KU-698-969

Wireless Data Demystified

John R. Vacca

McGraw-Hill
New York • Chicago • San Francisco • Lisbon
London • Madrid • Mexico City • Milan • New Delhi
San Juan • Seoul • Singapore
Sydney • Toronto

The McGraw-Hill Companies

To Dennis Pleticha, the network guy.

Library of Congress Cataloging-in-Publication Data

Vacca, John R.
Wireless data demystified / John R. Vacca.
p. cm.
Includes bibliographical references and index.
ISBN 0-07-139852-X (alk. paper)
1. Wire communication systems. I. Title.
TK5103.2.V33 2002
621.382—dc21 2002038692

Copyright © 2003 by The McGraw-Hill Companies, Inc. All rights reserved. Printed in the United States of America. Except as permitted under the United States Copyright Act of 1976, no part of this publication may be reproduced or distributed in any form or by any means, or stored in a data base or retrieval system, without the prior written permission of the publisher.

1 2 3 4 5 6 7 8 9 0 DOC/DOC 0 8 7 6 5 4 3 2

ISBN 0-07-139852-X

The sponsoring editor for this book was Stephen S. Chapman, the editing supervisor was Stephen M. Smith, and the production supervisor was Sherri Souffrance. It was set in Century Schoolbook by Victoria Khavkina of McGraw-Hill Professional's Hightstown, N.J., composition unit.

Printed and bound by RR Donnelley.

UNIVERSITY OF STRATHCLYDE
23 JUL 2008
UNIVERSITY LIBRARY

McGraw-Hill books are available at special quantity discounts to use as premiums and sales promotions, or for use in corporate training programs. For more information, please write to the Director of Special Sales, McGraw-Hill Professional, Two Penn Plaza, New York, NY 10121-2298. Or contact your local bookstore.

This book is printed on recycled, acid-free paper containing a minimum of 50% recycled, de-inked fiber.

Information contained in this work has been obtained by the McGraw-Hill Companies, Inc. ("McGraw-Hill") from sources believed to be reliable. However, neither McGraw-Hill nor its authors guarantee the accuracy or completeness of any information published herein and neither McGraw-Hill nor its authors shall be responsible for any errors, omissions, or damages arising out of use of this information. This work is published with the understanding that McGraw-Hill and its authors are supplying information but are not attempting to render engineering or other professional services. If such services are required, the assistance of an appropriate professional should be sought.

D
621.382
VAC

CONTENTS

FOREWORD

The future always brings more data and the necessity to move that data farther, faster, and less expensively. The biggest obstacle to developing data-intensive wireless applications is the need for speed. The expansion of wireless high-speed data networks and services will open an entirely new era of communications and connectivity.

Wireless high-speed data networks will be deployed on a global basis and thus lower the cost of high-speed connectivity. These incredible networks will benefit virtually every industry from banking and manufacturing to distribution and transportation. Wireless high-speed data networks will also provide tremendous benefit to defense and space efforts.

This book provides network designers, application developers, and product designers with a solid foundation in wireless high-speed technology and applications. The biggest challenge to managing or starting a career in information technology and telecommunications is keeping pace with emerging technologies and applications. This book examines every aspect of wireless high-speed data networks. The comprehensive discussion of data network platforms, next-generation high-speed wireless technology, and data satellites gives readers an unprecedented opportunity to improve their knowledge and advance their skills.

I highly recommend this book to students, professionals, enterprise knowledge centers, and university libraries.

—MICHAEL ERBSCHLOE
VICE PRESIDENT OF RESEARCH
COMPUTER ECONOMICS
CARLSBAD, CALIFORNIA

ACKNOWLEDGMENTS

There are many people whose efforts on this book have contributed to its successful completion. I owe each a debt of gratitude and want to take this opportunity to offer my sincere thanks.

A very special thanks goes to my editor Steve Chapman, without whose continued interest and support this book would not have been possible; and to acquisitions coordinator Jessica Hornick, who provided staunch support and encouragement when it was most needed. Thanks are given also to Stephen Smith, editing manager; Sherri Souffrance, senior production supervisor; Victoria Khavkina, desktop publishing operator; George Watson, copy editor; Peter Karsten, proofreader; and Charles Burkhour and Steven Gellert, senior computer artists, whose fine editorial work was invaluable. And a special thanks is given to Michael Erbschloe, who wrote the foreword for this book.

I thank my wife, Bee Vacca, for her love, her help, and her understanding of my long work hours.

Finally, I wish to thank the organizations and individuals who granted me permission to use the research material and information necessary for the completion of this book.

—John R. Vacca

INTRODUCTION

The surefire way to get ahead is to think ahead. So, while you are working in the here-and-now (whether revamping a client's Web site or upgrading the client's supply chain), there's no time like the present to examine technologies that haven't been widely adopted but could have a huge impact on enterprises during the next 10 years.

Take wireless data networks, for instance: Anywhere, anytime access to corporate data from your notebook, PDA, or mobile phone is an attractive service for a vendor to sell. While voice and wireless data carriers are beginning to roll out such "always-on" wireless data connections, faster, more reliable, and more ubiquitous always-on networks are on the way. (Many technical terms, abbreviations, and acronyms used in this book are defined in the Glossary.)

National carriers are in various stages of rollout: Verizon has rolled out a 2.5G CDMA service, which offers speeds of up to 384 kbps, on about 20 percent of its national network. Sprint plans to offer 3G services at peak speeds of 144 kbps in 2003, and more than 3 Mbps within 2 years after that. AT&T Wireless, Cingular, and VoiceStream are also in various stages of rolling out GPRS-enabled networks with top speeds of about 155 kbps.

NOTE The "national" rollouts, however, leave many geographic areas uncovered.

Many of the new services, and the devices to access them, will flow through a traditional two-tiered distribution model. Sierra Wireless, for example, is reselling its wireless products through Ingram Micro to make it look as much like a traditional model as possible, so vendors won't have to navigate the bureaucracy of a wireless data carrier.

Carriers will rely on vendors to build applications to drive wireless data usage. But the carriers, some of which also sell services, could be resistant to the channel having a piece of the pie. That could mean bumpy relations, until carriers figure out who to compete with and when to cooperate with the channel.

Vendors will also need to learn new tools, such as data compression, protocol optimization, and security software geared for wireless data networks. But the biggest challenge is writing applications for the lower bandwidth and intermittent availability of wireless data networks. To test developers' wireless data skills, you should walk around the back of their machines every couple of hours and yank out their Ethernet cords. If their applications keep running, then they pass.

Just how much money might be involved, how carriers will price wireless data services, and how the revenue will be shared hasn't been determined. Carriers could share part of a customer's monthly usage fees with a reseller. Another possible revenue stream for vendors is carriers reselling or developing their own wireless data applications.

But what about wireless data LANs? While carriers have delayed rollouts of wireless data wide-area networks, wireless data local-area networks (WiFi LANs) have surged in popularity. WiFi LANs provide network access only for approximately 300 ft around each access point, but provide for bandwidth up to 11 Mbps for the IEEE 802.11b protocol, and up to 100 Mbps for the emerging 802.11a protocol. Best of all, the technology is available now and affordable. Mainstream vendors offering WiFi products include Apple, Cisco Systems, Compaq, HP, Intel, Lucent, and 3Com.

WiFi LANs are an attractive way to extend corporate networks to other locations and are a cost-effective alternative to wired LANs. That's because they save the cost both of running cable and of updating user information as they move among physical locations. WiFi is especially popular in the manufacturing, distribution, and retail industries.

Vendors should know that WiFi LANs require skills conventional LANs don't, such as conducting site surveys to figure out how many access points are needed. The WiFi protocol is new enough that wireless data vendors can't count on interoperability among network interface cards or access points from different vendors. Security is also a concern, thanks to several hacks of the WiFi encryption protocol.

It's not as simple as it appears at first blush. A vendor hoping to sell and support WiFi networks needs to understand LANs, WANs, and the wireless data network over which the signals travel. In addition, the interoperability among components from multiple vendors is just not as good as you would expect on a wired network.

Vendors and integrators also need to factor in how applications running over WiFi mesh with other wireless data technologies. Data shouldn't be lost or leaked as users move among wireless data networks or between wired and wireless data environments. Developers also face a special challenge in designing applications that are usable on either a high-bandwidth WiFi LAN or a lower-bandwidth 2.5G or 3G network, where less data can be shared.

But the worst mistake is to do nothing. Savvy customers expect a vendor to offer both wired and WiFi options. If you can't properly address both types of LAN, regardless of which the customer chooses, you run the risk of losing the deal.

With that in mind, recent advances have made wireless data networks practical for voice, data, image, and video services in areas as small as an office and as large as the entire world. Wireless data systems reliably provide the flexibility demanded by today's increasingly mobile users and geographically distributed applications. This book provides you with a comprehensive technical foundation in mobile systems and wireless data products, services, and applications development as well as the knowledge required to implement wireless data systems that meet the needs of your enterprise.

Purpose

The purpose of this book is to show experienced (intermediate to advanced) mobile Internet professionals how to quickly install wireless data network technology. The book also shows, through extensive hands-on examples, how you can gain the fundamental knowledge and skills you need to install, configure, and troubleshoot wireless data network technology. This book provides the essential knowledge required to deploy and use wireless data network technology applications: integration of data, voice, and video. Fundamental wireless data network technology concepts are demonstrated through a series of examples in which the selection and use of appropriate high-speed connectivity technologies are emphasized.

In addition, this book provides practical guidance on how to design and implement wireless data network applications. You will also learn how to troubleshoot, optimize, and manage a complex mobile Internet using wireless data network technology.

In this book, you will learn the key operational concepts behind the mobile Internet using wireless data network technology. You will also learn the key operational concepts behind the major wireless data network services. You will gain extensive hands-on experience designing and building resilient wireless data network applications, as well as the skills to troubleshoot and solve real-world mobile Internet communications problems. You will also develop the skills needed to plan and design large-scale mobile Internet communications systems.

Also in this book, you will gain knowledge of concepts and techniques that allow you to expand your existing mobile Internet system, extend its reach geographically, and integrate global wireless network systems. This book provides the advanced knowledge that you'll need to design, configure, and troubleshoot effective wireless data network application development solutions for the Internet.

Through extensive hands-on examples (field and trial experiments), you will gain the knowledge and skills required to master the implementation of advanced residential wireless data network applications.

Finally, this intensive hands-on book provides an organized method for identifying and solving a wide range of problems that arise in today's wireless data network applications and mobile Internet systems. You will gain real-world troubleshooting techniques, and skills specific to solving hardware and software application problems in mobile Internet environments.

Scope

Throughout the book, extensive hands-on examples will provide you with practical experience in installing, configuring, and troubleshooting wireless data network applications and mobile Internet systems. Also throughout the book, hands-on demonstrations highlight key elements in wireless data networking. These include deploying WAP-enabled information systems and implementing a wireless data security video. In addition to advanced wireless data network application technology considerations in commercial organizations and governments, this book addresses, but is not limited to, the following line items as part of installing wireless data network–based systems:

- Plan and build a wireless data network system.
- Determine which digital multiaccess technology is appropriate for your organization's needs.
- Create circuit-switched and packet-switched core network infrastructures.
- Increase speed and bandwidth to form 3G wireless data networks.
- Implement mobile IP for Internet applications and services "on the move."
- Exploit the new features of next-generation mobile devices.

This book will leave little doubt that a new architecture in the area of advanced mobile Internet installation is about to be constructed. No question, it will benefit organizations and governments, as well as their mobile Internet professionals.

Intended Audience

This book is primarily targeted toward anyone involved in evaluating, planning, designing, or implementing wireless data networks. Users of cellular, pager, and other private and public wireless data networks who want to gain an in-depth understanding of network operations will also benefit. Basically, the book is targeted for all types of people and organizations around the world that are involved in planning and implementing wireless data networks and other mobile Internet systems.

Plan of the Book

The book is organized into five parts, with an extensive glossary of wireless data networks and other mobile Internet systems, 3G, 4G, and wireless data Internet networking terms and acronyms at the back. It provides a step-by-step approach to everything you need to know about wireless data networks as well as information about many topics relevant to the planning, design, and implementation of high-speed, high-performance mobile Internet systems. The book gives an in-depth overview of the latest wireless data network technology and emerging global standards. It discusses what background work needs to be done, such as developing a mobile Internet technology plan, and shows how to develop mobile Internet plans for organizations and educational institutions. More important, this book shows how to install a mobile wireless data broadband system, along with the techniques used to test the system and certify system performance. It covers many of the common pieces of mobile wireless data broadband equipment used in the maintenance of the system, as well as the ongoing maintenance issues. The book concludes with a discussion of future wireless data network planning, standards development, and the wireless data broadband mobile Internet industry.

Part 1—Overview of Wireless High-Speed Data Technology

Part 1 presents the fundamentals of wireless data networks: technology, platforms, services and applications, marketing environment, and standards for next-generation high-speed wireless data connectivity.

1. *Wireless Data Network Fundamentals.* This introductory chapter explores the uncertainty around the deployment of the higher-quality 3G wireless data networks. Organizations will likely have to live with the standards, coverage, reliability, and speed issues that exist today for at least the next several years.
2. *Wireless Data Network Protocols.* This chapter discusses how the the 5-UP will provide enhancements to the 802.11a standard that will enable home networking to reach its ultimate potential with scalable communications from 125 kbps through 54 Mbps. Robust,

high-rate transmissions are supported in a manner compatible with 802.11a, while allowing low-data-rate, low-cost nodes to communicate with little degradation in aggregate network throughput.

3. *Services and Applications over Wireless Data Networks.* This chapter discusses the wireless data moves in m-commerce. Not all m-commerce relies on location-based wireless data tracking.
4. *Wireless Data Marketing Environment.* This chapter discusses the state of the wireless data market environment. It also makes a lot of predications.
5. *Standards for Next-Generation High-Speed Wireless Data Connectivity.* This chapter discusses the state of the wireless data standard environment. Like Chap. 4, it also makes a lot of predications.

Part 2—Planning and Designing Wireless High-Speed Data Applications

Part 2 of the book is the next logical step in wireless data network application development. Part 2 also examines planning and designing wireless data and satellite applications, architecting wireless data mobility design, fixed wireless data network design, wireless data access design, designing millimeter-wave devices, wireless data services, and U.S.-specific wireless data design.

6. *Planning and Designing Wireless Data and Satellite Applications.* In this chapter, the integration of a terrestrial IP backbone with a satellite IP platform is addressed, with the main aim of enabling the resulting system for the global Internet to a differentiated service quality for mobile applications of a different nature. The detailed description of the functional architecture and the task performed by an interworking unit within the gateway interconnecting the two environments are highlighted.
7. *Architecting Wireless Data Mobility Design.* In this chapter, a new CDMA architecture based on CC codes is presented, and its performance in both MAI-AWGN and multipath channels is evaluated by using simulation. The proposed system possesses several advantages over conventional CDMA systems currently available in 2G and 3G standards.
8. *Fixed Wireless Data Network Design.* Fixed low-frequency BWDA radio systems at 3.5 and 10.5 GHz are presented as an attractive solution in this chapter. System architecture is presented from a signal processing and radio-frequency perspective.

9. *Wireless Data Access Design.* This chapter demonstrates that fixed wireless data have a significant role to play in the future of broadband communications, being used in areas in which the copper or cable infrastructure is not appropriate or by new operators that do not have access to these legacy resources. It also demonstrates that operators can economically and technically offer broadband services to users of 10 Mbps or more provided that they have a spectrum allocation of 100 MHz or more.
10. *Designing Millimeter-Wave Devices.* This chapter introduces and demonstrates a short-range LOS LMDS-like millimeter-wave and FSOW architecture for a BWA system that possesses many technological and operational advantages. These include ease of installation and alignment, low radiation power, and, effectively, a link free from major multipath, obstruction (trees, buildings, and moving objects), and adjacent cell interference.
11. *Wireless Data Services: The Designing of the Broadband Era.* This chapter provides an introduction to a variety of techniques used to provide robust image transmission over wireless data channels. Controlled redundancy can be added in the source coding and/or channel coding, and lossless compression techniques can be made more robust to transmission errors with little or no sacrifice in efficiency.
12. *U.S.-Specific Wireless Data Design.* This chapter presents the need for an optimized OTA transport, intelligent application protocol design, and payload compression as some of the key factors to consider in designing a mobile application for GPRS. It is only after evaluating these factors and the resultant compression ratio that the developer will be able to make a value decision as to the most efficient method to implement a particular solution.

Part 3—Installing and Deploying Wireless High-Speed Data Networks

This third part of the book discusses how to install and deploy wireless data satellite networks, implement terrestrial fixed wireless data networks, implement wireless data and satellite applications, apply the packet-over-SONET/SDH specification (POS-PHY level 3), deploy high-speed wireless data networking applications, and implement wireless data access.

13. *Deploying Mobile Wireless Data Networks.* This chapter discusses the deployment of wireless data network devices. Wireless data hand-held devices are a liberating technology for the mobile worker.

14. *Implementing Terrestrial Fixed Wireless Data Networks.* In this chapter, the implementation of terrestrial (nonsatellite) fixed wireless data technologies is discussed. As with wireline technologies, almost every specific service can be provided by terrestrial fixed wireless data technologies.
15. *Implementing Wireless Data and Mobile Applications.* How much functionality will reside on the devices? And how will the information on those devices be in sync with server information? This chapter answers the last question—too often the most overlooked component of going mobile.
16. *Packet-over-SONET/SDH Specification (POS-PHY Level 3): Deploying High-Speed Wireless Data Networking Applications.* This chapter introduces several emerging techniques currently under development for next-generation SONET/SDH systems. Taking into account these new techniques, the chapter elaborates on new SONET/SDH transport services likely to become reality within a few years.
17. *Wireless Data Access Implementation Methods.* This chapter quantifies the benefits of using antenna arrays (in the context of emerging mobile wireless data systems) as a function of the number of available antennas. Although absolute capacity and data-rate levels are very sensitive to the specifics of the propagation environment, the improvement factors are not.

Part 4—Configuring Wireless High-Speed Data Networks

Part 4 shows you how to configure wireless data, broadband wireless data networks, wireless data satellite networks, and residential wireless data access technology.

18. *Configuring Wireless Data.* This chapter presents architectural solutions for the following aspects, identified in the TRUST project: mode identification, mode switching, software download, and adaptive baseband processing. Finally, these solutions provide insight into the type of entities necessary to develop a feasible RUT based on SDR technology.
19. *Configuring Broadband Wireless Data Networks.* This chapter gives an overview of the challenges and promises of link adaptation in future broadband wireless data networks. It is suggested that guidelines be adapted here to help in the design and configu-

ration of robust, complexity/cost-effective algorithms for these future wireless data networks.

20. *Configuring Wireless Data Mobile Networks.* This chapter very briefly discusses the configuration of wireless data mobile networks. Configuring wireless data connectivity has implications for the specific mobile computing hardware you choose.

21. *Configuring Residential Wireless Data Access Technology.* This chapter very briefly discusses the configuration of residential wireless data access technology. The meaning of residential (home) networking configuration is changing because of the introduction of new wireless data access technologies that are allowing for more advanced applications.

Part 5—Advanced Wireless High-Speed Data Network Solutions and Future Directions

This fifth part of the book discusses residential high-speed wireless data personal area networks and presents a summary, recommendations, and conclusions.

22. *Residential High-Speed Wireless Data Personal Area Networks.* This chapter presents an overview of high-rate wireless data personal area networks and their targeted applications, and a technical overview of the medium access control and physical layers and system performance. The high-rate WDPANs operate in the unlicensed 2.4-GHz band at data rates up to 55 Mbps that are commensurate with distribution of high-definition video and high-fidelity audio.

23. *Summary, Recommendations, and Conclusions.* This last chapter outlines the new challenges to the key technological advances and approaches that are now emerging as core components for wireless data solutions of the future. A summary, recommendations, and conclusions with regard to the information presented in the book are also presented.

This book ends with an extensive glossary of wireless data networks, 3G, 4G, and mobile wireless data Internet terms and acronyms.

Overview of Wireless High-Speed Data Technology

CHAPTER 1

Wireless Data Network Fundamentals

Wireless data is a recent and valuable addition to the arsenal of corporate mobile computing tools, and has been the subject of much recent attention. It needs to be considered within the context of the business problems being solved and the existing corporate mobile infrastructure, with a realistic eye toward the capabilities of the public wireless networks of today and tomorrow.

Based on this author's extensive hands-on experience, this chapter, as well as the rest of the book, has been written to address popular misconceptions, minimize the hype, and provide insight to wireless data networks. Each of the chapters serves to help further the understanding of the wireless data world and to offer practical hands-on recommendations and perspectives.

The book content is intended to be equally useful whether you are in the throes of a major wireless data deployment or merely keeping an eye on the technology, waiting for it to mature further. The focus is also on providing information and analysis to organizations that will be users of wireless data, not to the telecom companies and carriers that will obviously be profoundly impacted by increasing wireless adoption.

So, without further ado, let's start with the most obvious questions: What are wireless data networks? And why consider them?

Wireless Data Networks Defined

To link devices like computers and printers, traditional computer networks require cables.[8] Cables physically connect devices to hubs, switches, or each other to create the network. Cabling can be expensive to install, particularly when it is deployed in walls, ceilings, or floors to link multiple office spaces. It can add to the clutter of an office environment. Cables are a sunk cost, one that cannot be recouped when you move. In fact, in some office spaces, running and installing cabling is just not an option. The solution—a wireless network.

Wireless data networks connect devices without the cables. They rely on radio frequencies to transmit data between devices, For users, wireless data networks work the same way as wired systems. Users can share files and applications, exchange e-mail, access printers, share access to the Internet, and perform any other task just as if they were cabled to the network.

How Fast Are Wireless Networks?

A new industry-wide standard, 802.11b, commonly known as WiFi, can transmit data at speeds up to 11 megabits per second (Mbps) over wireless data links. For comparison, standard Ethernet networks provide 10 Mbps. WiFi is more than 5 times faster than prior-generation wireless data solutions and its performance is more than adequate for most business applications.

What Is WiFi?

WiFi is a certification of interoperability for 802.11b systems, awarded by the Wireless Ethernet Compatibility Alliance (WECA). The WiFi seal indicates that a device has passed independent tests and will reliably interoperate with all other WiFi certified equipment. Customers benefit from this standard as they are not locked into one vendor's solution. They can purchase WiFi certified access points and client devices from different vendors and still expect them to work together.

When Do You Need Wireless Data Networking?

The following are a few examples of cases in which a wireless data network may be your ideal solution:

- For temporary offices
- When cabling is not practical or possible
- Supporting mobile users when on site
- Expanding a cabled network
- Ad hoc networking
- Home offices

For Temporary Offices

If you are operating out of an office space that is temporary, use a wireless data solution to avoid the costs of installing cabling for a network. Then, when you relocate, you can easily take your wireless data network with you and just as easily network your new facility. With a wired network, the money you spend on cabling a temporary space is lost when you leave. Moreover, you still need to build a new cabling infrastructure at your new site. If you expect to outgrow your current facilities, a wireless data network can be a shrewd investment.

When Cabling Is Not Practical or Possible

Sometimes landlords forbid the installation of wiring in floors, walls, and ceilings. Buildings may be old or walls solid or there could be asbestos in the walls or ceilings. Sometimes cabling cannot be laid across a hallway to another office. Or you have a space used by many employees where cabling would be messy and congested. Whenever cabling is impractical, impossible, or very costly, deploy a wireless data network.

Supporting Mobile Users When on Site

If you have branch office employees, mobile workers such as your sales force, consultants, or employees working at home, a wireless data network is an excellent strategy for providing them with network connectivity when they visit your premises. Once their laptops are equipped to communicate wirelessly with the network, they will automatically connect to the network when in range of your wireless data access point. You do not burden your IT staff to set up connections and you avoid having often-unused cabling strewn about your facilities just for remote users. You also use your office space more efficiently because you no longer provide valuable office space for workers who are infrequently on site.

Expanding a Cabled Network

You should use a wireless data network to extend an existing network, avoiding the cost and complexity of cabling. You will be able to connect new users in minutes rather than hours. Also, you will be able to provide network connectivity for your conference rooms, cafeteria, or lobby without any cabling hassles. In addition, you will even be able to expand the network beyond your building to your grounds, enabling employees to stay connected when outside. They will also be able to access the network as effortlessly and seamlessly as any worker linked by cabling.

Ad Hoc Networking

If you need to create temporary computer networks, such as at a job site, a conference center, or hotel rooms, wireless data solutions are simple, quick, and inexpensive to deploy. From virtually anywhere at a location or facility, employees will be able to share files and resources for greater productivity. Their wireless PC cards communicate directly with each other and without a wireless data access point.

Home Offices

You should also use a wireless data solution to network your home office, avoiding unsightly cables strewn about the workplace. Moreover, you can network your family, enabling everyone to share printers, scanners, and—if you are using an access router or bridged cable/digital subscriber line (DSL) modem—Internet access. You should also be able to link to the network from any room or even the backyard.

How Private and Secure Is Wireless Data Networking?

If you select a solution with sophisticated security technologies, your wireless data communications will be very safe. Leading wireless data solutions provide 128-bit encryption, and, for the highest levels of security, the most advanced systems will automatically generate a new 128-bit key for each wireless data networking session. These systems also will provide user authentication, requiring each user to log in with a password.

Coming in the Wireless Data Back Door

There are many juicy targets that are vulnerable to eavesdroppers and malicious intruders. In short, with an off-the-shelf directional antenna and a vanilla wireless NIC, you can sit in your car or other public places in many metropolitan areas and connect to hundreds of networks, typically those of sizable corporations. (The Glossary defines many technical terms, abbreviations, and acronyms used in this book.)

Anyone with a pulse has read innumerable accounts of Wired Equivalent Privacy's (WEP) weaknesses and knows that there are widely available script-kiddie-level tools, such as Air Snort, that can quickly crack WEP encryption. Less than half of the networks in the United States have WEP enabled, much less IPSec or some other measure that might be safe from third graders. Remember, wireless data networks are practically always installed inside the firewall, so whatever protections your firewall provides are moot if an intruder comes in wirelessly. It's bad enough if a war-dialing intruder finds an unprotected dial-in port and gets inside your firewall. An 802.11b-based intruder may be connected at 11 Mbps, not 56 kbps, making you a much juicier zombie or warez repository.

There are two causes for the preceding state of affairs, beyond the network managers who don't care if anyone in a quarter-mile radius can access their networks, and those forced to install a wireless data network without effective security despite their objections. First, many people underestimate the distance over which 802.11b radio signals can be picked up. Second, many wireless data networks are being set up informally by users who don't know or care what WEP is or what a firewall blocks out.

In either case, the solution is easy: For example, go down to Fry's or RadioShack and pick up a high-gain 2.4-GHz antenna and an Orinoco card. Connect the antenna to the wireless data card and install the wireless card in a laptop. Take it out to the parking lot or up on the roof and see whether you can find a wireless data network. If it doesn't measure up to your security policy, shut it down until it does. While you're at it, you may not want to limit your audit to the exterior of your building. You may be surprised to find internal wireless data networks that don't leak to the street.

If any of your enterprise's employees, including you, work at home on 802.11b networks, it might be smart to drive by their houses with your wireless data vulnerability tool kit and check them out. Those home firewalls and even the VPN clients you provide home users with may not suffice. You can be sure that most work-at-home employees haven't implemented Kerberos authentication and IPSec. It wouldn't be all that surprising if they also have file sharing enabled without strong passwords, providing opportunities for their neighbors and drive-by intruders to read, modify, delete, and otherwise "share" their files. You'd also be doing your friends and neighbors a service by checking out the vicinities of their 802.11b networks.

There's a minor groundswell underway among "Internet idealists" for explicitly sharing access to one's own wireless data network with the public. Usually, the point of this sharing is to provide unpaid high-speed Internet access to other members of the community. There's an issue regarding whether paying for a DSL or cable modem line gives you the right to open it up to an arbitrary number of other users. Many service providers' terms of service prevent the resale of access services, but it's not clear if such language would apply to given-away service.

In any event, the morality and legality of such sharing will be worked out by the usual methods before long. Before you open up a free public network to anyone with a wireless data card, you'd think long and hard about preventing the things that could get the ISP to shut access down, such as spam-meisters, hack-vandal activity, and other sorts of offensive content. You'd also think long and hard about fencing off your own hosts and devices from what a worst-case malevolent user might do. Then you'd forget about the project altogether.

Now, let's thoroughly examine the current state of the wireless data network infrastructure. It is composed of four parts, of which the first three are designed to address specific aspects of the global wireless data infrastructure. The first part gives an overview of wireless data networks—defining speeds, protocols, and types of networks. The second part discusses the worldwide allocation and rollout of the 3G wireless networks. The third part provides wireless data coverage maps so that one can better understand current coverage levels by network. The final part offers some practical insight and recommendations based on the current state of the networks.

Overview of Existing Networks

Although most of the discussion so far in this chapter has focused on wireless data WAN technologies, other types are presented as well (see Table 1-1).[1] Note that existing first- and second-generation (1G and 2G) technologies are typically much slower than a 56-kbps dial-up line. And yet-to-be delivered third-generation (3G) networks will not come anywhere close to the speed of the wired office LAN for which most corporate applications are designed.

In Table 1-1, the wireless generation is a function of speed and maturity of technology and is usually representative of a family of similar technologies, while 3G networks need to meet International Telecommunications Union specifications. Theoretical throughput is the best-case

TABLE 1-1

Network Speeds and Standards

Type of Network	Wireless Generation	Connectivity/Protocol	Theoretical Throughput
WAN	1G	Mobitex/Motient	9.6 kbps
WAN	2G	CDPD, CDMA, TDMA	19.2 kbps
WAN	2G	GSM	9.6 kbps
WAN	2.5G	Ricochet (filed Chapter 11)	100–150 kbps
WAN	2.5G	GPRS, 1XRTT	100–150 kbps
WAN	3G	CDMA2001x, TS-SCHEMA, W-CDMA, EDGE	384 kbps
LAN		Wired LAN	10–100 Mbps
LAN		801.11b	11 Mbps
PAN		Bluetooth	1–2 Mbps

attainable speed over the network and is typically 50 to 100 percent faster than real-world performance.

Wireless Data Types

There is a dizzying array of wireless data standards and technologies available to choose from. While wireless data services have been much slower to catch on than wireless voice services, they are slowly growing in acceptance, along with the speeds they provide and their availability.

Modem Data Modems transmit data from a serial line over an analog voice facility (analog cellular radio channel) as a series of tones. They work best over analog channels, because digital coding and compressing of audio damages or destroys the modem tones. Analog cellular channels (30 kHz) using the MNP-10 or ETC protocols can transmit at around 9.6 to 19.2 kbps. However, the modem at the other end also has to have similar capabilities. Because of this problem, some wireless carriers installed modem pools using pairs of back-to-back cellular and standard modems.

Digital Circuit-Switched Data Digital circuit-switched data attempts to replicate the modem experience with TDMA, GSM, or CDMA digital cellular or personal communications service. The problem is that modem tones cannot be reliably transmitted through a voice coder. Removing the voice coder requires a new protocol (of which some have been developed), and a modem pool is not an option. But this is different from an analog modem pool, because only a single modem is required as the switch receives the data in a digital format. A rough estimate of the data capacity of digital cellular can be gained by looking at the voice coder bit rates. Usually, this is about the amount of bandwidth available for data. TDMA uses 8-kbps voice coders, and up to three time slots can be aggregated (for a price). GSM uses 13-kbps voice coders and up to eight time slots can be aggregated (but this is usually done only for GPRS, which is a packet data standard). CDMA uses 8- or 13-kbps voice coders, but is more flexible in the amount of bandwidth that can be assigned to an individual customer.

Personal Communications Systems Personal communications systems (PCS) is a name given to wireless systems that operate in the 1800- to 1900-MHz frequency band. According to the initial concept, these systems were supposed to be very different from cellular—better, cheaper, simpler. However, the only technologies that were implemented were upbanded cellular standards; so, now consumers rarely know whether their cellular phone is operating in the cellular or PCS band:

- PCS1900—upbanded GSM cellular
- TIA/EIA-136—upbanded TDMA digital cellular
- TIA/EIA-95—upbanded CDMA digital cellular[4]

The major change from cellular to PCS is that all personal communications systems are digital. The few new concepts that were promoted were never implemented, including:

- J-STD-014—Personal Access Communications System (PACS), a combination of Bellcore WACS and Japan's Personal Handyphone Service (PHS)
- TIA IS-661—Omnipoint composite CDMA/TDMA
- TIA IS-665—OKI/Interdigital Wideband CDMA[4]

NOTE The PCS frequency allocation in the United States is three 30-MHz allocations and three 10-MHz allocations.

Analog Control Channel Data Some clever engineers have figured out ways to use the analog control channel (it is actually a digital channel, set up to service analog cellular systems) to transmit low-bit-rate data. This channel runs at only about 1 kbps and has to be shared with a large number of voice users. Aeris (http://www.diveaeris.com/) and Cellemetry (http://www.cellemetry.com/technical.html) are the prime users of this service.

WARNING URLs are subject to change without notice!

By faking a voice transaction, Aeris and Cellemetry can cause a small amount of data (4 to 16 bytes) to be sent to a central computer [which emulates a home location register (HLR)]. The advantages of this are high mobility (for asset tracking applications) and low capital costs, because the infrastructure is generally in place. These systems are largely used for industrial purposes, although some consumer applications exist, such as alarm monitoring systems.

Analog Packet Data: CDPD Cellular digital packet data (CDPD) uses an analog voice channel to send digital packet data directly from a phone to an IP network. It provides about 19 kbps for each channel, but this must be shared by multiple users. The strength of this technology is that the cost is kept low because it reuses much of the existing cellular infrastructure, but it takes channels away from voice users. Originally, it was planned that CDPD would transmit data when voice channels were idle, thus not consuming any capacity, but this proved to be too difficult to manage. CDPD systems service over 50 percent of the U.S. population and are found in several other countries, including Canada. CDPD has experienced some new life as a bearer protocol for Wireless Access Protocol (WAP), eliminating many of the delays experienced when circuit-switched data are used.

Data-Only Systems There are two major public data-only wireless systems available in the United States: Motient and Mobitex. According to Mobitex, its system covered 95 percent of the U.S. population in 2002 and provides coverage in Canada through a relationship with Rogers Wireless (Cantel). Motient (according to Mobitex) had coverage of 81 percent of the U.S. population at the same time. By comparison, CDPD covered only about 57 percent of the U.S. population. These data systems are similar in performance to CDPD, giving a shared bandwidth (per cell site) in the 9600 to 19,200 bps range.

Digital Cellular/PCS Packet Data The next big game in town is 3G wireless data. This implies speeds of 144 kbps for mobile terminals and 2 Mbps for stationary devices. Here, the world is divided into two camps: GSM/W-CDMA versus cdma2000/1xEV.

The GSM/W-CDMA strategy is to move first to general packet radio service (GPRS), which allows use of multiple time slots within a GSM channel (composed of eight time slots). Theoretically, this should allow speeds up to 115 kbps, but early devices are more in the 20-kbps range. W-CDMA will provide higher capacity, but it is too early to tell what realistic values are.

CdmaOne provided second-generation data rates of 14.4-kbps circuit data and up to 115-kbps packet data in theory. IS-2000/cdma2000 is being more widely implemented for data services. It is claimed to provide 144 kbps in its 1X mode. Future plans are for 1XEV-DO (a data-only system) that will provide 2 Mbps from the cell site and 144 kbps from the mobile unit (see sidebar, "3G Wireless Delivered by CDMA2000 1xEV-DO"). Yet another generation, known as 1xEV-DV (including voice services), is being designed to support about 2 Mbps in both directions.

3G Wireless Delivered by CDMA2000 1xEV-DO

CDMA2000 1xEV-DO technology offers near-broadband[7] packet data speeds for wireless data access to the Internet (see Fig. 1-1).[3] CDMA stands for code-division multiple access, and 1xEV-DO refers to 1x evolution-data optimized. CDMA2000 1xEV-DO is an alternative to wideband CDMA (W-CDMA). Both are considered 3G technologies.

A well-engineered 1xEV-DO network delivers average download data rates between 600 and 1200 kbps during off-peak hours, and between 150 and 300 kbps during peak hours. Instantaneous data

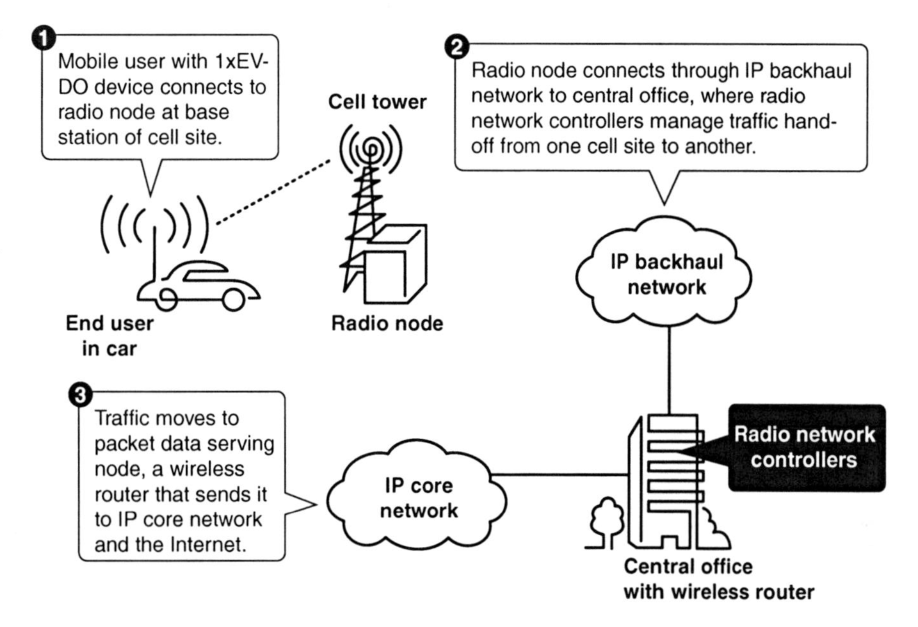

Figure 1-1 CDMA2000 1xEV-DO technology provides high-speed wireless access to the Internet over an all-IP network.

rates are as high as 2.4 Mbps. These data rates are achieved with only 1.25 MHz of spectrum, one-quarter of what is required for W-CDMA.

In an IP-based 1xEV-DO network, radio nodes perform radio-frequency processing, baseband modulation/demodulation, and packet scheduling. Radio nodes installed at a cell site can support hundreds of subscribers. Radio network controllers (RNCs) typically are located in a central office and provide hand-off assistance, mobility management and, terminal-level security via a remote authentication dial-in user service server. Each RNC can support many radio nodes and connects to a service provider's core data network through a standard wireless router called a packet data serving node. Finally, an element management system lets service providers manage 1xEV-DO radio networks.

1xEV-DO takes advantage of recent advances in mobile wireless communications, such as the adaptive modulation system, which lets radio nodes optimize their transmission rates on the basis of instantaneous channel feedback received from terminals. This, coupled with advanced turbo coding, multilevel modulation, and macrodiversity via sector selection, lets 1xEV-DO achieve download speeds that are near the theoretical limits of the mobile wireless data channel.

1xEV-DO also uses a new concept called multiuser diversity. This allows more efficient sharing of available resources among multiple, simultaneously active data users. Multiuser diversity combines packet scheduling with adaptive channel feedback to optimize total user throughput.

A 1xEV-DO network is distinguished from other 3G networks in that it is completely decoupled from the legacy circuit-switched wireless voice network. This has let some vendors build 1xEV-DO networks based entirely on IP technologies. Using IP transport between radio nodes and RNCs lowers backhaul costs by giving operators a choice of backhaul services, including frame relay, router networks, metropolitan Ethernet, and wireless data backhaul. IP-based 1xEV-DO networks take advantage of off-the-shelf IP equipment, such as routers and servers, and use open standards for network management.

1xEV-DO networks have the flexibility to support both user- and application-level quality of service (QoS). User-level QoS lets providers offer premium services. Application-level QoS lets operators allocate precious network resources in accordance with applications' needs. Combined with differentiated services–based QoS mechanisms, flexible 1xEV-DO packet schedulers can enable QoS within an entire wireless data network.

The International Telecommunications Union and Third Generation Partnership Project 2 recognize 1xEV-DO as an international standard. Subscriber devices based on the standard will become available in the first half of 2003 in North America. These devices will come in various forms, including handsets, PC cards, PDA sleds, and laptop modules.

Multimode 1xEV-DO terminals that support CDMA2000 1x voice will let subscribers receive incoming voice calls even while actively downloading data using 1xEV-DO. While 1xEV-DO is capable of supporting high-speed Internet access at pedestrian or vehicle speeds, it is can also be used from homes, hotels, and airports.3

NOTE It is hard to validate the preceding speed claims.

Technology is changing almost as fast as the marketing hype. Furthermore, carriers may decide that high-speed data is not as profitable as lower-speed data and voice services.

NOTE With voice coders running at 8 kbps, someone running at 800 kbps is taking approximately 100 times the resources.

Are voice coders going to pay 100 times the per-minute rate for voice services? Even if higher-speed data service is implemented, packet data channels are shared resources. Combined with overhead from multiple protocol layers, throughput may be limited to much less than the theoretical maximum.

I-Mode I-mode is a Japanese specification for providing Internet-like content to wireless devices.[5] It uses cHTML for data encoding, unlike WAP, which uses WML. Both protocols plan to migrate to xHTML, which should accommodate advances made by both protocols.

Wireless Application Protocol WAP is an application protocol designed to bring Web-like services to wireless data devices with extremely limited input and output capabilities. It uses a variant of HTML coding that, among other things, includes a binary compression scheme to make transmission of Web pages more efficient. Its biggest limitation is probably the fact that wireless devices with a numeric keypad and a tiny, low-resolution screen simply do not make great Web-surfing devices. However, no matter what its detractors say, it was a big advance in data, moving attention away from merely moving bits and bytes to actually supporting real-life applications for consumers and businesses. The specification was developed by the WAP Forum (http://www.wapforum.org/).

Wireless LAN Protocols Wireless LAN protocols have a somewhat easier job with wireless data. Terminals are usually stationary and systems are not expected to cover a wide area. Most of the standards use unlicensed spectrum, so anybody can set up one of these networks. IEEE 802.11 is definitely the premier standard here, allowing transmission at Ethernet speeds (10 Mbps), with higher speeds planned for the future. HomeRF is a competitor, but it seems to be treading on similar territory, and has perhaps missed the window of opportunity. Bluetooth is not truly a wireless LAN standard, but a Personal Area Network (PAN) standard. It provides a 1-Mbps channel to connect up to eight devices together. Rather than aim at connecting computers and printers (which is what 802.11 is usually used for), Bluetooth is more oriented toward personal cable replacement, perhaps connecting a phone, mouse, keyboard, and computer together. RF technology is also often used for wireless data networks. It provides good speeds, but is limited by the need to maintain line-of-sight between communicating devices.

Wireless Data IP Convergence Driven by the consumer hunger for anywhere, anytime communication, IP and wireless data are coming together. It's important to begin exploring this evolving landscape and what it means for the future of communications.

First let's define exactly what is meant by *IP* and what is meant by *wireless data* in this context. IP is short for Internet Protocol. Most data networks combine IP with a higher-level protocol called Transport Control

Protocol (TCP), which establishes a virtual connection between a destination and a source. IP by itself is something like the "snail mail" postal system. It allows you to address a package and drop it in a network without ever establishing a specific or direct link between you and the recipient. TCP/IP, on the other hand, establishes a connection between two hosts so that they can send messages back and forth for a period of time.

As previously explained, *wireless data* describes telecommunications in which electromagnetic waves (instead of some form of wire) carry the signal over part or all of the communication path. A wireless data device can connect to other devices like cellular phones, laptops, personal digital assistants (PDAs) with wireless modems, and wireless LANs. Generally, wireless data IP is a gathered body of data or packets over a wireless transmission path.

It's always challenging to ensure that technologies complement each other, and the convergence of IP and wireless data is no exception. While IP has the greatest potential for bringing together next-generation voice networks, wireless data technology is seen as one that will bridge the gap between the stationary and mobile workforces—giving end users the "always connected" capabilities they crave.

In this case, the mobile/wireless device landscape is complex. And this complexity leads to some specific issues the industry must address as it adds IP to the wireless data solution set:

- Which devices will be best suited to which applications (wireless IP phone, PDA, etc.)?
- Which devices will gain market segment leadership?
- Will users continue to use targeted, stand-alone devices or migrate to multifunction devices such as those that combine the functionality of a PDA and a cellular phone?
- What technological developments will ease existing device and connectivity constraints?
- Does the solution environment have enough wireless IP bandwidth available?[6]

Generally, striking the right balance will mean evaluating each mobile/wireless data application and its requirements separately. Applications need to be evaluated for the frequency and type of data transfer they require. If an application requires only periodic synchronization with a central repository, but also involves significant amounts of data entry on the client device, then most of the application logic should be on the client device. For example, sync-based content delivery can be effective for applications that handle sales force automation. It would be easy to store catalogs, client information, reference material, and other structured data files on the device and update them periodically when the user returns to the office.

On the other hand, applications that require either frequent or on-demand updates from a central repository, but don't require much input from the client, might be better off with a thin-client architecture on a device that connects more frequently—for instance, a cellular IP phone.

Of course, the greatest challenge will fall to developers of applications that require frequent, on-demand updates and rich graphical displays. These applications will need to add significant value to an organization to justify their development cost—and the high risk of failure inherent in meeting their design goals.

Unfortunately, the development picture for these wireless data applications will only become cloudier because of the ever-changing landscape and its impact on standardizing to a development environment and languages (for example, WML, XML, HTML, C-HTML, WAP, Java/J2ME, C—any derivative, HDML, XHTML, tag versus code). The marketplace's diversity, complexity, and constraints all make it hard for vendors to clearly see how to position themselves for success. For the same reasons (and because of today's economic slump), customers are reluctant to embark on extensive mobile/wireless data projects unless they see the potential for significant cost savings, productivity gains, or a clear competitive advantage.

Device ergonomics, bandwidth, coverage, and roaming constraints (plus the lack of heavy demand for these products) all make it hard to predict just when the market for wireless data IP solutions will take off. The more optimistic vendors point to standards that improve compression algorithms, intelligence controlling the display of the software residing on the device itself, and the growing demand for more information by both consumers and employees.

The eventual market segment opportunity will depend on the availability of more bandwidth and improvements to displays and mobile devices. End users are certainly attracted to the prospect of anytime, anywhere access to reliable information. That's why, despite the challenges, there's high interest in mobile devices, mobile access, and the potential of wireless IP for cellular phones. Vendors looking to penetrate this market segment will need to find a balance between establishing a track record of successful customer implementations and keeping themselves open to abrupt changes in the market segment.

The slowing U.S. economy has led to softer vertical and horizontal demand for wireless data devices. Moreover, this market segment is in for some real challenges in 2003 because of the ever-changing who's who in the wireless data world, the new applications being developed, and the potential for vendors of wireless hand-held devices to support wireless data IP.

The bottom line? Even though the general wireless data industry remains a favorite high-tech opportunity, it's not immune to temporary setbacks and slowdowns. Although wireless and IP are here to stay, vendors and manufacturers will come and go and application development will struggle to stabilize. In the long run, you'll all be accessible anytime, anywhere—and probably wishing you were still relying on your answering machines for near-real-time communications.

Ultra-Wideband Wireless Data Networks The most extreme claims about ultra-wideband (UWB) wireless data networks technology are that it could deliver hundreds of megabits of throughput per second, that its power requirements to link to destinations hundreds of feet away are as little as one-thousandth those of competing technologies such as Bluetooth or 802.11b, that transceivers could be small enough to tag grocery items and small packages, and that traffic interception or even detecting operation of the devices would be practically impossible. A slightly different way to look at the difficulty of detection and interception would be to claim that UWB devices wouldn't interfere with other electromagnetic spectrum users.

While the significant deployment of UWB devices is years away, each of the stupendous claims made for the technology has at least a modicum of supporting evidence. UWB devices operate by modulating extremely short duration pulses—pulses on the order of 0.5 ns. Though a system might employ millions of pulses each second, the short duration keeps the duty cycle low—perhaps 0.5 percent—compared to the near–100 percent duty cycle of spread-spectrum devices. The low duty cycle of UWB devices is the key to their low power consumption.

In principle, pulse-based transmission is much like the original spark-gap radio that Marconi demonstrated transatlantically in 1901. Unlike most modern radio equipment, pulse-based signals don't modulate a fixed-frequency carrier. Pulse-based systems show more or less evenly distributed energy across a broad range of frequencies—perhaps a range 2 or 3 GHz wide for existing UWB gear. With low levels of energy across a broad frequency range, UWB signals are extremely difficult to distinguish from noise, particularly for ordinary narrowband receivers.

One significant additional advantage of short-duration pulses is that multipath distortion can be nearly eliminated. Multipath effects result from reflected signals that arrive at the receiver slightly out of phase with a direct signal, canceling or otherwise interfering with clean reception.

NOTE If you try to receive broadcast TV where there are tall buildings or hills for signals to bounce from, you've likely seen "ghost" images on your screen—the video version of multipath distortion.

The extremely short pulses of UWB systems can be filtered or ignored—they can readily be distinguished from unwanted multipath reflections. Alternatively, detecting reflections of short pulses can serve as the foundation of a high-precision radar system. In fact, UWB technology has been deployed for 20 years or more in classified military and "spook" applications. The duration of a 0.5-ns pulse corresponds to a resolution of 15 cm, or about 6 in. UWB-based radar has been used to detect collisions, "image" targets on the other side of walls, and search for land mines.

So, when will we really see 3G? Let's take a look.

When Will We See 3G?

The deployment of 3G networks has not yet begun in earnest. Once the presumed viability of 3G became widely expected, each country initiated allocation of the 3G radio spectrum within its geography. You can see in Table 1-2 that this is an ongoing staggered process.[2] In some countries the licenses were simply awarded (freeing capital for immediate build-out), while in others auction prices reached staggering proportions, prompting industry analysts to question whether the auction winners will be able to afford to build the networks or find any way to profitably commercialize the services.

TABLE 1-2

Status of 3G Spectrum Awards

Country	Licenses Awarded to Date (of Total)	Method	Award Date
	Europe, Middle East, and Africa		
Austria	6	Auction	November 2000
Belgium	3	Auction	February 2001
Croatia	Not applicable	Not applicable	Not applicable
Czech Republic	2 (of 3)	Auction	December 2001
Denmark	4	Sealed bid	September 2001
Estonia	0 (of 4?)	Beauty contest	2002?
Finland	4 + 2 regional	Beauty contest	March 2000
France	2 (of 4)	Beauty contest	May 2001
Germany	6	Auction	July 2000
Greece	3 (of 4)	Auction	July 2001

TABLE 1-2

Status of 3G Spectrum Awards (Continued)

Country	Licenses Awarded to Date (of Total)	Method	Award Date
Europe, Middle East, and Africa (continued)			
Hungary	Not applicable	Not applicable	Not applicable
Ireland	0 (of 4)	Beauty contest	June 2002?
Isle of Man	1	Not applicable	May 2000
Israel	3	Auction	December 2001
Italy	5	Auction	October 2000
Latvia	Not applicable	Not applicable	Not applicable
Liechtenstein	1	Not applicable	February 2000
Luxembourg	0 (of 3)	Beauty contest	2002
Monaco	1	Not applicable	June 2000
Netherlands	5	Auction	July 2000
Norway	3 (of 4)	Beauty contest combined with annual fee	December 2000
Poland	3 (of 4)	Beauty contest (cancelled)	December 2000
Portugal	4	Beauty contest	December 2000
Slovakia	0 (of 3)	Beauty contest	2002
Slovenia	1 (of 3)	Auction	November 2001
Spain	4	Beauty contest	March 2000
South Africa	0 (of 5)	No contest	2002
Sweden	4	Beauty contest	December 2000
Switzerland	4	Auction	December 2000
Turkey	0 (of 4 or poss. 5)	Not applicable	2002?
United Arab Emirates	Not applicable	Not applicable	Not applicable
United Kingdom	5	Auction	April 2000

TABLE 1-2

Status of 3G Spectrum Awards (Continued)

Country	Licenses Awarded to Date (of Total)	Method	Award Date
		Asia Pacific	
Australia	6	Auction	March 2001
Hong Kong	4	Revenue share	September 2001
India	Not applicable	Not applicable	
Japan	3	Beauty contest	June 2000
Malaysia	Not applicable	Not applicable	2002?
New Zealand	4	Auction	July 2000
Singapore	3 (of 4)	Auction (cancelled)	April 2001
South Korea	3	Beauty contest	December 2000
Taiwan	0 (of 5)	Auction	2002
		Americas	
Canada	5	Auction	January 2001
Chile	Not applicable	Not applicable	Not applicable
Honduras	0 (of 1)	Auction	2002?
Jamaica	0 (of 2)	Auction	2002?
Uruguay	Not applicable	Auction	2002
United States	Not applicable	Not applicable	Not applicable
Venezuela	Not applicable	Not applicable	Not applicable

Licensing Costs for 3G

Over 100 3G licenses have now been secured worldwide via a combination of auctions, beauty contests, "sealed bid" competitions, and automatic awards. Table 1-3 shows the UMTS Forum's analysis of 3G licensing costs. Data are supplied for information only.[2]

The costs are either the highest auction cost for 2 × 5 MHz of spectrum or the corresponding administrative cost over the lifetime of the license. The costs for France are not yet completely known and have been estimated.

TABLE 1-3 Basic Data Concerning the Licensing in Those Countries That Have Issued 3G Licenses (and Two That Are Going to Issue Licenses)

Country	Date, year-month-day	Cost per $1,000,000 US per 2 × 5 MHz of Spectrum	Population, 1999	GDP, $1,000,000 US	Cost per Head of Population per 2 × 5 MHz of Spectrum	Cost as Percent of GDP
New Zealand	01-01-18	3	3,819,762	63.8	0.8	0.004
Switzerland	00-12-06	12	7,262,372	197	1.7	0.006
Norway	00-11-29	17	4,481,162	111.3	3.8	0.015
Singapore	01-04-11	21	4,151,264	98	5.1	0.020
Portugal	00-12-19	33	10,048,232	151.4	3.3	0.021
Slovenia	01-09-03	34	1,927,593	21.4	17.6	0.151
Denmark	01-09-15	43	5,336,394	127.7	8.1	0.032
Czech Republic	01-09-15	48	10,272,179	120.8	4.7	0.038
Belgium	01-03-02	50	10,241,506	243.4	4.9	0.020
Hong Kong	01-09-19	61	7,116,302	158.2	8.6	0.037
Austria	00-11-03	66	8,131,111	190.6	8.1	0.033
Australia	01-03-19	70	19,169,083	416.2	3.7	0.016
Greece	01-07-13	73	10,601,527	149.2	6.9	0.047
Poland	00-12-06	217	38,646,023	276.5	5.6	0.075
The Netherlands	00-07-24	238	15,892,237	365.1	15.0	0.062
South Korea	00-12-15	272	47,470,969	625.7	5.7	0.041
Spain	00-03-13	419	39,996,671	677.5	10.5	0.059
Italy	00-10-27	1224	57,634,327	1212	21.2	0.096
France	01-05-31	619	59,330,887	1403.1	10.4	0.042
United Kingdom	00-04-27	3543	59,510,600	1319.2	59.5	0.256
Germany	00-08-18	4270	82,797,408	1864	51.6	0.218
Finland	99-03-18		5,167,486	108.6		
Liechtenstein	00-02-15		32,207	0.73		

TABLE 1-3 Basic Data Concerning the Licensing in Those Countries That Have Issued 3G Licenses (and Two That Are Going to Issue Licenses) (Continued)

Country	Date, year-month-day	Cost per \$1,000,000 US per 2 × 5 MHz of Spectrum	Population, 1999	GDP, \$1,000,000 US	Cost per Head of Population per 2 × 5 MHz of Spectrum	Cost as Percent of GDP
Japan	00-06-30		126,549,980	2950		
Thailand	00-01-15		61,230,874	388.7		
Sweden	00-12-16		8,873,052	184		
Ireland			3,797,257	73.7		
Hungary			10,138,844	79.4		

GDP = gross domestic product.

NOTE The UMTS Forum is not responsible for the table's accuracy or completeness.

Apparently there are big differences in the amount of money paid or to be paid in the future, whatever measure of size is used. If the license prices per head of population and per 2 × 5 MHz are plotted after the date of issue of the licenses, there is a clear tendency that the price is declining with time (see Fig. 1-2).[2]

This is an effect of the declining business climate in the telecom sector. However, there is also an effect of the size of the market. If the costs per head of population and per 2 × 5 MHz of spectrum are plotted

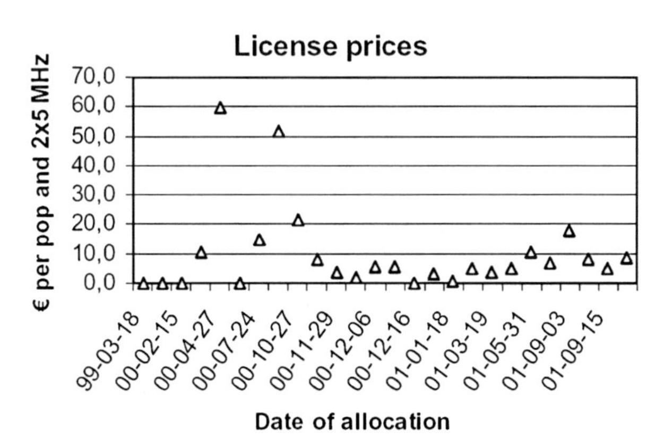

Figure 1-2 License prices are declining with time.

against the size of the country, a tendency becomes apparent that the price increases with the size of the market (see Fig. 1-3).[2]

This tendency is even more clear when the gross domestic product (GDP) is used instead of population (see Fig. 1-4).[2] With the exception of the two early licensings in the United Kingdom and Germany, all countries that have required a license price in the upper part of Fig. 1-4 have had problems in finding applicants for all licenses.

There is another effect of the auctions. If you make a division of the licensees in three categories—global players, regional players, and local players as shown in Fig. 1-5—it becomes clear that the small players in the market have small chance to compete for the expensive auction licenses.[2] The beauty contests, on the other hand, have allowed the local players to have about one-third of the licenses. This effect is by no means unexpected, but may be an important reason for many countries to choose the beauty contest as an allocation method.

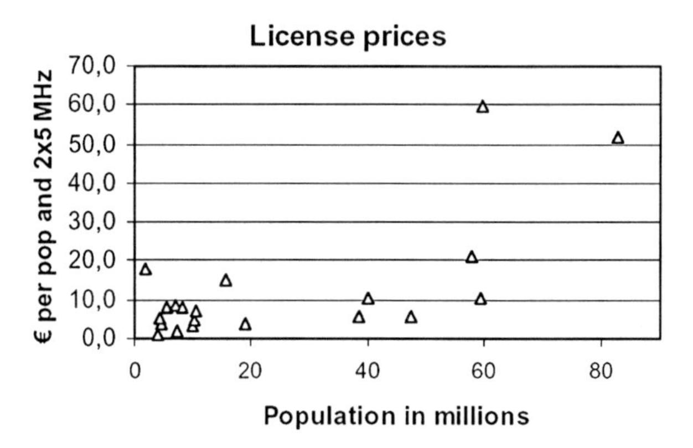

Figure 1-3 License prices are increasing with the size of the market.

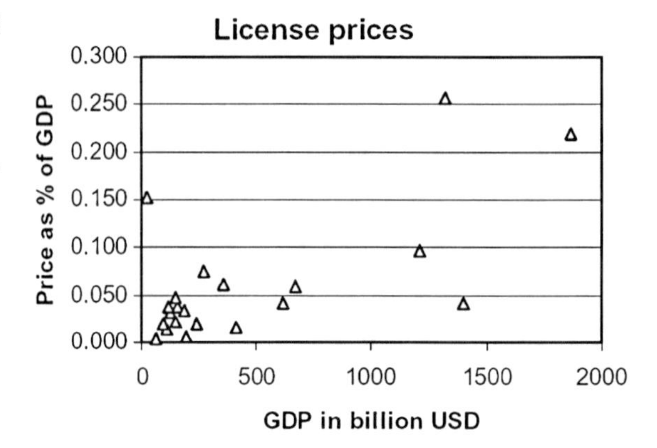

Figure 1-4 The GDP is used instead of population.

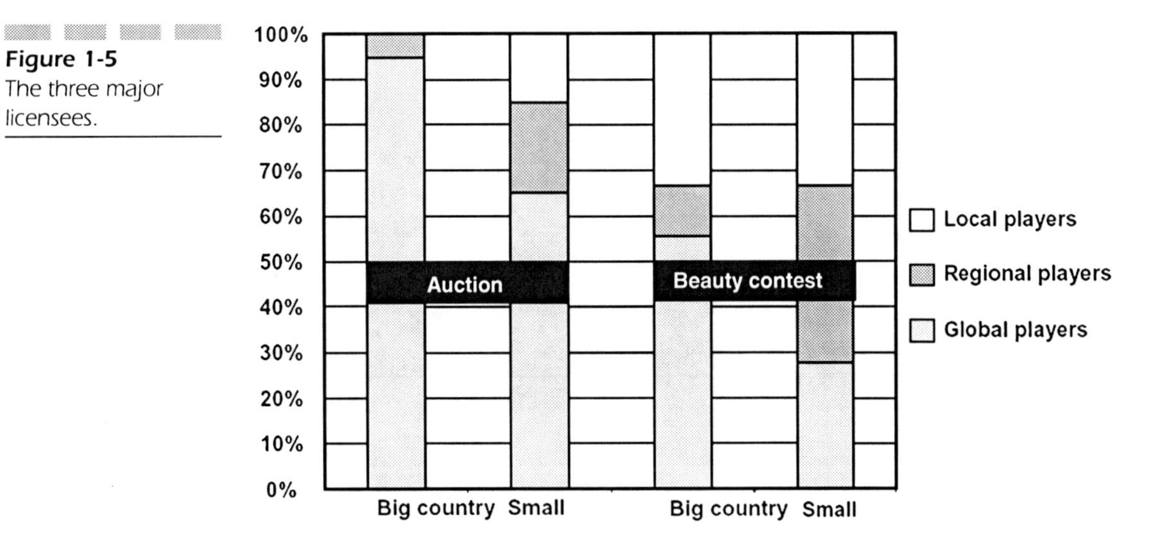

Figure 1-5
The three major licensees.

Standards and Coverage in the United States

The United States in particular faces heightened challenges related to a lack of standards and a vast geographic area. Both factors impact coverage for any given network. The maps in Figs. 1-6 to 1-8 show wireless data coverage in the United States for a variety of networks.[1]

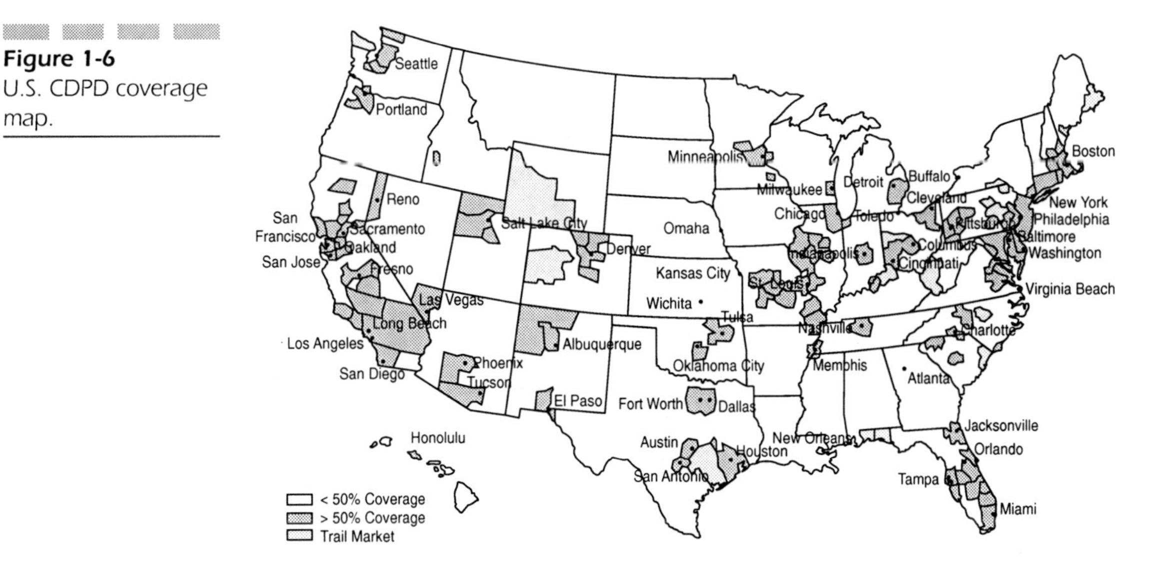

Figure 1-6
U.S. CDPD coverage map.

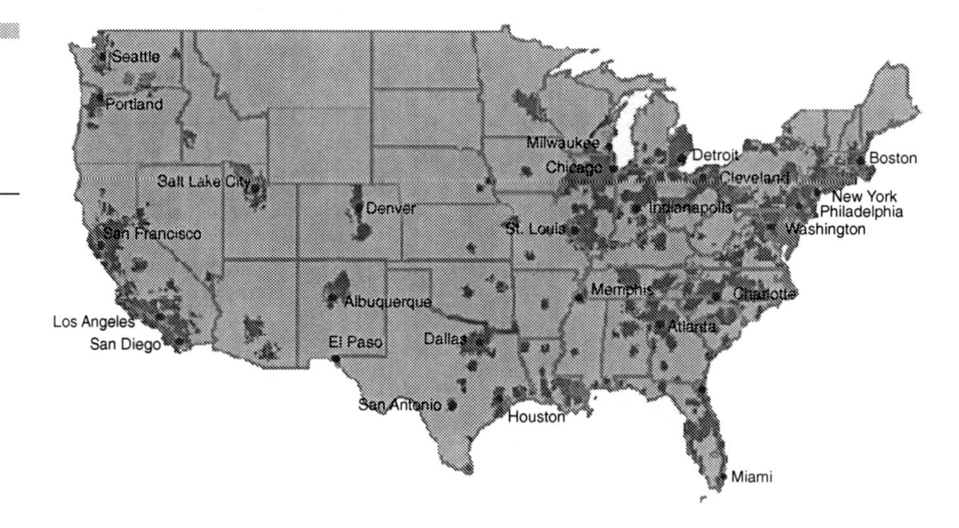

Figure 1-7 Coverage map for Cingular interactive network based on Mobitex.

Figure 1-8 Coverage map for Sprint's CDMA network.

Coverage in Europe

You should compare the level of U.S. coverage for any given technology with that offered in Europe where there is one standard. The maps in Figs. 1-9 and 1-10 are typical for European countries in terms of coverage.[1]

NOTE Coverage in Asian countries varies widely and no single map would be representative for that region.

Implications for the Short Term

Based on the preceding information, the following implications should be considered in planning your near-term wireless data investments:

- Value of 3G technology/spectrum is less than initially thought.
- Deployment is likely to be delayed.
- Standards are still uncertain.
- Coverage is incomplete.

Value of 3G Technology/Spectrum

The fees paid for 3G spectrum licenses have been trending downward, signifying the reduced perceived value of the licenses. This is due to bidding telecom firms' questioning how they can commercialize the service and make a profit based on the cost of the spectrum and building out the network. Look back at the fees paid over time in the United Kingdom, then Germany, then Australia.

Large carriers, including British Telecommunications and NTT DoCoMo, Japan's largest wireless provider, have postponed 3G offerings after technical glitches. Several European 3G auctions have collapsed. And some European operators are now asking for governments to refund the money spent to buy licenses to the 3G wireless spectrum, a dramatic about-face. While all purchasers still believe in the value of deploying the 3G networks, the potential revenue streams are being questioned, and overpaying for the spectrum could have implications on deployment time frames.

Deployment

In the United States, unlike Europe, the spectrum allocated for 3G is currently occupied and being used by the Department of Defense. In order to

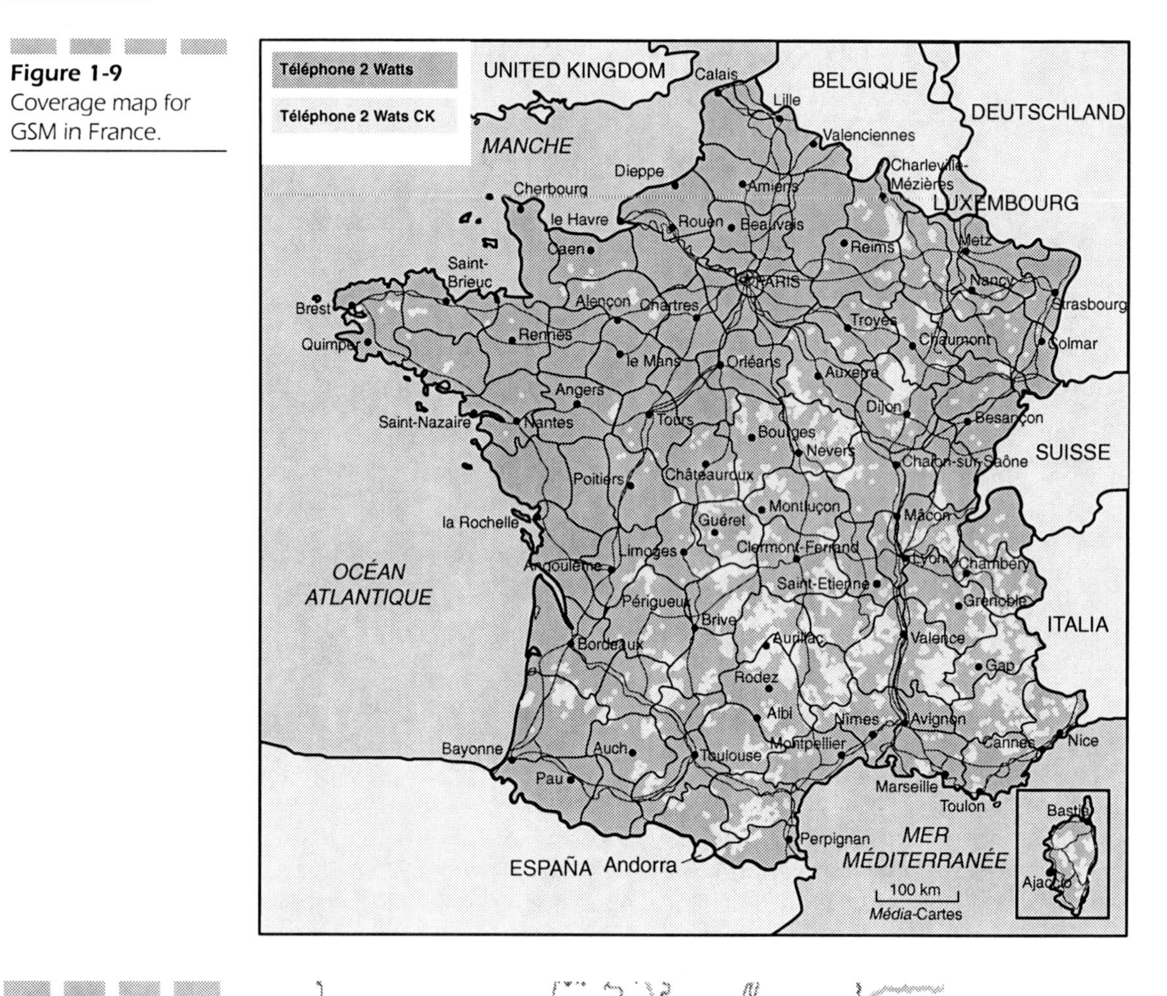

Figure 1-9 Coverage map for GSM in France.

Figure 1-10 Coverage map for GSM in Germany.

first auction off the 3G wireless spectrum, sufficient frequency must be allocated, and those occupying the current frequency must be compensated accordingly. Discussions are ongoing with the FCC and Commerce Department, and there is a commitment to resolve this in time for the auction. However, until this is resolved, the auction can't happen.

3G network equipment suppliers (Lucent, Siemens, Nortel, and Cisco) have recently experienced significant revenue shortfalls, and that is partly because of the slowdown in network infrastructure spending by the telecom firms on deployment of 3G networks. The network equipment suppliers' financial results provide a harbinger of 3G technology deployment time frames.

Standards

While the 3G spectrum auctions and early deployments get started, there are a host of competing 3G standards. Actual deployment of 3G networks worldwide could very well overcome the coherence of the existing outside-the-U.S. 2G standard of GSM. The global 3G picture may wind up looking more like the standards mix that exists in the United States today.

Coverage

As the maps in Figs. 1-6 to 1-10 suggest, and as you experience in your daily usage of cell phones, coverage is not complete. Planned 3G rollouts are scheduled to be completed in the 2005–2007 time frame. Additionally, sales of the infrastructure components to support the upgrade of technologies from 2 to 2.5G have remained somewhat sheltered from the downturn, revealing that network providers may suspect that 2.5G technology may suffice until all the 3G issues are worked out.

Perspective on Wireless Data Computing

In recent media coverage, wireless data computing has been presented as a revolutionary paradigm shift. Wireless data computing is perhaps a less dramatic advance. Cellular phones didn't fundamentally change the way people communicated—talking on the phone wasn't new, but the convenience and availability cell phones brought were. Wireless data will bring corporations equally powerful benefits—within a framework you already understand.

A more practical way to look at wireless data is to put it in perspective within the overall context of building and delivering mobile computing solutions. The process of bringing mobile technologies to bear on business processes is nothing new. It requires a disciplined review of the alternative technologies and architectures to determine those best suited to solving the business problem at hand. Wireless data hasn't changed this.

New Connectivity Option

Any enterprise mobile computing solution will involve successive layers of technology, as shown in Fig. 1-11.[1] Viewed from this perspective, you see wireless data as just another connectivity option. This is obviously a bit understated, as the option for wireless data connectivity definitely impacts your choices in the other layers.

The important point is that wireless data does not stand your whole IT operation on its head. It is merely a new connectivity option, one that may allow you to add business value by extending existing systems or further automating business processes.

Another way of adding perspective to this new wireless data option is to look back at how mobile computing has evolved in the past 10 to 15 years. Seeing new options in any given layer is neither rare nor surprising.

Rapid Change in Mobile Computing

In the 1990s, you saw the arrival of sophisticated customer relationship management applications as a prime target for mobilization. Toward the end of the decade, hand-held devices began the transformation from per-

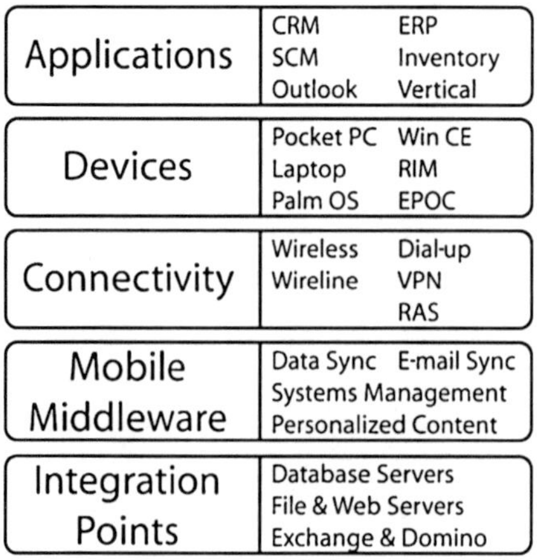

Applications	CRM SCM Outlook	ERP Inventory Vertical
Devices	Pocket PC Laptop Palm OS	Win CE RIM EPOC
Connectivity	Wireless Wireline	Dial-up VPN RAS
Mobile Middleware	Data Sync Systems Management Personalized Content	E-mail Sync
Integration Points	Database Servers File & Web Servers Exchange & Domino	

Figure 1-11 Component layers for mobile solutions.

sonal organizers into centrally managed extensions of the existing IT environment. Likewise, you've seen the types of back-end integration points for mobile computing grow from just relational database servers to include e-mail, file, and intranet servers as well.

Seen this way, wireless data represents a new option within one layer of the very dynamic and fast growing space of enterprise mobile computing. What is driving this change? Increased mobility, the ever increasing pace of business, and rapid advances in technologies. All these factors combine to make mobile computing ever more promising—and increasingly a basic requirement to competing successfully. Wireless data is the latest advance and it merits cautious investigation and investment.

The Pros and Cons of Wireless Data

Wireless connectivity for corporate information access offers a variety of potential business benefits driven by user convenience, timeliness of information, and increased ability to transact business. There are organizations out there that have aggressively adopted wireless computing technology and seen the following types of benefits:

- Increased sales
- Decreased costs
- Improved customer service
- Competitive advantage
- Rapid return on investment (ROI)[1]

However, keep in mind that supporting wireless data connectivity also has the potential to increase certain challenges. These challenges are central to mobile computing solutions in general—regardless of the connectivity option chosen. However, the relative immaturity of public wireless data networks does tend to exacerbate them. These challenges include:

- Coverage
- Reliability
- Standards
- Speed
- Costs[1]

In many cases, the unique benefits of wireless data can make it worthwhile to deal with the challenges. Your organization may find innovative ways to wirelessly enable existing business applications. You might find value in formally embracing hand-helds and speeding deployment—with or without wireless data. It all comes back to the business process being supported, and how that translates into the overall solution.

Keep in mind that next-generation wireless data networks will mitigate these challenges sooner or later, and that wireless will emerge as a truly strategic enabling technology. IT organizations are well served to cut their teeth on wireless data today in order to begin building core competencies for the future.

Wireless Data Impact on Other Layers

When building out your mobile solution with wireless data communications, you should take into account the effect on the other layers in the model:

- Applications
- Devices
- Integration points
- Mobile middleware

Applications The application layer should be driven first and foremost by the business problem that you are trying to solve and that led you to mobile computing in the first place. Therefore, it is unlikely that choosing wireless data is going to affect your choice of the application. However, wireless data might let you revisit existing processes and applications to see if there are opportunities to seize competitive advantage with new mobile initiatives.

Devices Regarding devices, all of the major types of mobile computing devices offer one or more options for wireless data connectivity. However, not all devices have options for all networks, so the decision to support a specific device is usually made hand-in-hand with the decision to support a particular type of wireless data connectivity. You can read more about devices, the networks they support, and key criteria for selecting devices in Chap. 20, "Configuring Wireless Data Mobile Networks."

Integration Points The back-end integration points are largely determined by the application layer. However, you should consider the existing back-end systems within your environment and look for ways to wirelessly enable them to solve business problems and build competitive advantage.

Mobile Middleware Ideally, the mobile middleware you choose will help overcome many of the challenges of going wireless. Your mobile infrastructure platform should support whatever devices, networks, and integration points you wish to mobilize. Thus, the choice to go wireless will indeed affect your choice of mobile middleware, which should be platform-agnostic and support all major standards.

More on the Middleware Layer Remember that basic purpose of a mobile middleware platform is to:

- Help authenticate mobile devices connecting to network resources.
- Optimize for low-bandwidth, intermittent connections.
- Provide secure access only to users authorized to receive information.
- Support all types of information—data, files, e-mail, Web content.[1]

Even if you are dealing with a very specific project for a specific device and network, it is important to plan for the future and choose a comprehensive platform. The alternative is buying and maintaining a portfolio of middleware solutions as you pursue future projects and support other devices and networks and types of information. This is not only more expensive and inefficient, but it creates integration nightmares.

Systems management for mobile and wireless devices also presents unique challenges. There are strong benefits to deploying one mobile middleware solution to meet the preceding requirements as well as providing specialized mobile system management capabilities.

Examples of Strong Wireless Value

The following are examples of the types of solutions that companies have deployed today where wireless data connectivity adds strong value to the overall solution:

- Risk management and insurance
- Electric meter reading
- Wireless data hand-held e-mail

Risk Management and Insurance

A large property and casualty insurer helps clients manage risk by sending risk engineers on site to profile and analyze client facilities. Data are captured on site on laptops, and synchronized back to a central database. An extranet site provides customers having access to their site with reports stored in this database. The company uses wireless data connectivity to make these reports (previously paper-based) available within hours—not weeks. This creates a huge service advantage. This project is typical of a trend in service-related industries, where providing information to customers about their own operations is just as important as providing the actual service performed.

Electric Meter Reading

A major electric utility company employs a large field-based workforce that captures billing information by physically visiting customer sites and reading the values from their electric meters. The historical way this information flowed into the billing process was that the reading was recorded on site on a paper form, which was then forwarded to the corporate office for data entry, and only then could a bill be sent. Using today's wireless data devices, the same utility captures the reading on site directly into a hand-held device, and at the end of the day, the staff member using the device wirelessly uploads the day's readings directly into the billing system database. This knocks several days off the time it takes to collect receivables, and results in more accurate billing—two things any CFO is eager to do.

Wireless Hand-Held E-Mail

Finally, for executives of a large vehicle manufacturer, the ability to keep in touch with key partners and customers from anywhere is an important competitive advantage. Being able to pick up and reply to e-mail while on the go is just as important to this company as doing the same with voice mail. Wireless e-mail opens the door to increased productivity for these mobile knowledge workers who are now able to do work in a taxi, waiting in the lobby for a meeting to start, between flights, or over breakfast in the morning. This easily applies to knowledge workers in a wide variety of industries.

Conclusion

This introductory chapter explored the uncertainty around the deployment of the higher-quality 3G wireless data networks. Organizations will likely have to live with the standards, coverage, reliability, and speed issues that exist today for at least the next several years. Of course, some companies have already proved it's possible to be successful with today's wireless data networks. Nevertheless, there is a reason to be optimistic and proceed cautiously with applying wireless data to your business model today, while we all wait for the exciting high-performance networks of the future.

In any event, the next 22 chapters will thoroughly discuss in finite detail all of the topics examined in this chapter, and much much more. Have a good read, and enjoy!

References

1. Synchrologic, 200 North Point Center East, Suite 600, Alpharetta, GA 30022, 2002.
2. UMTS Forum, 2002.
3. Vedat Eyuboglu, "CDMA2000 1xEV-DO Delivers 3G Wireless," Airvana, Inc., 25 Industrial Avenue, Chelmsford, MA 01824, 2002.
4. Cellular Networking Perspectives Ltd., 2636 Toronto Crescent, NW, Calgary, Alberta T2N 3W1, Canada, 2002.
5. John R. Vacca, *i-mode Crash Course,* McGraw-Hill, 2002.
6. Jim Machi, "Wireless IP: Another Convergence?" Intel Telecommunication and Embedded Group, Intel Corporation, 5000 W. Chandler Blvd., Chandler, Arizona 85226-3699, 2002.
7. John R. Vacca, *Wireless Broadband Networks Handbook,* McGraw-Hill, 2001.
8. John R. Vacca, *The Cabling Handbook*, 2d ed., Prentice Hall, 2001.
9. John R. Vacca, *High-Speed Cisco Networks: Planning, Design, and Implementation,* CRC Press, 2002.

CHAPTER 2

Wireless Data Network Protocols

Popular wireless data networking protocols such as Bluetooth, IEEE 802.11, and HomeRF were originally developed for the 2.4-GHz frequency band by organizations that made design tradeoffs based on values such as complexity, price, and performance. Because the protocols were developed independently and these values differed according to the markets and applications the organizations intended to serve, the various protocols do not easily interoperate with one another and can cause significant mutual interference when functioning in the same radio space. The problem becomes especially acute in environments such as residential networks where a single network may be required to serve a broad range of application classes.

A newer high-performance wireless data LAN standard, IEEE 802.11a, operates in the 5-GHz band and offers much higher speeds than previous WLAN standards, but does not adequately provide for unified networks that support multiple classes of devices with differing speed, performance, power, complexity, and cost requirements. These differing classes of devices will become increasingly important as LANs move beyond the limits of office-oriented computer interconnection services and into the realm of data, video, and audio distribution services for interconnected devices in offices and homes. (The Glossary defines many technical terms, abbreviations, and acronyms used in the book.)

Nevertheless, the data wireless marketplace is booming. New wireless data products are being introduced daily. The unlicensed industrial/scientific/medical (ISM) band at the 900-MHz and 2.4-GHz frequencies creates opportunities for high-quality wireless data products to be introduced. Wireless home networking initiatives are being announced and developed, including the BlueTooth, HomeRF, and IEEE 802.11 working groups and others. Industry leaders seek technologies for new digital cordless telephones with high-end features. There is a high level of expertise required to design high-speed and high-quality wireless data products in these spread-spectrum product market segments. Large consumer product manufacturers are turning to technology providers to obtain the latest wireless data technologies and shorten time to market. The system-on-a-chip (SoC) marketplace is "exploding" too. Application-specific integrated circuit (ASIC) complexity is estimated to reach 7.2 million gates by the end of 2003. This allows multiple functionality to be integrated into a single chip, lowering the cost and size of products based on such chips.

Because a single company becomes unable to design such high-integration components, and with demanding time-to-market constraints, system companies are turning to third-party ASIC designers. These third parties provide intellectual property (IP) in the form of subsystem ASIC designs as "building blocks" to their complete SoC designs. Companies like ARM, MIPS, RAMBUS, and others have already seized that opportunity and offer differentiated IP cores. The third-party IP market is estimated to grow from $5.9 billion in 2003 at a compound annual growth rate of 76 percent. There is a very special opportunity for companies than can offer special experience and intellectual property in the spread-spectrum area to companies that wish to integrate wireless data connectivity in their system-on-a-chip products in the form of wireless data IP cores. The marketplace for wireless data products that can use such cores is estimated at $11.6 billion in 2003 and is expected to grow to over $56 billion in 2007.

With the preceding in mind, let's now look at the 5-GHz Unified Protocol (5-UP). This protocol is a proposed extension to existing 5-GHz wireless data LAN (WLAN) standards that supports data transfer rates to over 54 Mbps and also allows a wide variety of lower-power, lower-speed devices carrying diverse traffic types to coexist and interoperate within the same unified wireless data network.

Unified Multiservice Wireless Data Networks: The 5-UP

The proliferation of cheaper, smaller, and more powerful notebook computers and other mobile computing terminals has fueled tremendous growth in the WLAN industry in recent years. WLANs in business applications enable mobile computing devices[6] to communicate with one another and access information sources on a continuous basis without being tethered to network cables.[3] Other types of business devices such as telephones, bar code readers, and printers are also being untethered by WLANs.

Demand for wireless data networks in the home is also growing as multicomputer homes look for ways to communicate among computers and share resources such as files, printers, and broadband Internet connections.[4] Consumer-oriented electronics devices such as games, phones, and appliances are being added to home WLANs, stretching the notion of the LAN as primarily a means of connecting computers. These multiservice home networks support a broad variety of media and computing devices as part of a single network. A multiservice home network is depicted in Fig. 2-1.[1]

Analysts project that the number of networked nodes in homes, including both PC-oriented and entertainment-oriented devices, will top 80 million by the year 2005. As can be inferred from Fig. 2-1, the multiservice home network must accommodate a variety of types of traffic. The ideal multiservice home LAN:

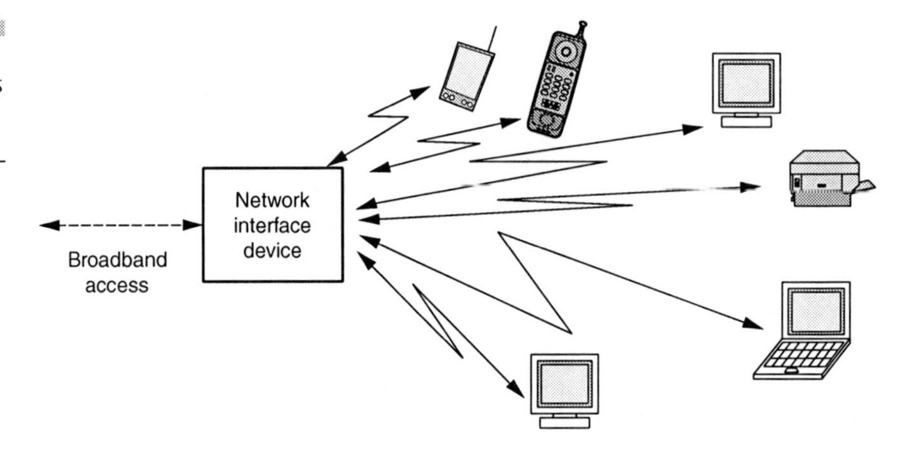

Figure 2-1
A multiservice wireless home network with broadband access.

- Supports differing traffic types such as low- and high-rate bursty asynchronous data transfer, telemetry information, multicast streaming audio and video, and interactive voice.
- Provides sufficient bandwidth to support an increasing amount of high-rate traffic both within the home and transiting the gateway.
- Allows multiple types of devices to operate on the network without interfering with one another.
- Efficiently supports diverse devices with differing price, power, and data rate targets.
- Efficiently allocates spectrum and bandwidth among the various networked devices.
- Can economically provide a single gateway through which services can be provisioned and devices can communicate outside the home.
- Provides coverage throughout the home, preferably with a single access point.[1]

Popular wireless data networking protocols such as Bluetooth, IEEE 802.11, and HomeRF meet some, but not all, of the multiservice home networking requirements. Furthermore, because the protocols were developed independently, they do not easily interoperate with one another and can cause significant mutual interference when functioning in the same radio space. The 802.11a WLAN standard offers speed and robustness for home networking that previous WLAN standards have not offered. Although access to this bandwidth for home networking is relatively recent, cost-effective chip sets have already been announced, such as Atheros' AR5000 802.11a chip set including an all-CMOS radio-on-a-chip (ROC). However, devices such as cordless telephones, personal digital assistants (PDAs), and networked appliances do not require all of the speed and features that 802.11a offers. An extension to these protocols that allows less expensive, lower-power, lower-data-rate radios to interoperate with higher-speed, more complex 802.11a radios is presented in this part of the chapter. The goal of this extension is to maintain high overall efficiency while allowing scalability: the ability to create dedicated radios with the capabilities and price points appropriate to each application and traffic type.

Background: 802.11 PHY Layer

Wireless data networking systems can be best understood by considering the physical (PHY) and media access control (MAC) layers separately. The physical layer of 802.11a is based on orthogonal frequency-division multiplexing (OFDM), a modulation technique that uses multiple carriers to mitigate the effects of multipath. OFDM distributes the data over a large number of carriers that are spaced apart at precise frequencies.

The 802.11a provides for OFDM with 52 carriers in a 20-MHz bandwidth: 48 carry data, and 4 are pilot signals (see Fig. 2-2).[1] Each carrier is about 300 kHz wide, giving raw data rates from 125 kbps to 1.5 Mbps per carrier, depending on the modulation type [binary phase shift keying (BPSK), quadrature PSK (QPSK), 16-quadrature amplitude modulation (QAM), or 64-QAM] employed and the amount of error-correcting code overhead (½ or ¾ rate).

> ***NOTE*** The different data rates are all generated by using all 48 data carriers (and 4 pilots).

OFDM is one of the most spectrally efficient data transmission techniques available. This means that it can transmit a very large amount of data in a given frequency bandwidth. Instead of separating each of the 52 subcarriers with a guard band, OFDM overlaps them. If done incorrectly, this could lead to an effect known as intercarrier interference (ICI), where the data from one subcarrier cannot be distinguished unambiguously from their adjacent subcarriers. OFDM avoids this problem by

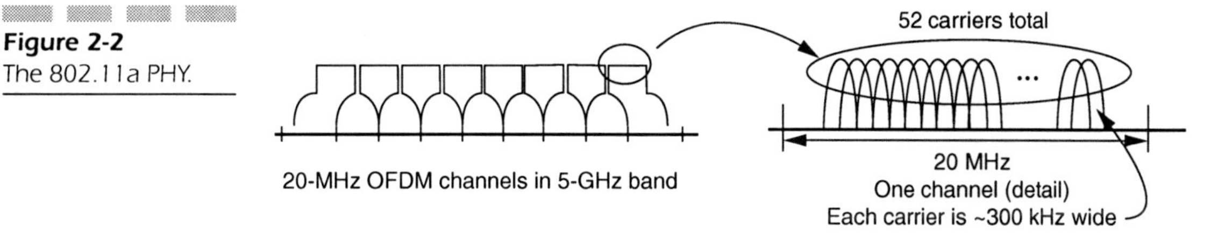

Figure 2-2
The 802.11a PHY.

making sure that the subcarriers are orthogonal to each other by precisely controlling their relative frequencies. In addition, coded OFDM is resistant to channel impairments such as multipath fading or narrowband interference. Because the coded information is spread across all the carriers, if a subset of the carriers is lost, the information can be reconstructed from the error correction bits in other carriers.

Background: 802.11 MAC Layer

Access methods for wireless data channels fall into three general categories: contention methods, polling methods, and time-division multiple access (TDMA) methods. The 802.11a is based primarily on contention methods, with some polling capabilities as well. Contention systems such as IEEE 802.11 use heuristics (random backoff, listen-before-talk, and mandated interframe delay periods) to avoid (but not completely eliminate) collisions on the wireless data medium. IEEE 802.11 also employs a beacon message that can be asserted by the access point and allows the access point to individually poll selected stations for sending or receiving data. The duration of the polling period is controlled by a parameter set by the access point and contained within the beacon message.

Contention systems are well suited to asynchronous bursty traffic. These systems work particularly well when the burst sizes are comparable to the natural packet size of the medium, or small multiples of the natural packet size. Slotted systems are well suited to isochronous applications that have a need for continuous channel bandwidth, although they may have extra overhead in comparison to contention systems when carrying asynchronous bursty traffic.

Another MAC layer consideration is whether there is a dedicated central controller such as an access point (AP) or base station. The 802.11a uses an AP, but has a fallback method for when there is no centralized controller (ad hoc mode). However, the operation of the network is more efficient with an AP present.

An Extension to 802.11a Is Needed

The 5-GHz 802.11a standard offers higher data rates and more capacity than 802.11b. However, to provide a complete solution for wireless data home networks, 802.11a needs to be extended to address remaining challenges. For example, the present standard does not support differing device/application types, nor does it enable a unified network that allows a single gateway or access point to support all the devices within a home. A cordless phone is a good example of such a device. It does not require a high data rate, but must provide high-quality sound and error-free transmission. As things stand now, there are only two ways to implement the phone in a standard 5-GHz wireless data network. You can make the phone a full 54-Mbps device and have it share time at a low duty cycle. This is an expensive solution for a cordless phone and draws high peak power while transmitting or receiving.

The second solution is to transmit at a data rate close to the cordless phone's natural rate, and make the rest of the network nodes wait for it to get off the air. This is highly inefficient and greatly reduces the overall throughput of the network.

The best solution is to allow the cordless phone to transmit at its natural rate at the same time other nodes are transmitting at their natural rates. Unfortunately, this type of operation is not supported under any of the existing 5-GHz wireless data network standards. An extension to 802.11a that allows overlaying transmissions using OFDM techniques has been proposed and is described later in the chapter.

The 5-GHz Unified Protocol

The 5-GHz Unified Protocol (5-UP) proposal extends the OFDM system to support multiple data rates and usage models. It is not a new standard, but an enhancement to the existing IEEE standard that would permit cost-effective designs in which everything from cordless phones to high-definition televisions and personal computers could communicate in a single wireless multimedia network with speeds up to 54 Mbps. The 5-UP achieves this by allocating the carriers within the OFDM signal on an individualized basis. As with the background on the existing standards, the 5-UP can be described by examining its PHY layer first, and then the MAC layer. Many of the elements of the MAC layer will be seen to be outgrowths of restrictions within the PHY layer.

5-UP PHY Layer

The 5-UP provides scalable communications by allowing different nodes to simultaneously use different subsets of the OFDM carriers. This is intuitive, and can be seen as an advanced frequency-division multiple access (FDMA) system. Most OFDM equipment can support this quite easily.

An example is shown in Fig. 2-3.[1] In this figure, the laptop, PDA, and voice over IP (VoIP) phone are simultaneously transmitting to an access point (not shown). The laptop device generates its OFDM signal using an inverse fast Fourier transform (iFFT). It would be simple for this device to avoid transmitting on some of the carriers by zeroing out some of the inputs to the iFFT and using only the remaining inputs to transmit data. Low-data-rate devices can then occupy the slots that were omitted by the laptop. In the case shown in Fig. 2-3, the PDA makes use of two of the omitted carriers, while the VoIP phone makes use of one.

At the receiving side, the radio would look similar to that shown for the laptop. All carriers can be simultaneously received by the access point and recovered through its single FFT-based receiver. The access point must then group the parallel outputs of the FFT into the separate streams. Finally, when the access point transmits to the other nodes, it can use a single iFFT to simultaneously create all the carriers. Each of the other nodes can receive only its subset of carriers, discarding the carriers intended for a different node.

The great advantage to this approach is that both the analog and digital complexity required in the radio scales with the number of carriers that can be transmitted or received. In the ultimate case of just one carrier, the radio becomes a single-carrier biphase shift-keying (BPSK) or quadrature PSK (QPSK) radio, transmitting at 1/52 the output power required to achieve the same range with a full 52-carrier radio. Table 2-1 highlights the relative analog and digital complexity required to achieve a given data rate.[1]

The 5-UP enables the building of radios with a broad range of complexity, which in turn results in a range of power and price points that serve a number of different data-rate requirements, allowing all to function simultaneously and efficiently in a high-data-rate system. Table 2-2 lists examples of the data rates and applications that can be met using various modulations and numbers of carriers.[1]

5-UP PHY Layer Constraints

While the evolution from an OFDM system to an advanced frequency-division multiple access (FDMA) system is intuitive, there are a number of constraints required to make it work. These constraints come from

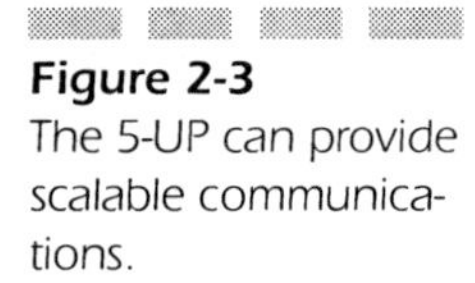

Figure 2-3 The 5-UP can provide scalable communications.

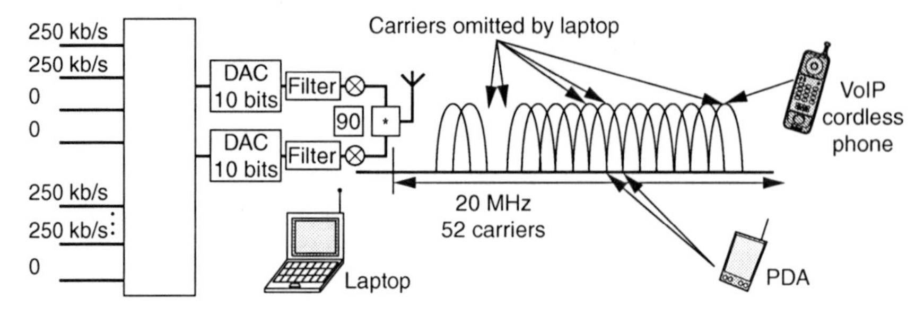

TABLE 2-1 Transmitter Power Based on Regulations for the Lower 100 MHz of the U.S. UNII Band

Data Rate	No. of Carriers	Modulation	Transmitter Power, Average, mW	Transmitter Power, Peak, mW (Approximate)	ADC/DAC	FFT Size
125 kbps	1	BPSK	0.8	1	4 bits	None
750 kbps	1	16-QAM	0.8	1.4	5 bits	None
1.5 Mbps	4	QPSK	3.2	4	5 bits	4
6 Mbps	8	16-QAM	6.4	8	6 bits	8
12 Mbps	16	16-QAM	12.8	16	7 bits	16
36 Mbps	48	16-QAM	40	48	8 bits	64
54 Mbps	48	64-QAM	40	48	8 bits	64

TABLE 2-2

Data Rate and Application Examples with Various Modulations and Numbers of Carriers

Data Rate	Applications	Carriers	Modulation
125 kbps	Cordless phone, remote control	1	BPSK
1.5 Mbps	High-fidelity audio	2 or 4	16-QAM or QPSK
12 Mbps	MPEG2 video, DVD, satellite, XDSL, cable modem, data network	12, 16, or 32	64-QAM, 16-QAM, or QPSK
20 Mbps	HDTV, future cable, or VDSL broadband modem	18 or 27	64-QAM or 16-QAM

the close spacing of the carriers (required to achieve high efficiency) and practical limitations in the design of inexpensive radio transceivers.

Narrowband Fading and Interference Control One disadvantage to using the carriers independently is that narrowband interference or fading can wipe out the complete signal from a given transmitter if it is using just one or a few carriers. Under those conditions, no amount of coding will allow the missing signal to be recovered.

Two solutions are well known to make narrowband signals more robust. The first is to employ antenna diversity. Radios can be built that can select between one of two antennas. If the desired carriers are in a fading null at one antenna, then statistically they are not likely to be in a null at the other antenna. Effective diversity gains of 8 to 10 dB are normally observed for two antenna systems.

A second way to provide robustness to narrowband fading and interference is to "hop" the subcarriers in use over time. This approach will work even for the case in which only one subcarrier is used at a time. For example, the node could transmit on subcarrier 1 in the first time period, and then switch to subcarrier 13 in the next period. Packets lost when the node is on a frequency that has interference or fading could be retransmitted after the next hop. Several such hopping nodes could be supported at the same time, hopping between the same set of subcarriers on a sequential basis. A similar arrangement could be used for nodes that use multiple subcarriers simultaneously, hopping them all in contiguous blocks, or spreading them out and hopping the entire spread of subcarriers from one channel set to another over time (see Fig. 2-4).[1]

A carrier allocation algorithm that is more intelligent than blind hopping can also be implemented. Narrowband fading and interference are likely to affect different nodes within a wireless data network differently because of the various nodes' locations. Thus, a given subcarrier may

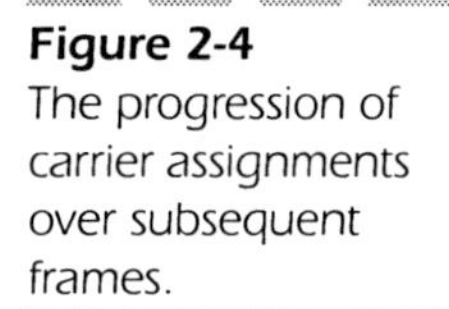

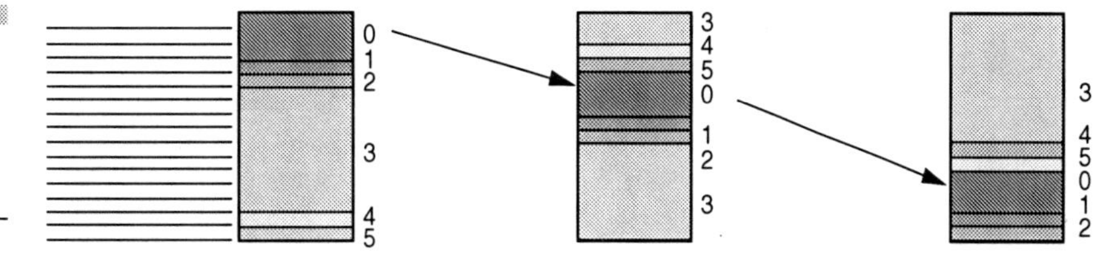

Figure 2-4 The progression of carrier assignments over subsequent frames.

work poorly for some of the nodes, but it might work well for other nodes. The subcarriers could therefore be intelligently allocated, swapping the assignments between nodes until all nodes are satisfied.

The 5-UP MAC

The 5-UP may readily be adapted to work with existing industry standard protocols such as 802.11a. Figure 2-5 shows a picture of the 5-UP frame as it would be embedded into an 802.11a system.[1] In the figure, the different rows represent different carriers, while the columns represent different slots in time.

To make the 5-UP work, three fundamental things are required. First, there must be a way to carve out time during which the 5-UP overlaid communication can take place. In the case of 802.11, this can be done by using the point coordination function (PCF) beacon. The original definition of 802.11 included two medium-access control mechanisms. These are the distributed coordination function (DCF) and the PCF. DCF is Ethernet-like, providing for random channel access based on a listen-before-talk carrier sense multiple-access (CSMA) technique with random backoffs. This is the most commonly used access mechanism in current 802.11 equipment.

The PCF access mechanism is based on centralized control via polling from the access point. In this access mode, all nodes are silent until they are polled by the access point. When polled by the access point, they can send a packet in return.

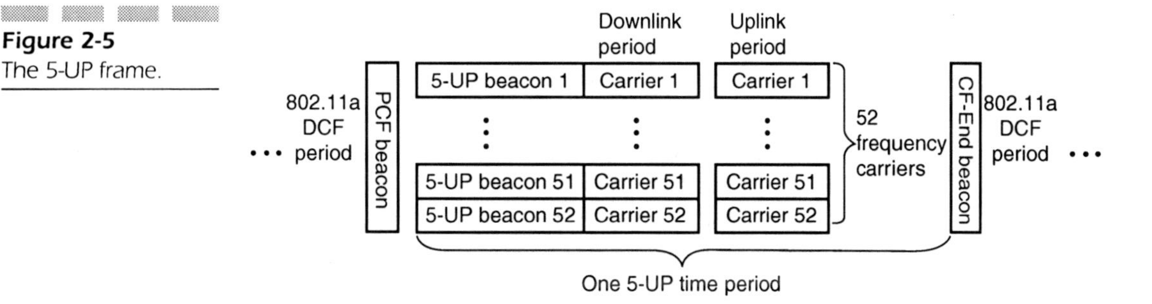

Figure 2-5 The 5-UP frame.

Two beacons are used to define the time during which the PCF access mechanism is in operation (the contention-free period) rather than the DCF mechanism. The PCF beacon announces to all the nodes that the polling access period is beginning. When nodes receive this beacon, they do not transmit unless they receive a poll from the access point that is addressed specifically for them. The end of the PCF (contention-free) period is signaled by a contention-free end beacon (CF-End).

In an 802.11 system, the contention-free periods are typically periodic, allowing for nearly isochronous communication of some portion of the traffic. The PCF beacon can be used to reserve a time period during which all legacy nodes will remain silent and the 5-UP can operate. Once the PCF beacon has been transmitted by the access point, all nodes must remain silent as long as they are not requested to transmit by a valid poll message. Because overlaid 5-UP traffic will not appear to be valid poll messages, legacy nodes will remain silent throughout the 5-UP period. The 5-UP-enabled nodes can then be addressed using the 5-UP without interference from legacy nodes.

After the 5-UP period has ended, the access point can send an 802.11 CF-End message, as defined in the standard, to reactivate the 802.11 nodes that were silenced by the initial PCF beacon. Following the CF-End message, communication would return to the nonoverlaid 802.11a method.

In this manner, the channel can be time-shared between traditional 802.11a operation and 5-UP operation. Legacy nodes will participate only in the 802.11a period, and will not transmit or receive any valid packets during the 5-UP period. Nodes that can operate only during the 5-UP period, such as nodes that can operate only on a subset of the carriers, will not be able to transmit or receive during the 802.11a period, but will be active during the 5-UP period. Finally, nodes that are able to handle both 802.11a and 5-UP messages can transmit or receive in either period. The access point can adjust the timing of the PCF and CF-End beacons to balance the traffic requirements of 5-UP and legacy 802.11a nodes.

The second requirement for embedding the 5-UP into the 802.11a protocol is to ensure that all devices know when they need to transmit in the 5-UP overlaid fashion and when to transmit according to the 802.11a methods. For nodes that understand the 5-UP only, or can use only a subset of the carriers, all communication outside of the 5-UP period will be indecipherable and will appear as noise. However, when the 5-UP period arrives, the 5-UP beacon transmitted at the beginning of this period will be intelligible. The 5-UP beacon is transmitted on each carrier individually such that even a single-carrier device can receive and understand it. This beacon includes information on the length of the 5-UP period and when the next 5-UP period is scheduled. Once synchronized, nodes that communicate only during the 5-UP period can sleep during the 802.11a periods.

Nodes that do not understand the 5-UP will know not to try to transmit during the 5-UP period, as described in the preceding. Nodes that understand both the 5-UP and the 802.11a protocol can understand all the packets that are transmitted, gaining information from both sets of beacons and potentially transmitting and receiving during both periods of operation.

Direct peer-to-peer communication or communication with the access point can be allowed in the nonoverlaid period. However, during the 5-UP overlaid period, only communication to or from the access point is allowed.

The third basic requirement is that 5-UP nodes must be able to request service, and must be instructed which carriers, hopping patterns, and time slots they should use. The 5-UP beacon is transmitted on each carrier such that even a single-carrier node can interpret this beacon no matter to which carrier it has tuned. The beacon includes information about which carriers and time slots are available to request service or associate with the network. As shown in Fig. 2-5, there are uplink slots (transmitting to the access point) and downlink slots (receiving from the access point). The node requesting service waits until it gets a response during a downlink slot. The response includes the carriers and time slots that will be allocated for traffic for that device. It also would indicate the hop pattern and timing if the network is operating in a hopping mode.

Some information, such as the time reference and when the overlaid communication period begins and ends, needs to be transmitted on each carrier; however, other information such as which time slot is assigned to which node for a given carrier is unique to each carrier. Information unique to a given node (sleep/wake information) needs to be transmitted on only one of the carriers assigned to that node.

Now, let's discuss how TIA/EIA standard IS-856 cellular data (1xEV) can be married with IEEE 802.11b wireless data to enable wide-area Internet access for service providers and users. In other words, the *lingua franca* of the Internet is TCP/IP, and wireless data devices are learning to speak this language. But what is the "wireless data Internet?" There are a number of different answers to this question. The question poses problems for equipment manufacturers, service providers, and users alike. You desire seamless access to the Internet, and in order to have that, all these different modes must operate transparently for users.

Wireless Data Protocol Bridging

Both 802.11 and the Telecommunications Industry Association/Electronics Industry Alliance (TIA/EIA) IS-856 are wireless data networking protocols. However, each meets different goals. Devices for short-range 802.11

wireless data networks are rapidly proliferating. Wireless data network providers (carriers) are eager to deploy high-speed wireless data protocols such as IS-856 that complement their wireless voice networks. The IS-856 standard is integrated into the protocols for code-division multiple access (CDMA) networks. Finding an effective means to connect 802.11 devices to increasingly available high-data-rate cellular networks answers the need of users for 802.11 devices to take advantage of the eventual ubiquity of high-speed cellular networks.

The 802.11 and IS-856 protocols have similar architectures. Wireless data stations are untethered. Both use similar modulation techniques for moving bits of data through the wireless medium. Both provide medium access control (MAC) to manage the physical and data link layers of the open systems interconnect (OSI) protocol model. Access points mediate access to other networks. Each has protocols for handing off between access points a station's logical connections as stations move into different coverage regions. Both are well adapted to support higher layers of the TCP/IP protocol stack.

However, significant differences exist as well. The differences arise from the different design goals these protocols serve. The 802.11 standard is designed to build short-range wireless local-area networks (WLANs), where the maximum distance between stations is on the order of 100 m. While IS-856 supports LANs, the range over which stations communicate is tens of kilometers. The IS-856 standard is designed to be an integral part of a cellular communication network that operates in licensed frequency bands assigned specifically for cellular communication. Networks of 802.11 devices use unlicensed frequency bands and must work in spite of the possibility of other nearby devices using the same radio spectrum for purposes other than data communication.

These differences, principally the difference in range, fostered the idea that these two wireless data systems could be combined to complement each other. Another factor behind this idea is the proliferation of 802.11-capable devices and the desire of their users to connect to the Internet via their Internet service provider (ISP). Thus, this part of the chapter up to this point has demonstrated how 802.11 networks and IS-856 networks can be bridged to facilitate user demand for this connectivity as they range through an IS-856 network with their 802.11 device.

Connecting the two protocols is quite straightforward. It can be done simply because these protocol designs complement each other in key ways. This part of the chapter provides overviews of how IS-856 and 802.11b manage the wireless data medium. Following the overview, the technique used to bridge the protocols is described. This part of the chapter concludes with some suggestions on how an ISP can take advantage of these techniques to offer wide-area access to its subscribers who are using 802.11 devices.

Overview of 802.11 Architecture

The introduction to this part of the chapter listed a number of similarities and differences between IS-856 networks and 802.11 networks. The differences are primarily due to the way in which each wireless data protocol is used. Networks of 802.11 devices are short-range wireless data networks. Today, typical applications for 802.11 protocols provide wireless data access to TCP/IP networks for laptop computers. The 802.11 protocols aren't limited to this kind of application. Any group of devices designed to share access to a common short-range communication medium can be built on 802.11's services. In the future, devices designed for particular tasks that incorporate communication with other nearby devices will be able to take advantage of 802.11's services in ad hoc networks. Some of these devices may simultaneously be part of the more structured environment of the Internet. This will have important implications when a single user or group of nearby users has a variety of devices that could interact for the benefit of their owners.

Devices able to take advantage of a wireless data network will use TCP/IP protocols as their means to exchange information with other devices. Because 802.11 defines MAC protocols, which correspond to the data link and physical layers of the OSI model, 802.11 is well suited to provide the basic connection on which the rest of the TCP/IP protocol stack depends.

This aspect of 802.11 enables it to fit neatly with IS-856 networks. For example, an IS-856 network could easily provide the backbone needed to connect a number of separate 802.11 networks into a single network domain. This idea is explored later when the particular architecture used for the IETF network is described.

IEEE MAC Protocol for Wireless Data LANs

One of the fundamental design goals for 802.11 is to provide services that are consistent with the services of 802.3 networks. This makes the peculiarities of wireless data communication irrelevant to higher layers of the protocol stack. The 802.11 MAC protocols take care of the housekeeping associated with devices moving within the 802.11 WLAN. From the point of view of the IP layer, communication via wireless data with 802.11 is no different from communication over an 802.3 data link, fiber, asynchronous transfer mode (ATM), or any other data link service. Because these different media are capable of different data rates, users can perceive differences in performance. But any well-designed application will operate successfully over all these media. This greatly reduces complexity for application designers. Reduced complexity results in more reliable and more robust applications, more rapid development by designers, and broader utility for users.

Designed for Multiple Scenarios

The fundamental organizational unit of an 802.11 network is called a basic service set (BSS). The members of a BSS are the wireless data stations that share a specific 802.11 WLAN. How a BSS connects to other networks defines the variants.

A BSS not connecting to another network is termed an independent BSS or iBSS (see Fig. 2-6).[2] An iBSS uses MAC protocols to establish how its members share the medium. There can be no hidden nodes in an iBSS. Each member must be able to communicate directly with all other members without relays. An iBSS is ideal for a collection of personal devices that move with the owner. For example, a PDA, laptop, cell phone, CD or DVD player, or video and/or audio recorder could be members of an individual's personal network of communication devices. An 802.11 network connecting them would provide an individual user with a rich array of ways to communicate with others. Another example might be a coffee maker, alarm clock, lawn sprinkler controller, home security cameras, home entertainment systems, and a personal computer.

Figure 2-6 Independent basic service set.

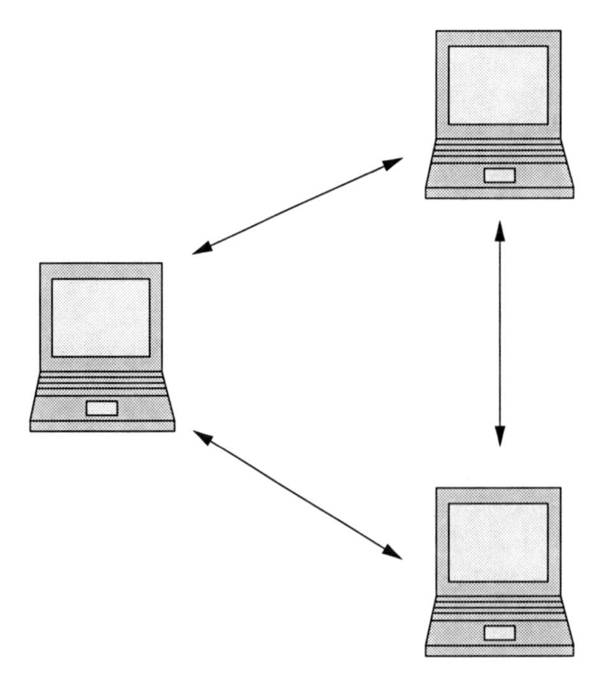

A network made up of these devices could turn on the coffee maker when the alarm goes off in the morning. It would allow a homeowner to water the grass from an easy chair, and make sure it is not watering the sidewalk, or turn the sprinklers on a burglar while calling the police and playing recordings of large dogs barking.

When a BSS connects with another network via an access point, it is termed an infrastructure BSS. Because this is the most common configuration today, the acronym BSS usually implies an infrastructure BSS. The access point is both a member of the BSS and mediates access to other networks on behalf the rest of the BSS. Generally, the members of the BSS beside the access point are personal computers. To facilitate coverage of a campus within the same 802.11 network, a group of BSSs, called an extended service set (ESS), define how access points hand off connections for members of the network as stations move between access points. The access points are connected by backbone links that provide the medium for the hand-off protocol (see Fig. 2-7).[2]

The 802.11 standard supports simultaneous existence of iBSS and BSS networks. It provides means for labeling networks and conditioning access so they can operate without interfering with each other. It is entirely reasonable that the computers mentioned in the iBSS examples in the preceding could participate simultaneously in a private 802.11 network and an infrastructure 802.11 network providing Internet access. While this idea has fascinating possibilities, further discussion is beyond the scope of this chapter.

Figure 2-7
Extended service set.

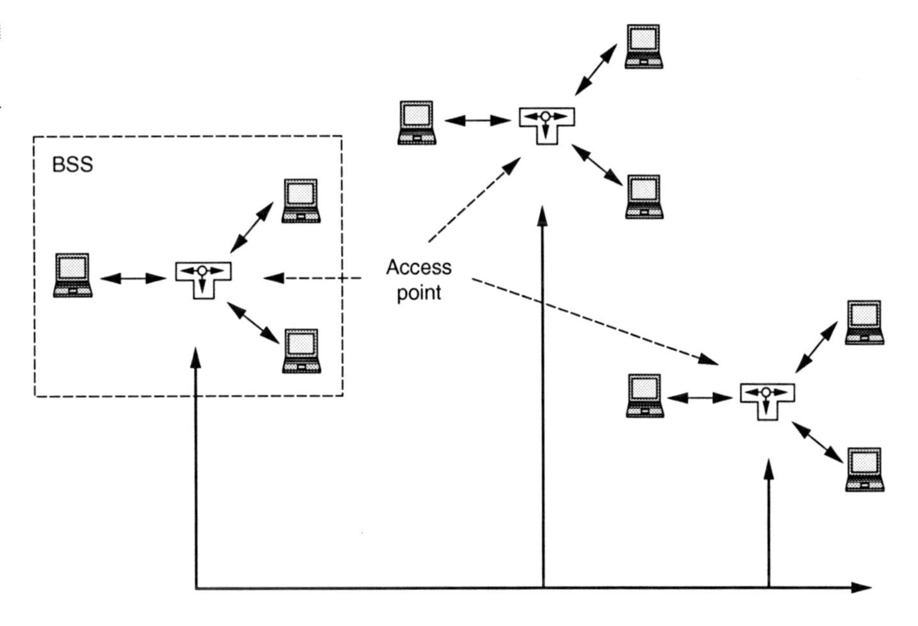

MAC Layer Protocols

The 802.11 standard consists of several MAC layer protocols to provide the variety of services necessary for the kinds of wireless data networks just described. A Beacon protocol enables a BSS or an iBSS to organize its communication. The Beacon information contains the network label information so 802.11 devices can discover the networks that exist within range of their antennas. The Beacon establishes the timing intervals of the network. Timing intervals mediate how stations access the medium. For an iBSS, once timing and network identity are determined, stations may exchange data. For a BSS, there are two additional groups of services to manage traffic.

Distribution Services and Station Services

The nine services for a BSS are grouped into distribution services and station services. There are five distribution services and four station services.

Distribution Services Distribution services manage traffic within a BSS and transfer traffic beyond the BSS. They provide roaming capability so a wireless data station can move between the BSSs in an ESS. The five services are association, reassociation, disassociation, distribution, and integration.

Association creates a logical connection between a wireless data station and the access point. Once association is established, the access point will deliver, buffer, or forward traffic for a wireless data station. The association service is used when a wireless data station first joins a BSS or when a sufficiently long enough period has elapsed with no communication between the access point and the wireless data station.

Reassociation is similar to association. A wireless data station uses reassociation when moving between access points. A wireless data station moving into an access point's coverage notifies the new access point with a reassociation request identifying the access point previously serving the wireless data station. The new access point then contacts the prior access point for any traffic that has been buffered for the wireless data station.

Either the wireless data station or the access point can use disassociation. A wireless data station sends a disassociation message when it is leaving the BSS. An access point may send a disassociation message to a wireless data station if it is going off line or has no resources to handle the wireless data station. In the latter circumstance, a wireless data station may attempt to associate with a different access point, provided there is one in range.

Access points use the distribution service to forward frames received from a wireless data station in its BSS. Frames may be forwarded to another station within the BSS, to another station within an ESS, or to a router for delivery to a destination outside the WLAN.

Integration and distribution provide a portal to non-802.11 networks. Integration takes an 802.11 frame and recasts it as a frame for a different type of data link service such as Ethernet.

Station Services While distribution services enable wireless data stations and access points to establish communication, station services grant permission to use a BSS and accomplish delivery of data in the BSS. The four services are authentication, deauthentication, privacy,[5] and data delivery.

Authentication, deauthentication, and privacy are potentially valuable. However, the current definition of these services cannot be relied on to protect access to the WLAN. In lieu of these limitations, there are alternative means, such as IPSec, to ensure the integrity of IP traffic sent across an 802.11 WLAN. More detailed discussion of these issues is beyond the scope of this chapter.

Of these services, data delivery is the most important. It provides reliable delivery of datagrams while minimizing duplication and reordering. It is the essential service for moving data across the WLAN. Data delivery, distribution, and management services are the essential services provided by the MAC layer of 802.11.

802.11: Versatile Wireless Data Environment

The MAC protocols provided by 802.11 permit the creation of a variety of short-range wireless data networks. These networks range from ad hoc collections of stations to integral subnets of a complex internetworking structure. The flexibility of 802.11 may well obviate the need for other protocol stacks for personal devices. Regardless, 802.11's easy adaptability for TCP/IP networking has proved its value for large communities. It is for one such large community that the Internet Engineering Task Force (IETF), combining the strengths of 802.11 and IS-856, proved to be especially valuable.

An Overview of IS-856 Access Network Architecture

This overview describes how the wireless data station and the access network provide transparent data transmission for the logical sessions between the wireless data station and the Internet. The description is based on a prototype implementation of the architecture. A scalable implementation would differ in some respects from the prototype, particularly with regard to methods for authentication and authorization of wireless data stations. The description notes those details and offers alternatives more suitable for commercial implementation.

CDMA cellular networks are spread-spectrum packet radio networks. Originally, the CDMA protocol was designed for efficient transmission of packets carrying voice data. Voice has different constraints from efficient data transmission. Voice transmission minimizes delay times at the cost of some data fidelity. The human ear is more tolerant of a little distortion than it is of delay. For data transmission, nearly the reverse is true. Errors in data bits increase packet retransmission, and that hurts overall network throughput.

In a CDMA network, the base station sends data to wireless data stations over the forward link. Wireless data stations use the reverse link to communicate to the base station. The IS-856 standard uses CDMA's reverse link packet structure, retaining compatibility with voice traffic. The forward link packet structure is different, but the modulation techniques are the same, preserving compatibility in the forward link. However, management techniques for voice traffic and for data traffic differ considerably. A voice call consists of a single CDMA connection during which the call begins and ends. Packet data transmission comprises multiple CDMA connections, so that the CDMA network is used only when the wireless data station must exchange data with the rest of the network. A single logical network session (a browser session or an e-mail exchange) will consist of a number of CDMA connections.

In the prototype IS-856 system all wireless data stations were known, so registration of the wireless data station in the network was simplified. In a commercial system, IS-856 systems would use the Remote Authentication Dial-In User Service (RADIUS) to manage the registration and configuration information a particular access network would need. RADIUS is not the technique used to register cellular phones in CDMA networks. The carrier would unify its accounting and billing for data upstream of the systems by using RADIUS with other systems used to account for voice traffic.

The RADIUS protocol is a means to authenticate connections to a data network and optionally provide configuration information to the device making the connection. When a user of a wireless data station begins a session with an ISP, the wireless data station and a network access server (NAS) exchange a series of messages that identify the user, and obtain parameters configuring the Point-to-Point Protocol (PPP) session used between the station and the access network. The network access server may rely on databases further upstream for authentication information it needs when the station attempts to connect.

Asymmetric Data Paths

To provide maximum data throughput for all wireless data stations in the network, IS-856 uses asymmetric data paths. This is not unlike the asymmetry between forward and reverse links in CDMA voice systems. By taking this approach to a packet data network, it is possible to provide higher forward link burst rates than reverse link data rates. The user model for wireless data stations assumes reverse link data demand is similar to demand at the terminals of wired networks. The forward link to the wireless data station is capable of transmitting bursts up to 2.4 Mbps. The reverse link provides a constant data rate of up to 153.6 kbps for each station. These data rates are comparable to those typically found on cable networks such as Time Warner's Road Runner service or Cox@Home.

Access Network and Wireless Data Stations

The carrier's access network mediates connections between wireless data stations and the Internet by providing access points in each sector. The access network is a private network, invisible and transparent from the point of view of devices connected to the wireless data station or from the Internet beyond the access network. Access networks manage the IP space for all wireless data stations in the carrier's service area. Besides transporting data, the access network includes monitoring and maintenance capabilities.

The access network and the wireless data station use PPP as their data link protocol. PPP is carried over the radio channel using the Radio Link Protocol (RLP) of IS-856. RLP minimizes data loss and packet retransmission in order to provide an interface to the wireless data medium with error rates that meet or exceed the requirements for adequate PPP performance.

In the prototype, each wireless data station manages a local subnet. This subnet is part of the IP space assigned to the prototype system, not part of the access network. In a commercial implementation using the same approach, the subnet managed by the wireless data station would be part of an ISP's IP space. Using the Dynamic Host Configuration Protocol (DHCP), the wireless data station distributes the IP space it manages, and transfers TCP/IP traffic between the devices, the wireless data station services, and the access point. Because the wireless data station handles the PPP connection, downstream devices don't need to. They simply function as they would ordinarily in a TCP/IP LAN. The wireless data station and access point cooperate to shield devices from the PPP session and to permit persistent TCP/IP sessions, independent of the CDMA connections. This helps optimize the use of the CDMA network resources in a way that is transparent to the user.

Access Network Architecture

Figure 2-8 shows the connection between a wireless station (WS in the figure) and the access network, as well as the access network's internal structure.[2] The access network consists of several subsystems. The principal systems are the consolidation router, modem pool controller (MPC), and access point. User Datagram Protocol/Internet Protocol (UDP/IP) is used within the access network to connect subsystems. These will be described next. While this is a description of a prototype architecture, most of the same components and functions must be present in a commercial system. Because this is a prototype, configuration information storage[7] and maintenance are simplified.

The consolidation router creates the boundary between the access network and the rest of the Internet. It provides routing information to the Internet for all wireless data stations managed by the access network. It also routes traffic within the access network, ensuring that private traffic stays within the access network. Routes for user devices to the Internet are derived from information maintained by the MPC.

The MPC is the heart of the access network. It houses the configuration server (CS), overhead manager (OHM), and a set of selector functions (SFs). The MPC uses the OHM and SFs to manage the state of wireless data stations within all of the cells served by the access network. The OHM's primary role is to assign an SF for use during a wireless data station session. In the prototype, the OHM also delivers configuration information it obtains from a static database in the configuration server. In a commercial system, the configuration server would interact with the RADIUS authentication, authorization, and accounting (AAA) server to obtain the necessary information for its database. When a wireless data station registers with an access network (via some access point), the access point notifies the OHM about the wireless data station. The OHM assigns an SF to manage the wireless data station connection. In a commercial implementation, the SFs may retrieve wireless data station parameters from either the configuration server database or directly from the AAA server. The SF cooperates with the wireless station to maintain PPP state. The SF encapsulates the PPP packet in RLP, and then forwards it via UDP to the access point. The SF also updates the consolidation router with current routing information for the wireless data station. When a wireless data station moves between access points by moving into a new sector, the SFs for each access point update the wireless data station routes for the consolidation router.

An IS-856 access point divides into two structures, a local router and modulation equipment connecting the access network to the cellular network. An access point shares its modulation equipment among a number of wireless data stations. Over time, the wireless data stations served by

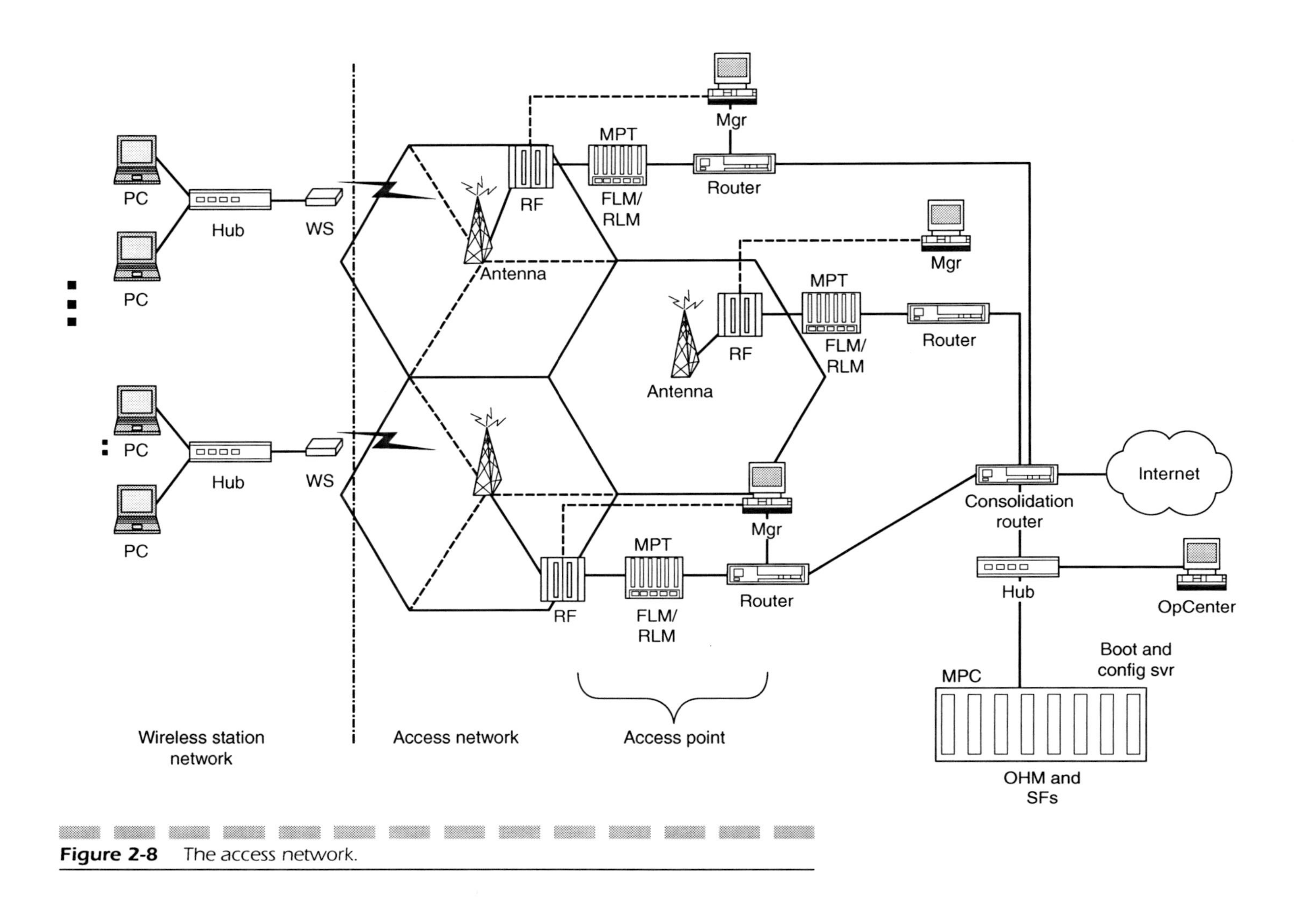

Figure 2-8 *The access network.*

an access point will change. The local router within the access point enables the modulation equipment to connect to the rest of the access network regardless of how resources are assigned to wireless data stations. The modulation equipment consists of pairs of forward link modules/reverse link modules (FLMs/RLMs) and an RF adapter. Collectively, this is called the modem pool transceiver (MPT). Each FLM or RLM is an IP device on the access network LAN. The RF adapter connects FLMs and RLMs to the RF system of the CDMA base station.

An FLM receives packets destined for wireless data stations. It provides the network and data link layer interface performing intermediate modulation of the data. After the intermediate-frequency (IF) stage, it hands the data stream to the RF adapter for broadcast in the cell sector. An RLM performs the inverse process. It receives an IF stream from the RF adapter, demodulates the data, and forms it into a packet, forwarding it to the SF.

Figure 2-9 shows how the access network uses UDP to encapsulate packets that are exchanged between the wireless data station and the Internet.[2] The IP datagram contains the user data flowing to and from the mobile node. The other protocol layers in the diagram show the encapsulation used to make the access network transparent to the Internet and to devices connected to the wireless data station. An IS-856 system preserves the PPP state between a wireless data station and an SF. This must be accomplished despite movement of wireless data stations between sectors and, consequently, between access points. The access network preserves this information by using UDP to wrap the entire packet down to the RLP layer. If a wireless data station changes access points, the SF updates its internal route to the new FLM/RLM. In this way, the SF and the wireless data station can maintain PPP state, regardless of how the wireless data station moves between sectors.

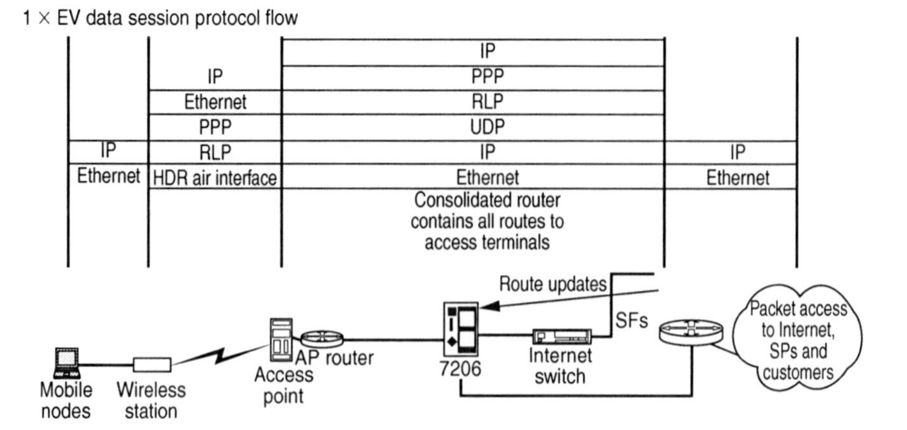

Figure 2-9 Access network protocol flow.

Forty-Ninth IETF Meeting Network

The IETF relies heavily on Internet communication for developing the protocols that are essential for the smooth operation of the Internet and for protocols for new services that can be provided over the Internet. The IETF meets three times yearly for face-to-face working group meetings to assist the work carried out by members over the Internet. An essential part of every IETF meeting is the increasingly misnamed "terminal room." The terminal room is a LAN created for the meeting to provide Internet access to attendees, and to members who cannot attend in person. Until recently, the LAN for each meeting provided wired access throughout the meeting areas of the hotel where meetings are held. The last few meetings have experienced an explosion in demand for 802.11 wireless data access as more attendees employ 802.11 wireless data networks at home. As a result, attendees have come to expect 802.11 coverage throughout the meeting areas of the main hotel.

As the number of people attending IETF meetings has grown, the meeting hotels have no longer been able to provide enough hotel rooms for all the attendees. Secondary hotels are used for the overflow. However, extending the meeting network to the secondary hotels has not been possible, putting attendees staying at the secondary hotels at a distinct disadvantage.

The design of the network for the forty-ninth meeting in San Diego demonstrated a solution for the access problem in the secondary hotels, provided that attendees in the secondary hotels had 802.11 cards for their laptop computers. By combining a prototype IS-856 network with 802.11 access points in these hotels, adequate access for those attendees was provided (see Fig. 2-10).[2]

An 802.11 BSS was installed in each secondary hotel. The 802.11 access point was connected to a prototype Qualcomm IS-856 wireless

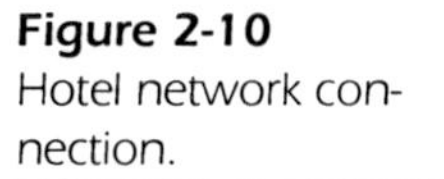

Figure 2-10
Hotel network connection.

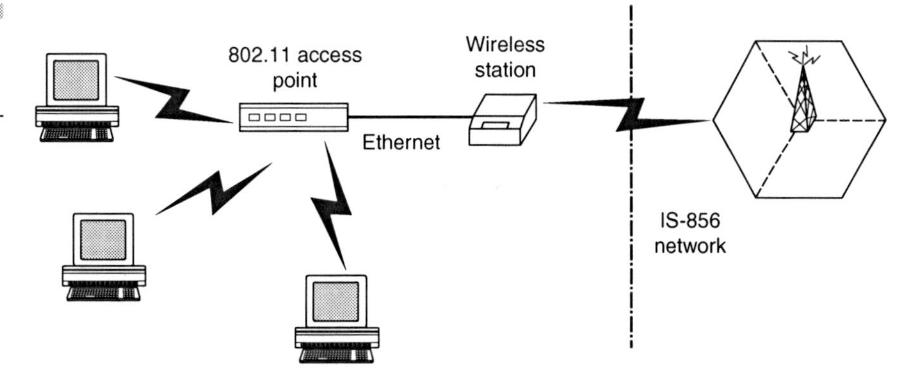

data station via a short 10baseT Ethernet cable. Each IS-856 wireless data station was assigned an IP address range from the prototype network it could distribute to the 802.11 cards of attendees' laptops. The 802.11 access point provided BSS housekeeping and the IS-856 wireless data network provided the backbone links connecting the 802.11 networks. It wasn't a true ESS, because users could not roam between BSSs and preserve their network address. However, in principle, there is nothing to prevent the forwarding necessary for an ESS.

During the meeting, some attendees were equipped with an IS-856 wireless data station for their individual use. This was done to compare the performance of individual use of the IS-856 network with the shared access provided by connecting an 802.11b network to the Internet via the IS-856 network. An 802.11b network provides data rates comparable to 10-Mbps wired networks. Because users of the 802.11 BSSs reported similar performance when a single IS-856 wireless data station was shared among multiple users, this experiment demonstrated that an IS-856 network provides an adequate backbone for an 802.11b ESS.

Finally, Fig. 2-11 shows a sample of the average data rates of both individual and shared IS-856 wireless data stations operating during

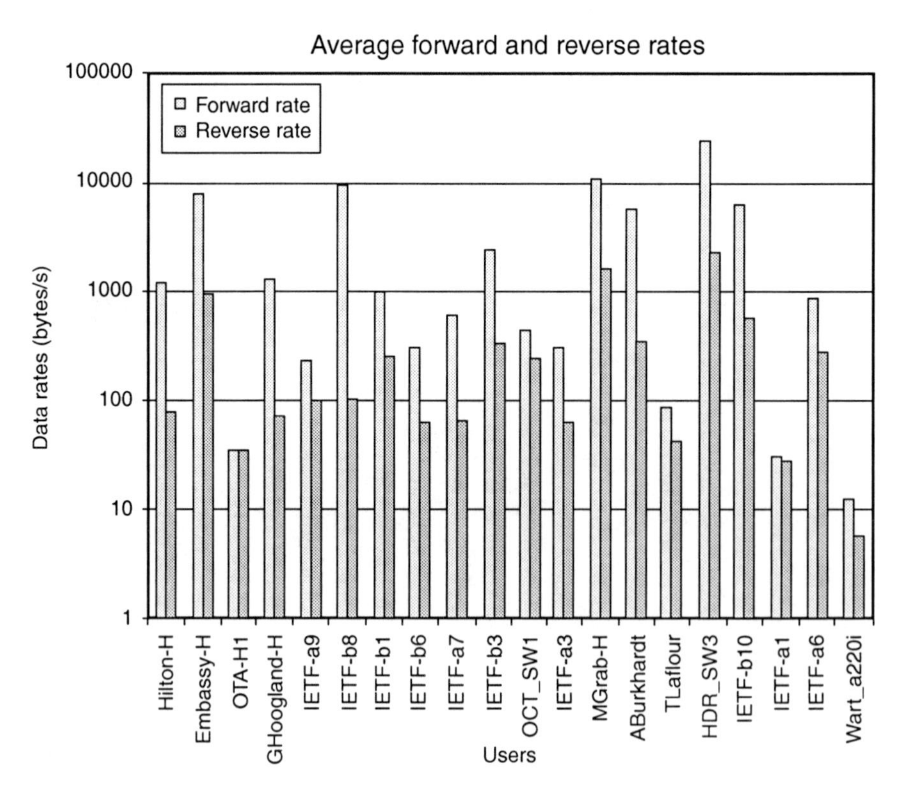

Figure 2-11
Aggregate data rates.

the meeting.[2] One can see from the chart that forward and reverse data rates are comparable. Users of the shared wireless data station in the 802.11 BSSs didn't seem appreciably affected by the difference in data rates between 802.11b and IS-856. A dozen or more users were sharing access to the IS-856 wireless data station via the bridge. The users were enthusiastic in their ability to access the net via the bridge. From their feedback on the performance of the prototype, using IS-856 wireless links as a backbone for an 802.11 ESS is promising.

Conclusion

This chapter discussed how the 5-UP will provide enhancements to the 802.11a standard that will enable home networking to reach its ultimate potential with scalable communications from 125 kbps through 54 Mbps. Robust, high-rate transmissions are supported in a manner compatible with 802.11a, while allowing low-data-rate, low-cost nodes to communicate with little degradation in aggregate network throughput. The 5-UP allows the construction of radios tuned to the performance requirements of any application from 125 kbps up, in increments of 125 kbps.

With 5-UP enhancements, each node can get a private, unshared channel with no collisions, fewer lost packets, no backoffs, and no waiting for the medium to free up. The 5-UP requires no big buffers because transmission rates can closely match required data rates, making 5-UP a natural for multimedia support and quality of service (QoS).

In summation, the 5-GHz Unified Protocol is a definitive step forward in the development of a new higher-functionality wireless data LAN standard for home networking that will allow all wireless data devices, regardless of their bandwidth requirements, to operate on the same network. The 5-UP will enable QoS, bandwidth reservation, and data rates up to 54 Mbps, while at the same time providing scalable cost, power usage, and bandwidth allocation.

This chapter also discussed how the 802.11 standard has provided a very popular method for individual wireless data access to the Internet. The quasi-ESS built with an IS-856 backbone offers interesting possibilities for practical systems. Both carriers and ISPs will face increasing demand from their customers for wireless data Internet access. There are at least two approaches that exploit the ease with which an 802.11 net can be bridged with an IS-856 backbone.

One possibility is that carriers will provide both ISP and infrastructure services. Carriers will succeed in this approach as long as they are adept at providing a wide range of services and support demanded by their consumer subscribers. As successful ISPs have discovered, service and support demand for consumer Internet access will extend well beyond the demands of simply providing wireless data Internet access.

Another possibility is that carriers will concentrate solely on building the infrastructure to transport data. ISPs will purchase wireless data access much as they purchase wired transport today. In this scenario, carriers would serve a more homogeneous set of customers consisting of ISPs with similar requirements. The ISP will focus on serving its specialized community of subscribers and take advantage of its knowledge of its customers to provide consumer subscribers with attractive services and support tailored to their tastes.

It is difficult to predict which of these scenarios will dominate the future of wireless data Internet access, or if some wholly different model will appear. It is certain, however, that the cost effectiveness of deploying IS-856 and the high consumer demand for 802.11-based devices will lead to the use of both wireless data protocols to satisfy demand for access to the wireless Internet.

Finally, this chapter discussed how the use of high-altitude platforms has been proposed for a joint provision of cellular communication services and support services for navigation satellite systems. Results obtained in the system design have shown that they are suitable to implement macrocells of large radius. In some cases, the number of sustainable physical channels is limited by the standard constraints, but can be improved by information on user location. Communication channels can then be used for the transmission of navigation messages to mobiles and exploited by users to notify the network of their position. The large coverage region and some navigation support services with better performance with respect to terrestrial stations make HAPs a promising infrastructure for a future system that will require the cositing of navigation and communication stations for the provision of integrated services.

References

1. Bill McFarland, Greg Chesson, Carl Temme, and Teresa Meng, "The 5-UP Protocol for Unified Multiservice Wireless Networks," Atheros Communications, Inc., *IEEE Communications,* 445 Hoes Lane, Piscataway, NJ 08855, 2002.

2. John W. Noerenberg II, "Bridging Wireless Protocols," Qualcomm, Inc., *IEEE Communications,* 445 Hoes Lane, Piscataway, NJ 08855, 2002.

3. John R. Vacca, *The Cabling Handbook,* 2d ed., Prentice Hall, 2001.

4. John R. Vacca, *Wireless Broadband Networks Handbook,* McGraw-Hill, 2001.

5. John R. Vacca, *Net Privacy: A Guide to Developing and Implementing an Ironclad ebusiness Privacy Plan,* McGraw-Hill, 2001.
6. John R. Vacca, *i-mode Crash Course,* McGraw-Hill, 2002.
7. John R. Vacca, *The Essential Guide to Storage Area Networks,* McGraw-Hill, 2002.

CHAPTER 3

Services and Applications over Wireless Data Networks

U.S.-based wireless data carriers are determined to develop enterprise services and applications even though trends overseas indicate the consumer market may be the most lucrative place for the technology. Leading the push is Sprint PCS Group, which recently unveiled plans to bring its Business Connect corporate application access service to Handspring Inc.'s (http://www.handspring.com/) Treo hand-held device.

As a result, Treo customers will have wireless data access to IBM's Lotus Software Division's Notes data and Microsoft Corp.'s Exchange data. A Sprint-branded version of the Treo that runs on Sprint's next-generation wireless data network will be available when the network launches in early 2003. Beyond basic access to Exchange and Notes data, however, the company plans to partner with IBM Global Services for customized enterprise solutions.

AT&T Wireless also plans a carrier-hosted service for companies looking to give employees wireless data access to their corporate applications. However, AT&T does recommend that corporate customers install Infowave's middleware behind their firewalls before installing a wireless data network. The company is also looking at how to bring wireless data LAN technology into its portfolio, as 802.11b, or WiFi, products continue to make inroads in the enterprise (see sidebar, "Faster Transmission Speeds for Wireless Data LANs").

Faster Transmission Speeds for Wireless Data LANs

Thinking of adding wireless data LAN installations to your resume? A number of strong products are available that are secure, easy to configure, and well suited to the small and midsize customer. And, increasingly, it's the small and medium business (SMB) customer that's looking at wireless data LANs as a way to cut down on cabling costs[3] and boost productivity among workers.

The wireless data LAN is a "nice and clean" extension to an office's wired LAN. Wireless data LANs are attractive to offices that want to enable workers to take laptops into a conference room. Wireless data has a place now.

A New Standard

Interestingly, small vendors have been able to come out with wireless data LAN gear that meets the faster 802.11a transmission rate ahead of larger vendors. The new standard performs at speeds as much as 5 times faster than 802.11b, the prevalent standard used by most wireless gear makers.

However, larger vendors including Cisco Systems[9] and 3Com indicate they also support the faster standard and are working on products, including access points, for commercial use. Larger vendors haven't entered the market because they don't want to disrupt the growing number of customers using 802.11b wireless gear. Still, they may be developing solutions that will allow access points using either standard to work together.

Small vendors that have come out with wireless data LAN gear that meets the 802.11a transmission rate (using chips from Atheros Communications) include SMC Networks, Proxim, and NetGear. One selling point for 802.11a gear is that customers can download larger files more quickly. SMC's access points sell for $365, supporting up to 64 wireless data clients simultaneously and operating in the 5-GHz frequency range at transmission speeds of up to 54 Mbps and 72 Mbps. SMC's adapter costs $145.

Another selling feature is that it's easier to find a clear channel with 802.11a. That's because the access points operate at the 5-GHz frequency and not the more crowded 2.4-GHz band used by 802.11b products.

Keep in mind that offices must have Internet connections fast enough to let them take advantage of the 802.11a access points. Also, the faster gear can be 5 to 10 percent more expensive. The 802.11a access points have more channels for areas that are spread out; the increased number of channels reduces the chance that access points will have to be set on the same channel. The result: better reception.

Opening New Doors

Another place for wireless data LANs is small offices without complete corporate networks. They want to use wireless data LANs, an affordable way for them to share high-speed Internet access and pass files back and forth, and they don't want to build a wired network because of all the construction that entails.

For example, NuTec Networks in Roswell, Georgia, sells almost exclusively to small and midsize offices. For them, wireless data LANs open doors to selling other devices, especially security products. For small offices, NuTec Networks recommends "rock solid" OriNoco 802.11b network kits from Agere Systems, a wireless networking company in Allentown, Pennslyvania.

For very small offices (fewer than 10 users), NuTec Networks installs the RG-1100 broadband gateway, which provides high-speed Internet access from DSL, ISDN, or cable-modem connections. It also provides enhanced 128-bit RC4 encryption and support for VPNs.

A kit for the laptop, which includes the Gold PC World Card, sells for $349. A kit with a USB client device for a desktop computer is also $349.

For larger offices, Agere offers the Access Point 500 for up to 30 simultaneous sessions and the Access Point 1000 for up to 60 simultaneous sessions. Both products have 128-bit encryption and power-over-Ethernet adapters and are compatible with RADIUS servers, which allow user authentication. The AP 500 is $495 and the AP 1000 is $895.

For first-time wireless data users, Heartland Business Systems (http://www.hbs.net/) plans to implement 3Com's Access Point 2000, which ships for $229. The 802.11b access point automatically configures itself and selects the clearest channel to operate on, according to 3Com. It has standard state-of-the-art security features that include 40-bit-wired equivalent privacy and 128-bit shared-key encryption.[1]

Verizon Wireless, meanwhile, is testing its initial third-generation services on a corporate audience. Soon, the company will team with Lucent Technologies Inc. to launch a high-speed data service trial in the Washington, D.C., area. Using code-division multiple access–based 1xEV-DO technology, the data transmissions could be as fast as 2.4 Mbps. And, by using virtual private networks, the service will provide users with secure access to corporate applications. The company will launch a similar trial with Nortel Networks Inc. in early 2003. (The Glossary defines many technical terms, abbreviations, and acronyms used in the book.)

NOTE Both Verizon and Sprint are also selling Audiovox Corp.'s new Thera Pocket PC hand-held, which offers integrated wireless data capabilities.

The flurry of enterprise wireless data activity here is in stark contrast to the wireless data world in Europe and Japan. Wireless data providers there are banking on multimedia and consumer applications and services being the driving forces for 3G. This concerns U.S. companies, which find their customers are looking overseas to see which technologies take hold and which fail.

There's considerable overhang because of the overpromise in Europe. Demonstrations of movies played on phones isn't helping the case that 3G is practical or necessary.

Companies want the utility, not the concept. Costs are a big factor in initial acceptance of new technologies and make it all the more likely it will be corporations and their needs that will spur the growth of 3G services, rather than features for the consumer.

Unfortunately, the rest of the world doesn't seem to agree. Mobile messaging service and mobile commerce are leading the pack. Issues such as authentication and digital rights management are closer to the bottom of the list.

Companies like NTT DoCoMo Inc., Telecom Italia Mobile, and Korea Telecom Freetel Inc. have echoed the sentiment that video messaging is where they see the future of 3G. They also indicate they see a future in wireless data advertising, an area that U.S. carriers have avoided. The advertising business model is pretty discredited here in the United States.

So, with the preceding in mind, can mobile commerce (m-commerce) find a place in your wireless data network? Actually, location-based wireless data services could help mobile data track down its killer application. Let's take a look.

Wireless Communications or Commerce?

You're wandering through Bucharest, Romania, lost and about to be late for an important appointment. You're worried that your assistant back home can't keep the network running in your absence, but your immediate attention is focused on survival.

There are no cabs or subway stations in sight, and the signs are all written in a language you don't understand. Does the street you're on lead to the office you're supposed to visit, or the part of town that the guidebook told you to avoid?

Hoping for an answer, you whip out your wireless data–equipped cell phone. What happens next depends on future developments in mobile data.

If you believe the vendors' optimistic predictions, your smart phone is an invaluable tool. It displays a map of the local area, complete with a route to your destination. A bus will be passing by in 5 minutes, so you have the option to buy a ticket electronically or call the closest unoccupied taxi, which can reach you in 2 minutes. You choose neither, as the phone also informs you that one of your colleagues is in a café less than a block away. Noting that the corporate network has reported no problems, you decide to join her and travel to the meeting together.

Alternatively, you may never get a chance to see the map or the network status report. Your phone is instead overloaded by pornographic spam, trying to entice you into the houses of ill repute that fill the neighborhood you have inadvertently entered.

The only obvious transport option is a cab company that's paid to partner with the cellular operator, and it can't reach you for at least half an hour. You're aware that your every move is logged and reported to the boss, who will want to know why you're not at the meeting already. The expensive-looking cell phone has attracted the attention of some unpleasant characters. You switch it off and start to panic.

Both scenarios rely on location-based wireless data technology that uses satellites or radar-style systems to determine a cell phone's position to within a few meters.[7] This technology is set to become more widespread in 2003, though its own initial location is quite surprising. In an industry usually led by Europe and Japan, the first country to offer location-based wireless data services and applications across all its cellular networks will be the traditionally tardy United States.

America's carriers aren't deploying the technology because they think it'll be profitable; they're doing so because of a regulation designed to help emergency services pinpoint 911 callers. Nevertheless, it will give mobile business a much-needed kick start.

If you're considering any kind of application for mobile data, location-based wireless data services could play a role. America's carriers also throw up new but predictable privacy[5] and security concerns: Should you be keeping track of your employees and potential customers? Or should you be worried about marketers, your service provider, and the FBI keeping track of you?

Corporate applications of location-based wireless data services are often described as "m-commerce," marketing-speak that prompts many of us to flinch in disgust. The phrase has become a catchall term for any business conducted using a cell phone, from checking your corporate e-mail to buying soda (both of which have yet to become mainstream, though they are offered by some carriers). But, while m-commerce is certainly overhyped, it isn't entirely empty; nor is it just pocket-size e-commerce.[8]

M-commerce proponents originally claimed that it would enable customers to buy anything, anywhere. They forgot that cell phones already allow people to do this, and in a way that doesn't involve navigating a menu system five layers deep or typing URLs on a 12-button keypad. When British operator Orange asked a group of volunteers to survive for a day by ordering their food through a Wireless Application Protocol (WAP) phone, it found that they quickly gave up on the wireless Web and called for pizza.

The operator blamed this on WAP's primitive state, but even a perfect user interface wouldn't stimulate much m-shopping. If people are prepared to wait several days for delivery, as most online shoppers are, their order probably isn't urgent enough to require a cell phone.

Most of us use cell phones to keep in touch with friends, colleagues, and contacts, not to buy things. This will likely be the case even when mobile data capability becomes widespread, though the type of communication may change: e-mail (with attachments), database access, and perhaps video or multimedia will supplement basic voice service. As companies have realized this, m-commerce has become a lot less fashionable.

In June 2001, analyst firm Ovum (http://www.ovum.com) asked 60 enterprises in the United Kingdom what they saw as the main application for wireless data. Of the nine available responses, not one enterprise mentioned mobile commerce.

Nearly half chose the ability to retrieve data from corporate networks, and all said they had data that mobile users could benefit from. Some jargon-happy vendors describe this as business-to-employee (B2E) m-commerce, but it's really just remote access.

Nevertheless, mobile commerce isn't dead. Operators are spending billions of dollars on third-generation networks, and they cannot recoup those investments in charges for bits or minutes. They hope to recover their expenditures through more innovative services that take advantage of a cell phone's great distinction—that it accompanies its user nearly everywhere.

Some of these are extensions of existing Web services. They rely on a phone's ability to keep in constant contact with customers, helping them to make time-sensitive decisions. Location-based wireless data technology is something new and unique to the mobile world, permitting genuinely innovative services: for example, a phone that can provide precise traffic and weather forecasts, guide police to a thief whenever it is stolen, and record a person's movements both on line and off line. This last one particularly worries many people, so the industry is emphasizing that location-based wireless data services don't (yet) mean an electronic tag of the kind currently applied only to convicts.

Data aren't stored long-term. Certain services might do this in the future. Parents might have a location-detection device sewed into their kids' backpack or shoes.

Triangulation

All location-based wireless data technologies rely on some variant of triangulation, which means calculating a phone's position by measuring its distance from two or more known points. In the simplest systems, these points are the base stations that sit at the center of every cell. Therefore, all the processing is done by the network, and doesn't require new phones.

Distances are generally measured by using a primitive form of radar: Each base station sends out a radio pulse, timing how long the response takes. Some systems also try to infer distance from signal degradation: The farther away the phone, the weaker its signal.

Neither method is particularly accurate because radio waves don't always travel directly between two points. They're reflected off walls, trees, and hills, which can make a phone appear to be farther from a base station than it really is.

For increased precision, most systems try to triangulate using at least three sites. The problem with this approach is that not all areas are within range of three different base stations, as networks are usually designed to minimize the overlap between cells. Many remote areas are served by only one, making any kind of triangulation impossible. A single measurement can ascertain how far away a user is from the tower, but not in which direction.

For 20 years, sailors and explorers have known that the most exact way to determine location is through the Global Positioning System (GPS), a constellation of 24 satellites run by the U.S. Air Force. Its weaknesses used to be that terminals cost thousands of dollars, and that the military introduced a random error to frustrate enemy users, which also affected civilian applications.

Both faults have since vanished: The error was switched off in 2000, and GPS receivers are now small and cheap enough to put inside a cell phone.

Only Qualcomm has shipped a GPS phone (it's used in Japan), but all the other major vendors plan to produce them in 2003. They claim that because the receivers only need to pick up the satellite signals, not transmit them back, they can be the same size as regular cell phones—not the bricks usually associated with satellite telephony. Most phones will eventually be equipped with GPS, whether customers want it or not.

GPS works in the same way as ground-based systems that measure time differences, though it's complicated by two factors. First, the satellites are moving, so they continuously transmit their own positions rather than sending simple radio pulses. Second, there's no return path from the receiver back to the satellite. The satellites overcome this by transmitting the precise time, measured by an onboard atomic clock.

A receiver can calculate its distance from each satellite by comparing the received time to its own clock, and then performing triangulation. The receiver needs to lock onto four satellites simultaneously: three to triangulate (because the system is three-dimensional, measuring altitude as well as map coordinates) and one extra to keep its clock synchronized with the network. This results in a location pinpointed to within 5 m (16 ft) and time measurements more accurate than the Earth's rotation.

Assisted GPS

Regular GPS receivers, when first switched on, can take several minutes to find four satellites, which isn't acceptable for location-based wireless data services. Instead, most will use a solution called Assisted GPS, which also keeps an active GPS receiver at every base station. This broadcasts the precise time (eliminating the need for a fourth satellite) and tells the phone where to look for the other three satellites.

NOTE A cell phone's battery would be drained too quickly if it kept the GPS receiver on all the time.

Assisted GPS is particularly useful in code-division multiple-access (CDMA) networks, because all of their base stations already include GPS receivers. It also has two other benefits: The base stations act as a backup when satellites aren't visible, and this can be more accurate.

NOTE CDMA systems need to know the precise time for synchronization purposes, and the GPS time signal provides the accuracy of an atomic clock at a much lower cost.

Most satellite systems require a clear line of sight between the satellite and the receiver. The GPS signals are slightly more resilient (they can pass through many windows and some walls), but they still won't work deep inside a building or underground. Assisted GPS can fall back to base station triangulation in these situations, providing at least some information whenever a user is able to make a call.

A few years ago, enthusiasts invented a system called Differential GPS. This uses a stationary GPS receiver to calibrate the system and correct errors, providing a simple work-around for the military's security. That's why they removed it.

A cellular network with Assisted GPS could easily be converted into a large-scale Differential GPS, though no carriers have yet announced plans for this. With the deliberate error gone, it could correct for natural errors due to atmospheric interference, potentially pushing the resolution to within 1 inch.

Assisted GPSs offload all of the positioning calculations to a server somewhere on the network. No matter what mobile operators claim, this isn't to reduce the weight or power consumption of the handsets [the math is trivial (for a microchip), and vendors say that they could build the triangulation capability into a phone]. Rather, it's so that operators have a service that they can bill for.

High-Resolution Maps

The most impressive location-based wireless data services to date are high-resolution maps, complete with a you-are-here sign. These are already available on some specialized GPS devices, without the services of a cellular carrier.

Their weakness is that the map data must be stored in memory, and must be reprogrammed manually for different areas. Online maps have been tested in Tokyo, but their high bandwidth requirements mean that they need 3G—still many years away for most of us.

However, many other applications are practical over today's narrowband connections. A few European carriers are already experimenting with them, using short message service (SMS) or even regular voice telephony.

In Helsinki, people can use their cell phones to get directions over both WAP and an automated voice system. In Geneva, directory assistance will tell callers the address of their closest nightclub or restaurant. Parisians can receive a text message whenever a friend is within a predetermined range, enabling the two people to arrange to meet in person.

As useful as these services are, they're not widespread. They don't require broadband,[4] but they do require new hardware in the cellular network—and for the most accurate fix, new phones. With budgets already stretched by upgrades to 2.5G, operators are reluctant to spend more on what is still sometimes seen as a less profitable technology.

Regardless of the business case, there's another compelling reason for location-based wireless data technology: public safety. When someone dials 911 from a landline, the emergency services can tell exactly where the caller is. Cell phones originally offered no such guarantee, relying on the often-distraught caller to describe the location.

Without location-based wireless data technology, emergency calls from cell phones are sometimes more of a nuisance than a lifesaver. Mountain rescuers complain that wireless data networks have encouraged foolhardy climbers to carry a cell phone instead of real safety equipment, endangering the team sent up to find them when they call for help.

Car crashes are often reported by hundreds of passing motorists, jamming the switchboard and risking further accidents. Talking on a cell phone is already a dangerous distraction for drivers, implicated in up to 30 percent of all accidents according to the National Highway Traffic Safety Administration (http://www.nhtsa.dot.gov). Looking around to make guesses about location can only make this worse.

In 1999, the FCC passed its Enhanced 911 (E-911) mandate, a two-part order intended to make the operators supply location information to emergency services. Phase I came into effect immediately and was easy to comply with. It says that operators must reveal which base station a caller is closest to, data that all cellular networks already track (so that they can route calls). This helps somewhat, but one cell tower can cover an area of up to 4000 km^2, or a million acres, so a more precise system was needed.

Phase II says that operators must provide specific latitude and longitude coordinates. The allowed margin for error depends on the type of location-based wireless data technology used: GPS and Assisted GPS are assumed to be twice as accurate as systems that rely on the network of base stations alone. Operators were supposed to comply with Phase II by October 2001, but like so much else in the wireless data world, this deadline has been pushed back. Just about every operator has applied for some sort of waiver.

Location-Based Wireless Data Services to Go

Waivers will hold back location-based wireless data services for a few months, but U.S. carriers still plan to offer them during the first half of 2003. Elsewhere, it will take longer, as operators aren't under the same regulatory pressure.

One reason is that Asian and European networks use smaller cells, making basic information about which one a caller is in more useful. Another is that Europe and Asia have firmer plans for 3G, so they can perform both upgrades at once.

The trend toward wireless data networks supplemented with Bluetooth or wireless data LAN technology is also more advanced in Europe. The maximum range of these is sometimes as little as 10 m (33 ft), making it easy to determine location without any kind of triangulation.

Users may be less concerned with deployment than with how the operators plan to get the necessary returns. Location technology gives them a powerful tool to spy on their customers, and some may be tempted to abuse it.

In July 2001, the first settlement in a lawsuit involving cell phones and health gave a group of users a $2.5 million payout—not because the phones had harmed them, but because the operators had invaded their privacy by sharing personal data in an alleged conspiracy to cover up the risks.

Most analysts predict a boom in mobile advertising, though they can only guess at the figures. Jupiter Research (http://www.jup.com) forecasts that revenue from advertisements sent to U.S. cell phones will reach $800 million by 2006. Ovum's estimate is even more bullish, a terrifying $3 billion, with a worldwide market 5 times that.

Though current revenue is zero, more than 91 companies have already joined the Wireless Advertising Association (WAA, http://www.waaglobal.com), a group lobbying Congress on behalf of the would-be industry. Even it recommends that advertisers adopt a "double opt-in" policy, rather like e-mail groups that require people to confirm their subscriptions. Burying a consent clause in a cellular contract isn't enough.

Carriers lured by the numbers should be cautious and heed warnings from Japan and Australia. Mobile operators there have already faced boycotts after their networks became overwhelmed by (non-location-based) advertising.

NTT DoCoMo found spamming very profitable, but was forced to block it anyway. Telstra went a step further, refunding charges incurred for listening to its telemarketers' voice mail. In both cases, customers found the ads so irritating that they simply left their cell phones at home.

So, why should you care about any of this? You should care because the need for satellite wireless data broadband is real, especially in areas where service providers are unable to reach customers with traditional terrestrial wireless data broadband options.

Reseller Opportunities with Two-Way Satellite Access

Terrestrial networks are rapidly expanding, evolving, and struggling to be the ubiquitous systems customers expect. Nonetheless, there's still a long road to travel. The last few months have undoubtedly demonstrated the opportunities and pitfalls involved with deploying wireless data broadband to the world.

Regardless of the amount of fiber laid, DSL access multiplexers (DSLAMs) installed, cable plants upgraded, and wireless towers constructed, many bandwidth-hungry consumers and businesses remain unreachable. Except that is, from above.

A new breed of satellite technologies and services allows providers to bring high-speed, always-on, two-way access to the planet's farthest reaches. For example, McLean, Virginia–based StarBand Communications (a joint venture of Israeli satellite powerhouse Gilat Satellite Networks, EchoStar Communications, and Microsoft) is the first company to launch two-way consumer service in the United States.

The company claims that, for $69.99 a month, it can provide download speeds up to 500 kbps with a floor of 150 kbps and upload speeds bursting to 150 kbps with an average of 50 kbps. For an additional $30 per month, customers can also subscribe to EchoStar's Dish network services using the same satellite dish shown in Fig. 3-1.[2]

StarBand Communications (http://www.starband.com/) enjoys an all-star cast of venture partners with exceptionally deep pockets, but it will have to prove it can perform. Since its launch in December 2000, StarBand has switched its model, choosing to focus on providing wholesale services to retail partners.

StarBand and its peers that are targeting consumer markets have a number of sizable hurdles to clear. Soaring equipment costs, trou-

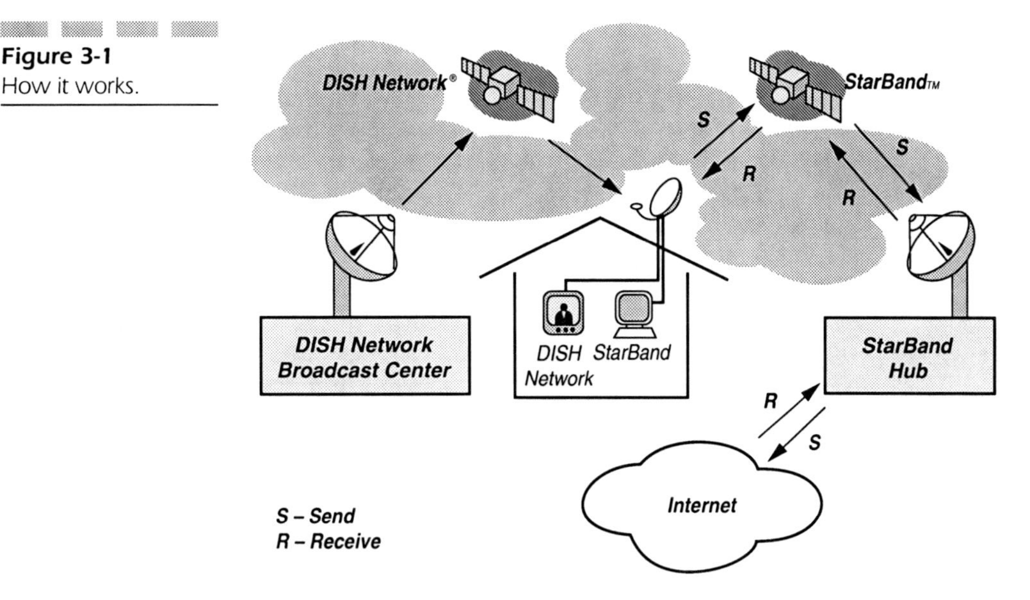

Figure 3-1
How it works.

blesome installations, satellite-transponder capacity issues, high latency, and increasing terrestrial competition combine to create a potential black hole for first-generation residential broadband satellite (RBS) providers. Companies that can navigate among the many perils will find that a lucrative market awaits, practically begging for service.

Target Markets

There is little question residential broadband satellite providers will play second fiddle to terrestrial and fixed wireless data services in the United States. In spite of this, there's still plenty of market action for high-flying satellite services.

At the end of the third quarter of 2001, a report from the Yankee Group estimated that there were 56,000 RBS subscribers in the United States. The report also forecast that, by year-end 2006, 6.7 million U.S. subscribers will be accessing the Net by satellite. These projections include one-way customers using a telco return channel. A report by Pioneer Consulting puts residential two-way users at 4 million worldwide by 2006.

As DSL and cable modem shops duke it out, often to the bloody death, the satellite industry is confident it can soar by and beam up a few million subscribers here and there. DSL subscribers suddenly missing a

DSL provider are looking straight up to satellite providers to fill the bandwidth void. Northern Sky Research LLC (http://www.northernskyresearch.com/) estimates that the number of addressable homes, defined as homes that currently have a computer or some form of Internet access and are unlikely to get a terrestrial connection, is north of 27 million in 2002.

Of course, this number will shrink as terrestrial networks continue to spin their webs. Addressable homes will decrease to 17.5 million in 2005. That number is certainly on the low end of the spectrum. It's probably a lot higher than that. The situation is similar to direct broadcast satellite (DBS) when it was rolled out. The initial assumption was that only rural users would access the system, but today people with other cable options are the largest percentage of DBS subscribers. By 2005, that revenue potential is upward of $25 billion for residential and enterprise access services, although the supply will rush to catch demand over the next 3 to 4 years, possibly constraining revenue's real growth potential in the short term.

Crowded Neighborhood

By taking advantage of leased Ku-band capacity on two orbiting geosynchronous equatorial orbit (GEO) satellites, GE-4 and Telstar 7, instead of launching their own birds, StarBand Communications claims to be the first provider of two-way residential broadband service in the United States. But others have been quick to show up in the game.

For example, DirecPC (http://www.direcpc.com/) has provided consumer satellite access for over 4 years and just rolled out its two-way service, Direcway, in June 2001. The company claims to have 227,000 subscribers on its system, including those using a dial-up connection for upstream requests. DirecPC is owned by Hughes Network Systems, which, through a number of subsidiaries and corporate designations, is ultimately controlled by General Motors Corp.

Hughes' and StarBand's founding partner, Gilat Satellite (http://www.gilat.com/Home.asp), happen to be bitter enemies in the very small aperture terminal (VSAT) market. These two industry giants control roughly 99 percent of the time-division/domain multiple-access (TDMA) VSAT market. Inside that space, the companies slice the pie in half, each owning approximately 50 percent. They aggressively fight for fractions of percentage points year after year.

Gilat, determined to remain Hughes' number-one foe, refused to sit still while its rival took control of the RBS market. With distribution partners in tow, Gilat, through its Spacenet subsidiary, formed StarBand in January 2000 to take on Hughes and DirecPC in a high-stakes battle 22,236 miles above earth's surface. With decades of VSAT experience and powerful partners, these two companies appear poised to become leading long-term providers in the RBS market.

But wait, not so fast. Much like the terrestrial combat being waged among powerful Baby Bells and aggressive competitive LECs (CLECs), several satellite start-ups are targeting the residential/short-message entity (SME) market, hoping to win business.

Denver-based WildBlue (http://www.wildblue.com/me/rel0815.html) is probably the wildest and furthest along among the hopefuls. Rather than lease capacity on existing satellites, the company is hatching its own birds. WildBlue 1 is being manufactured by Loral with a planned launch in early 2003 by Arianespace. WildBlue will take advantage of Ka-band capacity.

It's those two letters, Ka, that are driving WildBlue and others to take on such risky and expensive endeavors. WildBlue estimates it will cost at least $800 million to get its service off the ground. The average cost for development, construction, and launch of a large GEO is $360 million. Ka-band, using spot-beam technology, allows enabled satellites to produce up to 4 times the bandwidth that existing Ku-band birds can provide in the same amount of radio spectrum.

Multiple-spot beams narrowly focused on specific geographic regions rather than one large beam, as Ku employs, grant Ka-band/spot-beam technology the ability to reuse a large amount of frequency. Along with WildBlue and Hughes' Spaceway project, Astrolink Technologies, Teledesic, and Cyberstar have all announced plans to launch Ka-band GEO and LEO satellites beginning in 2004.

Subscribers

By choosing to rent instead of buy, StarBand and its partners have an early lead in the two-way market, but first-mover advantage not does guarantee long-term success. With the inherent technical issues imposed by sending packets 23,236 miles into space and back, StarBand must contend with relatively high customer premises equipment (CPE) costs, laborious installations though independent contractors, its multiple owners, and trying to turn a profit.

Early on, former controlling partner Gilat, which manufactures the VSAT and modems, realized it would need experienced distribution partners to move StarBand's service. EchoStar, under pressure from Wall Street to come up with a wireless data broadband offering, put an initial $60 million into the venture.

For a variety of reasons, EchoStar (http://www.dishnetwork.com/content/aboutus/index.shtml) was recently forced to pump another $60 million in StarBand and effectively take control of the company. Several insiders have speculated that EchoStar was not providing sufficient attention to the project early on. The company was forced to get involved when StarBand management, mostly brought over from Gilat, led the fledging company through a series of rollout missteps.

After the latest round, EchoStar's equity stake in the company is 32 percent. Through Echostar, StarBand has plans to launch its own satellite in the coming years, and at that point EchoStar's equity stake will climb to 60 percent.

The partnership allows StarBand to take advantage of 34,000 Dish retailers with the two companies comarketing services as a single, bundled product. Many of EchoStar's roughly 7 million current Dish DBS subscribers have StarBand written all over them. The EchoStar relationship is not exclusive, however. In addition to its StarBand investment, EchoStar has thrown another $60 million into StarBand's competitor, WildBlue.

For small and midsize ISPs looking to become StarBand retail partners, the company has not yet set up a formal program (see sidebar, "Small Wireless Data ISPs"). A number of StarBand's Dish resellers have their own programs that vary from shop to shop. StarBandDirect.com, an independent online StarBand reseller, originally intended to offer a reseller program that paid a $105 commission per StarBand sale, but was forced to drastically cut the commission percentage after experiencing huge installation problems that made the economics unfeasible.

Small Wireless Data ISPs

Factories with high-current electrical equipment, arc welding, and high-intensity lighting cause radio-frequency "pollution" that is extremely challenging to RF systems. For example, Cirronet (http://www.cirronet.com/), to meet the challenge, made good use of its robust wireless data technology in developing a wireless data system for ISPs. Cirronet's WaveBolt is very purposefully designed for the unique requirements of the small to medium-size wireless data ISP market:

- Its signal allows reliable operation in the crowded 2.4-GHz license-exempt spectrum worldwide.
- Its customer premises equipment (CPE) component (subscriber unit) is designed to be installed by customers, so no expensive truck roll is required.
- Its CPE device is under $400, including (Cirronet claims) the amortized cost of the base station.
- Its range is reasonable at 2 to 5 miles.[2]

The subscriber unit is remarkable in its simplicity: a small plastic "brick" with mounting tabs and a 125-ft cable. All of the RF electronics and a 90° flat panel directional antenna are contained in the subscriber unit.

The customer mounts the subscriber unit on the side of the structure closest to a base station and routes the cable inside, where the cable is connected to an ac adapter and a universal serial bus (USB) adapter, which are then connected to a computer's USB port. An earlier version of WaveBolt, called SsuRFnet, is connected to a computer's serial port, with a maximum speed of 115 kbps.

To the computer, the WaveBolt customer unit appears as a modem device and uses Windows' built-in dial-up networking. A single .INF file is installed on the computer to initialize the customer unit, and, once loaded, the customer is then able to connect to the base station.

Cirronet was one of the first companies to take advantage of a recent FCC Part 15.247 rule change allowing wider than 1-MHz hopping channels when using frequency-hopping spread spectrum (FHSS). WaveBolt uses a 2-MHz hopping channel to achieve 900-plus kbps.

Apparently, WaveBolt incorporates a fair amount of overhead, such as forward error correction [it's far more common to use a 1-MHz hopping channel to achieve 2 Mbps (2 bits/Hz); WaveBolt's bits/Hz ratio is about 0.5]. Such robust modulation allows better (apparent) receive sensitivity and range, as well as improved near-line-of-sight coverage.

Much has been made about the need for low-cost, customer-installable CPE that operates in the license-exempt spectrum. Cirronet is one of the first to offer such a system. Cirronet's WaveBolt should be seriously considered by any ISP planning to provide Internet access over wireless, which is almost every ISP.

Wireless ISP of the Month

A new wireless system being developed by AT&T, called Project Angel, has evolved into revenue service as AT&T Fixed Wireless Services (FWS), operating in a number of U.S. cities, including Dallas/Fort Worth and Anchorage.

FWS allows AT&T to operate as a CLEC and do so without the use of unprofitably priced ILEC telephone lines. Its markets are among the few that actually do offer a choice of local telephone service, and, by all accounts, it is being well received by customers. The service offers up to four lines of telephone service and "burst to 1 Mbps/best effort" Internet access.

As a CLEC, FWS is allowed to bundle long-distance services, and AT&T takes full advantage. Effectively, the entire state of Texas is within the local calling area of Texas FWS customers. The majority of its profits are from low-cost, high-margin services such as call waiting, caller ID, and voice mail. This approach is so successful that AT&T offers a version of FWS without Internet capability.

AT&T FWS was designed to operate in narrow spectrum slices, such as the narrow 10-MHz personal communications services bands. AT&T Wireless Services has reallocated its PCS spectrum for mobile services, and now plans to deploy FWS using wireless communications service (WCS) spectrum just above 2.3 GHz.

For ISPs that hope to provide voice services, there are several vendors offering wireless data equipment designed for voice services, as opposed to the more common "Oh, our equipment works fine for voice over IP," which isn't exactly the same thing.[2]

Microsoft, the third partner of the powerful venture team, markets StarBand service under the MSN umbrella, buying access and equipment from StarBand on a wholesale basis. Microsoft's relationship with RadioShack allows StarBand access to over 8000 retail locations nationwide. Microsoft put a reported $60 million into the company for a 26 percent equity stake.

NOTE While StarBand is currently focused on the residential market, it does have plans to target SOHO and small business users beginning in the first quarter of 2003.

Working Model

StarBand is fortunate to have rich founders, but can the business stand on its own two feet? One major factor affecting the bottom line is the price of CPE equipment. The external modem package with dish unit costs $499. Professional installation runs $199. So minimum setup costs to the customer are roughly $740 with shipping and taxes. This makes the CPE a high-ticket item when compared to terrestrial peers, and another factor affecting the service's status as a medium of last resort for residential wireless data broadband.

However, StarBand and other providers are taking a substantial hit to get CPE prices to the current level. Data from a number of vendors put CPE manufacturing costs between $1100 and $2000 per unit with eventual volume driving unit costs down to around $540 in 2006.

StarBand is subsidizing CPE costs anywhere from $600 to $1500 out of pocket per unit. At the current $69.99 price point, subscribers will have to remain on the system for at least 7 months for StarBand to break even on equipment alone. Because of its large up-front investment, StarBand has subscribers sign a 1-year contract.

Another piece of the RBS puzzle is the number of subscribers per satellite transponder. Transponders are the satellite-based electronics that receive, amplify, and relay signals to ground-based network operations centers (NOCs). Average lease costs run $150,000 to $200,000 per transponder per month. StarBand originally thought it would be able to squeeze 30,000 subscribers through one transponder. New technology could one day make this possible, but today the company claims it's serving only slightly more than 8600 subscribers per transponder.

Fundamentally, StarBand can't get to 20,000 subscribers per transponder with its network configured the way it is now. StarBand is probably looking at a sub-10,000 range. Network issues will become more apparent as StarBand grows from servicing 50,000 customers today to hundreds of thousands over the next few years.

According to a rough model, StarBand shows a positive gross margin on the basis of its current capacity, but this does not include CPE costs or selling, general, and administrative (SG&A) expenses. The company cited high marketing costs as a major factor in switching to a wholesale model.

Rollouts

Excessive customer acquisition costs are certainly one reason StarBand chose to abandon the direct-sales process. A troublesome rollout was probably another.

The entire rollout has been a complete disaster and the lack of certified installers and numerous problems forced StarBand to refund many of its customers' installation fees. In almost every aspect of the rollout, StarBand could have handled things much better.

Every single first-generation wireless data broadband technology has suffered through a series of rollout hiccups; it's the nature of the business at this life stage. However, one has to question whether StarBand rushed into a large-scale rollout it was not fully prepared to deal with. Early troubles with capacity on the company's terrestrial backbone connection, and cross-polarization problems that have seriously affected other services on GE-4 and Telstar 7, seem to reinforce this idea. The polarization problem, now resolved, was created by 300 improper installations.

Unfortunately for StarBand and many others, it's a catch-22. The market is demanding subscribers and revenues today and it wants cash flow tomorrow. This often means forcing customers to suffer for a bit as companies get their acts tighter on the fly. StarBand seems to be correcting problems rapidly and should be commended on its speed to market.

Early adopters are used to the large hassles of getting new technology to work, but once it's up and running, is StarBand worth the trouble and cost? As expected, the system demonstrates slight delays when initially requesting new Web pages, but images and text quickly fill in after that. The Net's two greatest strengths, free music and research capabilities, are easily handled by the service.

The initial delay is the much-discussed latency issue all geostationary orbits (GEOs) suffer. It takes half a second round trip to travel the 22,236 miles to the bird, relay to the NOC and then travel back. This does not include the packets' terrestrial trip from the NOC out to the chosen Web server and back.

Average round-trip ping times to yahoo.com in midevening are often in the 1200-ms range. This prevents StarBand users from employing interactive applications that require a low latency. StarBand's new 360 modem, along with compression technologies and caching, are helping to minimize the latency effect, but the lag will never be completely eliminated.

Finally, rain- and snow-fade also affect the system, but signals seem to punch through most of what the weather has to offer. Bottom-line: Without other wireless data broadband choices coming soon or at all for millions of potential broadband users, StarBand is a welcome service.

Conclusion

This chapter discussed the wireless data moves in m-commerce. Not all m-commerce relies on location-based wireless data tracking. Any kind of data service that doesn't require a broadband link can potentially be delivered to a cell phone, though not all of them will find customers. A cell phone's great advantage over a computer is that it accompanies the user nearly everywhere, so it's particularly suitable for applications that require immediate attention.

References

1. Steven K. Stroh, "Catch the Wireless Wave: Cirronet's WaveBolt Offers Small Wireless ISPs a Competitive Edge," P.O. Box 84, Redmond, WA 98073-0084 (Boardwatch, 1300 E. 9th Street, Cleveland, OH 44114), 2002.

2. Jack Ferguson, "Achieving Orbit: StarBand Communications Launches Two-Way Satellite Access with Reseller Opportunities," Boardwatch, 1300 E. 9th St., Cleveland, OH 44114, 2002.
3. John R. Vacca, *The Cabling Handbook*, 2d ed., Prentice Hall, 2001.
4. John R. Vacca, *Wireless Broadband Networks Handbook,* McGraw-Hill, 2001.
5. John R. Vacca, *Net Privacy: A Guide to Developing and Implementing an Ironclad ebusiness Privacy Plan,* McGraw-Hill, 2001.
6. John R. Vacca, *i-mode Crash Course,* McGraw-Hill, 2002.
7. John R. Vacca, *Satellite Encryption,* Academic Press, 1999.
8. John R. Vacca, *Electronic Commerce*, 3d ed., Charles River Media, 2001.
9. John R. Vacca, *High-Speed Cisco Networks: Planning, Design, and Implementation,* CRC Press, 2002.

CHAPTER 4

Wireless Data Marketing Environment

The year 2001 was good for the wireless data telecommunications marketing environment, featuring increased competition and innovation, a growing base of wireless data users, and lower prices, according to a report adopted by the Federal Communications Commission (FCC). The seventh annual report was presented to the FCC commissioners as an update on the state of the wireless data communications marketing environment.

According to the report, mobile telephony[4] services generated more than $63.6 million in revenues in 2001, as well as an increase in subscribers, from 109.5 million in 2000 to 210.6 million in 2001. Overall, wireless data service achieved a 40 percent penetration rate across the nation, while wireless data companies continued to expand their networks.

Some 260 million people, or almost 92 percent of the total U.S. population, have access to three or more different companies offering wireless data services, giving users expanded and competitive choices, according to the report. About 76 percent of the U.S. population lives in areas with six or more companies providing service, and 48 percent can choose from at least seven different companies for service.

Digital wireless data phone service continues to replace analog wireless data service across the nation, according to the report, with digital customers making up 63 percent of the industry, an increase from 51 percent in 1999 and 62 percent at the end of 2000. The increased competition has helped to lower service prices by about 23.4 percent, according to the report.

Wireless data Internet services have blossomed since late 2000, according to the report, with eight major mobile telephone carriers offering data services, including Internet access, short messaging service, and e-mail. Meanwhile, the use of traditional one-way pagers declined in 2001, a service offered by wireless phone companies.

The report, however, doesn't indicate that while the quantity of service providers has improved, the quality of phone service hasn't kept up. Many cellular carriers today are good enough. Companies like AT&T and Verizon have kludged together nationwide wireless data networks that generally serve consumers and the traveling public well.

But, there are still significant problems with in-building coverage and areas with no coverage. Voice cellular is sort of like a medium-priced Holiday Inn. It serves the purposes of most of the people.

Now, let's look at how getting into the wireless data market requires careful preplanning in a number of key areas. In other words, how do you sell wireless data? (The Glossary defines many technical terms, abbreviations, and acronyms used in the book.)

Marketing Wireless Data

Customers can rent cell phones for business trips to Switzerland or connect their laptops via satellite[5] when traveling in the polar regions. Advancements in wireless data point to both improved technologies and more opportunities for solution providers. Sixty-three percent of solution providers recently surveyed indicate they currently support, deploy, or service wireless data technology, up from 43 percent in 2001. On the customer side, 38 percent of IT professionals in the business-services market now use, or plan to deploy, wireless data technologies for their companies, according to Reality Research, Jericho, New York.

Still, it's not always easy to break into the wireless data world or create an extension to an existing business. The first step is to study the wireless data technology and service you want to deliver. In addition to doing your homework and easing into the market, it's important for solution providers to choose their wireless data vendors carefully and pay special attention to their claims with regard to product service and support.

Technically Speaking

Technical considerations, including access points, security, and connection roadblocks, are crucial in implementing wireless data solutions. It gets more complicated when you start deploying on multiple floors with multiple wireless data access points. Anyone who is going to resell a wireless data solution needs to think of the access point, like the antenna tower for a cell phone, and how to connect the remote computer.

Access points must be strategically placed for optimum usage and reception. Prices vary; the access point solution from 3Com, for example, sells for approximately $825. Corporate customers could buy an access point and a PC card for roughly $159 for their notebooks.

Another technical consideration is security. Solution providers need to ensure that appropriate firewalls and restricted access codes are in place. Possible limitations to the wireless data connection should also be noted. The consultant would have to look at such things as heavy metal construction and distance restrictions. In addition, it's wise to build a wireless data solution around an existing standard (802.11) to ensure longevity of use and compatibility with other products.

Marketing Plan

Finally, generating customer awareness of new and advanced technology is another challenge. A marketing plan, no matter how small, must be in place so customers have a better idea of how the technology can benefit them.

Since the discussion here is about an emerging technology, customer education is a big concern. You need to get the word out there, so the customer knows the value and so there's ease of use. Anything more complicated than a breadbox they won't use. Next, let's look at wireless data technology areas that are at the top of their class. In other words, let's look at how the wireless data marketing movement is doing.

The Wireless Data Marketing Movement

For those of you who haven't pursued wireless data as a growth-market opportunity because of uneven cell-phone coverage in your area or some other technical obstacle, the following is going to come as a bit of a shock: Your rivals have put glitches behind them and are putting distance between your organization and theirs.

New research reveals just how significant wireless data penetration has become. Today, as previously stated, 63 percent of solution providers say they support, deploy, or service wireless data technology for customers. Contrast that figure with 2001's, when only 43 percent of the solution-provider community was deploying wireless data solutions. In a short year, the market has transitioned from the brave, early adopters to the early majority. The strongest pocket of support resides among large solution providers (those with annual sales of $20 million or more), where some 72 percent are supporting or deploying wireless data technology today. There is certainly an air of optimism among the unwired. When surveyed about the outlook, the vast majority indicated they expect to be delivering more wireless data solutions and products to their customers in 2003.

No matter what you may have heard about Palm's bloated inventory of products or performance issues with wireless data LAN products, wireless data is clearly achieving critical mass in the solution-provider community, where 94 percent of solution providers expect the number of wireless data deployments to grow in 2003. That includes companies such as StellCom, a 17-year-old wireless data system integrator based in San Diego. According to StellCom, wireless data technology is going through a cycle similar to that of the PC. Like the PC, businesses once saw wireless data technology as attractive, but IT staffs viewed them as loose cannons. In 2001, businesses have stopped playing around. What's happening now is that wireless data is leaking more and more into the business processes, and companies are being forced to integrate the technology into their enterprise systems.

So, in this next part of the chapter, let's zero in on six wireless data technologies poised to propel the market:

- Wireless data LANs
- Satellites
- Mobile computing
- Hand-held computers
- Advanced RF technology
- Wireless data software

For example, wireless data LANs were chosen because of performance improvements and market potential.

NOTE According to market researcher IDC, the wireless data LAN market is expected to nearly triple in size during the next 4 years to $4 billion.

Hand-helds are being looked at here, simply because of the sheer number of new, higher-end devices that will literally transform the market. And, wireless data software is being looked at because of advances from Palm, Microsoft, and others, who are making new options possible. Here's a closer look at some of the products and innovations that could help you join the wireless data movement.

Wireless Data LANs

Though still an imperfect technology, wireless data LANs are, nonetheless, booming and remain at least one market segment that's expected to achieve its anticipated growth rate. IDC forecasts worldwide wireless data LAN semiconductor revenue alone to grow at a 30 percent compound annual growth rate during the next 4 years. And, 68 percent of networking solution providers already deploy wireless data LANs and WANs.

New innovations are one reason the market is heating up. For example, several manufacturers have unveiled new products that allow traditional indoor office technology to be used in the great outdoors. Enterasys Networks, for example, deployed its RoamAbout R2 wireless access platform in 2001 at Tulane University in New Orleans. The solution connects

80 buildings and a number of common outdoor areas across three campuses to a wireless data LAN (WDLAN) comprising approximately 1000 RoamAbout R2 access points. Enterasys officials indicate RoamAbout R2 is the first wireless data access platform that offers 54-Mbps performance based on the 802.11 standard and brings advanced Layer 3 and Layer 4 capabilities to the wireless data LAN.

For example, Cisco Systems,[7] one of the largest proponents of wireless data LAN technology, is another company moving outdoors. The company recently introduced its Aironet 350 wireless data LAN access point, a metal-cased hardware unit with an extended operating temperature range that's ideally suited for installation in harsh indoor and outdoor environments. Unlike the company's standard 350 access point, which operates in temperatures ranging from 0 to 50°C, the new device can operate in conditions ranging from −20 to 55°C. That's roughly the equivalent of a winter in Nome, Alaska, and a summer in Las Vegas. Similarly, D-Link's DWL%961000AP offers an operating temperature of −10 to 50°C and an outdoor range of nearly 1000 ft.

D-Link, a 15-year-old networking company in Irvine, California, is among those manufacturers trying to appeal to a broader base of customers. In the past, its solutions have been targeted mostly at small businesses and home users. Now, as the wireless data market has blossomed, D-Link is aiming its low-cost wireless data networking products, including its popular DWL-1000AP wireless data LAN access point, at larger customers in the midmarket. In 2001, D-Link launched a reseller program and is aiming to capitalize on WDLAN growth in both the private and enterprise markets. D-Link's DWL-1000AP can operate on both wired and wireless data LANs and offers a speed of 11 Mbps.

Wireless data LAN technology still has obstacles to overcome, including security risks and performance issues. Wireless data solutions are still slower than traditional wired LANs. But the mobility provided by wireless data LANs will continue to propel adoption rates and sales. Frost & Sullivan states in one recent report that it expects annual WDLAN shipments to reach 31 million units by 2008, despite speed and security issues. This bodes well, not only for other top WDLAN vendors such as Proxim, 3Com, and WaveLink, but for their solution-provider partners as well.

The security obstacles themselves can be considered opportunities for both vendors and solution providers. Meta Group recently reported that wireless data LAN vendors are scrambling to come to market with integrated virtual private network solutions bundled with their network offerings. Vertical industries such as financial services and health care, which require more stringent safeguards with respect to personal information, may soon emerge as key growth areas for such wireless data security solutions.

Blues for Bluetooth

Though the Bluetooth electronic communication standard appeared poised in 2000 to usher in a new era of electronic connectivity, the global manufacturing slowdown and continued interoperability difficulties have combined to stall the long-awaited rollout of Bluetooth-enabled products. Microsoft, the 500-pound gorilla of the computer world, has gone so far as to announce that its next-generation Windows release, XP, will not support Bluetooth. Wireless data LANs, meantime, claim more territory every day—including Microsoft's XP operating system.

From formidable to fine: That's the news in the one-way wireless data satellite market, depending on whom you talk to. So who's doing what? Next is a rundown of what some of the industry's top companies are doing. The future sure looks bright for the satellite wireless data market.

Satellite Wireless Data Markets

Many satellite equipment firms are worried about the economic slowdown, but Taiwan-based manufacturer Apex Communications isn't one of them. The reason? The dot-com phenomenon was largely a North American issue. Because a large part of Apex's satellite business is centered in Asia, the crash has not affected it to any significant extent.

In Asia, companies like Apex Communications are offering a number of wireless data distribution services supported by data communications over satellite. A couple of examples are an English language training system and a real-time stock exchange data broadcast, both operating in China, Taiwan, and the Philippines. The companies also use wireless data over satellite for distribution from the production site to FM stations in the major cities, and distribution from there to the end users via an FM subcarrier, and they will use digital radio in the future.

This said, Apexcom is hedging its bets by serving the two-way market, as well as the one-way. Right now, its flagship two-way product is the ACS2400 multimedia broadband VSAT system.[3]

Designed specifically to deliver IP over satellite, the ACS2400 supports high speed and wideband multimedia transmissions. This makes it well suited for a wide range of applications, such as digital audio and video broadcasting, wireless data, videoconferencing, IP telephony, distance education, and broadband Internet access.

All things considered, Apexcom remains optimistic about the future of the one-way wireless data industry. Its primary interest was, and remains, wireless data over satellite for well-defined niche market applications.

Brisk Business

Infolibria is ideally poised to capitalize on one-way wireless, data satellite broadcast and Internet accesses. The reason? This Waltham, Massachusetts, company makes streaming media data storage products[8] like MediaMall and DynaCache—products that can be used to receive one-way wireless data from satellites and then serve the data out on demand to users on a local-area network (LAN).

In terms of actual sales, Infolibria sells its products to companies like SES Americom, Panamsat, and Lockheed Martin. They, in turn, integrate Infolibria's solutions into their one-way wireless data offerings, and sell them to end users.

So, how is the market meltdown affecting Infolibria? They've seen a lot of their customers focus on managed solutions for the enterprise market. In addition, a lot of their customers' customers are looking at IP as a way of enhancing their internal communications and overall productivity; this means that business remains brisk, despite the economy. Perhaps this explains why Infolibria recently secured $52 million in new funding, with money coming from companies such as GE Capital, Mitsubishi, and Mellon Ventures.

Infolibria has been able to attract and retain a strong base of customers and partners, including AT&T, EMC, Lockheed Martin, and Mitsubishi. The company is confident that it will lead the way for streaming media adoption in the carrier and enterprise market. Not bad, given the current state of the venture capital market. Not bad, indeed.

What Recession?

If there's a recession on, then International Datacasting Corp. doesn't seem to have heard about it. In fact, when it comes to orders for wireless data satellite products, they've got one of the biggest backlogs they've ever had.

A case in point: IDC just announced new orders for $2.6 million US, including a new sale to the Canadian Broadcasting Corp. (CBC). CBC has ordered 12 FlexRoute digital audio uplinks to continue the conversion of CBC Radio's national distribution system from analog to digital. In addition, the U.S. company Sky Online has ordered a SuperFlex system to support its growing IP networking business in South America. IDC has also received orders for FlexRoute equipment from Korea's Dong-in Satellite Network, and for SuperFlex DVB/IP satellite receivers from Norway's Telenor.

However, there has been a general slowdown in the wireless data satellite market. The economic situation is making people more cautious with their money. They're still buying, but they're doing it somewhat more slowly than they did before.

So, why is IDC doing well in these troubled times? Well, it doesn't hurt that the company has staked its life on IP-based datacasting systems. To put it mildly, IP is the hottest standard on the market today. Even in tough markets, IP still sells. IDC is also benefiting from the world's continuing migration to digital technology. As long as there are analog satellite customers out there, the company still has a fresh crop of clients to harvest.

The bottom line: For IDC, these are still good times. Everyone else in the satellite equipment market should be so lucky, and so well positioned.

Opportunities for Growth

KenCast, Inc. (http://www.kencast.com/) isn't fazed by the economic downturn or the push for Internet services by two-way satellite. That's because KenCast sees opportunities for growth in a different way.

The reason? First, KenCast's Fazzt digital delivery system provides two-way Internet service by hybrid networks, using terrestrial Internet lines for access, query, and request while delivery is done by satellite.

Second, the secret is in the caching. Much content is delivered by Fazzt via satellite from content sources to increasingly large local caches at cable head ends,[2] telco central offices, and ISPs. Thus, local users with two-way wire access (DSL, cable, or telco plant) to the local cache can interact with it to retrieve the content they want.

What this means is that training videos, streaming files, and everything else can be immediately on hand for users, via two-way hybrid Internet, either from distantly located content or from a local cache. Except for rural and undeveloped areas without wire infrastructure, this is the most efficient way to provide two-way Internet service and the more commonly employed approach.

Hybrid Internet systems often use Fazzt to deliver by satellite in the Ku band and plan to do so in the Ka band. While Fazzt is particularly adept at recovery from rain attenuation signal loss in the Ku band, it is even more valuable in the Ka band, where rain attenuation is more of a problem.

To date, Fazzt is being used on over 600 systems worldwide, by everyone from the U.S. Air Force to movie and hotel data distribution. And sales are continuing to grow.

Doing It for Less

When it comes to one-way data broadcasting, Microspace Communications Corp. (http://www.microspace.com/media/press_releases/mcastpr.htm) is definitely a player. In fact, Microspace has a satellite broadcasting network with over 400,000 business-related satellite downlinks in 46 countries.

Central to Microspace's success is Velocity. Building on the company's initial 64-kbps FM^2 data service, Velocity provides users with MPEG-2/DVB video and high-speed wireless data transmissions via satellite. All they need are 36-inch receive-only antennas and MPEG-2 digital satellite receivers, both bought from third-party vendors. Microspace does the rest.

Initially launched on one GE 1 transponder, Velocity is now operating via three full-time transponders. Two are on GE 1, while the third is on Telstar 4. Compared to FM^2, Velocity delivers an awesome 8 to 10 Mbps of bandwidth per user. That's more than enough for business video or large file transfers from one site to many, simultaneously.

So how's business? Still growing. Microspace continues to add more capacity to keep up with customer demand. Despite what some people are saying, one-way satellite broadcasts are still alive and well.

One big opportunity is one-way emulating two-way traffic. This is done by broadcasting files via satellite to a company's entire range of sites simultaneously, and then letting users access those files on an on-demand, as-needed basis.

From Microspace's standpoint, it's getting all the benefits of more expensive two-way service. However, that's not how it appears to the company's accounting system.

Central to this concept is the incredible decline in the cost of server storage. In 1993, a 1-GB drive was $3500. Today, you can get 30 GB for $129.

As a result, Microspace is optimistic about one-way wireless data's future. Although Microspace can't do everything two-way, the company can do most of it, and for less money.

Keeping the Faith

For the past 30 years, Telesat has pioneered one-way satellite communications in Canada, including data, voice, and television. Today, it serves North America with its fleet of Anik and Nimiq C-/Ku-band satellites.

It's a very small market these days. Instead, the future and Telesat's opportunities lie in the two-way sector.

For instance, Telesat sold its one-way DirecPC business to Bell ExpressVu (the Canadian DBS company) in 2000. They're selling a consumer product already, so it makes sense for them to bundle DirecPC with it.

In addition, Telesat is focusing heavily on the two-way VSAT market. To date it has the Big Three automakers (Ford, GM, and Daimler-Chrysler) as Canadian customers. It also won the Fordstar contract away from Hughes. This means that it's providing maintenance to 6600 Ford sites across North America.

In the Ka-band space, Telesat has increased its stake in WildBlue's Internet-by-satellite venture, and expects to see a lot of businesses migrate from Ku band to Ka band. It continues to get demands from its clients that can't be addressed by a Ku-band footprint. To give the clients what they want, Telesat needs to increase its stake in the Ka band.

To address this demand, Telesat recently competed for, and won, the 118.7° W orbital slot from the Canadian government. It hopes to launch a C-/Ku-band satellite with a small Ka-band payload into this location by 2003.

Anik F3 will provide a variety of new services, including one-way broadcast, one-way streaming, and one-way caching services. It will also accommodate a number of the new two-way broadband services that are being planned. In other words, Telesat does have faith in two-way, but intends to keep its stake in one-way as well. Just in case.

Developing Alternatives

Like others in the satellite industry, equipment manufacturer Tripoint Global (http://www.tripointglobal.com/) is feeling the pain of the current recession. The one-way wireless data market is flat, quite frankly. Obviously, economic conditions are causing companies to rethink their communications plans, and when they get into these economic decisions, they start looking for alternatives that cost less than satellites, if they can find them.

From a sales standpoint, this means that it's just tough out there. There's just no other way to say it.

However, this doesn't mean that Tripoint Global is wringing its hands in fear. Instead, the company is trying to work with the market by developing one-way alternatives to two-way traffic.

For instance, it makes no sense for a national corporation to install two-way point-to-point sites when a one-way one-to-many broadcast approach can do the job for less. The key to making this work is *store-and-forward* technology. For instance, a company can download corporate intranet data (including videos and other materials) to all of its servers. Once there, the wireless data can be accessed locally on demand, just as if the user was on a live two-way link to headquarters.

As for those situations where two-way is a must, you should combine terrestrial return paths (including wireless data) with satellite to optimize performance with cost. The thought here is that things will have to integrate, because the cost basis won't allow them not to.

Still a Way to Go

Despite the wide range of opinions on one-way wireless data's future, enough good ideas seem to be out there to ensure that this medium stays alive and well for years to come. This isn't to say that two-way applications won't cut it in this market; they will. Nevertheless, at the same time, the ability of companies like KenCast and Microspace to emulate two-way service with one-way will open new markets for this established technology. So don't count one-way wireless data by satellite out yet; its days are far from numbered.

Thinking of going mobile? Before you do, you'd better take a look at the mobile wireless data market: those sleek and stylish laptops that win converts with features, lower prices, and more power.

The Mobile Wireless Data Markets

Consumers are accustomed to watching electronics get smaller and cheaper—except for that hulking monitor and full-size PC at home. Portable versions, the slim notebooks that pack a full computer's power in a small space, have cost much more than similar desktop models, so few consumers considered them as a second or replacement PC.

That's no longer the ease. Chips have gotten so fast, and hard drives so big, that the comparable desktop is an overmuscled hot rod—more machine than most people need. Falling prices mean lesser-powered and perfectly capable notebooks can be had for about $2000. No longer are they only executive jewelry or company issue—the cheap prices are turning them into a second home computer, allowing parents to send e-mail from the patio via a laptop while the kids polish their homework on the living-room PC. There's also an inherent coolness in notebooks. Consumers like a sleek, thin, silvery thing with all the processing power of a big box.

Crashing prices for the most expensive piece of a notebook computer (the fancy flat-panel screens) have brought portables within grasp of a whole new group of consumers. The education market, in particular, is booming for notebook makers. Dorm rooms can hardly hold the furniture, much less a big computer. Most parents find it surprising that they can get their kids a thin and light IBM laptop, with extra doodads, for $1700. The notebook saves space in the dorm and at home.

Companies were the first to buy the advantages of an ever-shrinking PC. Workers could take a PC project with them when they left the office. Consumers were the next to recognize the perks, such as watching a DVD movie or listening to a CD while on a business trip. Notebooks help blur the lines; it's about meshing work and play.

Yet shopping for laptops has its own complexities. Computer buyers typically face a tradeoff between price and power. Laptops complicate the question with a third element: weight. So, the three categories of mobile wireless data PCs can be defined as: heavy desktop replacements, the midrange thin and lights, and the truly thin and light ultraportables. Prices typically go up as weight goes down, so buyers first must decide how much they'll be on the go, and whether it's cross-country or across the living room.

To save on pounds, makers cut down on the number of drives that store data. The biggest notebooks, weighing 7 to 8 pounds, come with three of what the industry calls "spindles"—often a hard drive, a floppy, and an optical drive for CD-ROMs or DVDs. Most home users are fine with a bulkier, less expensive notebook because they'll just lug it from room to room. Dell and Compaq, among others, sell notebooks as desktop replacements for less than $1000. The cheaper models come with smaller hard drives, say only 10 gigabytes (GB), and 128 megabytes (MB) of memory (also known as RAM)—the minimum you'd want in a new computer. One model is the $999 Compaq Presario 700, which has an 850-MHz processor from AMD called the Duron, designed for less demanding work. It includes a 10-GB hard drive, 13-inch screen, and floppy and CD-ROM drives. Most consumers, though, are willing to spend a bit more, usually about $1300, for added power and capacity. That buys a Presario with a 900-MHz Duron processor, a 20-GHz hard drive, a DVD drive, and a 14-inch screen.

Middle of the Pack

Bigger notebooks have the widest selection of prices. A new midrange model is the $2299 TravelMate 740 from Acer. It sports a faster, 1-GHz chip and a 15-inch screen as well as added conveniences, such as an opening for optional drives and a fingerprint reader that blocks unauthorized users. At the high end is the A series from IBM. For a whopping $3499, you'll get a state of the-art 1.2-GHz processor, a 15-inch display, a 48-GB hard drive, and a DVD player that also can burn CDs. The machine includes two bays, or openings, that swap out a variety of optional drives, batteries, or even a new docking bay for a Palm hand-held PC, sort of a computer-on-computer.

Feature lists, though, can't tell you everything about laptops. Touch and feel are more crucial for laptops than for other PCs—weight is important, and so are looks. Another key factor: The first thing that consumers do is open the notebook and start typing. If they don't like the keyboard's feel, they usually close it and move on.

Most of the midweight notebooks, at 5 to 6 pounds, retain good-size keyboards and offer two drives, typically a hard and an optical. The

Toshiba Satellite 3005-S303 weighs in at about 5.5 pounds and starts at $1699. It comes with an 850-MHz Intel Mobile Pentium III processor, with the *mobile* meaning it's easier on batteries.

NOTE The mobile chip is rarer in cheaper machines.

The 3005 includes a 20-GB hard drive, a DVD drive, and a 14-inch display. As with most midweights, a floppy drive costs extra and plugs in from the outside.

Both of Apple's notebooks fall into the midsize group at about 5 pounds, with two drives. They are priced more competitively than Apple laptops of old and include more innovations than a typical Windows notebook; Apple, for instance, was the first to include built-in antennas for wireless data networking. They're limited to the smaller selection of software written for the Macintosh, but they are a good option for consumers who use the computer only for e-mail, Web browsing, and word processing. Apple's less expensive iBooks start at $1300, which buys a 500-MHz processor, 15-GB hard drive, and a 12-inch screen. A unique titanium case makes the Apple PowerBook (starting at $2200) a silvery, inch-thin package that compromises little compared with a desktop Macintosh. Packing that much muscle, however, makes the PowerBook a hot item, literally; like many notebooks, it runs too warm to hold on your knees (ouch). That helps explain why the industry prefers the term *notebook* to *laptop*.

Ultralights

Going to a smaller notebook can be tough. Ultraportable keyboards get scrunched, and another drive bay gets dropped (makers typically build in only a hard drive). Consumers too often are disappointed by all the tradeoffs.

Also, prices rise when laptops go on a diet. The latest chips and batteries are needed for decent performance and computing time, meaning most start at $2000. Though aimed at companies, consumers can get them with longer warranties that add about $200. An exception is Sony, which took its older, ultraportable SR series and plugged in a slow chip, an Intel 600-MHz Celeron processor. With a l0-inch screen and 10-GB hard drive, the SR33K is a wimp amid today's PC brawn—but it's a deal at $1000, after a $100 rebate, for those wanting a lap PC that can handle routine tasks.

Sure, half that price could buy a desktop, and one with more speed and capacity. But for most college students, the extra money is well worth the freedom it buys. For a few hundred dollars more, they can take their computers with them.

During the fever pitch of the bull market, everyone was dazzled by the promise of mobile commerce (m-commerce). The crystal balls at Jupiter Media Metrix, Ovum, and McKinsey revealed global mobile commerce revenues that were to be somewhere between $33 billion and $300 billion in 2006. In a world of WAP-enabled handsets and location-aware mobile wireless data networks, mobile commerce took center stage in the intimacy of the New Economy.

M-Commerce Wireless Data Network Markets

The hangover has been painful. WAP failed to deliver on its promise to make the desktop available on the mobile device. Beset by painstakingly slow access and nested menus reminiscent of DOS days, WAP has become persona non grata among North American and European wireless data consumers. Compounding the problem has been the failure of carriers to deploy location technology within the expected timetable. Just shy of the 2003 E911 Phase II deadline, every major carrier in the United States has requested waivers or extensions. The Public Safety Answering Point of San Francisco has conspicuously announced that none of the carriers in its region are able to provide the level of accuracy required by the FCC mandate. To make matters worse, one of the most promising publicly traded firms in the location technology industry, US Wireless, Inc., announced recently that it would seek bankruptcy protection under Chapter 11.

With WAP far from consumer consciousness and location technology beyond the horizon, is mobile commerce dead on arrival? Yes and no. Mobile commerce will probably never see $300 billion under the original paradigm where subscribers use their phones to go shopping on the Web. On the other hand, mobile commerce may be resurrected under a different paradigm—one in which retailers have the ability to send targeted ads and coupons to willing subscribers, not using WAP, but rather using simple text messaging [short message service (SMS)] and perhaps, someday, wireless data instant messaging.

You Can't Go Window Shopping with a Cell Phone If e-commerce[6] is a global shopping center, then mobile commerce is a corner convenience store. Early returns from Japan, and to a lesser extent Europe, have shown that mobile commerce is well suited to inexpensive, consumable items: ring tones, animated figures, virtual girlfriends, parking meter payments, and sodas. Simply put, mobile commerce today is superb for impulse purchases.

And yet, ironically, the WAP experience is anything but impulsive. To conduct a simple mobile commerce transaction, a wireless data subscriber must:

- Have a WAP-enabled cell phone with the WAP service activated.
- Place the phone into a WAP session and explicitly agree to pay a fee.
- Enter a URL using a torturous keypad entry scheme, or, if the subscriber is lucky, thumb through several layers of nested menus and "next" softkeys to find a book marked URL.
- Navigate through the destination WAP site to make a purchase, and on and on.[1]

Finally, to round out the mobile wireless data marketing environment, let's look at WDASPs. Going with a wireless data application service provider (WDASP) can take the sting out of getting your company's business in the wireless data Web.

WDASPs Offer Fast Track to Mobilizing Wireless Data Applications

Hotel chains and airlines do it with reservations; brokerage firms do it with stock trades. Trucking companies do it for signatures, salespeople with inventory. And if your organization isn't doing "it" (mobilizing its line-of-business operations, including product sales, support, and service), then it's missing a big opportunity. The slowing economy notwithstanding, it appears that going mobile isn't just for keeping in touch with grandma anymore.

On the contrary, the mobile "numbers" are huge. For starters, vendor Nokia indicates that 105 million Americans use cell phones. The number of hand-held computing devices should climb from 24.7 million in 2001 to 81.0 million by 2006, according to research firm IDC (http://www.idc.com). And consumers are expected to spend nearly $61 billion a year shopping from their cell phones by 2004, according to the Yankee Group (http://www.yankeegroup.com), a Boston research consultancy.

It's no wonder that the mobile wireless data marketplace puts a gleam in every marketer's eyes. Nor is it a big surprise that enterprises in several major industries are finding it worthwhile to offer customers anywhere, anytime access to the information in their back-end systems via mobile and wireless data devices.

In particular, hotel chains, airlines, and financial services companies see considerable upside potential in letting customers do business with them via Internet-capable phones, PDAs, and other wireless data devices such as the Research in Motion (RIM) (http://www.rim.com) Blackberry pager. For these industries, mobilizing their customers can mean increased revenue and better customer service.

In these economic times, however, many organizations are turning to the old standby, the outsourcer, for the resources they need to make their first, tentative sortie into the wireless data environment. In this case, the outsourcer is the so-called wireless data application service provider (WDASP), a small but growing cadre of for-hire companies that let enterprises get wet behind the ears, to mix metaphors, for a minimal outlay in personnel, time, and (more important) capital equipment investment.

The Six Continents Hotels chain (http://www.sixcontinents.com)—which owns, operates, or franchises more than 4300 hotels and about 600,000 rooms in hundreds of countries—is a typical example. It turned to WDASP Air2Web (http://www.air2web.com) so that guests could make and check on reservations with their cell phones and PDAs.

The Hilton (http://www.hilton.com) chain went with another WDASP, OpenGrid (http://www.opengrid.com), to build its wireless data customer service solution. And Bidwell & Company (http://www.bidwell.com), a privately held discount brokerage firm, turned to a third WDASP, 2Roam (http://www.2roam.com), to let clients access stock quotes and make trades from their cell phones and PDAs.

These companies chose the WDASP route for a variety of reasons. However, the key criteria behind going with a WDASP, executives at Six Continents and Bidwell acknowledge, was, cost-specifically, not having to initially invest in wireless data technologies. They say these costs—which include buying, deploying, and maintaining a wireless data application server and developing the software to communicate with multiple (and widely differing) wireless data networks and mobile devices—were too prohibitive to consider. However, organizations considering the move to a WDASP for their mobile commerce solutions have much to study before taking the plunge, according to analysts.

Conclusion

This chapter discussed the state of wireless data marketing. It also made a lot of predications. Let's take a look at what conclusions were drawn from these predications.

Pulling Ahead

Microsoft might think Bluetooth isn't ready for prime time and be unwilling to support it, but the software colossus has apparently decided wireless data LANs are here to stay: The long-awaited Windows XP operating system will support them. It's an easy decision to understand. Though Bluetooth seems stuck in idle, wireless data networking product sales grew 16 percent in the first quarter of 2001, even while much of the private sector was retrenching, according to a report from the Dell'Oro Group. Much of the pickup, according to analysts, was in the home and small business sector.

Though conceptually different and developed to answer different needs, wireless data LANs and Bluetooth exhibit overlapping functionality, and users employ them to perform many of the same tasks. They operate in significantly different ways, however. Two Bluetooth devices should be able talk to each other anywhere: in an office, in a gondola in Venice, or on the moon. But wireless data LANs can't communicate without the aid of a third party, a transmitter that receives messages from one device and then forwards them to the other. This hasn't proved to be the obstacle to widespread adoption it was once expected to be, with airports, hotels, and office buildings racing to install transmitters for business travelers' use.

Additionally, 802.11b-compliant wireless data LANs use the networking protocols and standards employed by traditional networks, eliminating the layer of translation software required by Bluetooth. Further enhancing the competitiveness of the wireless data LAN, a group of research engineers at Penn State announced in late July 2001 that broadband, wireless data, indoor, local-area communication networks that rely on non-line-of-sight infrared signal transmission can offer low error rates as well as safe, low (below 1 watt) power levels. The development relieves wireless data networkers of the problem of signal blocking by furniture and metal-core cubicle partition walls.

Too Close to Call

So what will be said about the year 2002, when 2003's state of the wireless data satellite market review hits the newsstands? Frankly, the situation is too close to call right now. On the upside, the demand for wireless data satellite services remains fundamentally solid. On the down, the recession is hurting new project funding, and slowing the growth of new markets.

The final prediction is that the survivors will be those with sufficient money reserves to weather the storm, innovative products and/or services to hold their own in the marketplace, and, above all, companies whose commitment to quality and service keeps their customers loyal. Beyond that, all bets are off.

Dead or Alive?

Of course mobile wireless data commerce is not dead! Mobile commerce has been crippled by your overzealous drive to make it conform to the shape of E911 and WAP. You have tried to impose particular technical solutions on top of impulsive human behavior, simply because the technology was present. Mobile commerce does not require the accuracy of E911 call processing; it may not require much location data at all! Nor does mobile commerce require WAP, the grandly planned flop, when in fact the brilliantly accidental IM and SMS will do quite nicely.

Advanced RF Technologies

The innovative NZIF architecture allows the handset manufacturers to benefit from leading-edge technologies. The highly integrated transceiver and baseband circuits will be a major step in preparing the way toward a dual-mode GSM/UMTS handset, by having the GSM/GPRS/EDGE portion already optimized in performance, size, and cost.

The different technologies presented here allow state-of-the-art products for 3G standards to be proposed. The cellular product evolution toward GSM/TDMA or any other multimode standards will need to concentrate most of the analog-sensitive functions within the RF part and allow the digital part to integrate more and more memory with increased MIPS requirements. This will lead to a different partitioning approach, offering many challenging tasks to the RF engineers. It is a first step toward a software radio.

The Long and Winding Road Ahead

With the economy in the tank, but competitive pressures remaining, WDASPs are likely to remain a viable choice for many organizations intent on mobilizing their e-business processes, at least for the near term. Many enterprises don't have a choice about mobilizing their services, however. The jury is still out about how to make money in the whole wireless data space. In the final analysis, going with a WDASP can help alleviate many of the risks of moving to a new technology.

References

1. Mark E. McDowell, "mCommerce—DOA? Or A-OK?," Invertix Corp., 5285 Shawnee Road, Suite 401, Alexandria, VA 22312, 2002.
2. John R. Vacca, *The Cabling Handbook*, 2d ed., Prentice Hall, 2001.
3. John R. Vacca, *Wireless Broadband Networks Handbook,* McGraw-Hill, 2001.

4. John R. Vacca, *i-mode Crash Course,* McGraw-Hill, 2002.
5. John R. Vacca, *Satellite Encryption,* Academic Press, 1999.
6. John R. Vacca, *Electronic Commerce*, 3d ed., Charles River Media, 2001.
7. John R. Vacca, *High-Speed Cisco Networks: Planning, Design, and Implementation,* CRC Press, 2002.
8. John R. Vacca, *The Essential Guide to Storage Area Networks,* Prentice Hall, 2002.

CHAPTER 5

Standards for Next-Generation High-Speed Wireless Data Connectivity

In the telecommunications world, wireless data is almost synonymous with hype. From Bluetooth to third-generation (3G), no new technology has performed as promised. Everything is either slower than anticipated or late to arrive—or both. Nevertheless, in computing, it's a different story.

With the preceding in mind, and to set the stage for the rest of the book, this chapter thoroughly discusses the present and future state of high-speed wireless data standards. The following are standards for next-generation high-speed wireless data connectivity:

- Wireless data LANs
- Fixed broadband wireless data
- Universal Mobile Telephone Standard (UMTS) and/or International Mobile Telecommunications (IMT-2000)
- J2ME
- RSVP
- Multistandards

The Glossary defines many technical terms, abbreviations, and acronyms used in the book.

Wireless Data LANs

Despite the worst recession the networking world has ever known, wireless data LANs have continued to spread faster than anyone predicted. Traditionally confined to warehouses and factories, wireless data LANs are now installed in offices, homes, and even public spaces. Almost all are based on the same standard, IEEE 802.11b (also known as WiFi or Wireless Ethernet), so the same hardware can be used throughout these different environments.

The number of IEEE 802.11b users grew from almost zero in early 2001 to more than 26 million at the end. That still isn't much compared to cell phones and wired Ethernet, but the growth will likely continue. The IEEE has two more versions on the way, 802.11a and 802.11g, which will increase data rates to the point where wireless data LANs can seriously challenge their copper and fiber equivalents.

However, it isn't clear which—if any—of these upgrades network managers should choose. The higher data rates come at the expense of compatibility, and all types of 802.11 still have serious weaknesses—most notably security, which might make you question whether to deploy a wireless data LAN at all. The IEEE is working to fix these, but so are rival groups and even governments. The result is a confusing array of standards, with no clear winner.

802.11b to 802.11a

The letters after the number 802.11 tell you the order in which the standards were first proposed. This means that the "new" 802.11a is actually older than the currently used 802.11b, which just happened to be ready first because it was based on relatively simple technology—direct sequence spread spectrum (DSSS), as opposed to 802.11a's orthogonal frequency-division multiplexing (OFDM). The more complex technology provides a higher data rate: 802.11b can reach 11 Mbps, while 802.11a can reach 54 Mbps.

Both of these figures are often quoted by vendors, but they're a bit misleading. Physical layer overhead cuts throughput by at least 40 percent, meaning the real rate of 802.11b is at most 6 Mbps. Often, it's a lot less.

All wireless data LANs use unlicensed spectrum; therefore, they're prone to interference and transmission errors. These errors mean that traffic has to be resent, which wastes bandwidth. A 50 percent error rate will reduce the real throughput by about two-thirds, to only 2 Mbps. And that's only half-duplex, shared by every node on the network.

To reduce errors, both types of 802.11 automatically reduce the Physical layer data rate. IEEE 802.11b has three lower data rates (5.5, 2, and 1 Mbps), and 802.11a has seven (48, 36, 24, 18, 12, 9, and 6 Mbps). The lower rates are used most of the time. The maximum is available only in an interference-free environment, and over a very short range.

Higher (and more) data rates aren't 802.11a's only advantage. It also uses a higher frequency band, 5 GHz, which is both wider and less crowded than the 2.4-GHz band that 802.11b shares with cordless phones, microwave ovens, and Bluetooth devices. The wider band means that more radio channels can coexist without interference. Each radio channel corresponds to a separate network, or a switched segment on the same network.

The precise number of channels varies by country because each regulator allocates a different amount of spectrum for unlicensed use. However, there are always more channels at the 5-GHz band. In the United States, the 2.4-GHz band is wide enough for only three, whereas 5 GHz has room for 11. The first 802.11a cards to ship support only eight of these, but it's still enough for most purposes. There's even a (so far) proprietary scheme developed by Atheros (http://www.atheros.com) that combines two 802.11a channels together to double the data rate.

Though 5 GHz has many advantages, it also has problems. The most important of these is compatibility: The different frequencies mean that 802.11a products aren't interoperable with the 802.11b base. To get around this, the IEEE developed 802.11g, which should extend the speed and range of 802.11b so that it's fully compatible with the older systems (see sidebar, "802.11g High-Speed Wireless Data Standard"). Unfortunately, interference means that it will never be as fast as 802.11a, and vendor politics have delayed the standard. It's not expected to be ratified until fall 2003.

802.11g High-Speed Wireless Data Standard

Recently, the IEEE 802.11 Task Group G approved its first draft for a wireless data local-area network (WDLAN) standard that provides data rates up to 54 Mbps in the 2.45-GHz frequency band. This new standard hikes the 11-Mbps data rate of the 802.11b standard to enable multimedia streaming over WLAN environments. To appreciate the importance of Draft 1.0, it is necessary to look at the history of 802.11g.

The 802.11g task group had its first official meeting in September 2000. By the time of the May 2001 session in Orlando, the task group had two competing proposals for the implementation of 802.11g. The May session turned into a two-way tug-of-war between Intersil (Irvine, California), which submitted an orthogonal frequency-division multiplexing (OFDM) modulation scheme, and Texas Instruments (Dallas), which submitted its own scheme known as packet binary convolution coding (PBCC). The vote was 58 percent for the OFDM proposal and 42 percent for the PBCC proposal, taking PBCC out of the running, but this was not the last time that the group would hear from Texas Instruments.

Because OFDM did not reach the 75 percent approval threshold, it was decided that the proposal should be voted on during the Portland, Oregon, session in July 2001. The plan was for the members to vote round-robin style until the 75 percent approval threshold could be met. Unfortunately, no voting took place during that session. Instead, the meeting was mired in a heated debate on bureaucratic procedures.

The next session was planned to take place in September 2001, but it was cancelled as a result of the events of September 11, further delaying the first draft of 802.11g. Because the session was not rescheduled, the delay meant that voting would not take place until November 2001.

The draft approved during the November session allows for the inclusion of both Intersil's OFDM modulation scheme and Texas Instruments' PBCC scheme. The draft also calls for the inclusion of a complementary code keying scheme, which is used in 802.11b. The compromise was necessary to move 802.11g forward and end the months of bickering within the task group.

The task group met in January 2002 to refine the draft in preparation for publication by the second half of 2003. The estimated final approval of 802.11g is scheduled for October 2003. Further details on the status of 802.11g are available on the IEEE 802.11 Web site at http://www.ieee802.org/11.[1]

Though thousands of companies sell 802.11b equipment, nearly all of it's based on chips and reference designs from only two vendors. Whoever's design is accepted as a standard is almost guaranteed a large market share among the original equipment manufacturers (OEMs). The largest 802.11b chip maker is currently Intersil (http://www.intersil.com), which proposed using OFDM in the 2.4-GHz band. Texas Instruments (www.ti.com), which aspires to make 802.11 chips, instead wanted its own enhanced version of DSSS. The final draft of the standard is a compromise, including both.

Delays in 802.11g's ratification have prompted many vendors to go straight to 802.11a, where a wider range of chip makers are working on reference designs. Among them are Atheros, National Semiconductor, Resonext, Envara, and even Cisco Systems, which acquired Radiata, the first company to demonstrate a working 802.11a prototype in 2000.

If you're going to upgrade anyway, you might as well upgrade to 802.11a. It might have been different if 802.11a products were still a year away, but they're here now.

Sharing the Airwaves

The range of various wireless data LAN technologies is also hotly debated. Most 802.11b networks can officially reach up to 100 m, or 330 ft, but this is only a rough guide: A higher-power transmitter can extend the reach, while interference and signal blocking can reduce it. The range reduction scenarios are more commonly encountered: Since wireless data LANs are usually used inside, safety rules limit a transmission's power, and walls or other objects interfere.

In any type of radio system, higher frequencies are more easily absorbed by everything from air to paper, leading to a shorter range. This led most people to assume that the new 802.11a and HiperLAN technologies, which use the 5-GHz band, would cover a much smaller area than 802.11b. According to tests conducted by chip maker Atheros, this isn't the case. Atheros is hardly impartial (it's the only vendor so far to have shipped 5-GHz chips), but it does have experimental results, and a theory to explain them.

According to Atheros' tests, 802.11a provides a higher data rate than 802.11b at every measured distance when used in a typical office environment. The explanation is that 5-GHz technologies use OFDM, which is designed to be resistant to multipath effects. The benefits of OFDM and the drawbacks of higher frequencies cancel each other out, making the range of 802.11a and 802.11b approximately the same.

What the 5-GHz lobby doesn't say is that 802.11g also uses OFDM, but in the same lower-frequency (2.4-GHz) band as 802.11b. This should give it a longer range than either of the other two technologies. No one has yet tested this because 802.11g is a newer standard that's still being thrashed out. However, if OFDM's benefits are extrapolated to the lower frequency, its range should be 50 percent greater than that of 802.11a and 802.11b.

Remember that coverage area depends on the range squared, so 802.11g could cover the same area as the other systems with less than half as many access points. Though Intersil and its other backers are currently focusing on backward compatibility, 802.11g's range could be its greatest selling point in the long term.

Of course, increased range isn't always a benefit. Because every user shares the available bandwidth, a larger range just spreads it out more thinly. This means that 802.11g is a good choice in environments containing few users, or where users don't need a high-speed connection. These include facilities such as warehouses, which until recently were wireless data LANs' main market, but probably not offices or homes.

Crowded areas such as conference centers and airports need the highest density of coverage they can get, and will eventually move to 802.11a. The large installed base means that they're likely to stick with 802.11b throughout 2003, and probably longer. IEEE 802.11g is compatible with this installed base, but it probably won't be available before dual-mode 802.11a and 802.11b systems. You'll be lucky to see .g products before the end of 2003.

The other problem with a longer range is that the signal is more likely to leak. If you haven't set up a secure system, intruders can crack into your network from farther away. If you have, it means that you're jamming somebody else's airwaves. Both are issues in skyscraper office buildings that house several companies.

This spreading can be overcome by using access points with directional antennas, which focus their transmission and reception on a specific area. The most common types radiate in an arc rather than a full sphere: They can attach to a wall and provide coverage on only one side of it. More complex antennas are available that can adjust to cover differently shaped regions, but these usually require trained radio engineers to set them up.

Directional antennas are frequency-specific, which could lead some users to choose 802.11g over 802.11a. The former is based on the same frequency as 802.11b, and hence could reuse the same antenna; the latter would need a new one. A dual-mode 802.11a/b access point requires two separate antennas. This applies to regular (omnidirectional) antennas too, but these are cheap to mass-produce; there's one built into every interface card, and vendors don't see any problem in miniaturizing them enough to produce dual-mode cards.

For users who don't need a directional antenna, upgrading from 802.11b to 802.11a shouldn't be a problem. Some vendors already sell "flexible" access points that are really just small chassis that link two or more CardBus slots to an Ethernet cable.[6] The slots can be used for any combination of 802.11 types, allowing the access point to be upgraded by using the same cards as laptops. Cards generally support only one radio channel at a time, so several cards of the same type can be used to set up a switched network.

Homeland Security

Though 802.11b is clearly the most popular wireless data LAN standard, neither of its successors is guaranteed the same acceptance. All share the same poor security and no support for QoS. The IEEE is working on many new standards to fix these weaknesses, but many users need security now. This has prompted vendors and even governments to step in with their own solutions.

All 802.11b products currently incorporate a system called Wired Equivalent Privacy (WEP), which encrypts all transmissions using 40-bit keys. However, most networks don't use it because it's switched off by default out of a naive belief that ease of use is more important than security. And even if they do use it, it's still easy to break into. Every user has the same key, meaning that the entire network is compromised if one laptop is stolen. It's also vulnerable to a fairly simple attack, which hackers have conveniently packaged into a freely downloadable program called Airsnort.

Some newer products incorporate a system known informally as WEP2. The IEEE recently renamed it Temporal Key Integrity Protocol (TKIP), in an attempt to disguise its ancestry. It uses 128-bit keys, but is fully backward-compatible with WEP, and thus vulnerable to the same attacks. TKIP may even be more vulnerable because it adds support for Kerberos passwords, which can often be guessed through a simple dictionary attack.

Many vendors are promoting an emerging standard called 802.1x as a solution. However, this covers only authentication, not full security, and it isn't yet complete. It does have security holes. Therefore, it is recommended that you protect all access points with a firewall and run all traffic through the same type of VPN used for remote access over the Internet.

HomeRF2 is another wireless data LAN standard that's already made it into shipping products. As the name suggests, this was intended as a cheap and simple standard for home networking, but unfortunately it's turned out to be neither. Thanks to the success of 802.11b, HomeRF2 products often cost more than those based on the more popular standard, though they do include both QoS and a better encryption system than WEP. Ironically, this could make them a good choice for enterprises that don't want their wireless traffic easily readable by the outside world.

European regulators are so dissatisfied with 802.11 that they aren't permitting 802.11a to be used at all. Instead, they've reserved their 5-GHz band for HiperLAN2, a system developed by the European Telecommunication Standards Institute (ETSI), the same group behind most cell phone standards. HiperLAN2 is almost identical to 802.11a at the Physical layer (it uses OFDM, and even has the same data rates), but higher up the protocol stack, it's closer to ATM than to Ethernet.

Some people prefer the name "hype LAN" because it's been talked about for so long without any real deployment. This criticism certainly fits the original standard (HiperLAN1), first set back in 1992, but never actually adopted by any equipment manufacturers. However, HiperLAN2 is real. European and Japanese vendors are working on it, with the first products expected to ship by 2003.

And So to 5G

NTT DoCoMo (http://www.nttdocomo.com) has already built a dual-mode system that combines HiperLAN2 with a cordless phone—it can even use the two simultaneously.[7] The advantage here isn't backward compatibility or even extra bandwidth: The phone has a maximum data rate of about 32 kbps, which doesn't add significantly to HiperLAN2's 54 Mbps. Rather, it's that the Japanese cordless phone standard uses very low transmission power, which prolongs battery life. A Web surfer can set up an asymmetric link that receives multimedia content via the LAN (reception requires less power than transmission) and sends mouse clicks back through the phone.

Ericsson (http://www.ericsson.com) is the only other vendor to have demonstrated a HiperLAN2 prototype in public. Like DoCoMo, Ericsson is more well known for cellular networks than wireless data LANs, which should give you some hints about HiperLAN's true intent. Despite the name, it's not really a LAN protocol at all: It's designed for broadband mobile data services, and could form the foundation of fourth-generation (4G) cellular networks.

HiperLAN's detractors sometimes claim that this emphasis on services means it will require an access point. This isn't true, though many service providers probably wish that it were. It is correct to say that HiperLAN can't operate as a true peer-to-peer system: Any network that enforces QoS needs one node to take charge and act as air traffic controller. However, this "master" node doesn't necessarily have to be mounted on a wall or connected to a wire. Bluetooth and HomeRF both include QoS for ad hoc networks between mobile devices, with nodes automatically falling into master and slave roles according to predefined criteria. There's no reason that HiperLAN2 can't do the same.

Critics of HiperLAN also claim that the technology is being boosted artificially by European regulators' insistence on it rather than 802.11a. While this is true, the regulators appear to be motivated less by protectionism and more by a desire to see a system that can use 4G services. Even the HiperLAN2 Forum says that it doesn't object to 802.11a, provided that the standard can meet its requirements for QoS, power control, and security.

The IEEE is now addressing these issues, which should secure approval within Europe for a future version of 802.11a. There's also a joint venture between ETSI and the IEEE called the 5-GHz Partnership Project (5GPP), which aims to merge 802.11a and HiperLAN2 into a single standard, tentatively known as the 5-GHz Unified Protocol (5-UP). By tying two or even three channels together, this standard would offer even higher data rates than the existing systems. Three channels will provide a real throughput of about 100 Mbps, more than most laptop PCs can handle.

These new systems should begin to appear in 2003. With high data rates, guaranteed QoS, and airtight security, they could pose a real challenge both to 3G and wired networks.

Now, to continue with the wireless data LAN theme, let's take a look at how dueling standards and security issues can't keep corporate America on the fence. Or can they?

Enterprise Wireless Data Standards Technology Comes of Age

Despite security concerns and competing standards, wireless data LANs are gaining traction in the corporate marketplace. There has been a tremendous resurgence in business recently.

It's getting to be quite interesting, actually. The wireless data LAN is being seen as a component that provides strategic benefit rather than just an access technology. As a technology, it has finally made it.

Wireless data LAN vendors sold 8.1 million 802.11b network interface cards and access points to businesses in 2001. That figure was up from 3.3 million in 2000, and sales will rise to 22 million units in 2003.

While security concerns were top-of-mind in 2001, customers seem satisfied with the way vendors are addressing those issues. Wireless data is not that different from any other access technology. It has to be a part of the whole enterprise security posture.

In many markets, wireless data LANs have moved from a "wow-driven" technology to a "needs-driven solution." Right now, health care, campus environments, and warehouse applications are the most active market niches.

It seems the war between different wireless data LAN standards may subside over time. Recently, some larger companies were sitting on the fence, waiting for the battle between noninteroperable standards to play out. But, many of those companies plan to move forward with 802.11b solutions in 2003.

Current wireless data LAN (WDLAN) products are based on IEEE's 802.11b standard (WiFi) and deliver 11 Mbps in the 2.4-GHz range. New products released in late 2001 and based on the 802.11a standard (or WiFi 5), deliver 54 Mbps in the 5-GHz range, so they're not interoperable with the 802.11b installed base. As previously explained, to further confuse matters, an IEEE committee has released a draft of yet another standard, 802.11g. Products based on that standard would deliver 54 Mbps in the 2.4-GHz range, so they would be compatible with the installed base of 802.11b products. Still, major networking vendors such as 3Com and Cisco have yet to release 802.11a products, and offerings based on 802.11g won't be available until 2003, at the earliest.

Many companies are going ahead with 802.11b deployments now, with plans to overlay one of the faster wireless data LAN technologies later on. Their customers realize this technology may be supplanted, but it won't disappear. So, the customers are relying on these companies to be their wireless data architects and integrators as the technologies evolve.

And, evolve they will. But, you need to make sure not to leave 3G cell networks out of the mix.

As carriers' cellular networks adopt that data-ready technology, corporate clients and vendors envision the day when workers' mobile phones can roam from a carrier's network to the corporate WDLAN as employees enter the office. That's where the integration gets interesting.

Getting Up to High Speed with Wireless Data LAN Standards

The recent introductions surrounding high-speed wireless data LAN products have more of the feel of a tailgate party than a formal coming-out event. For example, Microsoft has made a glittery debut with its Tablet PC software platform.

Intel, Proxim, and TDK are among the companies that recently unveiled their wireless data LAN base stations, network interface cards (NICs), and other devices based on the 802.11a standard. As previously discussed, the 802.11a standard supports use of the 5-GHz radio band and bandwidth of 54 Mbps—5 times that of today's 802.11b products. Some products will even handle video and other multimedia applications, as well as file transfers that would choke existing 802.11b products.

The chief 802.11a drumbeater is the Wireless Ethernet Compatibility Alliance (WECA), a trade group. WECA tests for compatibility among wireless data LAN products, granting them the WiFi brand when they pass muster. The group recently indicated that the brand name for the 5-GHz products will be WiFi5, and testing will start early in 2003.

Bringing Harmony to Wireless Data LAN Standards

By now, most enterprises realize how useful a wireless data network can be. But tumult in the standards arena has left many companies high and dry with networks that are incompatible with the latest developments. Because 802.11a and 802.11b operate at different frequencies, they are incompatible, meaning enterprises that have already deployed 802.11b networks, but want the faster speeds now available through 802.11a, have historically had no option but to completely rebuild their WDLANs. Security has also been one of the biggest problems with WDLANs.

As the 802.11 standards effort marches on, WDLANs will continue to gather speed and batten down security, but interoperability will remain an issue. Meanwhile, two products discussed in this part of the chapter, Proxim Harmony and Orinoco AS-2000, address interoperability and security shortcomings, respectively.

Proxim's Harmony allows 802.11b, 802.11a, and OpenAir wireless data devices to coexist and interoperate on the same network. That means end users can communicate with each other, regardless of what kinds of devices they use, and all devices can be centrally managed from a Web interface. Best of all, Harmony does not require that any additions be made to the network.

The central component of the Harmony solution is the access point controller, a stand-alone device that becomes the heart of the entire wireless data infrastructure. All wireless data access points are automatically discovered by the access point controller when placed on the network. The controller also enables administrators to centrally manage access points from a Web interface.

The access points are Layer 2 network devices that provide limited functionality. Essentially, the access point serves only as a bridge between the wired and wireless data networks, and all functionality is controlled by the access point controller.

With the Harmony architecture, users can roam subnets without any difficulty, which is not possible with most WDLAN implementations. That comes in handy when you are trying to deploy a VPN to secure wireless data traffic. Harmony also supports the 802.1x standard, which allows organizations to deploy secure, interoperable wireless data networks.

NOTE Typically, an organization must reorganize its network infrastructure, or at least ensure all wireless traffic resides on its own subnet.

A few organizations will find 802.11a networks useful. For those that have already invested in 802.11b, solutions such as Harmony offer a tempting alternative to starting from scratch.

Finally, there's an effort afoot to provide wireless data LAN roaming. How simple can it be?

Wireless Data LAN Standard Roaming

A group of leading vendors is working to iron out the technical and financial details needed to let mobile wireless data LAN users connect to almost any wireless data ISP (WDISP), in the same way cell phone users can roam and use multiple carriers to complete calls.

As previously discussed, the Wireless Ethernet Compatibility Alliance (WECA), which includes Cisco, IBM, Intel, 3Com, and Microsoft, is looking to forge relationships and network standards among WDISPs and eventually carriers that will enable roaming for 802.11b wireless data LAN users. These standards will let vendors share subscriber usage and billing data, so no matter how many different ISPs subscribers use to make a connection from a plane, train, or automobile, they get only one bill from their "home" ISP.

According to WECA members involved in the roaming project, the public access wireless data LANs now being deployed in airports, convention centers, and even restaurants will create a burgeoning web of wireless data LAN hot spots. These hot spots will let mobile workers with 802.11b-equipped computers connect over a shared 11-Mbps link to Internet-based services and corporate networks. Most wide-area wireless data links today are based on much slower cell phone nets.

What's being discussed here is interservice provider roaming. As you go from a corporate to a public net, you want to have user ID and a password for the ISP. But you don't want to have a different one for every wireless data ISP net that you might traverse. Within a corporate wireless data LAN, roaming among access points is handled as part of the 802.11b protocol.

The group is a mix of service providers, LAN equipment vendors, and PC makers, including Agere Systems, Dell, Enterasys, and Nokia and wireless data ISPs MobileStar and Wayport. Having roaming agreements is a great idea for any network. The utility uses a cellular phone network to connect field workers with laptops or PDAs to corporate data.

So are you clear on what you'd want from such a service as proposed by WECA? One service provider with one bill. As far as cost is concerned, it must be similar in cost to dial-up connections from a hotel room, including the hotel fees and long-distance charges for an average user session, but with faster throughput compared to dial-up.

Other issues could slow the roaming proposal. For example, the reach of such a wireless data LAN (WLAN) service will still be severely limited compared to the cell networks because 802.11b, sometimes called WiFi, is a local network with a radio range of roughly 150 ft. The public access WDLANs being created by the likes of wireless data ISPs MobileStar and Wayport initially will be found in urban, high-density areas. Most of these public WDLANs are targeted at white-collar business travelers. Blue-collar mobile workers likely will have to rely on low-speed, but widespread, cell networks, such as cellular digital packet data nets, for accessing data wirelessly. In addition, the service providers that go forward with wireless data LAN roaming will have to ensure they're offering a simple connection process and a single bill to make such roaming a desirable service for target users. And then there are security concerns.

Specifically, WECA is looking to define a tag that users could tack onto their subscriber name. The tag will alert any WDISP that the user requesting service is "owned" by some other provider. Data about the user and the service request will be passed to an independent clearinghouse, which would coordinate transactions among different parties—in this case, the WDISPs.

The arrangement will most likely use the Remote Authentication Dial-In User Service (RADIUS) protocol, which is widely used to coordinate authorization information between remote users and an authorization server. The clearinghouse would pass the user data to that user's WDISP, which then completes the authentication, bills the user, and makes the appropriate payment to the WDISP serving as the user's access connection. Users can then access their home WDISP services and, through the provider, their corporate net.

WECA members say the technology for sharing data between the ISPs is relatively straightforward and most of the complexity involves setting up standards for handling transactions between service providers. The billing systems are key to this. WECA is extending the RADIUS protocol with specific new attributes, such as user name, time spent online, and bytes in or out. WECA will also have information about where the user is, through a location code, so they can return site-specific services to that user.

A WDISP subscriber from the United States, gaining access via a wireless data LAN service in a Swedish airport, would receive information in English, for example. Keeping it simple will be the key to user acceptance. WECA has failed if this is difficult to use. Everyone has a vested interest in making this work.

There will be significant investment. The overall 802.11b market is expected to keep growing at a healthy rate despite the economic slowdown, according to a report by Cahners In-Stat (http://www.instat.com/partner.htm). By 2006, the firm estimates that companies will be spending nearly $7.5 billion on WDLAN equipment. Companies such as IBM, Compaq, and Dell are introducing notebook PCs with built-in 802.11b radios and antennas. Adapter card vendors have just started bringing out 802.11b cards for hand-held computers, such as those using the Microsoft PocketPC software.

The carriers are watching the project closely, according to WECA members. "There's a tremendous amount of work going on by all the carriers. They're all involved in WiFi products. They're very quiet about it, but they're all doing it."

WECA has no set schedule to complete its work, so it's difficult to say exactly when 802.11b roaming will become reality. The group will have a final document by 2003. Users can expect to see roaming being implemented more widely in the next few years, with the pace accelerating as carriers get into the action and as the number of WDLAN clients surges, each one representing a potential subscriber for wireless data services.

Now, let's look at the IEEE fixed broadband wireless data standard 802.16. For years, members of the fixed broadband wireless data sector have fought over standards. Fortunately, the IEEE 802.16 specification is being pushed forward to end the bickering.

Fixed Broadband Wireless Data Standard

Despite their promise, fixed broadband wireless data systems have fallen short in becoming a cost-effective method for delivering voice, video, and data services wirelessly to homes, offices, campuses, and other last-mile applications. Just look at the woes of the local multipoint distribution service (LMDS) market. Once thought of as the panacea for fixed broadband access, LMDS systems are struggling and the big players, like Nortel and ADC, are abandoning the LMDS ship.

So what's causing these problems? One answer can be found in a lack of standardization. Unlike their cable modem and DSL brethren, fixed broadband wireless data providers have been slow to settle on a single standard. Some have backed a vector orthogonal frequency-division multiplexing/data over cable service interface specification (VOFDM/DOCSIS) approach. Others have explored traditional modulation techniques, such as quadrature amplitude modulation (QAM). Still others have followed the proprietary path for development. This wide assortment has not only caused confusion in the market, but has also slowed the development of fixed broadband wireless data equipment.

Fortunately, a new solution is on the horizon. In an effort to bring standardization to the chaotic broadband wireless data sector, the IEEE has formed a task group, dubbed 802.16, to unite manufacturers under a single specification. OK, it's actually three specifications (802.16.1, 802.16.2, and 802.16.3) under one umbrella spec, but you get the point.

The 802.16 Architecture

To bring standardization to the broadband wireless data sector, the 802.16 group is currently working on three specifications. These include:

- IEEE 802.16.1, which defines the air interface for 10- to 66-GHz systems.
- IEEE 802.16.2, which covers coexistence of broadband wireless data access systems.
- IEEE 802.16.3, which defines the air interface for licensed systems operating in the 2- to 11-GHz band.[2]

All three 802.16 standards are designed with respect to the abstract system reference model. An 802.16 wireless data service provides a communication path between a subscriber site, which may be either a single subscriber device or a network on the subscriber's premises (such as a LAN-, PBX-, or IP-based network) and a core network. Examples of a core network are the PSTN and the Internet.

Three interfaces are defined in the 802.16 reference model. The first is the air interface between the subscriber's transceiver station and the base transceiver station. 802.16 specifies all of the details of that interface.

The second interface is between the transceiver stations and the networks behind them [also known as the subscriber network interface (SNI) and base station network interface (BNI)]. The details of these interfaces are beyond the scope of the 802.16 standards. The reason for showing these interfaces in the system reference model is that the subscriber and core network technologies (such as voice and ATM) have an impact on the technologies used in the air interface and the services provided by the transceiver stations over the air interface.

The final interface deals with the optional use of a repeater. The air interface specification allows for the possibility of repeaters or reflectors to bypass obstructions and extend cell coverage.

The Protocol Holds the Answers

Working from the bottom up, the lowest two layers of the 802.16 protocol model correspond to the Physical layer (PHY) of the OSI model and include such functions as encoding/decoding of signals, preamble generation/removal (for synchronization), and bit transmission/reception. In addition, the PHY of the 802.16 standard includes a specification of the transmission medium and the frequency band. Unlike the PHY, the Transmission layer is concerned with the encoding/decoding of signals, preamble generation/removal, and bit transmission/reception.

Above the Physical and Transmission layers are the functions associated with providing service to subscribers. These include transmitting data in frames and controlling access to the shared wireless data medium. These functions are grouped into a Media Access Control (MAC) layer.

The 802.16 MAC protocol defines how and when a base station or subscriber station may initiate transmission on the channel. Because some of the layers above the MAC layer, such as ATM, require specified service levels such as QoS, the protocol must be able to allocate radio channel capacity so as to satisfy service demands. In the downstream direction (base station to subscriber stations), there is only one transmitter and the MAC protocol is relatively simple. In the upstream direction, multiple subscriber stations are competing for access, resulting in a more complex MAC protocol.

On top of the MAC layer, the specification contains a Convergence layer that provides functions specific to the service being provided. A Convergence layer may do the following:

- Encapsulate protocol-data-unit (PDU) framing of upper layers into the native 802.16 MAC/PHY frames.
- Map an upper layer's addresses into 802.16 addresses.
- Translate upper-layer QoS parameters into native 802.16 MAC format.
- Adapt the time dependencies of the upper-layer traffic into the equivalent MAC service.[2]

In some cases, such as digital audio and video, a convergence layer is not needed and the stream of digital data is presented to the Transmission layer. Upper-layer services that make use of a PDU structure, however, do require a Convergence layer.

Bearer Services

Requirements for the 802.16 standard are defined in terms of bearer services that the systems must support. For example, an 802.16 interface must be able to support the data rate and QoS required by an ATM network or an IP-based network, or support the data rate and delay requirements of voice or video transmissions.

Separate bearer service requirements have been defined for 802.16.1. The 802.16.1 spec is designed to support three types of bearer services: circuit-based, variable packet, and fixed-length cell/packet. Circuit-based services provide a circuit-switching capability, in which connections are set up to subscribers across a core network. Variable packet services, on the other hand, include things like IP, frame relay, and MPEG-4 video. The fixed-length cell/packet is specifically aimed for ATM.

Requirements for these services are grouped into three categories. The first category is the data rate that must be supported. The second category refers to error performance. For most services an upper limit on the bit error rate (BER) is defined. For ATM, various specific QoS error parameters are also used.

The final category is maximum one-way delay. This delay can be defined as medium-access delay, transmit delay, and end-to-end delay. Medium-access delay measures the amount of time that the station, once the transmitter is turned on, must wait before it can transmit.

Transmit delay, on the other hand, refers to delay from SNI to BNI or BNI to SNI. It includes the medium-access delay plus the processing at the MAC layer for preparing transmission [from the subscriber transceiver station (STS) or base transceiver station (BTS)] and at the MAC layer for reception (at the BTS or STS).

End-to-end delay is characterized as the total delay between a terminal in the subscriber network and the ultimate service beyond the core network. This includes the transit delay.

Understanding the MAC

Data transmitted over the 802.16.1 air interface from or to a given subscriber are structured as a sequence of MAC frames. The term *MAC frame* as used in this context refers to the PDU that includes MAC protocol control information and higher-level data.

This is not to be confused with a time-division multiple-access (TDMA) frame, which consists of a sequence of time slots, each dedicated to a given subscriber. A TDMA time slot may contain exactly one MAC frame, a fraction of a MAC frame, or multiple MAC frames. The sequence of time slots across multiple TDMA frames that is dedicated to one subscriber forms a logical channel, and MAC frames are transmitted over that logical channel.

The 802.16 MAC protocol is connection oriented. Each MAC frame includes a connection ID, which is used by the MAC protocol to deliver incoming data to the correct MAC user. In addition, there is a one-to-one correspondence between a connection ID and service flow. The service flow defines the QoS parameters for the PDUs that are exchanged on the connection.

The concept of a service flow on a connection is central to the operation of the MAC protocol. Service flows provide a mechanism for upstream and downstream QoS management. In particular, they are integral to the bandwidth allocation process. The base station allocates both upstream and downstream bandwidth on the basis of the service flow for each active connection. Examples of service flow parameters are latency (maximum acceptable delay), jitter (maximum acceptable delay variation), and throughput (minimum acceptable bit rate).

Frame Format

The MAC frame consists of three sections: a header, with protocol control information and addresses; a payload, with data from a higher-level protocol; and a frame check sequence. Three header formats are defined by the 802.16 specification. There is a generic header format in both the uplink (toward the base station) and downlink (toward the subscriber) directions. These formats are used for frames that contain either higher-level data or a MAC control message. The third header format is used for a bandwidth request frame. The downlink header format consists of the following fields:

- *Encryption control (1 byte).* Indicates whether the payload is encrypted.
- *Encryption key sequence (4 bytes).* An index into a vector of encryption key information, to be used if the payload is encrypted.
- *Length (11 bytes).* Length in bytes of the entire MAC frame.
- *Connection identifier (16 bytes).* A unidirectional, MAC-layer address that identifies a connection to equivalent peers in the subscriber and base station MAC.
- *Header type (1 byte).* Indicates whether this is a generic or bandwidth request header.
- *ARQ indicator (1 byte).* Indicates whether the frame belongs to an automatic repeat request (ARQ)–enabled connection. If so, the ARQ mechanism found in a typical link control protocol is used, and a 2-byte control field is prepended at the beginning of the frame. The control bit structure contains a 4-byte retry number and a 12-byte sequence number. The retry number field is reset when a packet is first sent, and is incremented whenever it is retransmitted (up to the terminal value of 15). The sequence number field is assigned to each packet on its first transmission and then incremented.
- *Fragment control (2 bytes).* Used in fragmentation and reassembly.

- *Fragment sequence number (4 bytes).* Sequence number of the current fragment.
- *Header check sequence (8 bytes).* An 8-byte cyclic redundancy check (CRC) used to detect errors in the header.[2]

One of the more interesting aspects of the downlink header format is fragmentation. Fragmentation is used to divide a higher-level block of data into two or more fragments in order to reduce MAC frame size. This is done to allow efficient use of available bandwidth relative to the QoS requirements of a connection's service flow.

If fragmentation is not used, then the fragment control (FC) field is set to 00. If fragmentation is used, then all of the fragments are assigned the same fragment sequence number (FSN) and the FC field has the following interpretation: first fragment (10), intermediate fragment (11), last fragment (01). The MAC user at the destination is responsible for reassembling all of the fragments with the same FSN.

Uplink Headers

Now, let's look at the uplink header format. The uplink header format contains all of the fields of the downlink header, plus an 8-byte grant management (GM) field. This field is used by the subscriber to convey bandwidth management needs to the base station. There are three different encodings of this field, depending on the type of connection. There are also a number of subfields in the GM field. These include the slip indicator (1 byte), the poll-me field (1 byte), the grants per interval field (7 bytes), and the piggyback request (8 bytes).

The first two subfields, the slip indicator and poll-me field, for the GM field are associated with the unsolicited grant service (UGS). This service is designed to support real-time service flows. In essence, the base station, using MAC management messages, periodically grants an allocation of bytes to the subscriber on a given connection. The allocation is designed to keep up with real-time demands.

If a subscriber finds that its queue of data to send has exceeded a threshold, the subscriber sends a GM field with the slip indicator bit set and either requests a poll for bandwidth by setting the poll-me bit or requests that a given number of bandwidth grants be executed in the next time interval. The latter technique is used if this is a UGS with activity detection; this simply means that the flow may become inactive for substantial periods of time.

For other types of service, the GM field may be used to make a request for capacity. This is referred to as a *piggyback request* because the request is made as part of a MAC frame carrying user data rather than in a separate bandwidth request management MAC frame.

The bandwidth request header is used by the subscriber to request additional bandwidth. This header is for a MAC frame with no payload. The 15-byte bandwidth request field indicates the number of bytes of capacity requested for uplink transmission.

Don't Forget the PHY

The 802.16.1 PHY supports a different structure for the point-to-multipoint downstream channels and the multipoint-to-point upstream channels. These structures reflect the differing requirements in the upstream and downstream directions. In general, most systems will require greater downstream capacity to individual subscribers to support asymmetric data connections, such as Web applications over the Internet. For the upstream direction, the issue of medium access needs to be addressed, because there are a number of subscribers competing for the available capacity. These requirements are reflected in the PHY specification.

Under the 802.16 specification, upstream transmission uses a demand assignment multiple-access (DAMA)–TDMA technique. DAMA is a capacity assignment technique that adapts as needed to optimally respond to demand changes among the multiple stations. TDMA is simply the technique of dividing time on a channel into a sequence of frames, each consisting of a number of slots, and allocating one or more slots per frame to form a logical channel. With DAMA-TDMA, the assignment of slots to channels varies dynamically.

In the downstream direction, the standard specifies two modes of operation, one targeted to support a continuous transmission stream (mode A), such as audio or video, and one targeted to support a burst transmission stream (mode B), such as IP-based traffic. For the continuous downstream mode, a simple time-division multiplexed (TDM) scheme is used for channel access. Additionally, frequency-division duplexing (FDD) is used for allocating capacity between upstream and downstream traffic.

For the burst downstream mode, the DAMA-TDMA scheme is used for channel access. Three alternative techniques are available for duplexing traffic between upstream and downstream. The first is FDD with adaptive modulation. This technique is the same FDD scheme used in the upstream mode, but with a dynamic capability to change the modulation and forward error correction schemes.

The second is frequency shift–division duplexing (FSDD). This is similar to FDD, but some or all of the subscribers are not capable of transmitting and receiving simultaneously.

The final technique is time-division duplexing (TDD). Under this technique, a TDMA frame is used, with part of the time allocated for upstream transmission and part for downstream transmission. The availability of these alternative techniques provides considerable flexibility in designing a system that optimizes the use of capacity.

But, will 802.16 unify the industry? Let's take a quick look.

Unifying the Industry with the 802.16 Standard

The new fixed wireless data standard recently ratified by the Institute of Electrical and Electronics Engineers promises to bring some stability to a corner of the service provider market plagued by failures over the past few years. Aimed at unifying the industry behind one specification, the IEEE-endorsed 802.16 standard ultimately could reduce manufacturing costs and spur innovation in areas such as middleware and security. For integrators, 802.16 is a welcome relief. The standard will help settle interoperability issues and reduce prices.

Pricing also was a problem. Now that a standard is available, components can be manufactured in volume, which may eventually bring prices down. The 802.16 should reduce some uncertainty in the market and enable the industry to rally behind one group of standards.

Equipment interoperability and high prices have all but squelched fixed wireless data, which was once hailed as an alternative to uncooperative local phone companies or expensive fiber connections for the last-mile connection. Amid the economic slowdown in 2001, fixed wireless data carriers went into a tailspin: WinStar, Teligent, and Advanced Radio Telecom filed for bankruptcy protection; AT&T Wireless sold off its fixed wireless data business; and Sprint said it would hold its rollout to 13 markets.

The new standard seeks to define three classes of fixed wireless data: high-frequency spectrum from 10 to 66 GHz used by wireless data carriers, low frequencies from 2 to 11 GHz used by providers of server message block (SMB) and residential wireless data broadband services as well as wireless data campus networks, and unlicensed spectrum.

The first version focused on the 10- to 66-GHz frequencies. Among its benefits is an efficient allocation of spectrum, providing capabilities of up to 134 Mbps per channel at peak. This is particularly important for the carriers that have paid millions of dollars for spectrum licenses and want to get the most of out their spectrum, efficiently providing voice, data, and video over each channel. The IEEE committee expects to complete extensions to 802.16 for low frequencies by 2003. Following that, the group will work on a standard for equipment in unlicensed frequencies.

Next, let's look at that tangled family tree of wireless data technologies that's reaching for 144-kbps and 384-kbps convergent mobility. In other words, is the evolution of wireless data networks really moving toward 3G?

Universal Mobile Telephone Standard (UMTS) and/or International Mobile Telecommunications (IMT-2000)

The evolution of wireless data networks, from simple first-generation analog through 2G, 2.5G, and 3G, involves enormous complexity and rapid change. It also involves convergence, for 3G is the long-sought juncture at which the Global System for Mobile Communications (GSM) and code-division multiple-access (CDMA) evolutionary paths come together into a single, official, globally roamable system.

As defined by a standards body called the Third Generation Partnership Program (3GPP), a global wireless data standard, the Universal Mobile Telephone Standard (UMTS), should be firmly in operation by 2006. UMTS will have circuit-switched voice and packet-switched data. 3G networks must be able to transmit wireless data at 144+ kbps at mobile user speeds, 384 kbps at pedestrian user speeds, and an impressive 2+ Mbps in fixed locations (home and office). This flexibility derives from UMTS' two complementary radio access modes: frequency-division duplex (FDD), which offers full mobility and symmetrical traffic, and time-division duplex (TDD), which offers limited (indoor) mobility and handles asymmetric traffic, such as Web browsing.

Ultimately, UMTS itself will evolve into an "all IP" or "end-to-end IP" network, or at least a network in which IP is used as much as possible. UMTS is Europe's answer to an earlier (and ongoing) project, the ITU-T's International Mobile Telecommunications 2000 (IMT-2000), which stakes out frequencies for future use. Amusingly, the independent-minded allies of the United States and Japan refer to 3G as IMT-2000 (not UMTS), despite the fact that the Europeans, to keep the Americans in the loop, established a separate Third Generation Partnership Project Number 2 (3GPP-2) body.

Furthermore, the exact line separating 3G from its predecessors has blurred lately, especially since the highest-end 2.5G technology is called "3G" by both manufacturers and the IMT-2000. The great dream is that all of the high-end technology will interoperate with UMTS under the general term "3G."

A 3G phone is supposed to handle more than simple voice mobility. Cramming streaming color video, multimedia messaging, and broadband Internet surfing into a single device may make some of the first true 3G phones a bit bulky, a throwback to the 1980s.

Indeed, the cell phone should run as many timesaving intelligent agents as possible. When you must use the phone, you should be able to efficiently use voice, data, and touch-sensitive screen simultaneously. In fact, Cisco, Comverse, Intel, Microsoft, Philips, and SpeechWorks recently formed the Speech Application Language Tags (SALT) Forum to develop a device- and network-independent de facto standard to do just that.

Aside from the mobile phone, the wireless broadband–enabled laptop and PDA will also play a role in the wireless future. Some people may prefer making voice over IP (VoIP) calls from a laptop in a higher-bandwidth, fixed wireless data scenario (an animal different from a pure mobility play), while others may prefer a more portable, integrated cell phone/PDA.

The Family Tree

In the 1980s, some of you were using thick-as-a-brick analog phones. In the early 1990s, things began to change. At the moment, we're in a digital 2G wireless world. The Global System for Mobile Communications (GSM) is the world's most popular 2G mobile standard, having conquered Europe, Asia, Australia, and New Zealand, and is spreading through the United States, thanks to an aggressive marketing campaign by VoiceStream. GSM operates on the 900-MHz and 1.8-GHz bands worldwide except for the Americas, where it occupies the 1.9-GHz band.

Other 2G systems include the Integrated Digital Enhanced Network (iDEN), which Motorola [Arlington Heights, Illinois (http://www.mot.com)] launched in 1994. The iDEN runs in the 800-MHz, 900-MHz, and 1.5-GHz bands. GSM and iDEN use time-division multiple access (TDMA), which involves timesharing a channel somewhat like a T1 does.

Another 2G system, cdmaOne (also called IS-95A, which debuted in 1996), doesn't use timesharing. It uses a unique spread-spectrum technology, code-division multiple access (CDMA), which relies on a special encoding technique to let lots of users share the same pair of 1.25-MHz bands.

Qualcomm owns most of the CDMA-related patents. Major cdmaOne carriers include Verizon and Sprint in the United States and Bell Mobility and Telus in Canada. Unfortunately, typical data transmission rates for 2G networks range between 9.6 and 14.4 kbps. This isn't great for Web browsing and multimedia applications, but okay for SMS—short (160 Latin characters) text messages.

Between Two Gs

The major improvement 2.5G brings over 2G is the introduction of packet-switched data services that conserve bandwidth even though they're "always on." This means that when you use a data service over 2.5G, you occupy bandwidth only when you actually send and receive packets (shades of the Internet!). Voice calls on 2.5G, however, are definitely still circuit-switched, and use a constant bandwidth.

GSM Path On the GSM path, an effort was made to send packets over GSM circuit-switched voice channels, called high-speed circuit-switched data (HSCSD). More powerful, however, is General Packet Radio Service (GPRS), a GSM-based packet data protocol that can be configured to gobble up all eight time slots that exist in a GSM channel. With some software and hardware upgrades, GPRS can commandeer existing spectrum, servers, and billing engines. GPRS can support a 115-kbps data rate, though 50 to 60 kbps is more likely in practice, especially since the packets must contend for the same bandwidth as GSM circuit-switched voice, and providers will tweak the bandwidth according to the number of subscribers as they try to find a profitable mix of number of users versus bandwidth per user.

To enjoy both GPRS data and GSM voice, one must have a new subscriber terminal or "TE" (mobile phone, PDA, PC, or laptop card) that supports packets as well as voice. One also needs to upgrade software at the GSM base transceiver site (BTS) and the base station controller (BSC). The BSC also must have a new piece of hardware called a packet control unit (PCU), which helps direct data traffic to the GPRS network. Also, databases such as the Home Location Register (HLR) and the Visitor Location Register (VLR) should be upgraded to register GPRS user profiles.

Existing GSM mobile switching centers (MSCs) don't handle packets, so two new network elements, collectively referred to as GPRS support nodes, must be introduced. The serving GPRS support node (SGSN) delivers packets to mobile devices around the service area. SGSNs query HLRs for GPRS subscriber profile data, and they detect new GPRS mobile devices entering a service area and record their location.

The second new element is the gateway GPRS support node (GGSN), which is an interface to external packet data networks (PDNs) that work with protocol data units (PDUs). One or more GGSNs may support multiple SGSNs.

Motorola has championed GPRS with its Aspira GPRS network infrastructure, an offering that gives GSM network operators immediate wireless data services without having to rebuild the central infrastructure. Motorola's Aspira GPRS network subsystem [including the GPRS support node (GSN), MSC, and location register] is functionally separate from the base station subsystem. Operators can increase node capacity by just adding modular interface cards and downloading software and firmware processors. Motorola's packet controller unit (PCU), the interface between the voice GSM network and the packet network, performs radio functions and Aspira GPRS network functions and is built on a standard 16-slot compact PCI card cage.

In the field of GPRS software, Hughes Software Systems [Germantown, Maryland (http://www.hssworld.com)], employs 2800 programmers in India who work on ready-to-go GPRS solutions for carriers. Instead of deploying GPRS to everybody (200,000-plus–subscriber base), some providers are zeroing in on business campuses and more lucrative areas of likely adopters (20,000-subscriber base).

For even more bandwidth, GPRS can be upgraded to use a modulation technique called EDGE, which stands for enhanced data rates for GSM (or global) evolution. EDGE lets GSM operators use existing GSM radio bands to increase the data rates within GPRS' 200-kHz carrier bandwidth to a theoretical maximum of 384 kbps, with a bit-rate of 48 kbps per time slot and up to 69.2 kbps per time slot in good radio conditions. Existing cell plans can remain intact, and there is little investment or risk involved in the upgrade.

AT&T has announced it will move its entire network to GSM/GPRS and thence to EDGE. VoiceStream is also converting to GPRS, and Cingular Wireless indicates it will take the GPRS/ EDGE route, too. Cingular announced it was going to launch GPRS in Seattle, where AT&T recently trialed its GPRS service on Nokia phones.

Network operators wary of the seemingly lengthy GSM-GPRS-EDGE-UMTS path should take a look at Alcatel's [Calabasas, California (http://www.alcatel.com)] highly flexible Alcatel 1000 mobile switching center (MSC) for GSM and GPRS/ EDGE, which evolved from the Alcatel 1000 switch. Its UMTS features make it ready for 3G, as the switch is part of Alcatel's planned end-to-end UMTS solution leading toward all-IP multimedia services. The MSC comes in both a small stand-alone version for small networks and a high-capacity version based on an ATM switching matrix and UNIX servers. There are even specific functions for GSM satellite gateways.

In 1993, the International Union of Railways (UIC) decided to use GSM as a basis for a standardized radio communication system for railways within Europe. Now, let's discuss the MORANE project, which was set up in order to conduct trials on the system, as well as the GSM-R standard and its motivation.

GSM-R The European railways and the telecommunications industry have developed a new-generation digital radio communication system based on GSM (see sidebar, "GSM Talk"), called GSM-R. This new European standard offers an alternative to existing PMR/PAMR networks in the transportation domain. GSM-R is rapidly being deployed as the railway communications system of choice across Europe.

GSM Talk

GSM is very flexible and tolerant of improving technology and is a good hedge against obsolescence. The defining feature of GSM is the manner in which the network handles wireless data information, not any given technology. A common misconception is that GSM is equivalent to TDMA radio technology. While current GSM deployments use TDMA, it is not required. The entire network could be either built on or converted to CDMA as that technology expands. In short, buying into GSM does not lock you into a single vendor and works to assure a lasting investment.

An interesting feature of GSM is the subscriber identification module (SIM) chip, a chip that carries the unit's "personality," such as telephone number and access level. SIM chips have useful features for transit operations because the chip could be placed into a vehicle's radio by an operator to orient the vehicle's network to its assignment (such as train number, route, and block) for the day. If reassignment is desired the next day, the chip could simply be placed in another vehicle. GSM also has an inherent data capacity useful for many IT applications such as automated vehicle location (AVL), fare collection, and vehicle health and welfare monitoring.[4]

Recently, Deutsche Bahn joined the list, contracting to replace its analog telecommunications system with a GSM-R network for railway operations in Germany. The new integrated system will be used for train, vehicle, switching, operations, and maintenance communications. This range of functions is a hallmark of the GSM-R technology.

GSM-R has already entered commercial service in Sweden for Banverket operations (5000 miles). It is soon to enter operational service in European countries including Germany (20,000 miles), the Netherlands (2500 miles), Spain, Switzerland, United Kingdom, Italy, Finland, Belgium, and France. Currently, it is being considered for introduction by Indian Railways, Burlington Northern Santa Fe (which really could use it) in the United States, and several countries in Eastern Europe (Czech Republic, Hungary, Baltic countries, Slovenia, and Russia).

The GSM-R, which relies on the GSM worldwide standard for mobile communications, integrates all existing mobile radio services for railways as well as all transport and mass transit services. It offers all the basic features for an alternative to existing analog as well as digital private mobile radio/public access mobile radio (PMR/PAMR) radio systems.

The Drivers for Evolution European railways are major users of mobile radio systems. They use radio for a wide range of services, such as road radio communication, operation and maintenance, yard communications, and passenger information. Current systems vary throughout Europe, with different frequencies and technologies employed for different applications even within a single country. Most in-service equipment is based on analog technology and has exceeded its product life cycle. The railways faced the following questions:

- Which digital radio system should be used to replace aging analog radio systems currently in use?
- What technological solution will support the needs of your border-crossing traffic to coordinate with other systems?
- How can you ensure continuity of service and respect budgetary constraints if a new system is to be implemented?
- How can you guarantee future evolution of a new system?[4]

Considering these issues, the Union Internationale des Chemins de Fer (UIC) anticipated the need for a common wireless data frequency band and digital communications standard for border-crossing rail traffic. The UIC conducted a detailed technical and economical survey of digital technologies, and in 1993 decided to base the new system on GSM (Global System for Mobile Communications). This would ensure that railways could participate in the evolution of the public standard to include their specific needs. In addition, they might benefit from the economy of scale of the existing public market for the cost of their equipment.

The Standardization Work The decision to choose an open standard had some drawbacks, as not all specific requirements were covered by the GSM. Enhancements for special needs were necessary, which were researched and defined by the UIC's European Integrated Radio Enhanced Network (EIRENE) Project. The GSM system had to be modified to meet several types of requirements specific to railways:

- Those arising from railway operational needs such as special addressing facilities, numbering schemes, and man-machine interfaces.
- Those related to railway telecommunications needs such as broadcast and group calls, fast call set up associated with priority, and preemption mechanism.
- Those related to the train control European Rail Traffic Management System (ERTMS) application.[4]

In 1995, the UIC decided to establish a project to set up tests of the new system. Three railways, Societé Nationale des Chemins de Fer (France), Deutsche Bahn (Germany), and Firenze SMN (Italy), set up the consortium Mobile Radio for Railway Networks in Europe (MORANE) to conduct the trials. The overall aim of the project was to specify, develop, test, and validate the new GSM-R (GSM for Railways) system. In particular the MORANE project was intended to:

- Provide specifications for the new functionalities, the interfaces, and the system tests.
- Develop prototypes of the radio system (mobile and fixed part) and implement them on three trial sites in Germany, France, and Italy.
- Validate the prototypes with reference to the specifications and the user requirements.
- Investigate the performance of existing GSM and new GSM-R standards under railway-specific conditions.
- Contribute on a high level to the standardization for the future European Radio System for Railways.[4]

In order to provide the specifications and the related prototypes, major suppliers for GSM and for railway equipment were asked to join the project. Responsibilities were divided within the industry with respect to the different subsystems which have been identified for the new system. Research companies were included so that an independent test definition and evaluation were ensured.

The initial tasks performed were design specifications and system and equipment validation. Documents were elaborated to allow validation of the actual development results against the performance expected by the users.

The basic assumption for the development work was to use the standard GSM technology. The aim was to stay as close as possible within the standard evolution path of GSM in order to avoid specialized solutions for railways. The EIRENE project had already identified some basic telecommunication features, which they passed on to the European Telecommunications Standard Institute (ETSI) for standardization.

A further assumption for the specifications and developments for the MORANE prototypes was to base them on standard services already defined in GSM or on enhancements which could become open European standards. The infrastructure for trial sites was equipped with GSM-R equipment in order to:

- Evaluate the ability of the GSM-R system to operate in a railway environment.

- Enhance this GSM-based system in order to meet the railways' user needs.
- Validate and demonstrate to a large set of users the capability of the enhanced system to answer railways' radio communication needs.[4]

The project started in 1996. It was successfully finalized in 2000 with the approval of the new system by the users.

GSM-R Economic Aspects GSM-R offers to transport organizations end-to-end solutions for their radio communication networks. It allows digital communication for voice and data.

Road radio, yard radio, and operation and maintenance radio as well as vehicle radio are now available on an integrated and standardized platform able to evolve with the user's needs. The system is able to perform all the existing day-to-day operations of today's analog radios and offer a platform for evolution. It offers single or combined operation, as well as well as interagency operation possibilities.

Once GSM-R is implemented for a system, the high-performance data transmission it provides will allow new applications to be added on the existing system, such as IT systems for passenger information, on-board ticketing, diagnostics, and maintenance, as they are needed. Introduction of GSM-R offers most railroad organizations:

- Reduction of operating costs
- Reduction of maintenance costs (reduced spare parts and training costs)
- Increased spectrum efficiency
- High-speed data applications and service differentiation with existing systems
- Reduced capital expenditure by using standard equipment
- Increased flexibility of operation by using SIM cards[4]

The CDMA Route to 3G

The GPRS equivalent in the CDMA world is CDMA2000 1XRTT, which can assign more of the 1.2-MHz radio channel per user. It can also employ a more sophisticated modulation scheme to boost bandwidth for individual users, up to 144 kbps (bursting at 153.6 kbps, and up to 307 kbps in the future). It also involves a new phone and demands a change to some of the base station equipment, doubling voice network capacity and allowing data to be packetized and sent without the need to establish a traditional circuit.

CDMA2000 1XRTT lets service providers evolve gradually from 2G to 3G, since it's backwardly compatible with cdmaOne. By the same token, any combination of 2G and 3G capabilities can be deployed in a network simply by inserting new or upgraded "1X" cells at strategic network locations among the 2G cells.

While Europeans want GSM and its descendants to be the foundation of 3G, Americans are particularly fond of promoting CDMA derivatives. Indeed, Sprint already refers to CDMA2000 1XRTT as a "3G" service and will offer it throughout the entire Sprint PCS all-digital network by mid-2003. Even now, Sprint will sell you a backward-compatible silver SCP-5000 Sanyo phone with a full-color display for $399. Like Japanese carriers, Sprint will charge for screen savers and ring tones.

Another operator, Nextel, currently offers Motorola iDEN phones, but many in the industry feel that Nextel will also install a CDMA2000 1XRTT overlay on its nationwide integrated digital enhanced network. Just to muddy the waters further, CDMA2000 1XRTT has two new descendants, CDMA2000 1XEV-DO and CDMA2000 1XEV-DV. CMDA2000 1XEV-DO (evolution-data optimized) is about to be deployed in bandwidth-crazed South Korea, where 60 percent of the population have broadband access. Faster than CDMA2000 1XRTT, 1XEV-DO is essentially 3G in its prodigious handling of bandwidth, supporting fixed and mobile applications at 1.2 to 800 kbps on average and 2.5 Mbps peak.

In the United States, Verizon may do 1XEV-DO deployments at the end of 2003, as will an operator in Japan. Airvana [Chelmsford, Massachusetts (http://www.airvananet.com)] specializes in building 1XEV-DO end-to-end IP infrastructures. In North America, certain regional operators are looking to use 1XEV-DO to come up with not a mobile, but a fixed wireless data technology that delivers Internet access at about 200 to 250 kbps without line-of-sight transmission and without a truck roll. Semiurban areas with little DSL deployment are candidates.

Airvana is working with Nortel Networks [Richardson, Texas (http://www.nortel.com)] to jointly develop all-IP 1XEV-DO products. Nortel Networks expects its CDMA2000 1XEV-DO solution to be available in the second half of 2003.

Many network operators converting to CDMA2000 1XRTT or 1XEV-DO are using Lucent Technologies' [Murray Hill, New Jersey (http://www.lucent.com)] Flexent products designed to support 2G-to-3G evolution. Lucent knows a few things about CDMA, having installed 60,000 CDMA base stations among 70 customers over the years, giving it a 41 percent market share.

Moving to Lucent's "1X" architectures is made as painless as possible: For most Lucent Series II and Flexent base stations, adding circuit cards and upgrading network software are all that's needed to move from cdmaOne to CDMA2000 1X.

Lucent's platforms can be deployed in whatever frequency bands are allowed from 450 to 2100 MHz. They double voice capacity and support mobile Internet–based applications (144 kbps with CDMA2000-1X and up to 2.4 Mbps with 1XEV-DO), yet they can use existing 2G overhead/control channels for system acquisition, call establishment, and control. Lucent's CDMA comprehensive solutions include Flexent 3G-ready base stations; operations, administration, and management (OA&M) solutions; billing solutions; optical backhaul; and high-capacity MSCs.

Now, let's take a very detailed look at why the CDMA2000 system is the first of the new 3G mobile technologies to be deployed in a revenue-earning service. In other words, in this part of the chapter, let's look at CDMA2000 and describe its relationship to the TIA/EIA-95 systems it has evolved from.

CDMA2000 1X

As previously explained, the CDMA2000 third-generation wireless data standard was developed in response to the ITU call for third-generation wireless data systems and the continuing desire by wireless data operators to increase the performance and capabilities of their systems. In almost all of the Americas, where cdmaOne is extensively deployed, new spectrum is not being made available for third-generation systems. In many Asian countries, where cdmaOne is also extensively deployed, regulators are allowing third-generation systems to be deployed in existing spectrum, even when new spectrum is also being allocated for third-generation systems. As a result, a large number of the existing cdmaOne operators required that the third-generation air interface integrate well with their TIA/EIA-95 systems to provide a clean, economical, and transparent migration path. The cdmaOne operator community also challenged the designers of the cdma2000 air interface to double the voice call capacity over TIA/EIA-95. This was a significant challenge, as TIA/EIA-95 was already the highest-capacity wireless data air interface. The challenge was met and what resulted was CDMA2000, which can be deployed as an evolution of cdmaOne or as a new third-generation system.

The first commercial third-generation network was the CDMA2000 network which was launched in October 2000 by the Korean operator SKT. This was followed by introductions by LG Telecom and KT Freetel. A recent report indicated that 930,000 CDMA2000 subscribers were added in Korea during September 2001. By early 2003, most existing cdmaOne operators in North America and Asia will have commercially launched CDMA2000.

The initial version of the CDMA2000 air interface standard was developed by the Telecommunications Industry Association (TIA) standards body TR45; subsequent versions have been developed by 3GPP2, a consortium of five standards bodies: TIA in North America, TTA in Korea, ARIB and TTC in Japan, and CWTS in China. These regional standards bodies have converted the CDMA2000 specifications into regional standards. The TIA designator for the air interface standard is TIA/EIA/IS-2000. In May 2000, the ITU Radio Communication Assembly approved ITU-R M.1457, consisting of five IMT-2000 terrestrial radio interfaces, one of which is CDMA2000. The ITU terminology for CDMA2000 is CDMA multicarrier.

The CDMA2000 system is continually evolving. Commercial systems are using the first version of the CDMA2000 air interface standard. The initial version concentrated on providing higher performance for the dedicated channels. Revision A provided support for the new common channels and concurrent services. Work is nearing completion on Revision B and work is beginning on Revision C. In addition to these CDMA2000 revisions, a high-rate data-optimized companion standard, TIA/EIA/IS-856, also called lxEV-DO, has been developed. This was recently added to the CDMA multicarrier family of standards by the ITU.

The CDMA2000 air interface can be connected to either the ANSI-41 network or the GSM-MAP network. Existing cdmaOne operators are using the ANSI-41 network, which really consists of a circuit-switched portion (formally the ANSI-41 network) and a packet-switched portion. There is currently considerable work in the standards bodies to transition the circuit-switched portion of the network to a unified packet network using Internet protocols. This unified network is called the all-IP network.

Most CDMA operators have commercially deployed Revision A of the TIA/EIA-95 standard or the PCS variant, J-STD-008. Revision B of TIA/EIA-95 introduced many new features to the air interface; however, the main feature in commercial service is the higher-data-rate capability on the forward link, which can provide up to 115.2 kbps, not including overhead. With the deployment of CDMA2000, a full set of TIA/EIA-95-B capabilities is being deployed. This part of the chapter provides an overview of the CDMA2000 air interface and some of TIA/EIA-95-B capabilities that are being introduced with CDMA2000 deployments.

CDMA2000 1X Performance Enhancements The modulation and coding structure of the CDMA2000 forward link is quite similar to that of TIA/EIA-95; however, the following enhancements improve the forward link capacity over TIA/EIA-95:

- Fast forward link 800-Hz power control
- Transmit diversity (space-time spreading)
- Choice of a rate 1A or rate ½ error-correcting coding
- Turbo coding
- Independent soft handoff for the F-SCH[3]

The CDMA2000 reverse link changed considerably from TIA/EIA-95, primarily by using coherent modulation and code multiplexing different channels. This made for cleaner integration and better performance for higher-rate services. The details of the CDMA2000 reverse link design are beyond the scope of this chapter.

Whose 3G Is It, Anyway?

Americans pushed for descendants of cdmaOne to be the "official" 3G. The CDMA Development Group (CDG) reports that there are more than 34 CDMA2000-enabled handset models on the market and over 800,000 3G CDMA2000 subscribers worldwide.

The CDMA camp, however, suffered from a schism. The European Telecommunications Standards Institute (ETSI) and the Japanese operator NTT DoCoMo wanted Ericsson's wideband (W-CDMA) to serve as the basis for 3G, which demands a large swath of new spectrum. Qualcomm and the Korean carriers wanted backward compatibility and found they could achieve the same objective as Ericsson's W-CDMA by simply aggregating existing codes and channels. Qualcomm therefore promoted the series of incremental upgrades leading to the various flavors of CDMA2000, which they hoped would be adopted as an official 3G system. As things turned out, 3G UMTS is a combination of GSM technology and the W-CDMA radio interface.

Finally, for forward-looking operators, another amazingly flexible platform that's available now is the UltraSite from Nokia [Irving, Texas (http://www.nokia.com)]. It's a triple-mode site solution that supports HSCSD, GPRS, EDGE, and W-CDMA. UltraSite includes a compact, high-capacity base station housing GSM/EDGE transceivers or W-CDMA carriers, or a mix of them, expandable through cabinet chaining. UltraSite can be installed at new or existing GSM sites, to increase the cell capacity, or to enhance the data features of the site simply by adding new EDGE-capable transceivers or upgrading the site to W-CDMA. When Nokia and AT&T Wireless Services completed the first live EDGE data call using GSM/EDGE technology and a live GSM network environment recently, the call was made with a 1900-MHz Nokia UltraSite base station and a prototype Nokia EDGE handset connecting a laptop to the Internet.

Conclusion

This chapter discussed the state of the wireless data standard environment. Like Chap. 4, it also made a lot of predications. Let's take a look at what conclusions were drawn from these predications. WDLANs are covered first.

Mainstreaming WDLANs

Why wireless data LANs? Ask anyone who manages networks for an evolving organization. By eliminating LAN wiring, wireless data LANs reduce the cost of space planning and preparation; "moves, adds, and changes"; and equipment and peripheral upgrades—all this while also conferring short-range mobility on laptop and PDA users.

In the past, WDLANs were a hard sell simply because they were based on proprietary technology and didn't provide much practical bandwidth. Since 1997, however, a family of wireless data specifications grouped under what's referred to as 802.11 has undergone refinement by the Institute of Electrical and Electronics Engineers (IEEE, http://www.ieee.org) along with various manufacturers.

There are three main 802.11 transmission specifications: 802.11a, b, and c. All of them use the Ethernet transport protocol, making them compatible with higher-level protocols such as TCP/IP, with popular network operating systems, and with the majority of LAN applications.

In the United States, the most popular, "universal" WDLAN standard at the moment is 802.11b, now called WiFi. It operates in the 2.5-GHz frequency range and can transmit up to 11 Mbps. WiFi certification means interoperability: If necessary, you should be able to integrate WiFi gear from different manufacturers' products into one system.

The Business Case for WDLANs

NOP World-Technology's (http://www.nop.co.uk) recent "Wireless Data LAN Benefits Study" surveyed 300 companies, each with 100 or more employees using WDLANs. Their data reveal that using wireless data LANs lets end users stay connected 1.75 hours longer each day, amounting to a time savings of 70 minutes for the average user, increasing productivity by as much as 22 percent.

The study also shows that WDLANs save their owners an average of $164,000 annually on cabling costs and labor, more than 3.5 times the amount IT staffs had anticipated saving. Cost savings and productivity gains produce a per-employee annual estimated ROI of $7550.

WDLAN Client Adapters

The main component of a WDLAN is the WDLAN client adapters. Adapters (complete with adorable little antennas) will get your laptop, printer, PocketPC, PDA, or other device onto the WDLAN. They are network interface cards made essentially to the same specs as their wired brethren. Since mobile laptops are a natural for use on a WDLAN, 802.11b client adapters tend to be PCMCIAs, or PC cards as they're called these days.

There are also wireless-enabled PCI cards for desktop PCs (a good example being Linksys' [Irvine, California (http://www.linksys.com)] instant wireless data PCI card (Model WMP11). Linksys also provides a connector [their wireless data PCI adapter (Model WDT11)] that adapts a PCMCIA WDLAN adapter to work in a free PCI slot in a desktop PC.

If your desktop PC doesn't have a spare PCI slot, you can use a wireless data USB adapter, such as the ORiNOCO USB client adapter from Agere Systems [Santa Clara, California (http://www.orinocowireless.com)], the Linksys wireless USB adapter model WUSB11, the wireless USB client model WLI-USB-L11G from Buffalo Technology Inc. [Austin, Texas (http://www.buffalotech.com)], the USB client from Avaya [Basking Ridge, New Jersey (http://www.avaya.com)], and the USB wireless adapter model DWL120 from D-Link (Irving, California (http://www.dlink.com)]. These all cost between $95 and $150.

The new wireless networker from Symbol Technologies (Holtsville, NY (http://www.symbol.com) is an 802.11b CompactFlash I/II card for PocketPCs. It can also be used in the Casio E-125, Compaq iPAQ, and HP Jornada 520/540. The wireless networker is available through e-tailers for around $250.

Another interesting device in the same vein is Linksys' instant wireless network CF card (Model WCF11). It's a Type II CompactFlash card that connects directly to your PDA. With it, your little PDA can now send and receive data at speeds up to 11 Mbps and distances of up to 1500 ft. Compatible with Windows CE 2.1 and 3.0, it can also be quickly configured from your PC. It should be available by the time you read this.

If you're building a very small, impromptu WDLAN, all wireless data LAN adapter cards have an "ad hoc" mode. This enables them to communicate directly to each other, which means that you can quickly set up a peer-to-peer wireless data network.

Security or Lack Thereof

There's been some hysteria in the press recently over WDLAN security issues. First of all, these security anecdotes relate to "interior" LAN products as opposed to WAN versions of those products. Second, since various forms of encryption may decrease network performance by 20 percent, many manufacturers ship their products with this option defaulted to off. Your average IT technician then installs the equipment out of the box and discovers he or she can telecommute from the company parking lot.

In any case, "pretty good" security used to involve simple link-level security, such as provided by Wired Equivalent Privacy (WEP), a protocol that was part of the original 802.11b formulation. WEP uses RC4 encryption and 40-bit keys that must match between the mobile device and the AP. Continued attacks and deficiencies found in WEP, however, have encouraged companies to move up.

The IEEE 802.11 Task Group I (IEEE 802.11i) is currently working on a future version of the standard. The committee's 802.1X (or 802.1.x) standard provides a scalable, centralized authentication framework for 802-based LANs. It automatically creates and distributes new 128-bit encryption keys at set intervals. End users with 802.1X-friendly Windows XP clients, for example, can be authenticated by a RADIUS server and supplied with a WEP security key. The open standard is flexible enough to allow multiple authentication algorithms.

Cisco beefed up its Aironet WDLAN product security with Lightweight Extensible Authentication Protocol (LEAP). When you log in a LEAP system, clients dynamically generate a new WEP key.

Other companies, like Avaya, support all of this, but believe that the best way to secure traffic entering the network from a wireless data access point is to use the security and policy enforcement found in an IPSec-based VPN, which they also support in their products and in their VPN remote client desktop software.

Future Migration

The 802.11a is a form of wireless data ATM that runs in the relatively interference-free 5- to 6-GHz frequency band, has lower power consumption, and can transfer data at an impressive theoretical maximum of 54 Mbps. It also supports eight nonoverlapping channels, yielding 13 times the capacity of its more popular brother, 802.11b. Because of the frequency difference, however, it's not made easily compatible with other wireless data Ethernet technologies.

The 802.11a is, however, championed by companies such as Avaya. Their AP-3 access point, designed and built in cooperation with Agere Systems, has a dual CardBus architecture that allows for the cohabitation of 2.5-GHz and 5-GHz radio cards in the same box, which means you can slowly migrate your WDLAN from 802.11b to higher-bandwidth 802.11a by changing client cards when convenient.

The AP-3 has Spectralink VoIP support and can be remotely managed via a Web browser or standard SNMP management tools. The AP-3 also includes a new wireless distribution system (WDS) that enables a single radio in the AP-III to act as a repeater station or wireless data bridge to expand a network across a facility and between buildings. The AP-3 is also supported by Windows XP. It lists at $1295.

On the horizon is 802.11g, which operates at 2.4 GHz, supports speeds ranging from 11 to 54 Mbps, and is backwardly compatible with 802.11b. But many in the industry now feel that 802.11a has too great a head start and will end up the winner for high-bandwidth WDLANs.

Bluetooth

Bluetooth is a promising technology. The speed of its success may be hampered by regulatory obstacles, not to mention its cost.

As previously mentioned, Bluetooth is a short-range technology that allows radio-style transmissions between devices. At its simplest level, Bluetooth enables electronic devices within a building to communicate with one another. In the office, such equipment includes computer systems, printers, telephone systems, photocopiers, security systems, automatic coffee- or tea-making machines, dictation machines, and systems that control air conditioning and lights. In the home, likely devices could include personal computers; security systems; telephone systems; and heating, lighting, and environmental control systems. Bluetooth technology could even be used to send signals to appliances or entertainment systems.

The additional functionality (in the form of a radio transceiver) will result in additional costs. The current cost of adding such functionality is regarded by many as still relatively high—typically $10, depending on the volume of the purchase.

As the cost comes down, which it undoubtedly will, the number of devices that can be economically interconnected will increase rapidly. Some developers have already suggested that the unit cost of a Bluetooth-enabling device could be as little as $1 per unit. There is no reason why any electronically controlled device cannot be connected via Bluetooth. Of course, the ubiquitous mobile telephone will also be connected to most electronic devices via Bluetooth.

How Will Bluetooth Be Used? With myriad applications for Bluetooth technology, its ultimate usefulness lies in its ability to allow these electronic devices to interconnect. For example, it will allow the control of any device using a mobile telephone. On arrival for a conference at a hotel, one could be guided via a mobile phone to the correct conference room. The hotel's guest system would recognize the attendee's mobile phone number and guide the attendee accordingly.

Bluetooth technology provides tremendous flexibility because it has the potential to allow all electronic devices to be interconnected. Indeed, mobile telephones that incorporate Bluetooth technology provide a fruitful source of potential applications. Today when visitors walk into an office building, their presence is announced by a receptionist. Using Bluetooth, a mobile telephone could do this automatically with a message on a monitor announcing the visitor with no need for human intervention. Of course, this could also work the other way around. If someone didn't want to see the visitor, he or she could become unavailable.

Another possibility introduced by Bluetooth technology is the ability to subdivide components of electronic equipment. For example, a manufacturer could build a mobile telephone with a remote earpiece. The earpiece could communicate to the telephone network via the telephone base using a Bluetooth radio link.

One of the best-publicized effects of Bluetooth will be the aesthetic effects: namely, the removal of cables in offices and homes. Bluetooth technology replaces the need for such cabling. Bluetooth can also be combined with other technologies. It can be used in conjunction with triangulation technology, which determines the precise location of a mobile phone. In a building, such technology could be used to track the whereabouts of visitors. Alternatively, a Bluetooth device could be built into children's clothing so that if a child wandered away, the Bluetooth transmitter would signal a warning.

Bluetooth as a Standard Bluetooth technology has not been formally adopted as a standard by any standards body. It is, however, a de facto standard. Given the amount of support, it is highly likely to be a successful standard. Nine companies are the primary promoters of Bluetooth technology: 3Com, Ericsson Inc., IBM Corp., Intel Corp., Lucent, Microsoft Corp., Motorola, Nokia, and Toshiba Corp.

The official Bluetooth Web site (http://www.bluetooth.com) indicates that more than 3200 companies have indicated an interest in using Bluetooth. There are, however, alternative technologies. As previously mentioned, one is known as HomeRF, which stands for home radio frequency. In addition, IEC and ETSI have relevant accredited international standards, and IEEE has published 802.11b. Indeed, the IEC standards and the Bluetooth standard can potentially conflict in certain areas. The consequence of this is that, in theory, Bluetooth would conflict with European Union (EU) legislation.

The EU CE marking legislation is linked to the use and adoption of standards. Those European standards, which are adopted via CEN and CENELEC, are usually identical to ISO, IEC, or ETSI standards. In this instance, it is hard to see how CENELEC could ignore the existence of Bluetooth. In practice, it is to be expected that a standard actually adopted would not conflict with Bluetooth. Certainly, manufacturers should not to be concerned about this technical inconsistency between the theory and practice.

Another technology worth mentioning is a product produced by Time Domain Inc. (Huntsville, Alabama). It is based on ultrafast, ultralow-power transmissions in the very wide frequency bands. This technology, known as PulsOn, uses transmissions of 500 picoseconds and is said not to interfere with radio communications. It is, however, a far riskier technology. It has not been widely adopted. And, although it supposedly does not interfere with radios, the technology is, in fact, unlawful in some countries, because it transmits in frequencies reserved for terrestrial radio services.

Bluetooth and the Law Bluetooth transmitters will be subject to compliance in the EU with the Radio Equipment and Telecommunications Terminal Equipment and the Mutual Recognition of Their Conformity (Directive 99/5/EC of 9 March 1999, Official Journal L 091, 07/04/1999, pp. 10–28)—commonly referred to as the R&TTE Directive. This EU directive replaced an earlier directive (TTE-SES Directive 98/13/EC).

The R&TTE Directive is the CE marking directive that applies to radio equipment and telecommunications terminal equipment as defined in the directive. The definition of telecommunications terminal equipment encompasses Bluetooth devices. There are several exceptions, the most important of which is for radio equipment that is intended to be used solely for the reception of sound or television broadcasting. This exception does not include Bluetooth devices, since Bluetooth devices are intentional transmitters. Apparatus within the scope of the R&TTE Directive must:

- Meet the requirements specified in the Low Voltage (Electrical Safety) Directive [Directive 73/23/EC on the harmonization of the laws of the member states, relating to electrical equipment designated for use within certain voltage limits (OJ 1973, L77/29)].
- Meet the emissions and immunity protection requirements under the Electromagnetic Compatibility Directive [Directive 89/336/EC on the approximation of the laws of the member states, relating to electromagnetic compatibility (OJ 1989, L139/19)].
- If the apparatus is radio equipment, be constructed to avoid harmful interference.[5]

In addition, the R&TTE Directive allows the European Commission to make further rules relating to interoperability. In some cases, the apparatus must meet relevant harmonized European standards and bear the CE mark. Furthermore, manufacturers must maintain records confirming that the apparatus complies with the R&TTE Directive.

Many Bluetooth applications will have important legal ramifications. Most important, many uses of Bluetooth are contrary to United Kingdom and European data-protection laws. For example, when people enter a building (such as a shopping center), do they consent to their personal information (including their whereabouts) being transmitted throughout the building to all the shopkeepers?

Unfortunately, the EU has recently taken a strong stance on data-protection legislation, as can be seen from the EU Directive on the Protection of Individuals with Regard to the Processing of Personal Data and on the Free Movement of Such Data (95/46/EC OJ No. L281/31 of 23.11.95). Interestingly, it seems that Europeans are, in practice, far more relaxed about the use of their personal data than the law permits. And, because people will most certainly want access to Bluetooth technology, it is highly likely that the legal technicalities (such as infringement of data protection) will be overlooked both by users and providers of the technology. In practice, this would certainly be the best course to adopt, because the dangers of being left behind in the next technological revolution are far greater.

Bluetooth and Cryptography Telecommunication transmissions are susceptible to being overheard. Accordingly, there will be a need for some encryption to be built into Bluetooth devices. Bluetooth, by design, however, is secure. The United States has given a blanket exemption to all types of encryption technology designed for Bluetooth. Under new U.S. regulations, some items are exempt from a technical review prior to export. Section 15 Part 740.17(b)(3) (vi) of the Code of Federal Regulations states:

> Items which would be controlled only because they incorporate components or software which provide short-range wireless encryption functions may be exported without review and classification by [the United States' Commerce Department's Bureau of Export Administration] and without reporting under the retail provisions of this section.

The Preamble to the new U.S. regulations provides the following additional guidance:

> In section 740.17(b)(3) (Retail Encryption Commodities and Software), License Exception ENC is revised to authorize, without prior review and classification or reporting, those items which are controlled only because they incorporate components providing encryption functionality which is limited to short-range wireless encryption, such as those based on the Bluetooth and Home Radio Frequency (HomeRF) specifications. Examples of such products include audio devices, cameras and videos, computer accessories, handheld devices, mobile phones and consumer appliances (refrigerators, microwaves and washing machines).

Unfortunately, no similar blanket exemption has been issued by the United Kingdom or European authorities, which do not treat Bluetooth technology any differently from other wireless data technologies. Therefore, whether encryption is allowed in either Europe or the United Kingdom will depend rather upon the level of encryption. So far, the United Kingdom has been fairly restrictive in prohibiting the importation of strong encryption technology for private use.

Will Bluetooth Succeed? Ultimately, there is no doubt that Bluetooth will succeed. Many companies have put much money into this new technology. However, it is the speed at which it will become a success that is still open to debate. Current projections have indicated that from the few thousand Bluetooth-enabled devices that were delivered in 2000, several million will be delivered in 2003. How long it will take before Bluetooth-enabled devices become mass-market items is not yet clear.

One unfortunate development is that, even though Microsoft is one of the founders of Bluetooth technology, the company announced in the summer of 2001 that it would not yet be integrating Bluetooth device drivers within its standard Windows operating systems. It is presumably waiting for others to do so first. Given the prevalence of the Windows operating system, this is unfortunate, because it means that to operate Bluetooth-enabled products from a computer, a separate driver would be required. Although an independent driver undoubtedly will be developed in the marketplace, the lack of a driver forces individuals or companies to purchase a separate software driver.

A major issue to address is how quickly the public will take up the new technology. No manufacturer is likely to increase a product's unit cost by including Bluetooth technology until it becomes cost-effective to do so. Manufacturers are currently struggling to compete in the marketplace with less expensive devices that do not incorporate Bluetooth technology. Clearly, a key issue is how much a product's current technology costs compared to the cost of a product integrating Bluetooth. The assumption, of course, is that the cost of Bluetooth-enabled devices will decrease rapidly as mass manufacture becomes common. Certainly by 2006, one would expect the vast majority of electronic devices to be Bluetooth-enabled.

CDMA2000

As previously explained, CDMA2000 is a high-performance third-generation wireless data system that builds upon the highly successful TIA/EIA-95 system. CDMA2000 provides twice the voice capacity of TIA/EIA-95 and provides significantly enhanced capacity and higher rates for wireless data services. The forward link structure of CDMA2000 is compatible with TIA/EIA-95, thus permitting a single frequency to support both TIA/EIA-95 mobiles and CDMA2000 mobiles. While these and many other compatible aspects provide a graceful transition path for cdmaOne operators to third-generation systems, CDMA2000 can also be deployed as a totally new third-generation system.

GSM

Railroad organizations aiming at renewing their analog, costly to maintain equipment, should consider the alternative of using the GSM-based standard as the answer to their present and future needs. The GSM-R equipment previously discussed is developed and available off the shelf from several vendors provided that the frequency range is within the overall GSM 900 frequency range. This system is widely deployed in Europe in the demanding railway environment. This system is able to answer to most of the needs of the mass transit sector. The selection of the frequency range for usage in different countries should be carefully evaluated, and vendors like the SYSTRA Group can offer their services to help transport organizations in evaluating and optimizing their radio communication requirements to select the most appropriate solution and benefit from the most advanced digital technique.

RSVP

This chapter also investigated the problems of existing RSVP in providing real-time services in wireless mobile data networks. The chapter also gave short overviews on how to interoperate IntServ services over DiffServ networks and how to map IntServ QoS parameters into a wireless data link. The chapter then identified several schemes proposed for solving these problems under both micro- and macromobility. Even though they set up RSVP resource reservation paths efficiently, most of these solutions have no QoS mechanism sufficient to prevent service disruption at a new cell during handoff. Therefore, it was proposed that a dynamic resource allocation scheme be initiated for reducing service disruption of real-time applications due to frequent mobility of a host.

Multistandards

Finally, this chapter has presented a study of different radio access technologies and selection criteria for multistandard terminals. The evaluation is based on technology, omitting the market aspects of different regions. The analyzed selection criteria are range, capacity, and delay. The analysis of the range was theoretical and based on a channel model that was equal for all technologies. The calculations show that Bluetooth starts with the shortest range of about 5 m. The WDLAN extends up about 150 m, whereas the evaluated cellular technologies GERAN and UTRAN range from 100 to 2500 m. It appears that range information gives better selection criteria when combined with environment information than when combined with the service to be used.

For capacity, in the absence of deployment models for the number of access points to satisfy a given number of users, and with the significant differences between the deployment of licensed versus ISM bands, it was not possible to compare capacity characteristics under loaded system conditions for all technologies. To do so, it would be necessary to observe a given number of end users interacting with a number of access points.

The delay caused by radio access has more significance in regard to QoS provided to a user. The access and transmission delays vary from one technology to another significantly, with WDLAN providing the lowest figures and GERAN being at the other extreme. In delay, again the lack of deployment models between licensed and ISM bands hinders the comparison.

Based on the results, it can be concluded that it would be reasonable to support WDLAN and a cellular technology such as GERAN and/or UTRAN in a mobile terminal. In such a terminal, the WDLAN access should be preferred over the cellular technology for high-data-rate applications. If these two technologies can be combined in one terminal, there is no need for supporting others from the service perspective. These two provide sufficient QoS in all usage scenarios.

The parameters utilized in this study are only a subset of the overall complexity, and in the future, more detailed analyses of other criteria are also necessary. It has been pointed out that, for example, cost, size, and interference are additional items to study before final decisions can be made. Possible approaches would include fixing the cost of the system and comparing the quality in different solutions, or taking uniform quality and comparing the cost.

References

1. "High-Speed 802.11g Creeps Forward," *CE Magazine,* Canon Communications LLC, 11444 W. Olympic Blvd., Los Angeles, CA 90064, 2002.
2. William Stallings, "Standardizing Fixed Broadband Wireless," *Wireless Communications and Networks,* Prentice-Hall, 2001.

3. Edward G. Tiedemann, Jr., "CDMA2000 1X: New Capabilities for CDMA Networks," *IEEE Vehicular Technology Society News,* 445 Hoes Lane, Piscataway, NJ 08855, 2002.
4. Robert Sarfati, "The Evolution of GSM-R: The New European Wireless Standard for Railways," *IEEE Vehicular Technology Society News,* 445 Hoes Lane, Piscataway, NJ 08855, 2002.
5. Dai Davis, "Bluetooth: Standards and the Law," *CE Magazine,* Canon Communications LLC, 11444 W. Olympic Blvd., Los Angeles, CA 90064, 2002.
6. John R. Vacca, *The Cabling Handbook*, 2d ed., Prentice Hall, 2001.
7. John R. Vacca, *i-mode Crash Course,* McGraw-Hill, 2002.

PART 2

Planning and Designing Wireless High-Speed Data Applications

CHAPTER 6

Planning and Designing Wireless Data and Satellite Applications

As you know, wireless data networks are composed of two components—access points and client devices. The components communicate with each other via radio-frequency transmissions, eliminating the need for cabling.

So, what do you need to plan, design, and build a wireless data network? Let's take a look.

Access Points

A wireless data network is planned, designed, and built around one or more access points that act like hubs, which send and receive radio signals to and from PCs equipped with wireless data client devices. The access point can be a stand-alone device, forming the core of the network, or it can connect via cabling to a conventional local-area network (LAN). You can link multiple access points to a LAN, creating wireless data segments throughout your facility. (The Glossary defines many technical terms, abbreviations, and acronyms used in the book.)

Client Devices

To communicate with the access point, each notebook or desktop PC needs a special wireless data networking card. Like the network interface cards (NICs) of cabled networks,[3] these cards enable the devices to communicate with the access point. They install easily in the PC slots of laptop computers or the PCI slots of desktop devices, or link to USB ports. A unique feature found on the wireless data PC card of a leading vendor features a small antenna that retracts when not in use. This is extremely beneficial, given the mobility of laptop computers. You can also connect any device that doesn't have a PC or PCI card slot to your wireless data network by using an Ethernet client bridge that works with any device that has an Ethernet or serial port (printers, scanners etc.).

Once the access point is plugged into a power outlet and the networked devices are properly equipped with wireless data cards, network connections are made automatically when the devices are in range of the hub. The range of a wireless data network in standard office environments can be several hundred feet.

Wireless data networks operate like wired networks and deliver the same productivity benefits and efficiencies. Users will be able to share files, applications, peripherals, and Internet access.

Planning and Designing a Wireless Data Network

Now, what type of features should you plan and design into a wireless data network? In other words, you need to plan, design, and build the following features and solutions:

- Standards-based and WiFi certified
- Simple to install
- Robust and reliable
- Scability
- Ease of use
- Web server for easy administration
- Security
- A site survey application
- Installation

Standards-Based and WiFi Certified

As previously explained, WiFi is a robust and proved industry-wide network standard that ensures your wireless data products will interoperate with WiFi-certified products from major networking vendors. With a WiFi-based system, you will have compatibility with the greatest number of wireless data products and will avoid the high costs and limited selection of proprietary, single-vendor solutions. Additionally, select a wireless solution that is standards based and fully interoperable with Ethernet and Fast Ethernet networks. This will enable your wireless data network to work seamlessly with either your existing cabled LAN or one that you deploy in the future.

Simple to Install

Your wireless data solution should be plug and play, requiring only minutes to install. Plug it in and start networking. For even greater ease of deployment, your solution should support the Dynamic Host Configuration Protocol (DHCP), which will automatically assign IP addresses to wireless data clients. Rather than install a DHCP server in a standalone device to provide this timesaving capability, select wireless data hubs that feature DHCP servers built into them.

If you are adding a wireless data system onto your existing Ethernet network, an access point that can be powered over standard Ethernet cabling makes a great choice. This enables you to run the access point using low-voltage dc power over the same cabling you use for your data—eliminating the need for a local power outlet and power cable for each access point device.

Robust and Reliable

Consider robust wireless data solutions that have ranges of at least 300 ft. These systems will provide your employees with considerable mobility around your facility. You may choose a superior system that can automatically scan the environment to select the best radio-frequency (RF) signal available for maximum communications between the access point and client devices. To guarantee connectivity at the fastest possible rate, even at long range or over noisy environments, make sure your system will dynamically shift rates according to changing signal strengths and distance from the access point. Additionally, select wireless data PC cards for your laptop computers that offer retractable antennas to prevent breakage when the devices are moved about.

Scalability

A good wireless data hub should support approximately 60 simultaneous users. This should enable you to expand your network cost-effectively simply by installing wireless data cards in additional computers and network-ready printers. For printers or other peripherals that do not support networking, you should connect them to your wireless data network with a wireless USB adapter or an Ethernet client bridge.

Ease of Use

A wireless data network should be as effortless for users to operate as a cabled network. To ensure maximum performance and reliability at all times, chose a system that can automatically scan the local environment to select the strongest available radio-frequency channel for communications.

If you plan to connect multiple wireless data hubs to an existing cabled network, consider a solution that features automatic network connections. When a user roams beyond the boundaries of one wireless data hub into the range of another, an automatic network connection capability will seamlessly transfer the user's communications to the latter device, even across router boundaries, without ever reconfiguring the IP address manually. This is particularly useful for businesses with multiple facilities that are connected via the wide-area network (WAN). As a result, users will be able to move about your facility and beyond freely and remain connected to the network.

Web Server for Easy Administration

You will simplify administration of your wireless data network if you select an access point with a built-in Web server. This allows you to access and set configuration parameters, monitor performance, and run diagnostics from a Web browser.

Security

Choose a wireless data solution that offers multiple security layers, including encryption and user authentication. A secure solution will offer at least 40-bit encryption, and advanced systems can provide 128-bit encryption. For both ease of use and the strongest protection, select a superior solution that automatically generates a new 128-bit key for every wireless data networking session without users entering a key manually. Also, consider a system that features user authentication, requiring workers to enter a password before accessing the network.

A Site Survey Application

Your wireless data networking solution should include a site survey utility. The utility can help you determine the optimal location of wireless data hubs and the number of hubs you need to support your users. It will help you to deploy a wireless data solution effectively and efficiently.

Installation

Do you need a technician to install your wireless data network? Generally, you can install a wireless data network yourself. A wireless data solution is an effective strategy if your organization lacks networking experience. Some advanced systems can be set up in a minute or so. Installation and deployment procedures are discussed in specific detail in Part 3, "Installing and Deploying Wireless High-Speed Data Networks" (Chaps. 13 to 17).

Now, let's look at why the planning and design of a large-scale wireless data LAN poses a number of interesting questions. This part of the chapter describes the approaches developed and taken in the planning and design of wireless data networks.

A large-scale wireless data LAN must be planned and designed so that all of the target space has radio coverage (there are no coverage gaps). It must also be designed so that its capacity is adequate to carry the expected load. These requirements generally can be met by using the proper combination of access point locations, frequency assignments, and receiver threshold settings.

Large-Scale Wireless Data LAN Planning and Design

Wireless data LANs (WDLANs) were originally intended to allow local-area network (LAN) connections where premises wiring systems were inadequate to support conventional wired LANs. During the 1990s, because the equipment became available in the PCMCIA form factor, WDLANs came to be identified with mobility. They can provide service to mobile computers throughout a building or throughout a campus.

Generally, wireless data LANs operate in the unlicensed industrial, scientific, and medical (ISM) bands at 915 MHz, 2.4 GHz, and 5 GHz. The original WDLAN standard IEEE 802.11 (with speeds up to 2 Mbps) allows either direct-sequence or frequency-hopping spread spectrum to be used in the 2.4-GHz band. It also allows operation at infrared frequencies. The high-rate WDLAN standard IEEE 802.11b provides operation at speeds up to 11 Mbps in the 2.4-GHz band and uses a modified version of the IEEE 802.11 direct-sequence spread-spectrum technique. A newer high-rate standard, IEEE 802.11a, uses orthogonal frequency-division multiplexing (OFDM) to provide for operation in the 5-GHz UNII band at speeds up to 54 Mbps. IEEE 802.11b equipment is readily available in the market, and IEEE 802.11a equipment is expected to become available by early 2003.

WDLANs typically include both network adapters (NAs) and access points (APs). The NA is available as a PC card that is installed in a mobile computer and gives it access to the AP. The NA includes a transmitter, receiver, antenna, and hardware that provides a data interface to the mobile computer. The AP is a data bridge/radio base station that is mounted in a fixed position and connected to a wired LAN. The AP, which includes transmitter, receiver, antenna, and bridge, allows NA-equipped mobile computers to communicate with the wired LAN. The bridge, which is part of the AP, routes packets to and from the wired network as appropriate.

Each AP has a radio range, for communication with NAs, from approximately 20 to more than 300 m, depending on the specific product, antennas, and operating environment. The APs can be interfaced to IEEE 802.3 (Ethernet) wired LANs.

Most wireless data LANs allow "roaming"; that is, mobile computers can accept a handoff as they move from the coverage area of one AP to the coverage area of another, so service is continuous. In order for this handoff to be successful, it is necessary that the tables of the bridges contained in each AP be updated as mobiles move from one AP coverage area to another. In wireless data LANs, direct peer-to-peer (mobile-to-mobile) communication can be provided in one of two ways. In some wireless data LANs, it is possible for a mobile to communicate directly with another mobile. In others, two mobiles, even though they are both within range of each other, can communicate only by having their transmissions relayed by an AP.

The use of direct-sequence spread spectrum (DSSS) in IEEE 802.11 and 802.11b spreads the signal over a wide bandwidth, allowing transmissions to be robust against various kinds of interference and multipath effects. IEEE 802.11b WDLANs operate at raw data rates of up to 11 Mbps and occupy a transmission bandwidth of approximately 26 MHz. Exact spectrum allocations for 2.4-GHz ISM differ from one country to another. In North America the band is 2.400 to 2.4835 GHz.

IEEE 802.11 and 802.11b use the carrier sense multiple access (CSMA) with collision avoidance (CA) medium access scheme, which is similar to the CSMA/CD scheme used in IEEE 802.3 (Ethernet) LANs. With wireless data transmissions, the collision detect (CD) technique used in wired LANs cannot be done effectively, since the transmitter signal strength at its own antenna will be so much stronger than the signal received from any other transmitter. Instead, CSMA/CA adds a number of features to the basic CSMA scheme to greatly reduce the number of collisions that might occur if only CSMA (without CD) were used.

Planning and Design Challenges

The challenges in building such a large wireless data network are significant. They include planning and designing the network so that coverage blankets, for example, a campus, and adequate capacity is provided to handle the traffic load generated by the campus community. The WDLAN plan and design is defined as including two components: selection of AP location and assignment of radio frequencies to APs.

In laying out a multiple-AP wireless data LAN installation, one must take care to ensure that adequate radio coverage will be provided throughout the service area by carefully locating the APs. Experience shows that the layout must be based on measurements, not just on rule-of-thumb calculations. These measurements involve extensive testing and careful consideration of radio propagation issues when the service area is large, such as an entire campus.

The layout and construction of buildings determine the coverage area of each AP. Typical transmission ranges go up to 300 m in an open environment, but this range may be reduced to 20 to 60 m through walls and other partitions in some office environments. Wood, plaster, and glass are not serious barriers to wireless data LAN radio transmissions, but brick and concrete walls can be significant ones; the greatest obstacle to radio transmissions commonly found in office environments is metal, such as in desks, filing cabinets, reinforced concrete, and elevator shafts.

Network performance is also an issue. An AP and the mobile computers within its coverage area operate something like the computers on an Ethernet segment. That is, there is only a finite amount of bandwidth available, and it must be shared by the APs and mobile computers. The IEEE 802.11b protocol, using CSMA/CA, provides a mechanism that allows all units to share the same bandwidth resource.

The Carrier Sense Multiple-Access/Collision Avoidance (CSMA/CA) protocol makes radio interference between APs and NAs operating on the same radio channel a particular challenge. If one AP can hear another AP or a distant NA, it will defer, just as it would defer to a mobile unit transmitting within its primary coverage area. Thus, interference between adjacent APs degrades performance. Similarly, if a mobile unit can be heard by more than one AP, all of these APs will defer, thus degrading performance.

Design Approach

In selecting AP locations, one must avoid coverage gaps, areas where no service will be available to users. On the other hand, one would like to space the APs as far apart as possible to minimize the cost of equipment and installation. Another reason to space the APs far apart is that coverage overlap between APs operating on the same radio channel (cochannel overlap) degrades performance. Minimizing overlap between APs' coverage areas when one is selecting AP locations helps to minimize cochannel overlap.

NOTE One should not overprovision a wireless LAN by using more APs than necessary.

The rules of thumb are inadequate in doing this type of planning and design. Rather, each building plan and design must be based on careful signal strength measurements. This is particularly challenging because the building is a three-dimensional space, and an AP located on one floor of the building provides signal coverage to adjacent floors of the same building and perhaps to other buildings as well.

After the APs have been located and their coverage areas measured, radio channels are assigned to the APs. Eleven DSSS radio channels are available in the 2.400- to 2.4835-GHz band used in North America; of these, there are three that have minimal spectral overlap. These are channels 1, 6, and 11. Thus, in North America, APs can operate on three separate noninterfering channels. Furthermore, some NAs can switch between channels in order to talk with the AP providing the best signal strength or the one with the lightest traffic load. Use of multiple channels can be very helpful in minimizing cochannel overlap, which would otherwise degrade performance.

One approach is to assign one of these three channels to each of the APs and to do so in a way that provides the smallest possible cochannel coverage overlap. Making these frequency assignments is essentially a map coloring problem, and there are various algorithms that give optimal or near-optimal assignment of the three radio channels, given a particular set of AP placements and coverage areas.

The design must also consider service to areas with high and low densities of users. If many users of mobile computers are located in a small area (a high-density area), it may be necessary to use special design techniques in these areas. Most parts of a campus will be low-density areas. However, there will be some areas, particularly classrooms and lecture halls, that will be high-density areas, with high concentrations of users, mostly students.

Two design layout techniques that are useful in high-density situations are increasing receiver threshold settings and using multiple radio channels. Some wireless data LAN products allow one to set receiver threshold, thus controlling the size of the coverage area of the AP. A coverage-oriented design should use the minimum receiver threshold setting, maximizing the size of the coverage area of each AP. When capacity issues are considered, however, one may wish to use higher AP receiver threshold settings in high-density areas, reducing the coverage area of each AP.

The use of multiple radio channels can allow the use of multiple APs to provide coverage in the same physical space. For example, one might use three APs operating on three different channels to cover a large lecture hall with a high density of users. The exact capacity improvement is dependent on the algorithm used by the mobile unit to select an AP. A load-balancing algorithm will provide the greatest capacity increase. An algorithm that selects the strongest AP signal will not provide as great an increase.

Thus, one would like to carry out a plan and design that is coverage-oriented in most (low-density) areas, minimizing the number of APs, but capacity-oriented in some (high-density) areas, assuring adequate capacity to serve all users in these areas. The coverage-oriented design in the low-density areas minimizes the cost of APs, but the use of extra APs with higher receiver thresholds in high-density areas can be used to provide extra capacity.

Planning and Design Procedure

Because radio propagation inside a building is frequently anomalous and seldom completely predictable, the planning and design of an indoor wireless data installation must be iterative. The planning and design procedure includes five steps:

- Initial selection of AP locations
- Test and redesign, which is adjusting the access point locations based on signal strength measurements
- Creation of a coverage map
- Assignment of frequencies to APs
- Audit, which is documenting the AP locations and a final set of signal strength measurements at the frequencies selected[1]

In the next part of the chapter, a technique for carrying out the first step is described, along with the initial selection of access point locations. This initial plan and design is tentative and is intended to be modified in the second step of the planning and design process.

After the initial selection of AP locations is complete, APs are temporarily installed at the locations selected. The coverage areas of these APs and the overlaps between coverage areas are measured. Typically, coverage gaps and/or excessive overlaps are found. On the basis of the measurement results, the AP locations are adjusted as needed, more measurements are done, more adjustments are made, and so on, until an acceptable plan and design is found. The process is an iterative one. It may be necessary to repeat this planning and design-test-redesign cycle several times to find an acceptable solution.

After the final AP locations have been selected, a coverage map of the planning and design area is created. This coverage map may be created by using AutoCAD or other computer-based techniques.

After AP locations have been finalized, frequencies are assigned to the APs in a way that minimizes cochannel coverage overlap. Then, a complete set of coverage measurements (audit) is made for the entire building with the APs operating at the selected frequencies, and the results of these measurements are documented. At this point, the design is considered complete. The coverage map is updated to reflect final AP locations, coverage measurements, and frequency assignments.

Determining the Access Points' Initial Locations

Now let's look at a procedure for the initial selection of AP locations in a low-density area. In selecting locations for the APs, one should place them so that there are no coverage gaps in the target space, and the coverage overlaps between and among APs are minimized. While the first point is obvious, the second is more important than is immediately apparent. If too many APs are used, the cost of equipment and installation will be higher than necessary, and the performance of the network may also be degraded if the final design involves a great deal of cochannel coverage overlap. The amount of cochannel coverage overlap is determined by both AP placement and AP frequency assignment.

The coverage area is defined in terms of a specified received signal strength. This threshold level is selected in order to provide an adequate signal-to-noise ratio (*S/N*) and some additional margin. If, for example, in designing an IEEE 802.11b WDLAN, one measures an ambient noise level of −95 dBm and a 10-dB *S/N* is needed to ensure excellent performance, one might decide to allow an extra 5 dB of margin to allow for noise levels higher than −95 dBm. In this case, one would select a threshold of −80 dBm.

When high-density spaces exist, it is suggested that the AP placement first be done for these spaces and that the remaining low-density spaces then be designed, filling in the gaps between high-density spaces.

AP Placement In this part of the chapter, an idealized notion of AP coverage is introduced. This description is offered only to provide some insight into the layout approaches that can be used in different types of buildings.

The coverage volume of the AP is idealized as three coaxial cylinders, as shown in Fig. 6-1.[1] The middle cylinder, representing coverage on the floor on which the axis point is located, has radius R. The AP is located on the axis of this cylinder. The upper and lower cylinders, representing coverage on the floors above and below the one on which the AP is located, have radius R', which is less than R. The height of each of the three cylinders is the height of a floor in the building. These three cylinders can be thought of as a single object, which moves about as the location of the AP moves.

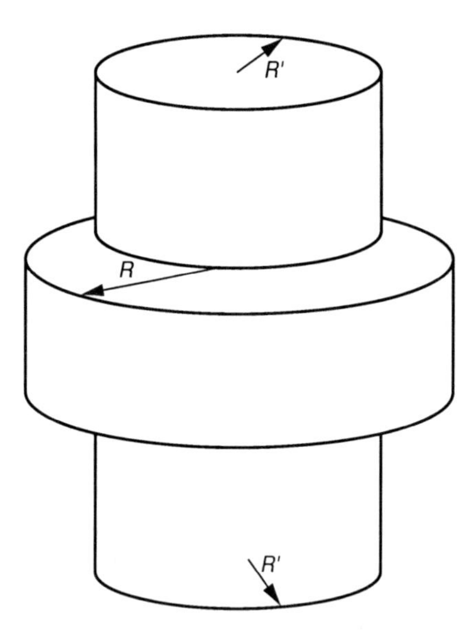

Figure 6-1 Idealized access point coverage.

The problem of locating APs within a building can be viewed as a problem of locating these shapes within the building in such a way that all spaces are filled with as little overlap as possible. While coverage volumes are not actually perfect cylinders, one can find the average coverage radius inside a building and use this as the radius of an idealized cylindrical coverage volume. This can be achieved by defining an acceptable signal strength threshold (−80 dBm) and determining the average distance from the AP at which signals fall below the threshold.

Procedure The initial selection of AP locations begins with a complete set of signal strength measurements within the building. Signal strength measurements should be made in all areas of the building, with particular attention to the building's construction so that the characteristics within each part of the building are understood. These measurements have two purposes: to divide the building into spaces that are relatively isolated from each other from a signal propagation perspective and to determine the typical coverage radius of an AP. Signal strength measurements should be taken to determine the same floor coverage radius R and the adjacent floor coverage radius R' of an AP.

Access points can be placed within a building in an array that is either linear or rectangular. An example of a linear array is shown in Fig. 6-2, and an example of a rectangular array is shown in Fig. 6-3.[1] Each of these shows how APs can be located in a single-floor building or in a building with only one floor needing WDLAN coverage. It is necessary only to locate the APs in a way that provides coverage throughout the floor and

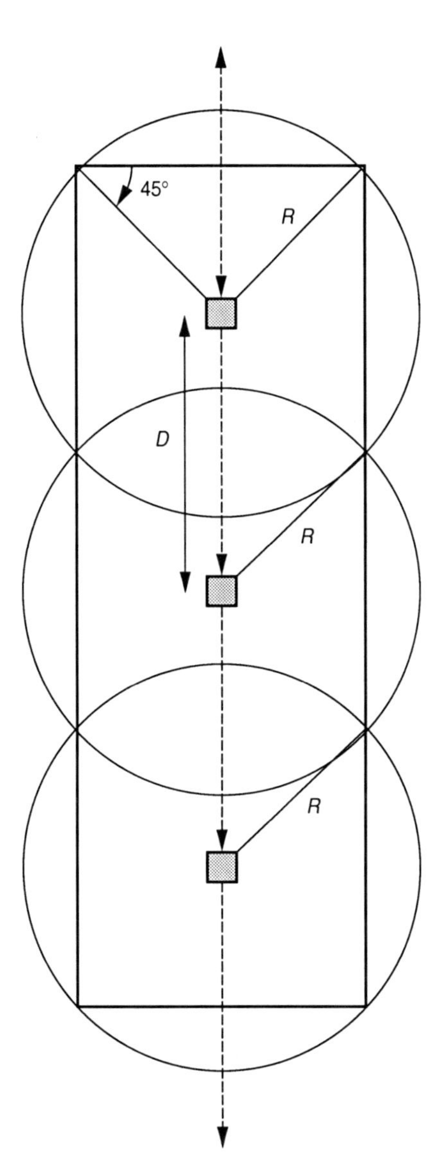

Figure 6-2
A linear array of APs in a single-floor building.

also minimizes as far as possible the overlap between and among AP coverage areas. A linear array is used when the building is narrow relative to *R*, and a rectangular array when the building width is large relative to *R*.

On the other hand, in a building that requires coverage on more than one floor, adjacent floor coverage must be considered in locating each AP. Usually, a staggered approach is used. As one moves along the length (or width) of a building, one places APs first on one floor and then on an adjacent floor. In this case, the coverage of an AP's adjacent floor coverage

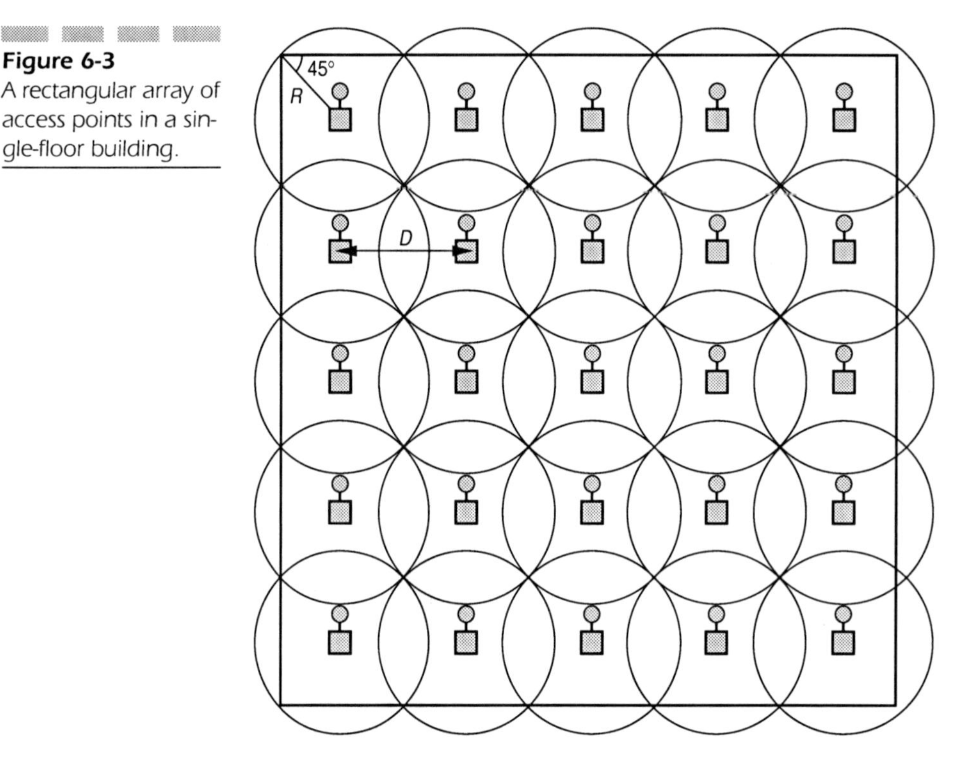

Figure 6-3 A rectangular array of access points in a single-floor building.

must dovetail with the coverage of the next AP's same floor coverage. As in a single-floor building, a linear array is used when the building is narrow relative to R, and a rectangular array when the building width is large relative to R.

Let's now illustrate by using four scenarios one will encounter when planning and designing an indoor wireless data network. Each is determined by whether the building is single-story or multistory and by the width of the building relative to R and R'. In each case, the appropriate layout approach is given and the figure that illustrates it is listed. Solid lines show coverage on a floor; dashed lines show adjacent floor coverage.

Scenario 1 A single-floor linear array is illustrated in Fig. 6-2.[1] This is a single-story building (or a building that requires wireless data coverage on only one floor) whose width (smallest outer dimension) is not large relative to R. D denotes the distance between adjacent APs.

Scenario 2 A single-floor rectangular array is illustrated in Fig. 6-3. This is a single-story building (or a building that requires wireless data coverage on only one floor) whose width (smallest outer dimension) is large relative to R. D denotes the distance between adjacent APs.

Scenario 3 A multifloor linear array is illustrated in Fig. 6-4.[1] This is a multistory building whose width (smallest outer dimension) is not large relative to R and R'. D' denotes the distance between adjacent APs on different floors.

Scenario 4 A multifloor rectangular array is illustrated in Fig. 6-5.[1] This is a multistory building whose width (smallest outer dimension) is large relative to R and R'. D denotes the distance between adjacent APs on the same floor, and D' denotes the distance between adjacent APs on different floors.

Frequency Assignment

After the AP locations have been finalized and a coverage map has been created, frequencies are assigned to the APs. In the United States and Canada, three nonoverlapping channels (channels 1, 6, and 11) are used. Thus, one can assign one of these three frequencies to each AP, doing so

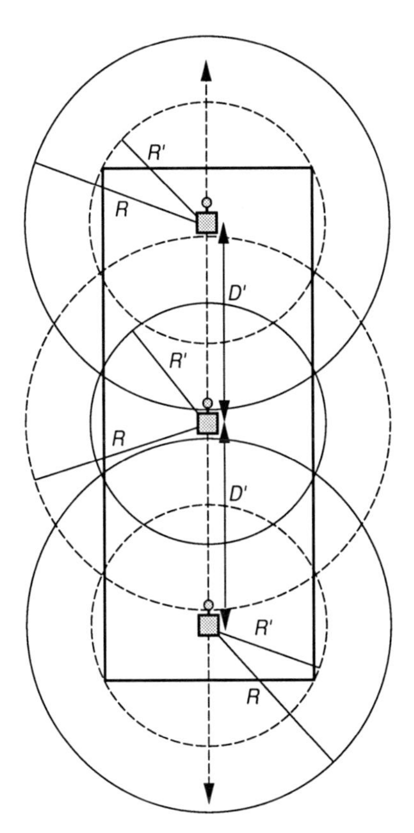

Figure 6-4 A linear array of access points in a multifloor building.

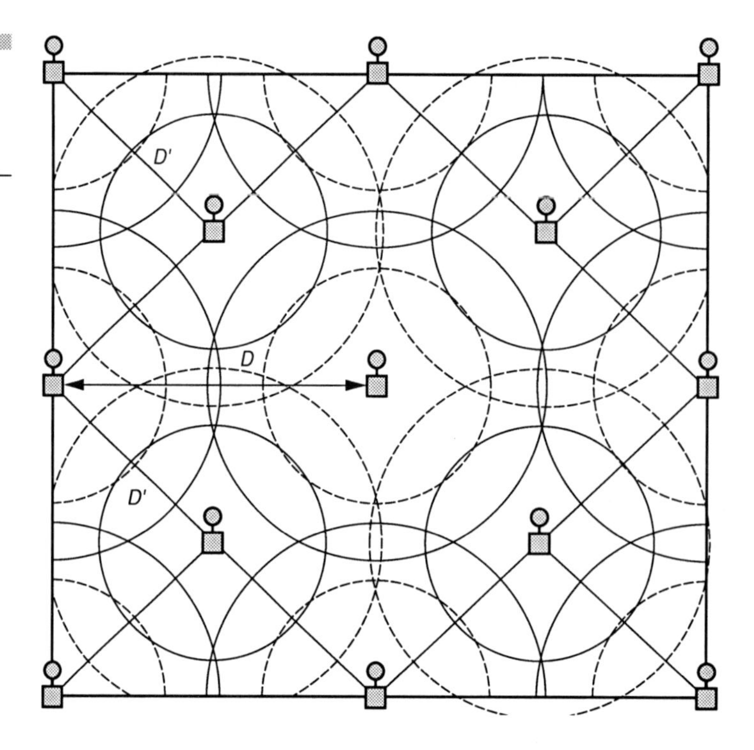

Figure 6-5 Rectangular array of access points in a multifloor building.

in a way that minimizes cochannel overlap. Assignment of frequencies is essentially a map coloring problem with three colors.

A variety of algorithms can be used to assign AP frequencies when the AP coverages are known. One can do this exhaustively by checking the cochannel overlap for all possible frequency assignments, and this is a reasonable approach if a computer is being used. Other, less time-consuming algorithms are also possible, and some of these can give near-optimal results. Another approach is to use the building coverage map that has been created to visualize the coverage overlaps and assign frequencies so that cochannel APs have only small coverage overlaps.

It is recommend that you assign AP frequencies in high-density areas before low-density areas. If, for example, one uses three APs to cover a high-density space, three different channels should be assigned to these APs. These frequency assignments will subsequently need to be considered in assigning frequencies to nearby APs covering low-density areas. This is true because APs covering the high-density space will usually have some coverage overlap with APs covering only low-density areas.

Now, let's look at how the planning and design of effective interworking between a multimedia terrestrial backbone and a satellite access platform[5] is a key issue for the development of a large-scale IP system designed for transporting multimedia applications with QoS guarantees.

This part of the chapter focuses on the planning and design of a gateway station that acts as an interworking unit between the two segments of the systems. The guarantee of differentiated QoS for applications within the envisaged global IP system is achieved effectively by assuming that the IP IntServ model in the satellite access system is combined with a DiffServ fixed-core network, in which the RSVP aggregation protocol is implemented. Thus, the design activity of the IWU mainly focuses on the following issues: seamless roaming between the two heterogeneous wireless data and wired environments, efficient integration between the two IP service models (IntServ and DiffServ), and suitable mapping of terrestrial onto satellite bearer for traffic with different profiles and QoS requirements.

Planning and Designing the Interworking of Satellite IP-Based Wireless Data Networks

Within the Internet community, strong expectations for a global system that is able to offer a differentiated quality of service (QoS) come from customers and applications. Such expectations both make the traditional Internet model based on the "same service to all" concept inadequate and, at the same time, move research and development activities toward the deployment of large-scale IP networks (implementing the concept of the global Internet).

Thus, on one hand, a commonly employed solution is to extend the potentialities of the Internet through service differentiation mechanisms, in order that some groups of customers and applications can obtain a superior level of service just by accepting different agreements with the carrier and higher costs. Such an enormous interest in IP QoS has brought about the rapid development of two standards for IP with quality assurance: one, an integrated services model coupled to the Resource Reservation Protocol (IntServ/RSVP), the other a differentiated services (DiffServ) model. On the other hand, it is clear that in order to offer the negotiated service quality to mobile end users in an enhanced broadband platform[4] for the global Internet, the Internet with QoS guarantees a new generation of multimedia satellite platforms that must converge toward integrated platforms.

NOTE The research reported in this part of the chapter deals with the issue of integrating IP with QoS assurance into a multimedia terrestrial-satellite infrastructure.

The Internet Engineering Task Force (IETF) proposes access networks working with the IntServ/RSVP architecture and core networks based on the DiffServ architecture. Such a proposal is driven by the essential difference between IntServ and DiffServ models: While the former is interested in offering end-to-end QoS guarantees to a single flow, the latter aims at scalability in large networks.

The envisaged solution guarantees many advantages. In particular, it provides a scalable end-to-end service with reasonable QoS guarantees across the core network, while an explicit reservation of resources is available on the access links where the bandwidth may be scarce.

The difficulties and the consequent awkward research issues that lie behind the deployment of effective interworking between the terrestrial and satellite segments are mainly tied to the contrasting features of the two cited IP models and the different natures of the environments involved (one common feature: The satellite bandwidth is still a precious resource, and the propagation delay strongly influences any design decisions). The proposed effective design of the whole terrestrial-satellite multimedia system will focus on the following design options:

- The design of a "reservation protocol" compatible with both enhanced-IP models (DiffServ and IntServ) that is able to handle heterogeneous connections with the required QoS on both the fixed and satellite sides.
- The implementation of a "mapping" among service classes of both models to carry out effective IntServ-DiffServ integration.
- The implementation of a mapping of fixed network bearer services over the bearer services offered by the satellite access network in order to perform effective integration of terrestrial and satellite segments.[2]

It goes without saying that a gateway station, interconnecting satellite and terrestrial segments, has a role of prime importance within the highlighted architecture. This makes its design particularly delicate.

The aim of this part of the chapter is therefore to address the research issues pointed out hitherto and present a proposal for the design of the interworking unit (IWU) operating within the terrestrial and satellite segments of an integrated system architecture for fourth-generation IP wireless data systems. The role of integrated QoS-aware IP models (DiffServ and IntServ) within the designed infrastructure is also highlighted.

IP Networks with QoS Guarantee

The research IETF carried out on QoS provisioning in IP networks led to the definition of two distinct architectures: integrated services (IntServ) (with its signaling protocol RSVP) and differentiated services (DiffServ).

The IntServ framework defines mechanisms that control the network-level QoS of applications requiring more guarantees than those available when the traditional best-effort IP model is exploited. Provision of end-to-end QoS control in the IntServ model is based on a per-flow approach, in that every single flow is separately handled at each router along the data transmission path.

The IntServ architecture assumes that explicit setup mechanisms are employed to convey information to the routers involved in a source-to-destination path. These mechanisms enable each flow to request a specific QoS level. RSVP is the most widely used setup mechanism.

Through RSVP signaling, network elements are notified of per-flow resource requirements by using IntServ parameters. Subsequently, such network elements apply admission control and traffic resource management policies to ensure that each admitted flow receives the requested service. It is thus clear that RSVP implements its functionality by means of signaling messages exchanged among sender, receiver, and intermediate network elements. A sender host uses the Path message to advertise the bandwidth requirements of its information flow downstream along the routing path. It also stores the path state in each node along the way. By using the Resv message, the receiving host reserves the amount of bandwidth necessary to guarantee a given QoS level. The Resv message retraces exactly the path to the sender host, reserving the resources in the intermediate routers (it creates and maintains reservation state in each node along the path used by the data) and is finally delivered to the sender host, so that it can set up appropriate traffic control parameters. The following factors have prevented a large deployment of RSVP (and IntServ) in the Internet:

- The use of per-flow state and per-flow processing raises scalability problems for large networks.
- Only a small number of hosts currently generate RSVP signaling. Although this number is expected to grow dramatically, many applications may never generate RSVP signaling.
- The needed policy control mechanisms (access control, authentication, and accounting) have become available only recently.[2]

In contrast to the per-flow orientation of RSVP, the DiffServ framework defines mechanisms for differentiating traffic streams within a network and providing different levels of delivery service to them. These mechanisms include differentiated per-hop queueing and forwarding behaviors (PHBs), as well as traffic classification, metering, policing, and shaping functions that are intended to be used at the edge of a DiffServ region. The DiffServ framework manages traffic at the aggregate rather than per-flow level. The internal routers in a DiffServ region do not distinguish the individual flows. They handle packets according to their PHB identifier based on the DiffServ codepoint (DSCP) in the IP packet header. Since DiffServ eliminates the need for per-flow state and per-flow processing, it scales well to large networks.

IETF is currently interested in two types of DiffServ traffic classes: uncontrolled and controlled. The first class offers qualitative service guarantees, but is unable to offer quantitative guarantees. An example of an uncontrolled traffic class is the assured forwarding (AF) PBH. The controlled traffic class uses per-flow admission control to provide end-to-end QoS guarantees. An example of controlled traffic class is represented by the expedited forwarding (EF) PBH.

IntServ/RSVP and DiffServ can also be used as complementary technologies in the pursuit of end-to-end QoS. IntServ can be used in the access network to request per-flow quantifiable resources along a whole end-to-end data path, while DiffServ enables scalability across large networks and can be used in the core network. The main benefits of this model are a scalable end-to-end IntServ framework with QoS guarantee in the core network, and explicit reservations for the access network where bandwidth can be a scarce resource.

Border routers between the IntServ and DiffServ regions may interact with core routers using aggregate RSVP in the DiffServ region to reserve resources between edges of the region. In fact, per-flow RSVP requests from the IntServ region would be counted in an aggregate reservation. The advantage of this approach is that it offers dynamic admission control to the DiffServ network region, without requiring the level of RSVP signaling processing that would be required to support per-flow RSVP. Details of this approach will be given later.

The Satellite-Terrestrial Integrated Framework

Let's now address the support of end-to-end IntServ over a DiffServ core network. Figure 6-6 illustrates the whole reference architecture, whose main components are a DiffServ network region and some IntServ network regions.[2]

The DiffServ network region is a terrestrial core network that supports aggregate traffic control. This region provides two or more levels of service based on the DSCP in packet headers. The IntServ network regions are segments outside the DiffServ region that may consist of generic IntServ access networks. In this case, let's consider an IntServ satellite access network on one side and any DiffServ terrestrial network on the other. The specific satellite network used here as a reference is the EuroSkyWay (ESW) geosatellite system (see Fig. 6-7), which is an enhanced satellite platform for multimedia applications.[2]

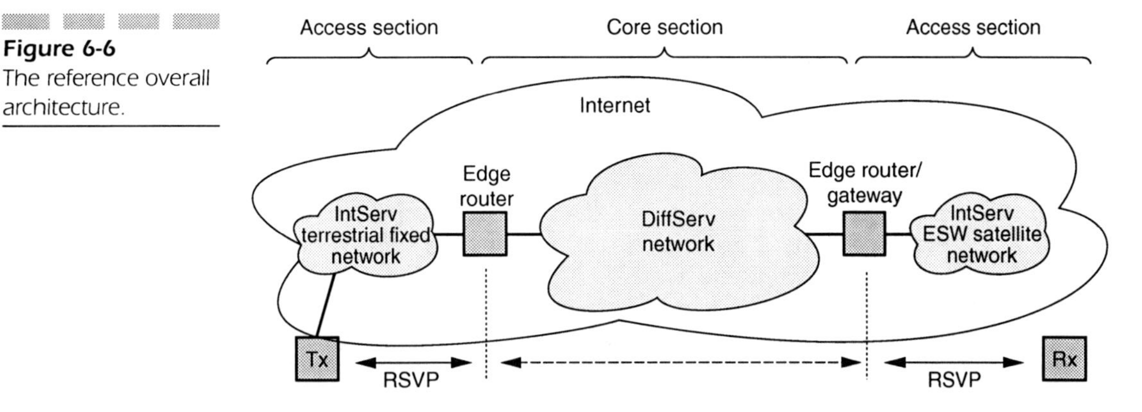

Figure 6-6 The reference overall architecture.

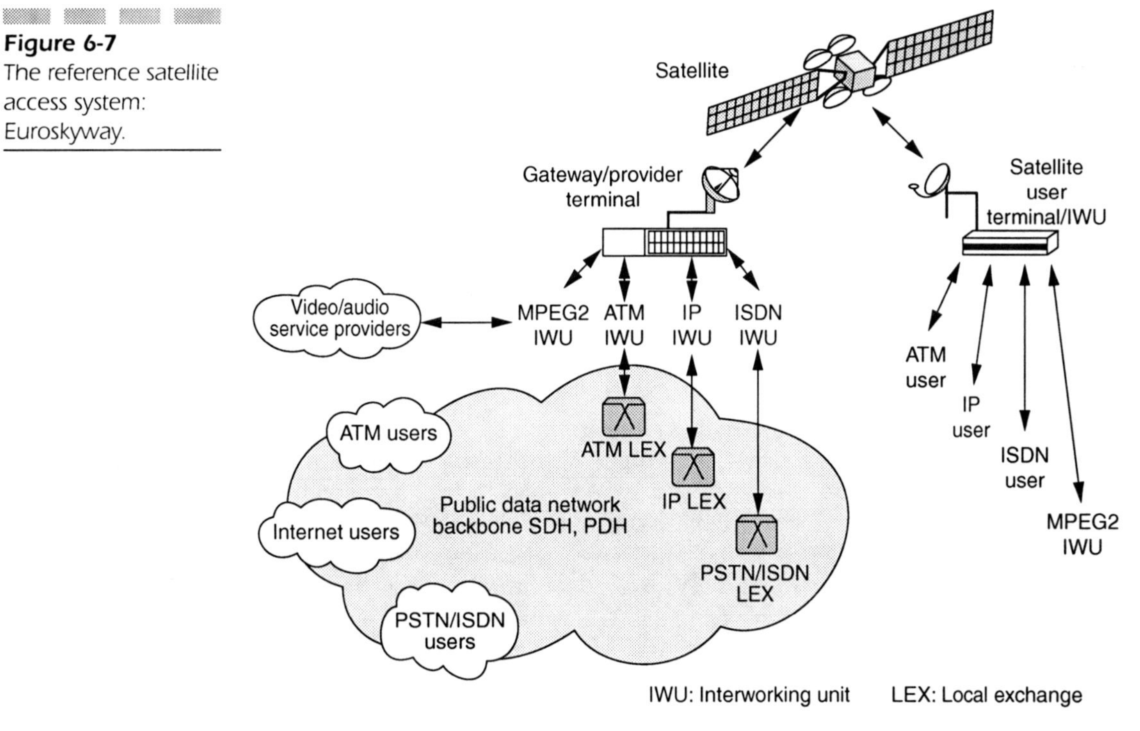

Figure 6-7 The reference satellite access system: Euroskyway.

Edge routers (ERs), which are adjacent to the DiffServ network region, act like IntServ-capable routers on the access networks and DiffServ-capable routers in the core network. In this approach, the DiffServ network is RSVP-aware and ERs also function as border routers for the DiffServ region. This means that ERs participate in RSVP signaling and act as admission control agents for the DiffServ network. As a result, changes in the capacity available in the DiffServ network region can be communicated to the IntServ-capable nodes outside the DiffServ region via RSVP. This feature gives the proposed architecture the further advantage of providing dynamic resource provisioning in the DiffServ core network, in contrast to static provisioning.

As for the satellite access network, its main components are illustrated in Fig. 6-7: a satellite with onboard processing (OBP) capability; a gateway station, interconnecting satellite and terrestrial segments; satellite terminals of different types; and a master control station. In particular, the master control station is responsible for call admission control (CAC); the reference system uses statistical CAC to increase satellite resource utilization. The satellite has OBP capability and implements traffic and resource management (TRM) functions.

The satellite network can be seen as an underlying network, aiming to interface a wide user segment by using different protocols, such as IP, asynchronous transfer mode (ATM), X.25, frame relay, narrowband integrated services digital network (N-ISDN), and MPEG-based ones (so-called overlying networks, OLNs). A valid example of this type of system is the EuroSkyWay satellite system.

The transparency of the satellite network is based on the use of IWUs, present at both the satellite terminal and the gateway/provider terminal level, but with different features. Because of the difference between the existing terrestrial network protocols, one IWU for each network protocol is envisaged.

Since the goal here is to enable seamless interoperation between Intserv and Diffserv segments of the reference architecture, this part of the chapter focuses on the functionality of an IWU conceived for the interconnection of the satellite system and the Internet core network.

For the sake of simplicity, but without losing generality, a single sender is considered here: Tx communicating across the reference network with a single receiver, Rx. Tx is a host in the terrestrial Intserv access network, and Rx is a mobile terminal of the satellite ESW system.

It's assumed that RSVP signaling messages travel end-to-end between hosts Tx and Rx to support RSVP/Intserv reservations outside the Diffserv network region. It's required that these end-to-end RSVP messages be carried across the Diffserv region without being processed by any of the routers in the Diffserv region. The remainder of this part of the chapter presents details of the procedures implemented for providing an effective interconnection between the DiffServ and IntServ regions of the reference network architecture.

Aggregate RSVP

Aggregate RSVP is an extension to RSVP being developed in order to enable reservations to be made for an aggregation of flows between edges of a network region, rather than for individual flows as supported by the current version of RSVP. In other words, Aggregate RSVP is a protocol proposed for the aggregation of individual RSVP reservations that cross an "aggregation region" and share common ingress and egress routers into one RSVP reservation from ingress to egress.

An aggregation region is a contiguous set of systems capable of performing RSVP aggregation. Routers at the ingress and egress edges of an aggregation region are termed *aggregator* and *deaggregator,* respectively. They dynamically create the aggregate reservation, classify the traffic to which the aggregate reservation applies, determine how much bandwidth is needed to achieve the requirement, and recover the bandwidth when the individual reservations are no longer required.

The establishment of a smaller number of aggregate reservations instead of a larger number of individual reservations allows reduction of the amount of state to be stored in the nodes on the path and of the signaling messages exchanged in the aggregation region. Such amounts are independent of the number of individual reservations.

The aggregation region is where the DiffServ model is adopted. Therefore, DiffServ mechanisms are used for classification and scheduling of traffic supported by aggregate reservations inside the aggregation region. One or more DSCPs are used to identify a traffic of aggregate reservations, and one or more PHBs are used to require a forwarding treatment to this traffic from the routers along the data path. By using DiffServ mechanisms (rather than performing per-aggregate reservation classification and scheduling), the amount of classification and scheduling state in the aggregation region is even further reduced. It is independent of the number of aggregate reservations.

There are numerous options for choosing which DiffServ PHBs might be used for different traffic classes crossing the aggregation region. This is the "service mapping" problem that will be described later in the chapter.

The edge routers at the ingress and egress sides of the DiffServ core network act as aggregator and deaggregator. In the reference architecture, the edge router in the terrestrial access IntServ network acts as an aggregator, while the edge router in the satellite IntServ destination network acts as a deaggregator. Let's call end-to-end (E2E) reservations the reservation requests relevant to individual sessions, and E2E Path/Resv messages their respective messages. Let's also refer to an aggregate reservation as a request relevant to many E2E reservations. The relevant messages are logically called aggregate Path/Resv messages.

To manage aggregate reservations, one has to be able to hide E2E RSVP messages from RSVP-capable routers inside the aggregation region. To this end, the IP protocol number in some E2E reservation messages is changed from its normal value (RSVP) to RSVP-E2E-IGNORE upon entering the aggregation region, and restored at the egress point. This enables each router within the aggregation region to ignore E2E reservation messages; messages are forwarded as normal IP datagrams. Aggregate Path messages are sent from the aggregator to the deaggregator using RSVP's normal IP protocol number.

As for QoS control, by means of traditional RSVP, the QoS control services are invoked by exchanging several types of data, carried by particular objects, including information that is sent from the sender to intermediate nodes and to the receiver, and describes the data traffic generated by that sender (Sender TSpec). This also includes information from the receivers to intermediate nodes and to the sender (FlowSpecs) that describes the desired QoS control service, the traffic flow to which the resource reservation should apply (Receiver TSpec), and the parameters required to invoke the service (Receiver RSpec). Furthermore, the ADSPEC object carries information collected from network elements toward the receiver. This information is generated or modified within the network and used at the receivers to make reservation decisions. This information might include available services, delay and bandwidth estimates, and operating parameters used by specific QoS control services.

The description of the flow generated by the source is made through the use of suitable parameters that are communicated to the receiver host. These are the token bucket parameters (token bucket rate r, token bucket size b, peak data rate p, maximum packet size M, and minimum policed unit m). As a consequence, the traffic profile is specified in terms of token bucket parameters.

In order to generate aggregate Path and Resv messages, the token bucket parameters (in the SENDER_TSPECs and FLOWSPECS) of E2E reservations must be added. Furthermore, the ADSPEC object must be updated, as described later in the chapter.

The Gateway and Its Functional Architecture

The gateway station plays a fundamental role within the reference network architecture shown in Fig. 6-6. Thus, the attention in this part of the chapter is directed toward the effective design of this device. As outlined already, it has a twofold functionality: interworking between the terrestrial and satellite network segments, and aggregating/deaggregating.

This functionality is located in the IWU module of the gateway, which is therefore also seen as an IP node. The internal structure of the gateway device is depicted in Fig. 6-8; it is split into some building blocks that are included in the control plane or data plane.[2]

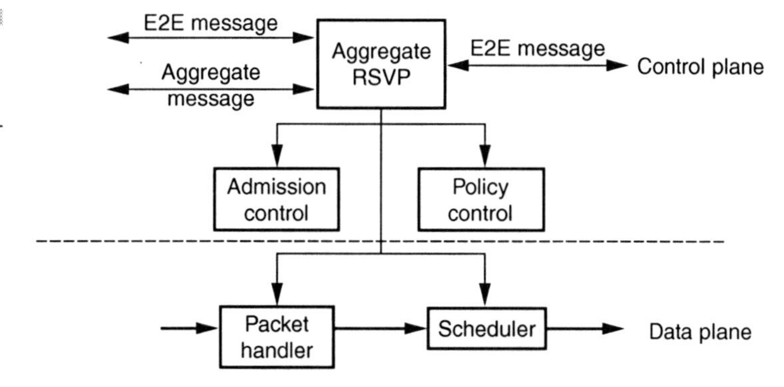

Figure 6-8 The gateway internal structure.

The Aggregating/Deaggregating Function of the Gateway: Operations at the Control Plane Level

The control plane of the gateway contains the functionality for establishing and clearing data paths through the network. As already mentioned, the gateway acts as an RSVP-capable router with the functionality of deaggregator at the egress of the DiffServ core network. As such, it is responsible for managing E2E Path and Resv message exchange. Specifically, the gateway is involved in the reception of E2E Path messages from the aggregator and the handling of E2E Resv messages coming from the Rx terminal. In the remaining part of this part of the chapter, the sequence of operations performed is described to set up an end-to-end RSVP QoS connection between the terrestrial Tx terminal and the satellite Rx terminal of Fig. 6-6.

Operations at the Data Plane Level

The data plane that's proposed here contains the functionality for transmission of traffic generated by user applications. As already shown in Fig. 6-8, the data plane includes two functional blocks: the packet handler and the scheduler. The packet handler is responsible for management of the aggregated traffic at the gateway input; it changes the aggregated DiffServ traffic into individual IntServ flows.

Figure 6-9 shows the packet handler functionality in detail.[2] Initially, any incoming aggregated traffic is policed in order to assess its conformance to the declared token bucket parameters. Out-of-profile traffic can be dropped, reshaped, or handled as best-effort traffic.

Subsequently, the DSCP classifier processes the DSCP value of the aggregated traffic and forwards the packet to one of the queues; a queue is provided for each type of DSCP value (best effort, BE, AF, and EF).

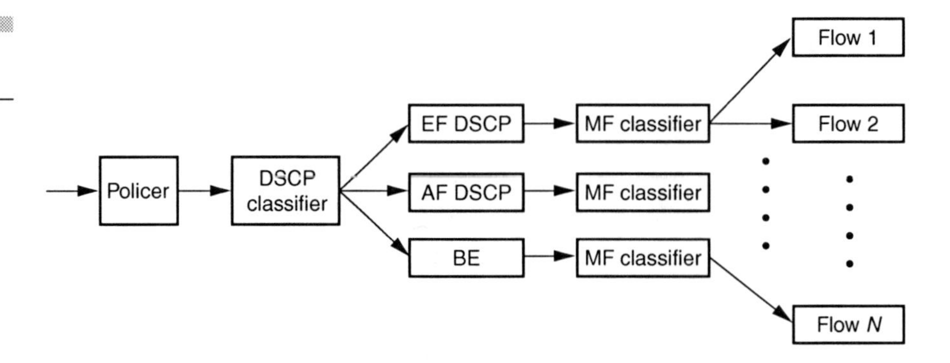

Figure 6-9
The packet handler.

At this point, traffic is still aggregated. The next step involves the separation of the flows of which it is composed. This separation is performed by the multifield (MF) classifier, which is able to classify single flows based on a combination of some IP header fields, which are source address, destination address, DS field, IP protocol, and source and destination port. The MF classifier assigns the packets of its queue to an IntServ service class and then forwards them to the appropriate queue.

Individual IntServ flows, whose packets are separately queued according to the flow type, represent the outgoing traffic from the packet handler. The packet scheduler is responsible for the transmission of packets queued in the packet handler according to a defined scheduling policy. It determines the different packet management at the network layer based on the desired QoS. Since the reference satellite system uses multifrequency time-division multiple access (TDMA), the scheduler assigns, on a frame basis, queued packets in the correspondent slots of the satellite connection.

Functionality of Interworking between the Terrestrial and Satellite Network Segments

Details on the most important functions of the gateway are given in the following part of this chapter. Let's take a look.

E2E Path ADSPEC Update at the Gateway Since E2E RSVP messages are hidden from the routers inside the aggregation region, the ADSPECs of E2E Path messages are not updated as they travel through the aggregation region. Therefore, the gateway is responsible for updating the ADSPEC in the corresponding E2E Path to reflect the impact of the aggregation region on the QoS that may be achieved end to end. To do so, the deaggregator should make use of the information included in the ADSPEC from an Aggregate Path, since Aggregate Path messages are processed inside the aggregation region and their ADSPEC is updated by routers.

In this reference system, however, it is not sufficient to update the ADSPEC, including just the impact of the aggregation region, since the gateway should also take into account the impact of the satellite path on the achievable end-to-end QoS. To perform this update, the gateway distinguishes two cases, according to the IntServ service class involved in the reservation procedure.

In the case of IntServ CLS, the ADSPEC includes only the break bit used to indicate the presence of a node incapable of managing the service along the data transmission path. Consequently, the gateway has to modify the break bit only if the satellite network does not support the CLS service.

In the case of IntServ GS, the ADSPEC update depends on how this service is mapped over the satellite link. If GS is mapped over satellite permanent connections, the *D* term in the expression *DB* includes only the duration of a frame during which the source host has to wait, in the worst case before transmitting a burst in the slots assigned to it. If GS is mapped over semipermanent connections, the *D* term also includes the further delay due to the per-burst resource request. In general, the following terms contribute to the *D* terms for a satellite connection:

- Time it takes for the burst transmission request to reach the traffic resource manager (TRM) and return (270 ms)
- Maximum waiting time of a request on board (TimeOut)
- One frame duration, as the request received during a frame by the TRM is analyzed during the next frame (26.5 ms)
- One frame duration due to TDMA (26.5 ms)
- One terminal configuration time interval (100 ms) and an onboard switching time interval (54 ms)[2]

Logically, for a permanent connection, *D* is equal only to the frame time given by TDMA. For the semipermanent connection, all the terms listed are present. Thus, D_{perm} = 26.5 ms, and D_{semip} = 477 ms + TimeOut. The *C* term relevant to the satellite link is invariant. If the requested delay is lower than DBS, the requested bandwidth over the satellite is greater than *p*; if a delay greater than DBS is sufficient, a smaller bandwidth is requested.

Before concluding, it is worth highlighting a further concept. Time delay is a major QoS function; thus, it is interesting to give some details on the end-to-end time delay the proposed architecture can offer to the supported applications and the influence this delay has on system performance. The first consideration is that, in the DBS previously considered, the time required to set up connections (mainly including round-trip delay times and a negligible time delay for processing) has to be included as well. In the following, some curves are given in which the total end-to-end delay is present on the abscissa axes.

Since a GEO satellite is used as an example satellite network (EuroSkyWay), it is clear that the proposed architecture is unsuitable for voice traffic and highly interactive real-time applications because of the long processing, path establishment, service mapping, and propagation time delay. Nevertheless, a wide range of low-interactive, real-time packet-based applications that allow for some end-to-end total delay time can be supported by the platform described in this part of the chapter, while achieving a good performance level.

Simulations have been conducted by loading the system with GS traffic only and with a number of sources greater than the maximum number actually accepted by the CAC. This is performed to stress the CAC system and verify the achievable loading level of the system.

The curves in Fig. 6-10 that are relevant to the system load are sketched by fixing b = 128 kb, r = 256 kbps, and burstiness B = 3, and for different values of the TimeOut expressed in terms of the number n of TDMA frames a resource request from a GS burst tolerates being buffered on board before being satisfied.[2] The curves show that the sustained load (and the number of accepted sources as well, curves not shown) increases with the overall requested delay for the source traffic. In fact, when the maximum end-to-end allowed delay (and consequently the delay bound) increases, the requested bandwidth R decreases, and the number of both the accepted sources and total exploited satellite channels increases.

Shown is just a sample situation. Anyway, the load behavior for different GS burstiness values has been analyzed with the aim of verifying how this parameter influences system performance. As expected, the system shows worse behavior when the source's burstiness increases. An increase in burstiness implies an increase in the requested bandwidth

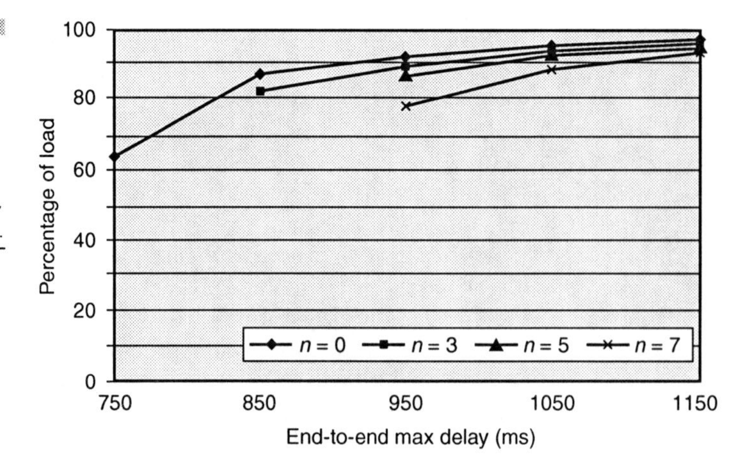

Figure 6-10 Percentage of load versus end-to-end maximum delay, for various values of the TimeOut (b = 128 kb, r = 256 kbps, B = 3).

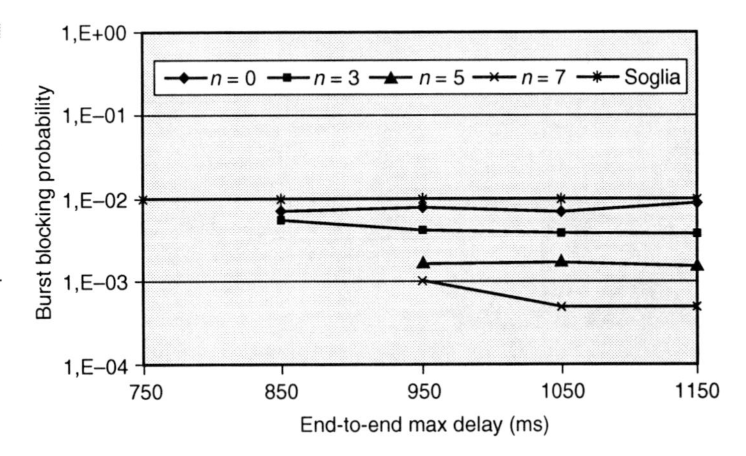

Figure 6-11 Burst blocking probability versus maximum request delay, for various values of the TimeOut (b = 1024 kb, r = 256 kbps, B = 3).

and a consequent decrease in the total exploited channels. Nevertheless, system performance still remains high for an allowed end-to-end delay range like that shown in Fig. 6-10.

The transport of the IP GS class over semipermanent satellite connections may introduce burst losses due to the statistical multiplexing performed by the CAC. Therefore, a metric of interest is the burst blocking probability (BBP), which measures the probability that a burst waiting for resources must be blocked (and then lost) as a result of unavailability of satellite channels. Specifically, the BBP curves for b = 1024 kb, r = 256 kbps, and B = 3 are shown for different values of the TimeOut in Fig. 6-11.

The curves in Fig. 6-11 show that the loss caused by the mapping of GS flows on semipermanent satellite connections can be kept below the bound of 0.01 established by the CAC mechanism.[2] Furthermore, by observing the curves in Fig. 6-11, it can be noted that the BBP decreases when the TimeOut increases, since the greater the maximum waiting time allowed on board for a resource request, the smaller the probability that a buffered request is discarded. In general, the BBP remains below the bound (0.01) established by the CAC, unless the TimeOut is zero, independent of the burstiness value. Also, in this case, the BBP remains below the bound established by the CAC, unless the TimeOut is zero. A similar behavior has always been found under any traffic profile and loading condition.

Conclusion

The design of a large-scale IEEE 802.11b WDLAN should be done in a way that ensures complete coverage of the target space and adequate capacity to carry the anticipated traffic load. The design must consider both the selection of AP locations and the assignment of frequencies to the APs.

AP locations should be selected so that all of the target space has radio coverage (there are no coverage gaps). AP locations should be selected and frequencies assigned in order to minimize cochannel coverage overlap. In high-density areas, coverage overlap can be used (with different frequencies) to provide increased capacity. Another technique useful in serving high-density areas is increasing receiver thresholds in order to reduce APs' coverage areas.

In this chapter, the integration of a terrestrial IP backbone with a satellite IP platform has been addressed with the main aim of enabling the resulting system for the global Internet to provide a differentiated service quality to mobile applications of a different nature. The detailed description of the functional architecture and the task performed by an interworking unit within the gateway interconnecting the two environments were highlighted. The resulting design of the IWU allows the effective interconnection of the terrestrial (DiffServ-based) and satellite (IntServ-based) segments, and the consequent potential achievement of both a good level of system resource utilization and the possibility of matching the QoS requisites of a wide range of applications.

References

1. Alex Hills, "Large-Scale Wireless LAN Design," *IEEE Communications Magazine,* 445 Hoes Lane, Piscataway, NJ 08855, 2002.
2. Antonio Iera and Antonella Molinaro, "Designing the Interworking of Terrestrial and Satellite IP-Based Networks," *IEEE Communications Magazine,* 445 Hoes Lane, Piscataway, NJ 08855, 2002.
3. John R. Vacca, *The Cabling Handbook*, 2d ed., Prentice Hall, 2001.
4. John R. Vacca, *Wireless Broadband Networks Handbook,* McGraw-Hill, 2001.
5. John R. Vacca, *Satellite Encryption,* Academic Press, 1999.

CHAPTER 7

Architecting Wireless Data Mobility Design

Once you have planned and designed your wireless data network to deliver corporate information, you still need to understand two basic mobile computing architecture models (synchronization and real-time access), and choose the most appropriate one. This is equally true for both wireline and wireless data connectivity. Let's examine the two models, the challenges that created the need for wireline synchronization, and reasons why synchronization is even more important in the wireless data world. (The Glossary defines many technical terms, abbreviations, and acronyms used in the book.)

Real-Time Access

The mobile computing device connects to the network whenever the user needs information, a query is sent to a communications server, and the requested information is located and transmitted back to the device for viewing. The user can interact with the information on the server only when a connection is available.

Synchronization

The mobile computing device connects occasionally to the network when possible, and synchronization middleware keeps information on the device in sync with that on the server. The user can interact with information on the device any time regardless of connection availability, and sync up when possible. Synchronization is also referred to by many as *offline access* or *store-and-forward technology.*

Why Synchronization?

Many people mistakenly assume that wireless data applications must automatically have a real-time access or thin client model. In fact, synchronization technologies originally developed for wireline-based mobile computing are even more applicable in the wireless data world of heightened challenges.

The factors identified in Table 7-1 have led corporations to demand mobile middleware solutions with synchronization capabilities.[1] These factors are relevant to discussions of both wireline and wireless data mobile computing.

TABLE 7-1

Challenges to Real-Time Access Model

Challenge	Notes
Coverage	Users need to track down a phone line or network port to connect, or to find cell tower coverage. Big impact on convenience and usability.
Speed	A function of throughput and latency. Users have to endure idle time while the query is transmitted, while the server searches for information, and while the information is transmitted back. Likewise for stored changes to be applied.
Communications costs	Store-and-forward sync can offset the added costs of mobile computing. The real-time model means repeat downloads of information, must send query to server and retrieve data each time the same info is accessed.
Reliability	The mobile worker is dependent on the reliability of network connections to accomplish tasks. Work can continue when connections drop with synchronization.
Standards	Wired standards are well established, with a variety of options. Wireless data standards are still emerging—increasing total costs to support mobile computing.

In the past, most corporations had already pursued mobile computing, leveraging wireline connections such as dial-up, WAN, VPN, or high-speed dedicated lines to remote locations. Many of these implementations relied on synchronization middleware from vendors such as Synchrologic, designed to help overcome the challenges of real-time access.

Data Wireless Makes Synchronization Even More Appropriate

The same architecture considerations must be weighed today in wireless data solutions. While the availability of wireless data networks undoubtedly adds convenience to the end user, in addition to potentially increasing the timeliness of information the user interacts with, the challenges of mobile computing still exist in the wireless data world. In fact, they are generally exacerbated and have a far more pronounced effect. Thus, the same factors that made synchronization a great technology for managing mobile computing with wires make it even more appropriate for many wireless data applications.

Corporations will have to make well-reasoned choices between wireline and wireless data communications, and between synchronization and real-time access. In fact, many organizations today are choosing to

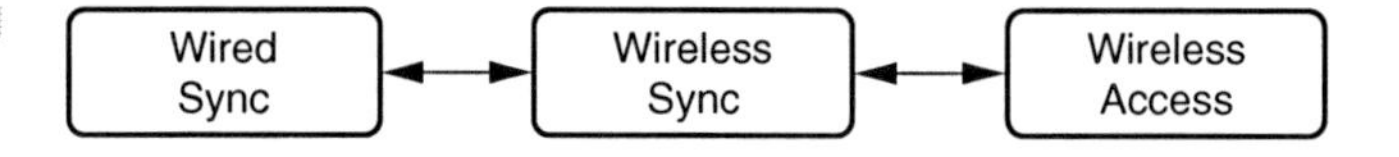

Figure 7-1 Options for mobile systems architecture.

pursue hybrid models that support multiple options in order to serve different users in different geographies at different times.

Today, enterprises are seen as demanding a mobile computing infrastructure that supports all mobile computing devices, provides comprehensive mobile infrastructure functionality, and supports both architecture models. Figure 7-1 summarizes the spectrum of mobile computing architectures, or "ways of working."

How Do You Choose Which Model for Your Wireless Data Application?

To select the most appropriate model, you will need to consider the following questions:

- Do users live and work in areas of ubiquitous wireless data coverage?
- Will work site building structures cause interference to wireless data?
- How important is guaranteed access to information stored[4] locally?
- How often does the referenced information change?
- How much more will real-time access cost for communications?
- Does real-time access add business value beyond synchronization?
- Will users wait for the query and response period?
- How granular is the information brought down with each query?
- Is instant access more or less valuable than up-to-the-second data?
- How long will users wait to download large attachments?
- Is lack of access to data acceptable if coverage is not available?[1]

Again, many times a hybrid model might be appropriate where the solution must support both real-time and synchronization architectures, for different groups of users, or for users to use selectively at appropriate times. Now, let's look briefly at synchronization.

Synchronization as Default Option

The bottom line is that you need to look at the business drivers behind the mobile computing project, consider the types of factors discussed in the preceding, take a look at how dynamic the information is, and carefully weigh the usage patterns for the application. Only then can you select the appropriate model. For practical reasons, synchronization should be considered the default option, unless a compelling reason for real-time access is forthcoming in your review.

Now, let's take a comprehensive look at the concerns of your peers with regard to architecting mobility. The discussion that follows will help shed light on the issues you face in addressing mobile computing.

Critical Steps in Supporting Mobile Enterprise Computing

What follows is the top action items organizations are pursuing today to build competitive advantage and deal with increasingly critical mobile and wireless data computing issues. Topics include:

- Application mobilization
- Controlling communications and support costs
- Managing and supporting mobile devices
- How to cut through the wireless data hype
- Lowering TCO of mobile devices
- Understanding the big picture[2]

Develop a Mobile Strategy Now

Increasing market pressures coupled with the rapid-fire growth of mobile computing have created a booming population of mobile and remote workers. The drive to stay competitive has tasked today's enterprise with exploiting any and every means necessary in optimizing service levels, increasing sales, boosting efficiency, and cutting costs. Mobile computing provides the enterprise with several compelling competitive advantages, including:

- Faster, decentralized decision making

- Increased responsiveness to customers
- Increased sensitivity to market changes
- Lowered commuting costs/time for staff
- Increased staff morale and productivity
- Reduced travel costs company-wide
- Decreased facilities costs[2]

Enterprise demand for support of mobile computing initiatives now requires extending the full complement of enterprise resources to do business anywhere, at any time. The enterprise that proactively pursues a comprehensive mobile computing strategy will be successful in building competitive advantage. True business agility requires flexible technologies and the ubiquitous proliferation of computing power.

Failure to architect and build a mobile strategy today will have the same effect as ignoring the invasion of PCs back in the 80's. By developing a mobile strategy that includes adopting standards, developing mobile infrastructure, and embracing mobile devices, your enterprise can effectively use mobile computing to stay competitive.

Keep an Eye on Wireless Data

Wireless data holds much promise for mobile computing. From real-time access for mission-critical applications to automated dissemination of competitive information, wireless data will dramatically affect the mobile computing landscape. But, mobile and wireless data are not interchangeable terms. Wireless data is one component of mobile. Though wireless data has substantial potential and some interesting uses today, myriad applications may prove more usable via wireline connections. Wireless data computing is a tricky endeavor, with numerous pitfalls ready to snare the enterprise that moves without careful consideration.

Wireless Data Today Today's wireless data networks are characterized by competing standards and protocols. No single network technology or operator will meet all your wireless data network needs. Most of the current wireless data networks known as 2G networks (GSM, CDPD, Mobitext, Motient) are built on analog and cellular digital networks—infrastructure designed to support voice communications. Data transmission is a more complex endeavor and the current public networks are ill-suited to efficiently support acceptable wireless data transmission rates. The 3G networks are better equipped to handle data transmission, but are not slated for completion for years to come. Figure 7-2 cites a Yankee Group study into enterprise concerns regarding wireless data adoption.[2]

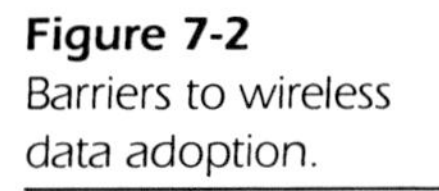

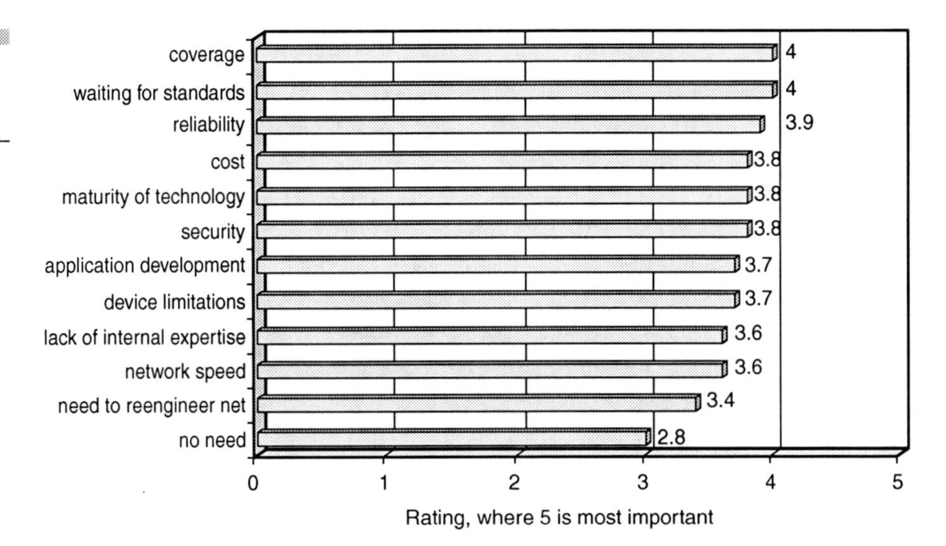

Figure 7-2
Barriers to wireless data adoption.

Accommodate the Occasionally Connected User: Real Time or Synchronization?

The fundamental challenge in mobilizing your enterprise is determining how a variety of mobile device users interact with data and information currently located on company servers. As previously discussed, in addressing the challenge, the enterprise must decide between two scenarios: real-time access and synchronization.

The Case for Synchronization While constant, real-time access is appealing, and well-suited for certain functions, synchronization may often be the more practical, smarter solution for the majority of enterprise needs. Synchronization offers the following benefits over real-time access:

- Reduced queries and network traffic
- Reduced user idle time
- Compression of staged data
- Reduced concurrent server processing loads
- Controlled communication costs[2]

As wireless data protocols mature, the lines will begin to blur between real-time and store-and-forward architectures, and organizations will deploy both options in a complementary fashion. The convergence of synchronization and real-time mechanisms is crucial in accommodating varying bandwidth and connection scenarios, and in graceful switching between modes of operation. The goal of 24–7 network availability can be shattered by the unreliability of a dial-up connection, and remote access to applications could prove useless when software is not in sync with desktop PCs.

The best way to accommodate the reality of occasionally connected users is to build a flexible infrastructure. Whether connected to corporate networks in a real-time environment, or working with localized applications in a deferred access environment, the optimal solution is to afford end users the luxury of being indifferent to, if not unaware of, whether or not they are connected. Users must avoid strategic investments in transitioning mobile wireless data technologies and focus instead on developing back-end logic that is device/network-agnostic and developing expertise in mobile application usability.

The current state of wireless data technology, coupled with the inconvenience of staying perpetually connected via wireline, has created the reality of the occasionally connected user. Your enterprise should build a mobile infrastructure, flexible enough to support both real-time access and synchronization.

Deploy E-mail and PIM to Hand-Helds

Since its debut, e-mail's tenure as the most killer enterprise application has remained relatively unchallenged. E-mail is the application wireless data adopters are asking for the most. As e-mail continues to be a vital method of communication, the ability to synchronize anywhere at any time and have access to the corporate intranet will provide significant productivity gains.

The advent of robust groupware applications like Microsoft Exchange and Lotus Notes has complemented enterprise e-mail systems with personal information management (PIM) data including calendars, contacts, to-do lists, and memos, providing one unified package for corporate workers to manage their busy lives.

Mobilizing groupware applications to hand-helds is more complex than a first glance would suggest. Early solutions featured a handheld–to–desktop synchronization model. With these products, the full burden of installation, support, and troubleshooting rested entirely on the shoulders of those least likely to be able to perform these functions—end users. Because these applications are not server-based, the flood of mobile devices through the corporate backdoor is further complicated.

A new breed of e-mail and PIM synchronization solutions now allows the exchange of data directly between hand-helds and more functional server-based groupware applications. Important features of a server-based e-mail and PIM synchronization solution include:

- One-step synchronization
- Connection transparency for users
- Complex filtering
- Encryption
- Flexible conflict resolution[2]

Enabling anywhere access to e-mail and groupware servers is the critical first step in empowering mobile workforces. Mobile workers are immediately more productive and the sense of disconnect associated with being away from the office is minimized.

Plan for Multiple Devices

If your organization is like most, there is already a mix of laptops, Palm devices, and pocket PCs in use by staff, and probably purchased by the company if only via expense reports. This is creating challenges for most IT shops. These range from network/data integrity issues, to an overtaxed help desk fielding support calls on devices it may or may not know exist, to inefficient systems management tools that don't work well for occasionally connected devices.

It was hard enough just trying to support laptops, and then hand-helds invaded your company through the back door. As the popularity of these devices continues to grow, the variety of models increases, and feature lists on these devices continue to grow, so too does the threat they pose to your corporation's information integrity as well as the costs, skills, and time required to support these devices. The time to take action is now.

Device Diversity One need only look at the success of Palm and other hand-held PDA manufacturers to gauge the blossoming proliferation of mobile devices used as companion devices to the venerable laptop. That flood of mobile devices is only going to continue: According to Meta Group, by 2005, each corporate knowledge worker will have four to five different computing and information access devices that will be used to access various applications.

While Palm's market share in the consumer hand-held space has remained dominant, various device manufacturers are gaining ground—especially in the enterprise market. According to IDC, by 2006, Microsoft's market share will surge to 41 percent, compared with 52 percent for Palm.

With the popularity of hand-helds reaching a fevered pitch, it's easy to forget about the laptop. For most, the laptop is the mobile workhorse. The effects of the recent slackening in growth of the PC market have not been mirrored in the laptop and notebook PC market. According to Gartner Group, the worldwide mobile PC market grew by 43.8 percent in the third quarter of 2001 compared with the third quarter of 2000.

The promise of Bluetooth and the growing popularity of wireless data LANs will further encourage enterprise adoption of laptops over the coming years. Many experts predict that over 60 percent of PC shipments will be laptop or notebook units within a few years. Falling prices and improved features, coupled with the increased market pressures, are forcing enterprise mobilization and contribute to the adoption of mobile PCs and hand-helds instead of traditional desktop workstations.

One of the most important points to take from this discussion is the notion that multiple device proliferation will continue, as Fig. 7-3 suggests.[2] In a study conducted by Yankelovich Partners, a high-technology research firm, when given the choice of carrying a wireless phone, two-way e-mail device, PDA, or pager, only 45 percent of professionals indicate they still wanted only one device.

You should not be bound by selective mobile infrastructure solutions that exclude certain devices. Business drivers should determine which devices are appropriate for which user groups, not your mobile platform software. In this way, you maximize business value and return.

Figure 7-3 Total computing device usage world forecast (in thousands).

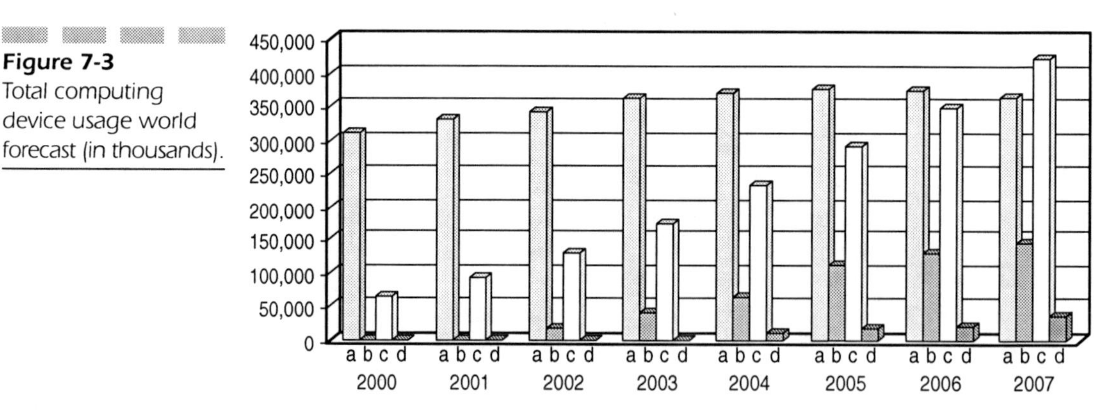

Structure and Automate Content Distribution

Getting access to file-based content and intranet pages is another area that can be challenging to mobile workers. Automated content distribution makes things much easier for the end user, enabling access to unstructured data, including spreadsheets, word processing documents, presentations, and graphics files. Ideally, this information is replicated throughout the mobile network. Without any effort, users have access to important files when on the road—even if a network connection is unavailable.

While data synchronization helps proliferate structured data, there is typically a wealth of information that is unstructured and saved in a variety of popular file formats. Many companies rely on e-mail to publish these files to end users. This methodology unnecessarily exposes the enterprise to several risks, including:

- *E-mail viruses.* No longer relying on e-mail attachments to move files through your network reduces the risk of exposure to viruses.
- *File versioning pitfalls.* Users don't have to sift through e-mails to find the latest version of a file.
- *Mailbox administration.* As attachments are less necessary, the stress on groupware servers due to large mailbox sizes is reduced.[2]

With automated content distribution, the most current files are automatically maintained and delivered to the appropriate personnel with no user intervention required. Content distribution also enhances the effectiveness of your corporate intranet. By making the site available off line as well as on line, your intranet becomes a more effective, relied-upon communication tool. Components of a robust content distribution mechanism include:

- Publish and subscribe architecture
- Web publication
- Remote device backup
- Overwrite versus rename
- File differencing
- Delivery logging
- Subscription management
- File versioning[2]

Through implementing a content distribution protocol, your mobile users get access to the most current, time-sensitive information found in files. This includes management reports, operations statements, pricing and product information, contracts, company forms and policies, and competitive information.

Implement a Robust Asset Management Solution

Critical to the mobilization of your network data is the deployment of a solution that will enable your IT staff to remotely manage device hardware and software inventories. Gathering devices or burning CDs can be expensive and wastes valuable resources when remote software distribution can automate software installs, upgrades, and removals. Likewise, fielding support calls from the mobile staff without an image of their device is extremely difficult.

Traditional LAN-based asset management solutions fail in the reality of the occasionally connected user. These solutions presume high-bandwidth, always connected devices, with high network reliability. The reality for mobile workers is different—they connect their devices with the network only occasionally, and typically over low-bandwidth, frequently dropped connections. Traditional systems management vendors have been slow to support mobile users, and as a result, these capabilities are typically sourced from the new breed of mobile infrastructure solution vendors. Of course, integration with the existing systems management solution is important here.

The systems management application you select should be comprehensive and flexible, providing customizable tools for systems maintenance, support, and troubleshooting. Must-haves in an asset management solution include:

- Comprehensive user profiling
- Condition-triggered alert mechanisms
- Flexible real-time logging
- Hierarchal log construction
- Console-based log views
- Encryption
- Full-device refresh
- Checkpoint restart
- Transaction rollback
- File compression

- Default user profiling
- Offline synchronization[2]

With a powerful systems management solution in place, you are poised to deliver top-quality support to end users. You should avoid undue strains on your scarce IT resources, and lower the total cost of ownership of mobile devices.

Mobilize Applications through Synchronization

For the enterprise dedicated to building competitive advantage through extending the reach of its network, the mobilization of core enterprise applications is of utmost importance. The following applications can yield significant benefits through mobilization for your field-based and frequently traveling workforces:

- Sales force automation
- Customer relationship management
- Enterprise resource planning
- Field service applications
- Supply chain management
- E-business applications[2]

An advanced data synchronization package, capable of supporting mobile PCs and hand-helds alike, is the only way to ensure mobile workers have constant access to critical corporate data. Careful consideration should be given to selection of data synchronization technology. Your solution should fit inside your existing applications elegantly and cleanly, freeing your technical staff from writing complex conduit code and an extensive integration effort. Make sure that the synchronization logic you define can be leveraged across multiple devices, and that you are not saddled with different administrative tools to support PCs, hand-helds, and other devices. Important things to look for in data synchronization tools include:

- Multiple platform support
- Multiple database support
- Open application development
- Field-level synchronization

- Offline synchronization
- Flexible change capture
- Graphical rules wizard
- Store-and-forward architecture
- Flexible conflict resolution tools
- Nonintrusive to applications[2]

With an advanced data synchronization engine, you will be able to easily mobilize your core enterprise applications without extensive integration and conduit coding. This should make mobile workers more productive by allowing them to do business anywhere.

Beware Consumer-Focused Vendors

In evaluating partners to help architect and build your mobile strategy, be sure your vendors have demonstrated enterprise experience. Many "enterprise" solutions are repackaged consumer solutions and lack the features required for success in the enterprise environment. Your solution should be built from the ground up with the enterprise in mind.

Administrative Control Your solution should contain flexible administrative controls that enable your mobile infrastructure to change as dictated by business dynamics. At a minimum, your administrative console should allow you to:

- Define the user base
- Define synchronization activities
- Set default configurations for sessions
- Configure the amount of user control allowed
- Subscribe users to activities
- Prioritize the order of activities
- Review extensive system logs
- Set alerts and notifications
- Remotely troubleshoot and address problems[2]

Make sure your mobile solutions partner has demonstrated experience in the enterprise market. This will ensure that the mobile solutions partner can offer secure, scalable, and flexible tools, so your mobile infrastructure can grow with your business.

Avoid Point Solutions

Integral to the development and execution of a robust mobile strategy is a commitment to infrastructure development and extension. Your IT team should not waste time integrating piecemeal mobile solutions. Because infrastructure is the basic, enabling framework of the organization and its systems, a holistic approach to its design, deployment, and management is pivotal to organizational success.

A comprehensive mobile infrastructure, remotely managed through a well-equipped administrative console, allows corporations to deploy groupware to hand-helds, mobilize enterprise applications, control assets, and manage and deliver content. The alternative is a mix of incompatible point solutions with proprietary systems management and support consoles—overtaxing enterprise resources and jeopardizing network integrity. Your infrastructure should be capable of supporting a variety of devices and platforms including:

- Laptop and tablet PCs
- Remote desktop PCs
- Palm OS devices
- Windows CE/pocket PC devices
- Industrial hand-held devices
- Point-of-sales systems
- Bar code readers
- Portable data terminals[2]

Figure 7-4 shows the components of a mobile infrastructure robust enough to support all of the mobile initiatives previously mentioned.[2] When considering mobile infrastructure products, map their solutions against the above model. Even if you don't require all of this functionality today, building an infrastructure capable of supporting these functions will help you avoid the pitfalls of point solutions and recognize the long-term benefits of a comprehensive infrastructure solution. These benefits include:

- Lowered training costs for administrators
- Easier for users—less effort to "get all their stuff"
- Flexible and easy for administrators
- Decreased support costs
- Decreased integration costs
- Support for all your devices

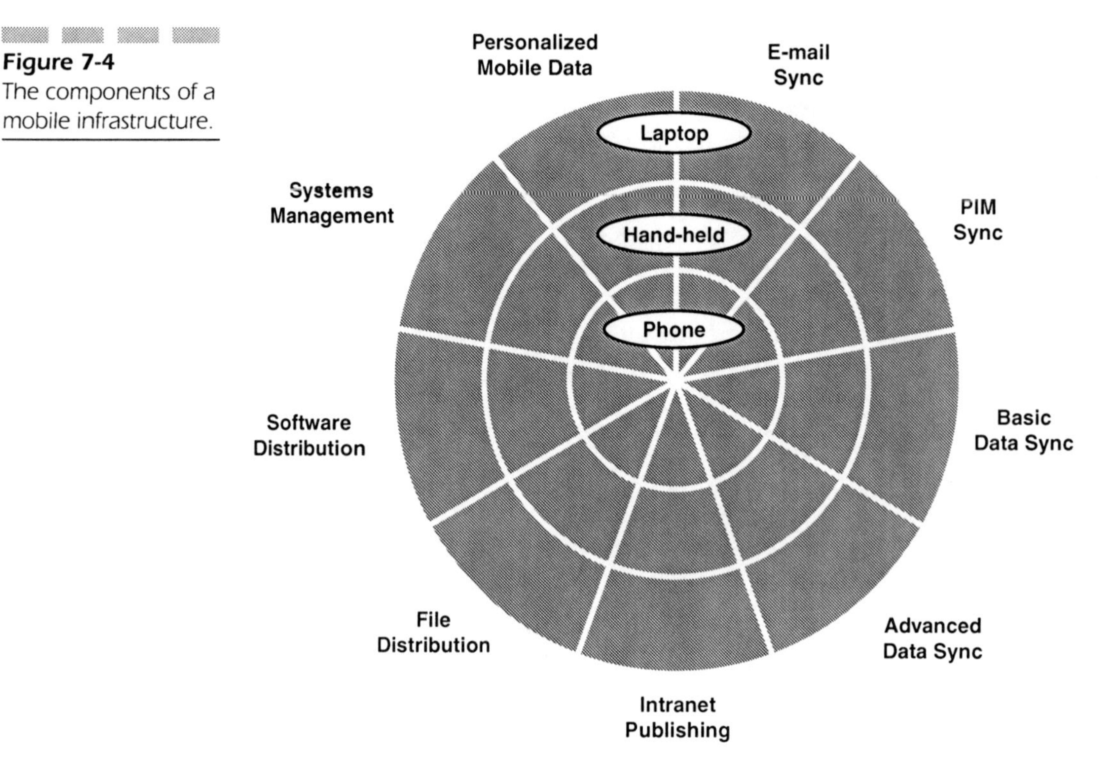

Figure 7-4 The components of a mobile infrastructure.

- Lowered software license costs
- Less time evaluating/negotiating vendors
- One point of contact for support/troubleshooting
- Increased user and administrator productivity[2]

Next, let's look at the ongoing research effort to construct a new multicarrier CDMA architecture based on orthogonal complete complementary codes, characterized by its innovative spreading modulation scheme, uplink and downlink signaling design, and digital receiver implementation for multipath signal detection. There are several advantages of the proposed CDMA architecture compared to conventional CDMA systems pertinent to current 2G and 3G standards. First of all, it can achieve a spreading efficiency (SE) very close to 1 (the SE is defined as the amount of information bits conveyed by each chip), whereas SEs of conventional CDMA systems equal $1/N$, where N denotes the length of spreading codes. Second, it offers MAI-free operation in both up- and downlink transmissions in an MAI-AWGN channel, which can significantly reduce the cochannel interference responsible for capacity decline of a CDMA system. Third, the proposed CDMA architecture is able to offer a high bandwidth efficiency due to its unique spreading modulation scheme and orthogonal carries. Lastly, the proposed CDMA architecture is particularly suited to multirate signal transmission because of the use of an offset stacked spreading modulation scheme, which simplifies the rate-matching algorithm relevant to multimedia services and facilitates asymmetric traffic in up- and downlink transmissions for IP-based applications. On the basis of the preceding characteristics and the obtained results, it is concluded in this part of the chapter, that the proposed CDMA architecture has a great potential for applications in future wideband mobile communications beyond 3G, which is expected to offer a very high data rate in hostile mobile channels.

Multicarrier CDMA Architecture

The great success of worldwide second-generation (2G) mobile communications has a tremendous impact on the lifestyle of people in the world today. In recent years, the voice-oriented services provided by the 2G mobile communication infrastructures in many countries have attracted increasing numbers of users. In Taiwan, more than 80 percent of the population subscribes to GSM mobile phone services. More than 200 million people in China use mobile phones, and each year more than 30 million people become new subscribers. The increasing trend in the penetration rate is expected to continue, especially in many developing countries.

The triumph of the 2G systems has also paved the way for the deployment of a new generation of mobile communications currently on the way in many developed countries. In May 2001, Japan initiated the world's first testing of commercial services for 3G mobile communications based on wideband code-division multiple-access (W-CDMA) technology, which can deliver various multimedia services on top of the voice-oriented and slow-rate data services available in the current 2G systems. In Taiwan, the government closed the bidding process on five 3G licenses at the end of 2001; it is expected that island-wide 3G services will be made ready in 2003, which is in phase with other countries in the world.

The maturing of 3G mobile communication technologies from concepts to commercially deliverable systems motivates one to think about the possible architectures for future generations of mobile communications. Nobody is very sure what the mobile communications beyond 3G will look like; what is certain at this moment is that the systems beyond 3G ought to deliver a much higher data rate than is achievable in currently almost-ready 3G systems. Some people expect that the possible data rate for 4G systems should be roughly in the range of 10 to 100 Mbps. In light of this objective, the question is how to guarantee such a high data rate in highly unpredictable and hostile mobile channels, and what types of air link architecture are qualified to deliver such high-data-rate services.

Considering the constraints on the available radio spectrum suitable for terrestrial mobile communications (from a few hundreds of megahertz to less than 100 GHz), one would argue that probably the most relevant and feasible way to achieve the goals promised by 4G systems is to work out some enabling technologies capable of improving, as much as possible, the air link bandwidth efficiency of the systems. In this part of the chapter, this issue is tackled comprehensively by proposing a new CDMA architecture that has great potential for future mobile communications.

It is well known that all current CDMA-based 2G and 3G standards (IS-95, cdma2000, and W-CDMA) use traditional direct-sequence CDMA techniques based on an identical principle: that each bit is spread by one single spreading code comprising N contiguous chips to attain a certain processing gain or spreading factor. The bandwidth of all those systems is determined by the chip width of the spreading codes used. Thus, it is natural to define a merit parameter called *spreading efficiency* (SE) in bits per chip to measure the bandwidth efficiency of a CDMA system. Therefore, it is clear that the SEs of all conventional CDMA-based mobile communication systems (IS-95, cdma2000, and W-CDMA) are equal to $1/N$, which is far less than 1. This in turn explains why those systems cannot offer better bandwidth efficiency.

In the past few years, industry research has focused on proposing a possible solution to improve the SE of a CDMA system with the help of a new spreading technique based on complete complementary (CC) codes, taking into account various implementation constraints of a practical CDMA system as follows:

1. The new CDMA architecture ought to be technically feasible with currently available digital technology.
2. The new system should not introduce too much multiple-access interference (MAI) to ensure higher capacity potential than that of conventional CDMA systems.
3. The proposed system should preferably have an inherent ability to mitigate multipath problems in mobile channels.

The multicarrier CDMA architecture based on orthogonal CC codes is one such proposal that can satisfy all the previously mentioned requirements. This part of the chapter demonstrates the capability to achieve high bandwidth efficiency and low bit error rate as a result of its innovative signaling design in both downlink and uplink channels. Several peculiarities pertaining to the new architecture in its receiver design are discussed here. To be more specific, a traditional RAKE receiver is no longer useful in the proposed CDMA architecture, and a new adaptive recursive filter is particularly introduced to detect a signal in a multipath environment. The technical limitations associated with the new system are also addressed in this part of the chapter.

This part of the chapter is outlined as follows. It describes basic properties of CC codes and shows some examples. This part of the chapter also introduces an operational model of the new multicarrier CDMA system using CC codes. It also explains how the proposed system can achieve MAI (free operation in both up- and downlink) in a multipath-free channel. In addition, this part of the chapter also shows the structure of the new recursive filter receiver for multipath signal reception and evaluates the performance of the proposed system under a multipath environment. Finally, it discusses the various aspects of the proposed system and possible future work.

Complete Complementary Codes

The core of the proposed new CDMA architecture is the use of orthogonal complete complementary codes, the origin of which can be traced back to the 1960s, when pairs of binary complementary codes were used whose autocorrelation function is zero for all even shifts except the zero shift. The concept has been extended to the generation of CC code families whose autocorrelation function is zero for all even and odd shifts, except the zero shift, and whose cross-correlation function for any pair is zero for all possible shifts. The work paved the way for practical applications of CC codes in modern CDMA systems, whose explicit architecture is proposed and studied in this part of the chapter.

There exist several fundamental distinctions between traditional CDMA codes (Gold codes, *m*-sequences, Walsh-Hadamard codes, etc.) and the CC codes concerned in the proposed CDMA system:

1. The orthogonality of CC codes is based on a "flock" of element codes jointly, instead of a single code as in traditional CDMA codes. In other words, every user in the proposed new CDMA system will be assigned a flock of element codes as its signature code, which ought to be transmitted, possibly via different channels, and arrive at a correlator receiver at the same time to produce an autocorrelation peak. Take CC codes of element code length $L = 4$ as an example, as shown in Fig. 7-5 (which lists two families of CC codes: one is for $L = 4$, the other for $L = 16$).[3] There are in total four element codes (A_0, A_1, B_0, and B_1) in this case, and each user should use two element codes (either A_0, A_1 or B_0, B_1) together, thus being capable of supporting only two users, as shown in Figs. 7-6 and 7-7, where both up- and downlinks of the proposed CDMA system are illustrated.[3] In this simple example, both flock size and family size are identical: 2. Table 7-2 shows the flock and family sizes for various CC codes with different element code lengths (L).[3]

2. The processing gain of CC codes is equal to the "congregated length" of a flock of element codes. For CC codes of lengths $L = 4$ and $L = 16$, their processing gains are equal to $4 \times 2 = 8$ and $16 \times 4 = 64$, respectively.

3. Zero cross-correlation and zero out-of-phase autocorrelation are ensured for any relative shifts between two codes. Let's consider $A_0 = (+++-)$, $A_1 = (+-++)$ and $B_0 = (++-+)$, $B_1 = (+---)$, being two

Figure 7-5 Two examples of complete complementary codes with element code lengths $L = 4$ and $L = 16$.

Element code length $L = 4$		Element code length $L = 16$	
Flock 1	A_0: + + + −	Flock 1	A_0: + + + + + − + − + + − − + − − +
			A_1: + − + − + + + + + − − + + + − −
			A_2: + + − − + − − + + + + + + − + −
			A_3: + − − + + + − − + − + − + + + +
	A_1: + − + +	Flock 2	B_0: + + + + − + − + + + − − − + + −
			B_1: + − + − − − − − + − − + − − + +
			B_2: + + − − − + + − + + + + − + − +
			B_3: + − − + − − + + + − + − − − − −
Flock 2	B_0: + + − +	Flock 3	C_0: + + + + + − + − − − + + − + + −
			C_1: + − + − + + + + − + + − − − + +
			C_2: + + − − + − − + − − − − − + − +
			C_3: + − − + + + − − − + − + − − − −
	B_1: + − − −	Flock 4	D_0: + + + + − + − + − − + + + − − +
			D_1: + − + − − − − − − + + − + + − −
			D_2: + + − − − + + − − − − − + − + −
			D_3: + − − + − − + + − + − + + + + +

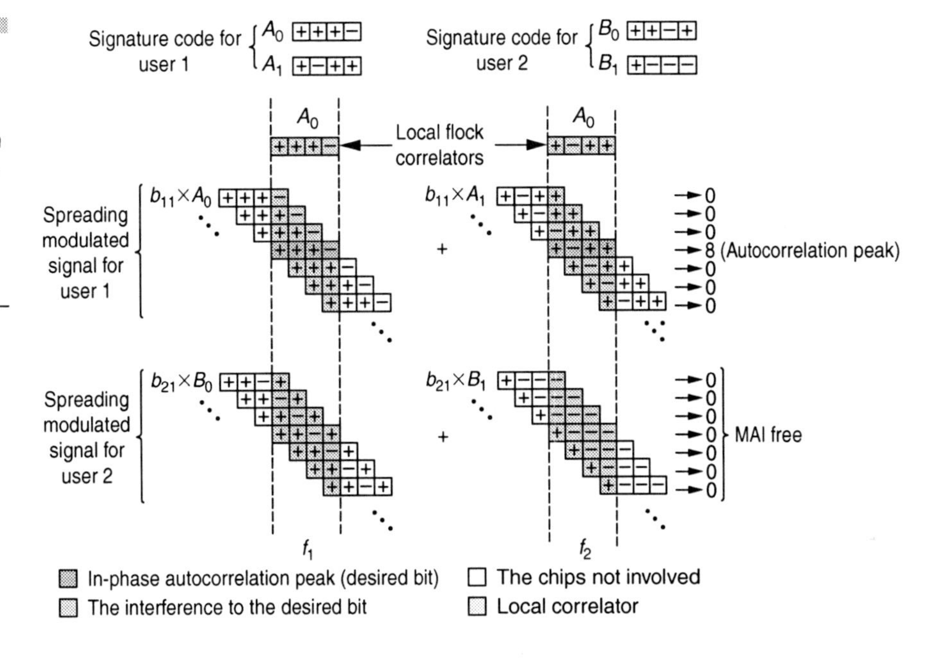

Figure 7-6 Downlink signal reception in a two-user CDMA system in an MAI-AWGM channel using CC codes of length $L = 4$, where user 1 is the intended one.

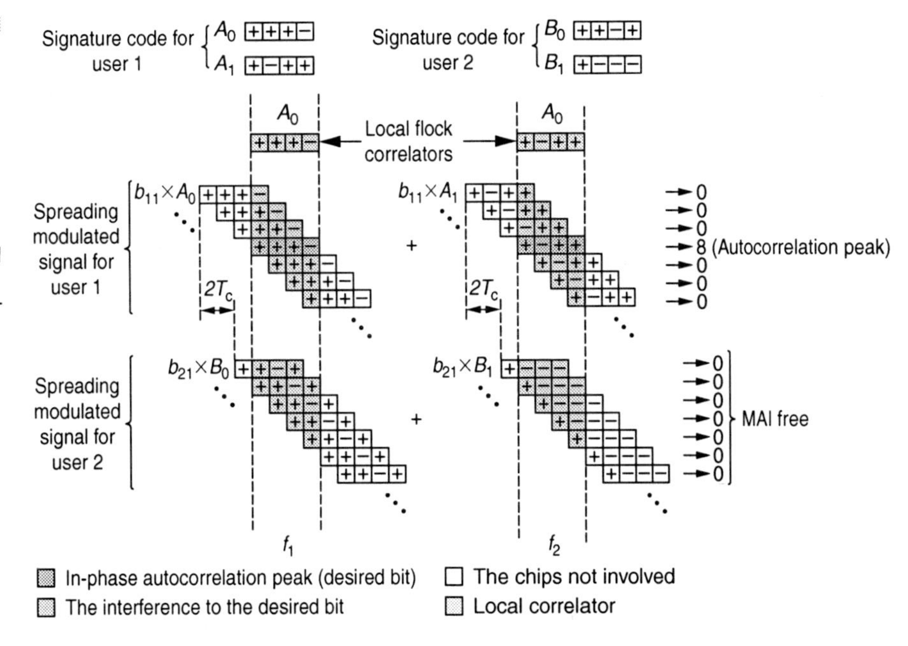

Figure 7-7 Uplink signal reception in a two-user CDMA system in an MAI-AWGM channel using CC codes of length $L = 4$, where user 1 is the intended one.

TABLE 7-2

Family Sizes and Flock Sizes for Complete Complementary Codes with Various Element Code Lengths L

Element Code Length ($L = 4^n$)	**4**	**16**	**64**	**256**	**1024**	**4096**
PG ($L\sqrt{L}$)	8	64	512	4096	32,768	262,144
Family size ($\sqrt{L}$)	2	4	8	16	32	64
Flock size ($\sqrt{L}$)	2	4	8	16	32	64

flocks of CC codes for a CDMA system of two users, A and B. Let $A_0 \otimes A_0$ and $A_1 \otimes A_1$ denote the shift-and-add operations to calculate autocorrelation function for A_0 and A_1, and $B_0 \otimes B_0$ and $B_1 \otimes B_1$ for B_0 and B_1 likewise. Then you have $A_0 \otimes A_0 + A_1 \otimes A_1 = (0, 0, 0, 8, 0, 0, 0)$, and $B_0 \otimes B_0 + B_1 \otimes B_1 = (0, 0, 0, 8, 0, 0, 0)$. Similarly, you can obtain the cross-correlation function between A and B as $A_0 \otimes B_0 + A_1 \otimes B_1 = (0, 0, 0, 0, 0, 0, 0)$, or $B_0 \otimes A_0 + B_1 \otimes A_1 = (0, 0, 0, 0, 0, 0, 0)$, illustrating the ideal cross-correlation property of CC codes.

4. Since each user in the proposed CDMA system is assigned a signature code comprising a flock of element codes, those element codes should be sent to a receiver using different carriers. In other words, every signature code is split up into several segments (or element codes) that ought to be transmitted to a receiver via different frequency channels.

Performance under Multiple-Access Interference

The conceptual diagrams for the proposed new CDMA system in a multipath-free channel are shown in Figs. 7-6 and 7-7, where down- and uplink spreading-modulated signals for a two-user system are illustrated. Each of the two users therein employs two $L = 4$ element codes as its signature code, which is exactly the same as listed in Fig. 7-5. The information bits (b_{11}, b_{12},...) and (b_{21}, b_{22},...), which are assumed to be all +1s for illustration simplicity in the figures, are spreading-modulated by element codes that are "offset stacked," each shifted by one chip relative to one another. When compared to traditional spreading modulation used in conventional CDMA systems, the new system has the following salient features. The most obvious is that the bit stream in the new system is no longer aligned in time one bit after another. Instead, a new bit will start right after one chip delay relative to the previous bit, which is spread by an element code of length L. Another important characteristic attribute of the new CDMA system is that such an offset-stacked spreading modulation method is particularly beneficial for multirate data transmission in multimedia services, whose algorithm is termed *rate-matching* in the current 3G mobile communication standards. The unique offset-stacked spreading method used by the proposed CDMA system can easily slow down data transmission by simply shifting more than one chip (at most L chips) between two neighboring offset-stacked bits. If L chips are shifted between two consecutive bits, the new system reduces to a conventional CDMA system, yielding the lowest data rate. On the other hand, the highest data rate is achieved if only one chip is shifted between two neighboring offset-stacked bits. Doing so, the highest spreading efficiency equal to 1 can be achieved, implying that every chip is capable of carrying one bit of information. Since the bandwidth of a CDMA system is uniquely determined by the chip width of spreading codes used, higher SE simply means higher bandwidth efficiency. Thus, the proposed new CDMA architecture is capable of delivering much higher bandwidth efficiency than a conventional CDMA architecture under the same processing gain.

It should be stressed that the "inherent" ability of the new CDMA system to facilitate multirate transmissions is based on its innovative offset-stacked spreading technique, which cannot be applied to traditional spreading codes. The current 3G W-CDMA architecture has to rely on a complex and sometimes difficult rate-matching algorithm to adjust the data transmission rate by selecting appropriate variable-length orthogonal codes according to a specific spreading factor and data rate requirement on the services. On the contrary, the proposed new CDMA system is able to change the data transmission rate on the fly, without the need to search for suitable codes with a particular spreading factor. What to do is just to shift more or less chips between two neighboring offset-stacked bits to slow down or speed up the data rate. That's it; no more rate-matching algorithms!

Another important feature of the rate-change scheme adopted by the new CDMA architecture is that the same processing gain will apply to different data transmission rates. However, the rate-matching algorithm in the Universal Mobile Telecommunications System (UMTS) W-CDMA standard is processing-gain-dependent; the slower the transmission rate, the higher the processing gain, if transmission bandwidth is kept constant. To maintain an even detection efficiency at a receiver, the transmitter has to adjust the transmitting power for different-rate services, which surely complicates both transmitter and receiver hardware.

The offset-stacked spreading technique also helps support asymmetrical transmissions in up- and downlinks, pertaining to Internet services in the future of all IP mobile networks. The data rates in a slow uplink and a fast downlink can be made truly scalable, such that "rate on demand" is achievable by simply adjusting the offset chips between two neighboring spreading-modulated bits.

Figures 7-6 and 7-7 illustrate that the proposed new CDMA architecture can offer MAI-free operation in both downlink (synchronous channel) and uplink (asynchronous channel) transmissions, because of the use of CC codes. It is assumed that the relative delay between the two users in Fig. 7-7 takes the multiples of chips. If this assumption does not hold, it can be shown as well that the resultant MAI level is far less than that of a conventional CDMA system. It should also be pointed out that the rate change through adjusting the number of offset chips between two neighboring stacked bits does not affect the MAI-free operation of the proposed CDMA system.

The MAI-independent property of the proposed CDMA architecture is significant in terms of its potential to enhance its system capacity in a multipath channel. It is well known that a CDMA system is an interference-limited system whose capacity is dependent on the average cochannel interference contributed from all transmissions using different codes in the same band. The cochannel interference in a conventional CDMA system is caused in principle by nonideal cross-correlation and out-of-phase autocorrelation functions of the codes concerned. In such a system, it is impossible to eliminate the cochannel interference, especially in the uplink channel, where bit streams from different mobiles are asynchronous such that orthogonality among the codes is virtually nonexistent. On the contrary, the proposed new CDMA system based on CC codes is unique because excellent orthogonality among transmitted codes is preserved even in an asynchronous uplink channel, making truly MAI-independent operation possible for both up- and downlink transmissions. The satisfactory performance in a multipath environment, as shown in Figs. 7-15 and 7-16, is also partly attributable to this property.[3]

It should also be noted that the two element codes for each of the user signature codes in Figs. 7-6 and 7-7 have to be sent separately through different carriers, f_1 and f_2. Therefore, the proposed new CDMA architecture is a multicarrier CDMA system. It is also possible for you to use orthogonal carriers, spaced by $1/T_c$ (where T_c denotes the chip width), to send all those element codes for the same user separately to further enhance the bandwidth efficiency of the system.

The bit error rate (BER) of the proposed CDMA system, under MAI and additive white gaussian noise (AWGN), is evaluated using computer simulations. The obtained BER performance of the new CDMA system is compared to that of conventional CDMA systems using Gold codes and m-sequences under identical operation environments. For each of the systems concerned here, a matched filter (single-correlator) is used at a receiver. Both down- and uplink are simulated considering various numbers of users and processing gains. Figures 7-8 and 7-9 typify the results obtained.[3] The former shows the performance for the BER in the downlink (synchronous) channel with a processing gain of 64 for CC codes being comparable to that for Gold code and m-sequence of length 63. The latter gives a BER in the uplink (asynchronous) channel with interuser delay equal to 3 chips. Both Figs. 7-8 and 7-9 show the BER only for the first user as the intended one; similar BER results can be obtained for the others. It is observed from Fig. 7-8 that at least 2-dB gain is obtainable from the proposed CDMA system compared to conventional systems using traditional CDMA codes. One of the most interesting observations for the new CDMA system is its almost identical BER performance (Fig. 7-8), regardless of the number of users, where two curves representing different numbers of users (one and four, respectively) in the system virtually overlap each other, exemplifying the MAI-independent operation of the proposed CDMA system. On the other hand, the BER for a CDMA using traditional codes is MAI-dependent; the more active users present, the worse it performs, as shown in Fig. 7-8. Next, let's look at the performance of the proposed CDMA system under multipath channels.

Signal Reception in Multipath Channels

It is well known that a conventional CDMA receiver usually uses a RAKE to collect dispersed energy among different reflection paths to achieve multipath diversity at the receiver. Therefore, the RAKE receiver

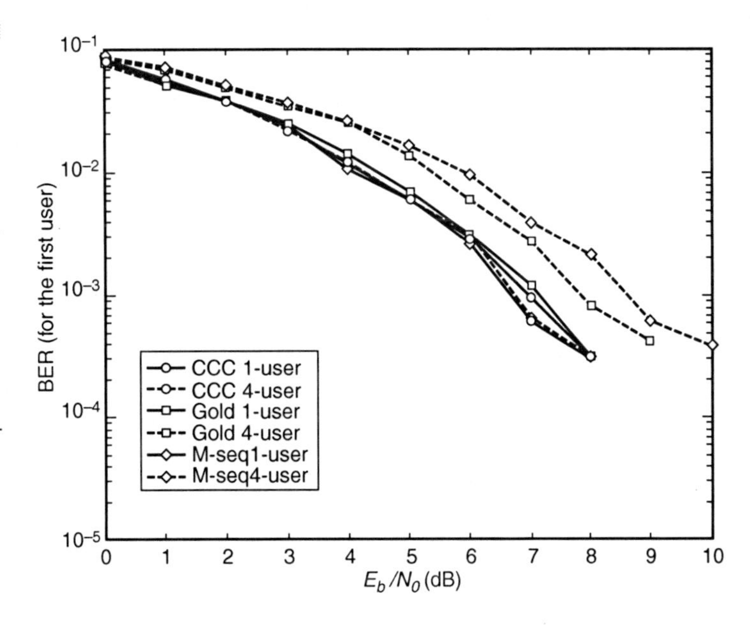

Figure 7-8 Downlink BER comparison for CC-code-based CDMA and conventional CDMA systems in an MAI-AWGN channel using a matched filter receiver. Lengths of Gold code/ m-sequence and CC code are 63 and 4 × 16, respectively.

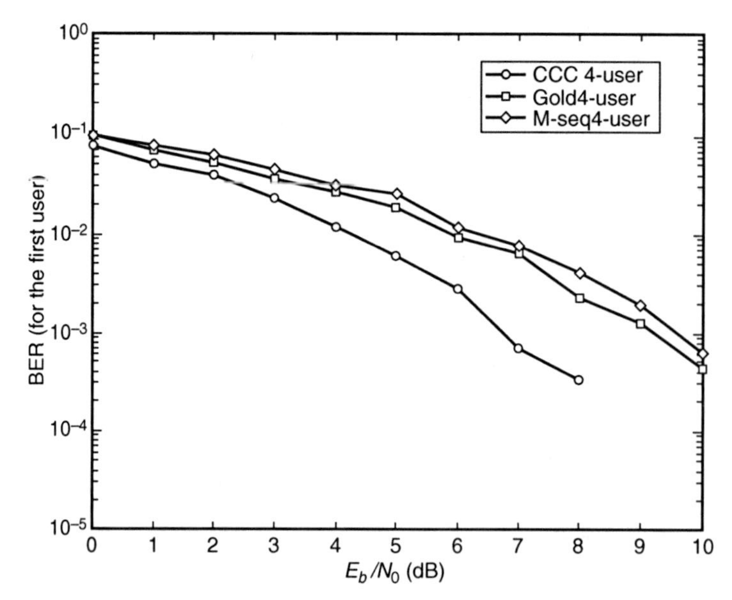

Figure 7-9 Uplink BER comparison for CC-code-based CDMA and conventional CDMA systems in an MAI-AWGN channel using a matched filter receiver. Lengths of Gold code/*m*-sequence and CC code are 63 and 4 × 16, respectively.

is a must for all conventional CDMA systems, including currently operational 2G and 3G systems. However, in the new CDMA architecture presented in this part of the chapter, the RAKE receiver becomes inappropriate because of the nature of the unique spreading modulation technique employed in the system. To illustrate how the proposed CDMA system makes the RAKE receiver obsolete, let's refer to Fig. 7-10, where a simple multipath channel consisting of three equally strong reflection rays is considered with interpath delay of one chip.[3] The RAKE receiver has three fingers to capture three paths and combine them coherently. Three columns in Fig. 7-10 show the output signals from three fingers; the shaded parts are the chips involved in the RAKE combining algorithm. Because of the offset-stacked spreading in the proposed CDMA system, there are in total five bits (b_1, b_2, b_3, b_4, and b_5) relevant to the RAKE combining procedure, where it is assumed that $b_3 = +1$ is the desirable bit. Therefore, b_1, b_2, b_4, and b_5 are all interfering terms, whose three possible error-causing patterns are $(b_1, 2b_2, 2b_4, b_5) = (1, -2, -2, -1)$, $(-1, -2, -2, 1)$, and $(-1, -2, -2, -1)$, respectively. Note that among total 16 possible combinations of binary bits b_1, b_2, b_4, and b_5, only three of them cause errors. Therefore, the error probability turns out to be $3/16 = 0.1875$ (if each path has the same strength).

From this example, you can see that the use of a RAKE receiver in the proposed CDMA system still causes BER $= 0.1875$ (with three identically strong paths), which is obviously not acceptable. Therefore, an adaptive recursive multipath signal reception filter, based on CC codes, is designed particularly for the CDMA system, as shown in Fig. 7-11,

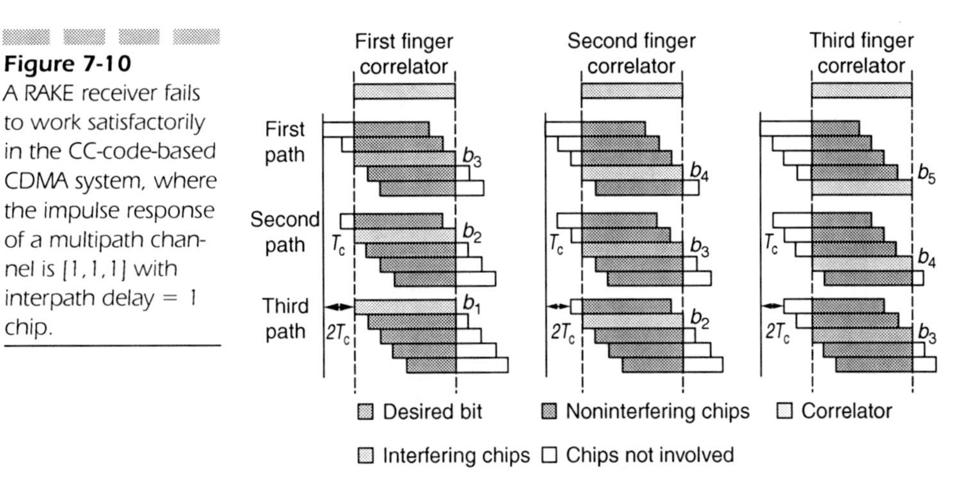

Figure 7-10 A RAKE receiver fails to work satisfactorily in the CC-code-based CDMA system, where the impulse response of a multipath channel is [1,1,1] with interpath delay = 1 chip.

where the receiver consists of two key modules: The lower part is to estimate the channel impulse response, and the upper part is to coherently combine signals in different paths to yield a boosted-up decision variable before the decision device.[3] For this adaptive recursive filter to work, a dedicated pilot signal should be added to the proposed new CDMA system, which should be spreading-coded by a signature code different from those used for data channels in the downlink transmission and time-interleaved with user data frames in the uplink channel transmission, as shown in Fig. 7-12.[3] The rationale behind the difference in the pilot signals for down- and uplink channels is explained as follows. The downlink transmission is a synchronous channel from the same source (a base station); thus, one dedicated pilot signaling channel is justified, considering that a relatively strong pilot is helpful for mobiles to lock onto it for controlling information. On the other hand, uplink transmissions are asynchronous from different mobiles. Therefore, it will consume a lot more signature codes if every mobile is assigned two codes, one for data traffic and the other for pilot signaling. Thus, the pilot signaling has to be time-interleaved with user data traffic in the uplink channels. Time-interleaved pilot signals in the uplink channels can also assist base stations to perform adaptive beam forming, required by a smart antenna system. The signal reception in both down- and uplinks can use the same recursive filter (see Fig. 7-11) for channel impulse response estimation as long as the receiver achieves frame synchronization with the incoming signal. In fact, the pilot signals in both down- and uplink channels consist of a series of short pulses, whose durations (T_{d1} and T_{u1}) should be made longer than the delay spread of the channel and whose repetition periods (T_{d2} and T_{u2}) should be made shorter than the coherent time of the channel to adaptively follow the variation of the mobile channel.

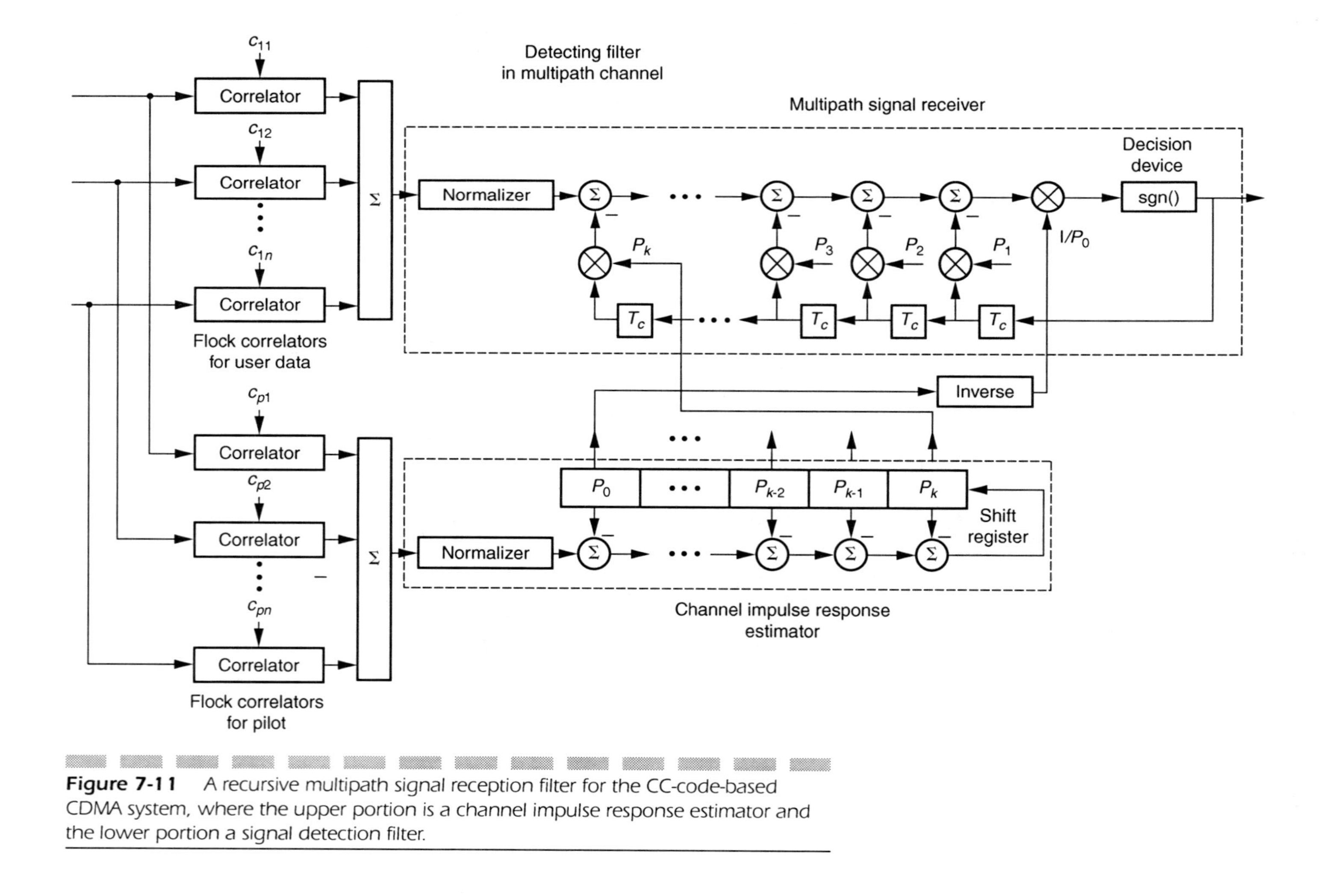

Figure 7-11 A recursive multipath signal reception filter for the CC-code-based CDMA system, where the upper portion is a channel impulse response estimator and the lower portion a signal detection filter.

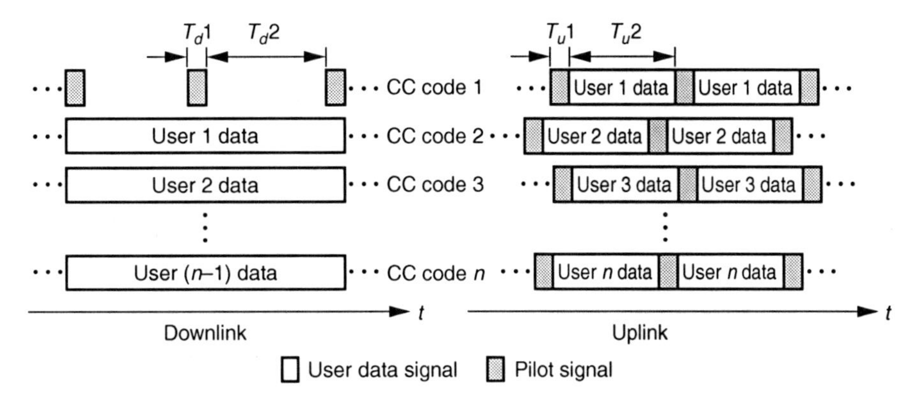

Figure 7-12 Down- and uplink channel signaling design for the CC-code-based CDMA system, where the downlink uses a dedicated pilot channel and the pilot signal in the uplink is time-interleaved with data traffic.

The detail procedure for the recursive multipath signal reception filter to estimate the channel impulse response and detect signal is illustrated step by step in Figs. 7-13 and 7-14, where it is assumed that a three-ray multipath channel is concerned, with mean path strengths 3, 2, and 1, respectively.[3] It is also assumed that exactly the same CC codes as in Figs. 7-6 and 7-7 are used, with one signature code ($c_{p1} = A_0$ and $c_{p2} = A_1$) used for the pilot channel and the other ($c_{11} = B_0$ and $c_{12} = B_1$) for the user data channel, if only downlink transmission is considered in this example. In this illustration, it is presumed that each pilot pulse consists of five continuous 1s, which is longer than the channel delay spread (= 3 chips) in this case. The input sequence (3 5 6 6 6) to the left side of multipath channel estimator in Fig. 7-13 is the received pilot signal after being convoluted with multipath channel impulse response and local flock correlators. It is seen from Fig. 7-13 that the channel impulse response can be estimated accurately and saved in the output register at the end of the algorithm. The obtained channel estimates will then be passed on to the multipath signal receiver in the upper portion of Fig. 7-11 to detect the signal contaminated with multipath interference, whose procedure is shown in Fig. 7-14, where it is assumed that the originally transmitted binary bit stream is (1, −1, 1, −1, 1). The input data to the multipath signal receiver, (3 −1 2 −2 1), is the received signal of the transmitted bit stream after going through multipath channel and local flock correlators.

The proposed recursive multipath signal reception filter possesses several advantages:

1. It has a very agile structure, the core of which is made up of two transversal filters, one for channel impulse response estimation and the other for data detection.
2. Working jointly with the pilot signaling, it performs very well in terms of accuracy in channel impulse response estimation, as shown in Fig. 7-13 and the obtained BER results. The multipath channel

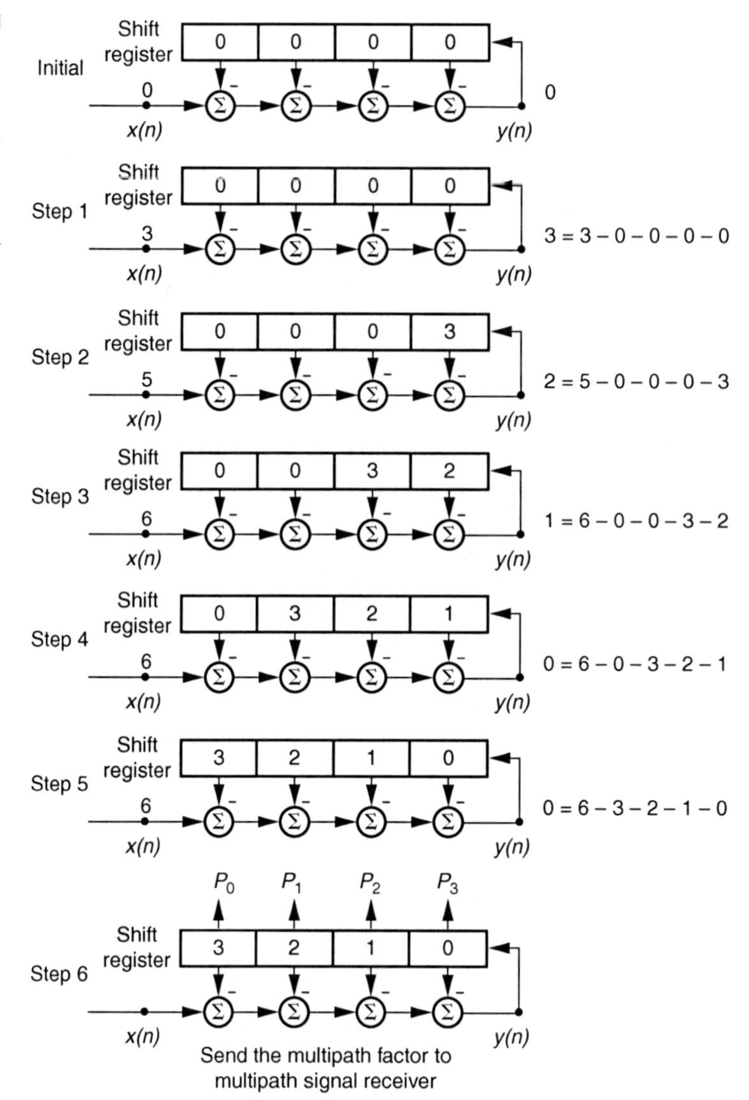

Figure 7-13 A step-by-step illustration of channel impulse response estimation using a recursive multipath signal reception filter.

equalization and signal coherent combining are actually implemented jointly in the proposed scheme under a relatively simple hardware structure.

3. It operates adaptively to the channel characteristic variation without needing prior knowledge of the channel, such as interpath delay and relative strength of different paths. On the contrary, a RAKE receiver

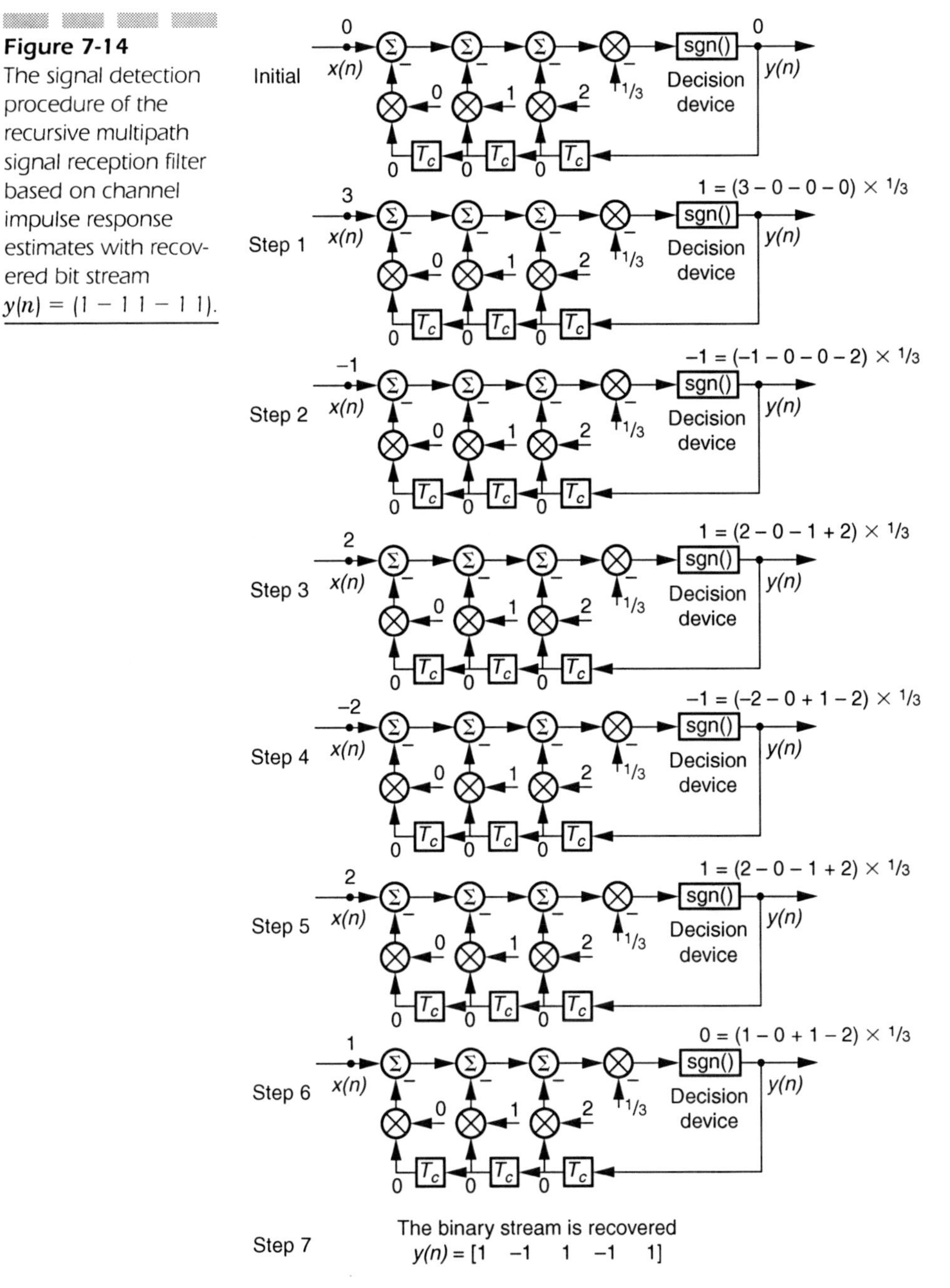

Figure 7-14
The signal detection procedure of the recursive multipath signal reception filter based on channel impulse response estimates with recovered bit stream $y(n) = (1 - 1\ 1 - 1\ 1)$.

in a conventional CDMA system requires the path gain coefficients for maximal ratio combining, which themselves are usually unknown and thus have to be estimated by resorting to other complex algorithms.

The performance of the proposed new CDMA architecture with the recursive filter for multipath signal reception is shown in Figs. 7-15 and 7-16, where two typical scenarios are considered: one for downlink performance and the other for uplink performance, similar to the performance comparison made for the MAI-AWGN channel in Figs. 7-8 and 7-9.[3] It is observed from the figures that, in terms of the BER in a synchronous downlink channel, three different codes perform similarly, whereas in an asynchronous uplink channel, the Gold code and m-sequence performances are much worse than the CC code, because the orthogonality among both Gold codes and m-sequences is destroyed by asynchronous bit streams from different mobiles. Nevertheless, the CC-code-based CDMA system outperforms conventional CDMA systems using either Gold code or m-sequence by a comfortable margin that can be as large as 4 to 6 dB, because of its superior MAI-independent property.

Bandwidth Efficiency

Previously in this chapter, it was demonstrated that the CDMA architecture based on CC codes and an adaptive recursive multipath signal reception filter is feasible and performs well. The system offers MAI-free

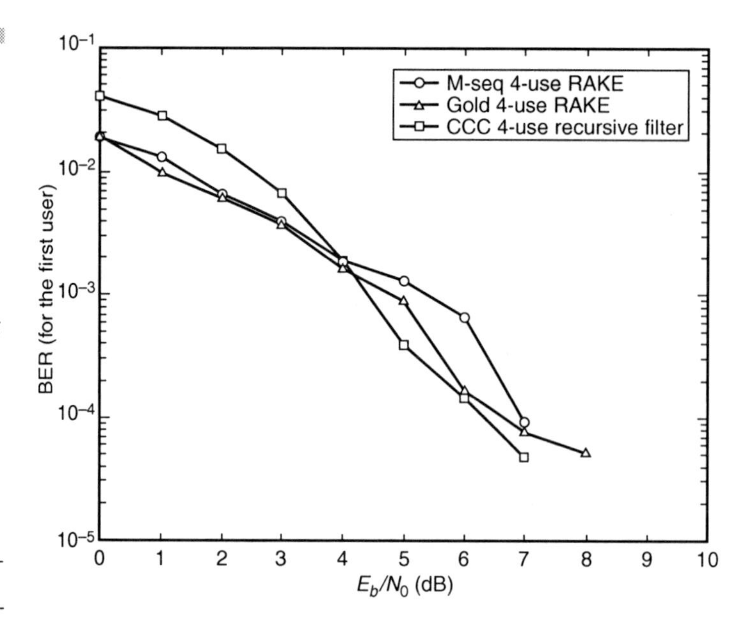

Figure 7-15 Downlink (synchronous) BER for CC-code-based CDMA and conventional CDMA systems in a multipath channel, with normalized multipath power; interpath delay = 3 chips; multipath channel delay profile = [1.35, 1.08, 0.13]; PG = 63/64; Gold code/m-sequence with MRC-RAKE; CC-code-based CDMA with the recursive filter.

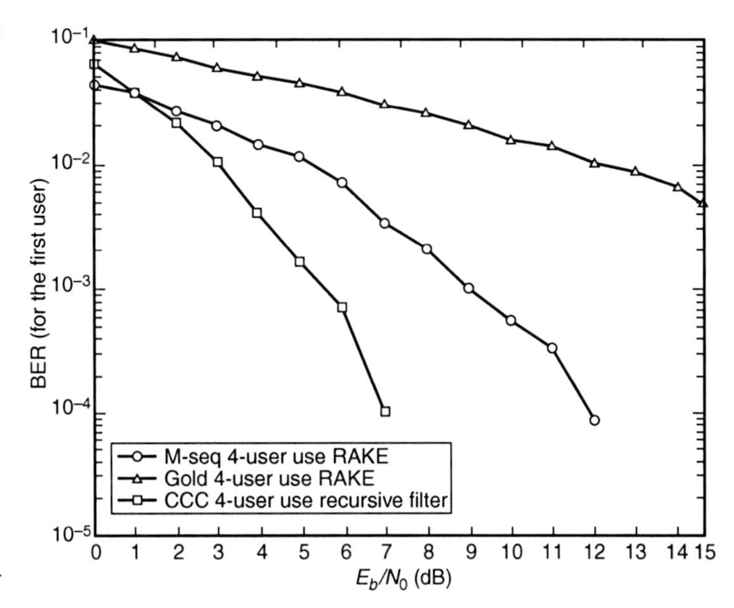

Figure 7-16
Uplink (asynchronous) BER for CC-code-based CDMA and conventional CDMA systems in a multipath channel, with normalized multipath power; interpath delay = 3 chips; interuser delay = 2 chips; multipath channel delay profile = [1.35,1.08, 0.13]; PG = 63/64; Gold code/*m*-sequence with MRC-RAKE; CC-code-based CDMA with the recursive filter.

operation for both down- and uplink transmissions in an MAI-AWGN channel. Another interesting property of the new CDMA system is its agility in changing the data transmission rate, which can be finished on the fly without needing to stop and search for a code with a specific spreading factor, as required in the W-CDMA standards. Therefore, the rate-matching algorithm in the proposed system has been greatly simplified.

Yet another important point that has to be addressed is the bandwidth efficiency of the proposed CDMA architecture. Spreading efficiency in bits per chip has been used to measure the bandwidth efficiency of a CDMA system because the bandwidth of a CDMA system is determined by the chip width of the spreading codes used. Table 7-3 compares the SEs of three systems: conventional CDMA and CC-code-based CDMA with and

TABLE 7-3

Spreading Efficiency (in Bits per Chip) Comparison of a Conventional CDMA System and a CC-Based CDMA System with and without Orthogonal Carriers

PG	8	64	512	4096	32,768	262,144
Conventional CDMA	1/8	1/64	1/512	1/4096	1/32,768	1/262,144
CC-code-based CDMA			1/8	1/16	1/32	1/64
CC-code-based CDMA (orthogonal carriers)	1			1/8	1/16	1/32

without orthogonal carriers.[3] It is clear that the CC-code-based CDMA systems have a much higher SE figure than a conventional CDMA does, especially when the processing gain is relatively high.

However, there exist some technical limitations for the proposed CC-code-based CDMA system, which ought to be properly addressed and can become the direction of possible future work for further improvement.

Obviously, a CC-code-based CDMA system needs a multilevel digital modulation scheme to send its baseband information, because of the use of an offset-stacked spreading modulation technique, as shown in Figs. 7-6 and 7-7. If a long CC code is employed in the proposed CDMA system, the number of different levels generated from a baseband spreading modulator can be a problem. For instance, if the CC code of $L = 4$ is used, as shown in Table 7-2, five possible levels will be generated from the offset-stacked spreading: 0, -2, and -4. However, if the CC code of $L = 16$ in Table 7-2 is involved, the possible levels generated from the spreading modulator become 0, -2, -4, ..., -16, comprising 17 different levels. In general, the modulator will yield $L + 1$ different levels for a CC-code-based CDMA system using length L element codes. Given the element code length (L) of the CC code, it is necessary to choose a digital modem capable of transmitting $L + 1$ different levels in a symbol duration. An $L + 1$ quadrature amplitude-modulated (QAM) digital modem can be a suitable choice for its robustness in detection efficiency. It should be pointed out that the simulation study concerned in this part of the chapter assumes an ideal modulation and demodulation process. Thus, the research takes into account the nonideal effect of multilevel carrier modulation, and demodulation remains a topic of future study.

Finally, another concern with the CC-code-based CDMA system is that a relatively small number of users can be supported by a family of the CC codes. Take the $L = 64$ CC code family as an example. It is seen from Table 7-3 that such a family has only eight flocks of codes, each of which can be assigned to one channel (for either pilot or data). If more users should be supported, long CC codes have to be used. On the other hand, the maximum length of the CC codes is in fact limited by the maximal number of different baseband signal levels manageable in a digital modem, as mentioned earlier in this chapter. One possible solution to this problem is to introduce frequency divisions on top of the code divisions in each frequency band to create more transmission channels.

Conclusion

In this chapter, a new CDMA architecture based on CC codes was presented, and its performance in both MAI-AWGN and multipath channels was evaluated by simulation. The proposed system possesses several advantages over conventional CDMA systems currently available in 2G and 3G standards:

1. The system offers much higher bandwidth efficiency than is achievable in conventional CDMA systems. The system, under the same processing gain, can convey as much as 1 bit of information in each chip width, giving a spreading efficiency equal to 1.
2. It offers MAI-free operation in both synchronous and asynchronous MAI-AWGN channels, which attributes to cochannel interference reduction and capacity increase in a mobile cellular system. This excellent property also helps to improve the system performance in multipath channels, as shown by the obtained results.
3. The proposed system is inherently capable of delivering multirate/multimedia transmissions because of its offset-stacked spreading modulation technique. Rate matching in the new CDMA system becomes very easy, just shifting more or fewer chips between 2 consecutive bits to slow down or speed up the data rate—no more complex rate-matching algorithms.

This chapter also proposed a novel recursive filter, particularly for multipath signal reception in the new CDMA system. The recursive filter consists of two modules working jointly; one performing channel impulse response estimation and the other detecting signal contaminated by multipath interference. The recursive filter has a relatively simple hardware compared to a RAKE receiver in a conventional CDMA system, and performs very well in multipath channels. The chapter also addressed technical limitations of the new CDMA architecture, such as a relatively small family of CC codes and the need for complex multilevel digital modems. Nevertheless, the proposed CDMA architecture based on complete complementary codes offers a new option to implement future wideband mobile communications beyond 3G.

The increasing amount of roaming data users and broadband Internet services has created a strong demand for public high-speed IP access with sufficient roaming capability. Wireless data LAN systems offer high bandwidth but only modest IP roaming capability and global user management features.

This chapter described a system that efficiently integrates wireless data LAN access with the widely deployed GSM/GPRS roaming infrastructure. The designed architecture exploits GSM authentication, SIM-based user management, and billing mechanisms and combines them with public WDLAN access.

With the presented solution, cellular operators can rapidly enter the growing broadband access market and utilize their existing subscriber management and roaming agreements. The OWDLAN system allows cellular subscribers to use the same SIM and user identity for WDLAN access. This gives the cellular operator a major competitive advantage over ISP operators, who have neither a large mobile customer base nor a cellular kind of roaming service.

Finally, the designed architecture combines cellular authentication with native IP access. This can be considered the first step toward all-IP networks. The system proposes no changes to existing cellular network elements, which minimizes the standardization effort and enables rapid deployment. The reference system has been commercially implemented and successfully piloted by several mobile operators. The GSM SIM-based WDLAN authentication and accounting signaling has proved to be a robust and scalable approach that offers a very attractive opportunity for mobile operators to extend their mobility services to also cover indoor wireless data broadband access.

References

1. "Wireless Architecture Options," Synchrologic, 200 North Point Center East, Suite 600, Alpharetta, GA 30022, 2002.
2. "CIO Outlook 2001: Architecting Mobility," Synchrologic, 200 North Point Center East, Suite 600, Alpharetta, GA 30022, 2002.
3. Hsiao-Hwa Chen, Jun-Feng Yeh, and Naoki Suehiro, "A Multicarrier CDMA Architecture Based on Orthogonal Complementary Codes for New Generations of Wideband Wireless Communications," *IEEE Communications Magazine,* 445 Hoes Lane, Piscataway, NJ 08855, 2002.
4. John R. Vacca, *The Essential Guide to Storage Area Networks,* Prentice Hall, 2002.

CHAPTER 8

Fixed Wireless Data Network Design

If you can't wait for DSL or cable modem[3] to be installed at your corporate headquarters or if it seems like broadband[4] will never be available at your remote sites, the design of a fixed wireless data network is becoming a viable alternative for last-mile Internet access.

Fixed wireless data has some advantages over wired broadband: It can be installed in a matter of days. Once the line of sight is established, the connection isn't susceptible to the types of weather-related or accidental outages that can occur with wired networks.

But there are important design issues that network executives will need to resolve before signing up for fixed wireless data, including security and possible performance degradation from interference with other service providers.

For example, on the island of Anguilla, a British territory 6 miles north of St. Martin in the Caribbean, Weblinks Limited (http://www.weblinksadvertising.co.uk/contact_frameset.html) has installed a wireless data Internet system that covers the entire 16-mile-long island, offering services to a growing number of e-commerce[6] companies. On a hurricane-prone and remote island like Anguilla, fixed wireless data offers several benefits over DSL and cable modem. A fixed wireless Internet system, such as Weblinks' in Anguilla, consists of centralized transceiver towers and directional antennas mounted at each end-user location to maximize range and minimize the number of towers needed to cover a large area (see sidebar, "Wireless Data Internet Infrastructure"). (The Glossary defines many technical terms, abbreviations, and acronyms used in the book.)

Wireless Data Internet Infrastructure

Independent service providers are building private networks based on a combination of optical and fixed wireless data technology, exclusive peering arrangements, and Internet data centers to support the B2B marketplace. The arrival of the twenty-first century in Latin America coincided with the migration of the region's Internet from a communications/recreation medium to a platform for mission-critical applications and e-business. With this change, the region's Internet infrastructure is evolving from its dependence on U.S.-based hosting facilities and incumbent owned and operated transport to a mix of fiber-optic and fixed wireless data private networks with Internet data centers (IDCs).

Until a few years ago, the dot-coms that pioneered Latin American Web content looked to local garages or U.S.-based Web-hosting firms for their infrastructure needs, since high-quality solutions did not yet exist in the region. The distance between U.S. hosting facilities and Latin American users, combined with subpar infrastructure tying the two regions, resulted in poor performance and high-latency connections. Such concerns were not critical, however, because of the informational nature of the first Web sites. The ready-made U.S. solutions, which transported international traffic over satellite networks[5] or directed in-region traffic "hot-potato" style through multiple hubs and network access points (NAPs), suited both providers and users.

Even today, many connections throughout the region suffer delay as a result of poor routing. For example, a user in Buenos Aires accessing a site hosted in California connects to an Internet service provider (ISP) that in turn connects to an Internet backbone provider. Upon leaving the ISP network, the connection travels across the Internet "cloud." The network providers inside the cloud have no incentive or ability to optimally route the connection. Their motivation is to minimize the costs by routing across inexpensive and usually overly utilized links or by passing the session off to another less expensive and lower-quality network as soon as possible. This process, known as hot-potato routing, increases the number of hops and degrades the quality of the session.

If a user connects to a local ISP in Argentina or Brazil to access content that is hosted in the same city or country, the user's traffic is often routed to the United States, where it will be redirected at a public NAP back to its destination in South America. That occurs because of the limited partnerships at public access points and lack of peering agreements between local providers.

The ISP's backbone provider is likely an incumbent telecommunications provider with a legacy voice-based network. The legacy network's routers and links can add significant latency and packet loss to the session. The provider's network is also likely to include single points of failure that pose the risk of session failure.

The precise number of hops, amount of packet loss, and amount of latency varies with each session and the network topologies of the connection. Generally, packets passing from sites in the United States to Buenos Aires would generate 500 ms or more of round-trip latency. Compounded by multiple packets making up a Web page, such latency can produce 8 s or more delay in page downloads.

Today's Pan-Regional Internet Backbone

The Internet is entering the second phase of its evolution in Latin America. By 2000, the region emerged as the fastest-growing Internet market in the world. Companies no longer use the Web merely to market their products and services; many are developing highly complex, transaction-enabled sites. Market researcher International Data Corporation foresees e-commerce in the region growing to more than $9 billion by 2004. Merrill Lynch predicts the Web hosting market in Latin America will reach $2.4 billion in revenue by 2006.

In light of this e-commerce growth, it is clear solutions presented by foreign hosting firms via satellite transmissions and public NAP routing no longer meet the needs of the region's businesses. This situation is opening the door for ISPs to build private networks and IDCs in the region. Today, the local hosting sector is meeting these new demands through an optical backbone that enables quality of service, private peering relationships, content distribution, and managed hosting.

Problems posed by hot-potato routing and NAP bottlenecks resulted in insufficient transport for the mission-critical applications of the second phase of the Latin American Internet. The reliability and performance of each connection were greatly affected by the logical proximity and network availability of the links. Furthermore, much of the international traffic was transmitted via satellite connections, which are expensive and lack scalability. Other options existed, like submarine cables, but these were primarily consortium ventures controlled by incumbent carriers and were voice-centric in nature.

As a result of these challenges, a huge demand for data-centric traffic capacity grew in the region. And the increasing concerns for the latency and packet-loss issues posed by satellites drove several global network providers, including 360networks, Emergia, and Global Crossing to build their own fiber-optic connections within the region, connecting to the United States and other international fiber networks. These new fiber cables have enabled new entrants in Latin America to construct pan-regional fiber backbones.

Through an international fiber-optic backbone, carriers found a highly scalable solution that allowed them to add customers quickly and cost-effectively. A provider or customer can now get an STM-1 (155-Mbps) connection with 10 times the capacity on a fiber network for the same cost as 15 Mbps of satellite capacity a year ago. But, the customer value of these new backbones comes through the control new providers are able to guarantee through private peering arrangements at IDCs and content delivery features that better manage the flow of traffic around the globe.

As a result of the growth in number of local hosting facilities and improved intracountry networks, about 50 percent of the traffic in

Brazil today stays local instead of traveling over pan-regional or international networks before reaching its destination. The physical proximity also assists companies with some of the psychological challenges of transitioning mission-critical applications to the Web. The ability to touch and see Web hardware provides reassurance to organizations that are moving highly important information on line. However, there is a reluctance to outsource mission-critical applications remotely as a major attraction for local hosting. A local solution allows the company to bring a potential client to see first hand the secure location of a hosting platform.

The physical proximity to the Latin American user base can also help with necessary local dedicated links. Many application service-provider designs, for instance, call for dedicated local loops between the IDC and offices with high user concentrations. While such links would be prohibitively expensive from the United States, they become affordable when run from a local location.

In this scenario, when the Buenos Aires user requests content, located, for example, in a Miami or Mexico IDC, the request travels through the user's ISP to a private optical network. The optical-network provider's routers then broadcast the requested IP address because the content is hosted on the same pan-regional network (see Fig. 8-1).[1] The fiber-optic infrastructure provides a fast, reliable connection to the content located in the Miami or Mexico IDC.

The optimal solution is for a hosting provider to operate an optical network with multiple paths and access points in each of its markets. Any traffic that enters the provider's network is quickly moved over private connections to the server. In this scenario, any user located near an access point can access any Web server anywhere on the network at the same high speed.

The hosting provider's pan-regional presence can be utilized to provide a distributed architecture for Web content as well, using technologies such as shared caching, dedicated caching, and server mirroring. This array of choices provides for a wider range of distributable content, including applications and secure content.[1]

Security Concerns

Another key issue with wireless data Internet is security. A poorly secured system lets eavesdroppers access sensitive information.

If you plan to transmit credit card numbers, Social Security numbers, and passwords over a wireless data network, then you'd better be sure

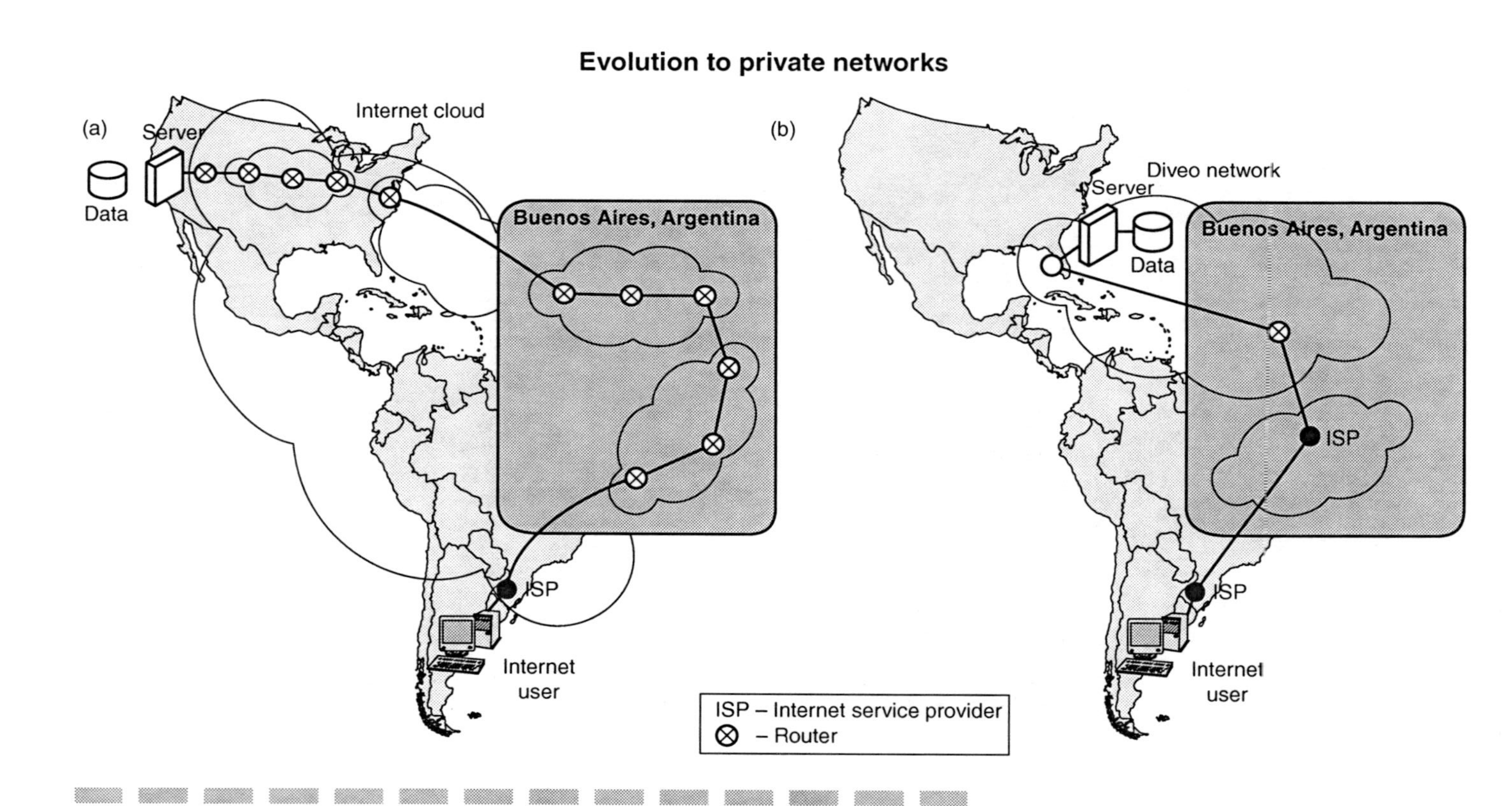

Figure 8-1 Map (*a*) illustrates the traditional hot-potato routing of Internet traffic, while map (*b*) shows the routing of Internet traffic over private optical networks with Internet data centers.

the system supports adequate security mechanisms. The IEEE 802.11 wired equivalent privacy (WEP) might not be good enough.

Researchers at the University of California at Berkeley have found flaws in the 802.11 WEP algorithm and claim it is not capable of providing adequate security. A problem with the 802.11 WEP is that it requires the use of a common key throughout the network for encrypting and decrypting data, and changing the keys is difficult to manage. This makes the system vulnerable to breaches in security, and network executives should be cautious when implementing 802.11 networks.

Network executives should ensure that wireless data service providers implement enhanced security beyond 802.11 WEP (such as IEEE 802.1x). Some vendors, such as Cisco,[7] implement security mechanisms that utilize a different key for each end user and automatically change the key often for each session. This greatly enhances information security.

Finally, let's look at an overview of a fixed low-frequency broadband wireless data access system for point-to-multipoint voice and data applications. Operating frequency bands are from 2 to 11 GHz, and the base station can use multiple sectors and will be capable of supporting smart antenna technology. The product system requirements, design of the radio subsystem specification, and an analysis of microwave transmission related to current radio technologies are presented. Examples of BWDA technology are provided.

Fixed Broadband Wireless Data Radio Systems

Global integration and fast-growing business activity in conjunction with remote multisite operations have increased the need for high-speed information exchange. In many places around the world, the existing infrastructure is not able to cope with such demand for high-speed communications. Wireless data systems, with their fast deployment, have proven to be reliable transmission media at very reasonable costs. Fixed broadband wireless data access (BWDA) is a communication system that provides digital two-way voice, data, Internet, and video services, making use of a point-to-multipoint topology. The BWDA low-frequency radio systems addressed in this part of the chapter are in the 3.5- and 10.5-GHz frequency bands. The BWDA market targets wireless data multimedia services to small offices/home offices (SOHOs), small and medium-sized businesses, and residences. Currently, licensed bands for 3.5-GHz BWDA systems are available in South America, Asia, Europe, and Canada. The 10.5-GHz band is used in Central and South America as well as Asia, where expanding business development is occurring. The fixed wireless data market for broadband megabit-per-second transmission rates, in the form of an easily deployable low-cost solution, is growing faster than that for existing cable and digital subscriber line (xDSL) technologies for dense and suburban environments.

This part of the chapter also describes the BWDA network system, the radio architecture, and the BWDA planning and deployment issues for 3.5- and 10.5-GHz systems. Table 8-1 summarizes the system characteristics for each frequency range according to various International Telecommunication Union—Radiocommunication Standardization Sector (ITU-R) drafts, EN 301 021, IEEE 802.16, and other national regulations.[2] A maximum of 35 Mbps capacity is achievable for 64 quadrature amplitude modulation (QAM) over 7-MHz channel bandwidth. Coverage ranges for line-of-sight links are given for 99.99 percent availability.

The BWDA System Network

A BWDA system comprises at least one base station (BS) and one or more subscriber remote stations (RSs). The BS and RS consist of an outdoor unit (ODU), which includes the radio transceiver and antenna, and an indoor unit (IDU) for modem, communication, and network management (see Fig. 8-2).[2] The two units interface at an intermediate frequency (IF); optionally, the RS ODU and IDU can be integrated. The BS assigns the radio channel to each RS independently, according to the policies of the media access control (MAC) air interface. Time in the upstream channel is usually slotted, providing for time-division multiple access (TDMA), whereas on the downstream channel, a continuous time-division multiplexing (TDM) scheme is used. Each RS can deliver voice and data using

TABLE 8-1

The 3.5- and 10.5-GHz System Characteristics

Product	3.5 GHz	10.5 GHz
Frequency, GHz	3.4–3.6	10.15–10.65
Tx/Rx spacing, MHz	100	350
Channelization, MHz	3.5, 5, 7	3.5, 7
RS upstream modulation	QPSK/16 QAM	QPSK
RS downstream modulation	16/64 QAM	16 QAM
RS upstream capacity, Mbps	5–20	5, 10
RS downstream capacity, Mbps	12–34	12, 23
Coverage radius, km	19	8

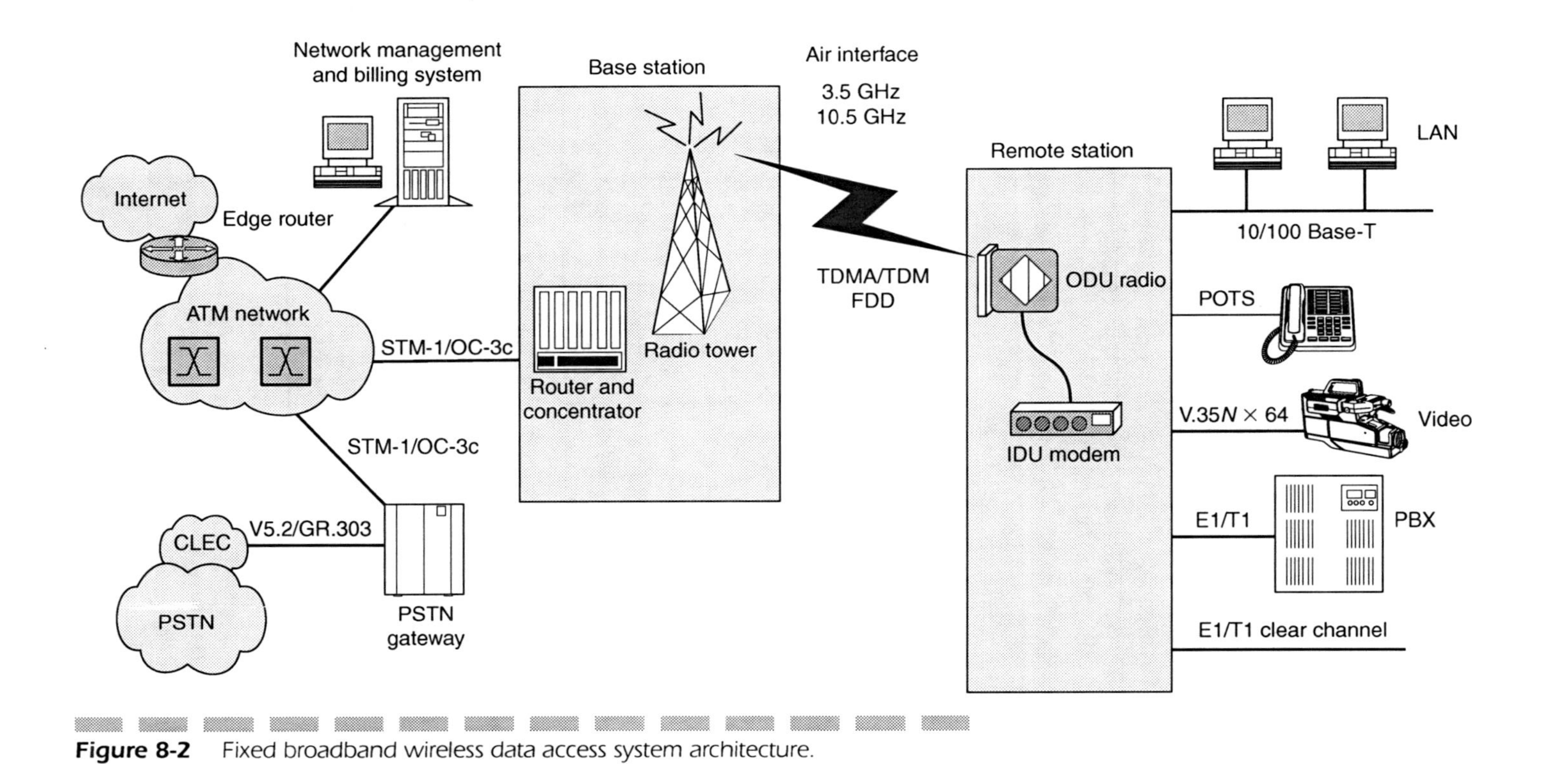

Figure 8-2 Fixed broadband wireless data access system architecture.

common interfaces, such as plain old telephone service (POTS), Ethernet, video, and E1/T1. Depending on the type of service required by the client, remote stations can provide access to a 10/100Base-T local-area network (LAN) for data access and voice over IP (VoIP) services to (1) a LAN and up to eight POTS units for small businesses or (2) a LAN and an E1/T1 channel connected to a private branch exchange (PBX) for small and medium enterprises.

The BS grooms the voice and data channels of several carriers and provides connection to a backbone network (IP or asynchronous transfer mode, ATM) or transport equipment via the STM1/OC-3c (155.52 Mbps) high-capacity fiber link. The ATM network gives access to the public switched telephone network (PSTN) gateway through competitive local exchange carriers (CLECs) using V5.2/GR.303 standards, or to an edge router for accessing the Internet data network through Internet service providers (ISPs). The ATM network interface is also connected to the network management system via Simple Network Management Protocol (SNMP) for performing tasks such as statistics and billing, database control, network setup, and signaling alarms for radio failures. Configuration of the radio network link is made possible through a Web browser http link via TCP/IP.

Each BS has a certain available bandwidth per carrier that can be fully or partially allocated to a single RS either for a certain period of time [variable bit rate (VBR) or best effort] or permanently [constant bit rate (CBR)]. BWDA systems are envisioned to work with a TDMA rather than a code-division multiple-access (CDMA) scheme in order to counteract propagation issues. Also, for non-line-of-sight (NLOS) environments, BWDA systems with a single carrier with frequency domain equalizer and decision feedback equalizer (FD-DFE) or orthogonal frequency-division multiplexing (OFDM) technologies are applicable. Small and medium-size businesses require fast and dynamic capacity allocation for data and voice packet-switched traffic. This TDMA access scheme can be applied to either frequency-division duplexing (FDD) or time-division duplexing (TDD). Both duplexing schemes have intrinsic advantages and disadvantages, so the optimum scheme to be applied depends on deployment-specific characteristics (bandwidth availability, Tx-to-Rx spacing, frequency congestion, and traffic usage). Targeting the business market, for example, are Harris ClearBurst MB (http://www.harris.com/harris/whats_new/pacnet.html) products, which are designed for FDD. In symmetric two-way data traffic, FDD allows continuous downstream and upstream traffic on both low- and high-band channels. Moreover, it has full flexibility for instantaneous capacity allocation, dynamically set through the MAC channel assignment.

The Radio-Frequency System

RF subsystems consist of the base station and remote station ODUs. This part of the chapter will provide a global understanding of the different RF technologies employed for high-performance low-cost radio design. In addition to meeting all the functional, performance, regulatory, mechanical, and environmental requirements, the radio system must achieve most of the following criteria:

- Cost-effectiveness
- Maintenance-free
- Easily upgradable
- Quick installation
- Attractive appearance
- Flexibility
- Scalability[2]

An example of a BWDA radio system is shown in Fig. 8-3: a base station ODU, part of the ClearBurst MB product.[2] Its radio enclosure contains two sets of identical transceivers with high-power amplifiers and RF diplexers for redundancy. A dual flat-panel antenna is directly integrated with the enclosure. A single coaxial cable is used to connect to the indoor base station router unit. The base station radio units can be mounted on

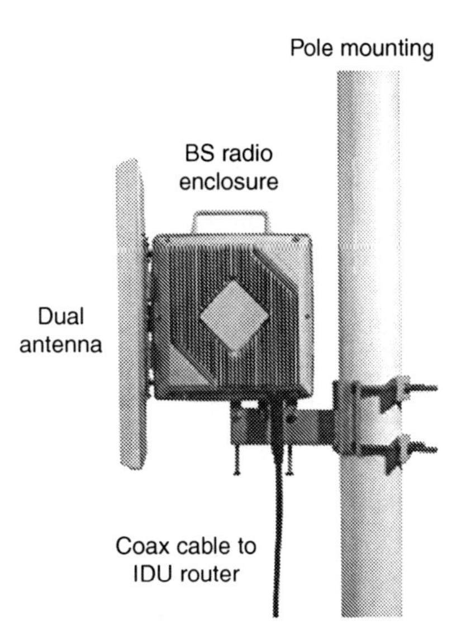

Figure 8-3
The Harris base station outdoor radio unit.

a pole, a tower, or a wall. The remote station ODU is an unprotected unit, where a single transceiver with a medium-power amplifier is used. The enclosure is directly connected to the flat-panel antenna. In addition, an alignment indication connector is also provided for antenna installation and alignment with the base station.

An ODU radio consists of transmitter and receiver circuits, frequency sources, a diplexer connected to the antenna, and a cable interface to connect to the indoor modem unit. Moreover, a minimum of "intelligence" is required in the radio to control the power level throughout the transceiver. Development of software-controlled radios is presently underway, but the issue of cost-effectiveness remains. Typically, for small businesses or residential markets, cost is the main factor that comes into play; hence, a design made simpler by limiting radio intelligence may translate into less demanding requirements for the radio processor. Software-controlled radios present many advantages, such as reducing hardware complexity, but it is up to the design engineers to compromise among the high performance, low cost, and flexibility of the product.

A low-cost, low-performance radio solution appropriate for the high-volume residential market is shown in Fig. 8-4 as a "dumb" transceiver.[2] This architecture uses a minimal number of hardware components, integrated with or without software control capabilities. Following the RF diplexer, the receive (Rx) path includes a low-noise amplifier, bandpass filters (BPFs) for image-reject and channel-select filtering, a downconverter mixer, and an open loop gain to allow a wide input dynamic range. The transmitter (Tx) consists mainly of an upconverter associated with some filtering and a power amplifier (PA). The local oscillator (LO) may provide for fixed or variable frequency to the mixers. A fixed LO would give a variable IF; hence, by using a wider BPF bandwidth, the receiver would not be immune to interference. Adding a microcontroller to the radio provides control of the phase-locked loop (PLL) for the transceiver synthesizer and can put the PA into mute mode. Single up/downconversion stages further reduce the overall cost, but at the expense of lower radio performance. Two separate IF cables simplify the interfacing.

Figure 8-4
A dumb transceiver: block diagram.

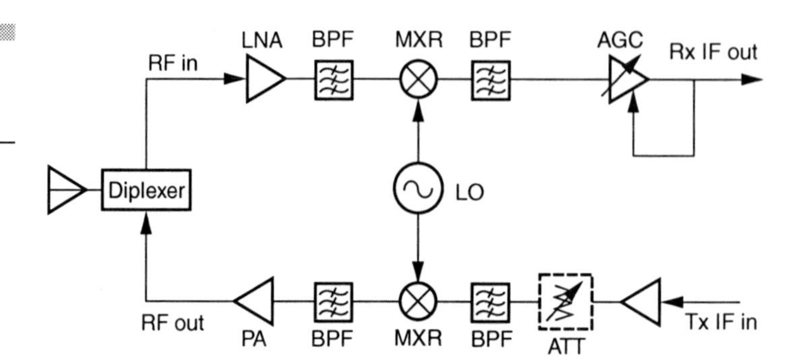

An intelligent transceiver involves more digital and software-controlled circuitry, and hence higher cost. Figure 8-5 shows a transceiver block diagram which includes closed-loop gain control, cable, and fade margin compensation on the transmit and receive paths, that is, power detection circuits on Rx IF, Tx chain, and PA.[2] The transmitter mutes on a synthesizer out-of-lock alarm in order to avoid transmitting undesirable frequencies, and also on no received signal. The microcontroller provides for the receive signal strength indicator (RSSI) level for antenna alignment, and for control and monitor channels. A single cable is used for all input and output IFs, the telemetry signal, and the dc biasing from the IDU. Software control also allows for calibrated radios, which results in no gain variation or frequency shifting of the signal with respect to temperature variation. Technology advancement in the past few years in the RF integrated circuit market allows for greater chip integration using commercial off-the-shelf (COTS) devices and simplified hardware board-level design. This architecture achieves better performance, especially for higher-modulation schemes, and therefore is suitable for higher-capacity radios targeting the business market.

The modulation scheme chosen for the radio system depends on several product definition factors, such as required channel size, upstream and downstream data rates, transmit output power, minimum carrier-to-noise ratio (C/N), system availability, and coverage. Table 8-2 gives the characteristics for quadrature phase-shift keying (QPSK) and QAM signals typically used for BWDA systems for 7-MHz channel bandwidth.[2]

A system can require symmetric or asymmetric capacity, depending on its specific application. For a symmetric capacity system, upstream and downstream traffic are equivalent, whereas for an asymmetric system, the downstream link usually requires more capacity. Hence, higher-level modulations with higher capacity are better suited to downstream transmissions. Using n QAM modulations for downstream transmission becomes advantageous, whereas QPSK can be used in the upstream

Figure 8-5 An intelligent transceiver: block diagram.

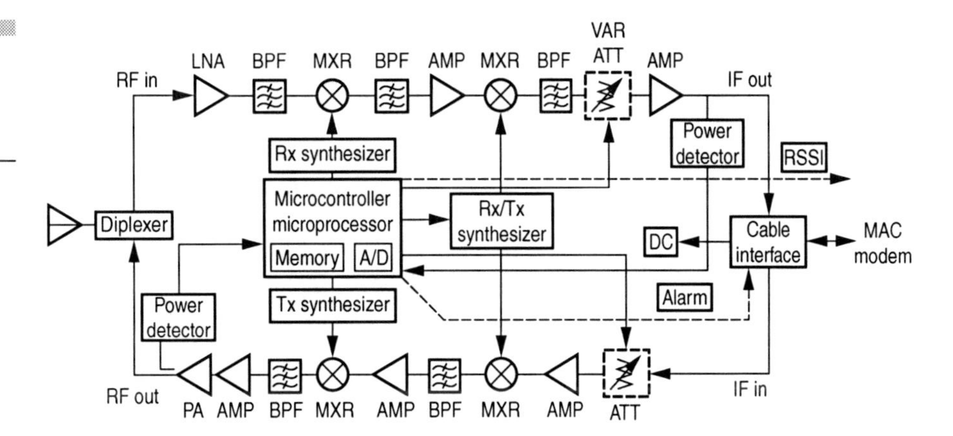

TABLE 8-2

QPSK and QAM Modulation Characteristics for 7-MHz Channel Bandwidth

Modulation	Data Rate, Mbps	*C/N* for 10^{-6} BER, dB
QPSK (upstream)	10.24	13.5
16-QAM (downstream)	23.68	16.6
16-QAM (upstream)	20.48	17.6
64-QAM (downstream)	35.52	22.8

direction. Since lower-level modulations perform better in more constrained environments, they can be used not only in burst, low-power, low-capacity, or upstream transmissions, but also can be adjusted dynamically in link fading conditions.

Radio Transmission System and Deployment

The maximum cell size for the service area is related to the desired availability level. At 3.5 and 10.5 GHz, the average cell radius for line-of-sight (LOS) 99.99 percent availability is 19 and 8 km, respectively. Principal factors affecting cell radius and availability include the rain region, the antenna and its height, foliage loss, modulation, Tx power, Rx sensitivity, and sectorization. These effects are generally related to the service area, such as dense urban, suburban, and low density. As an aid to determining these parameters, a powerful point-to-multipoint RF transmission engineering tool is used to estimate the maximum distance between the BS and RS, while maintaining the desired link performance and availability in a single- or multihub environment. Taken into account are the margins required to combat multipath fading, rainfall attenuation, and interference. The effect of the rainfall attenuation is negligible at 3.5 GHz, but noticeable at 10.5 GHz.

The base station hub is divided into a number of sectors to accommodate all received signals and cumulative traffic from the remote stations. The number of cell sectors affects the cost per cell and complicates cell planning, but also increases the capacity of the system. Each BS unit typically serves 1000 and 100 remote stations at 3.5 and 10.5 GHz, respectively. The deployment consists of a four-sector/90° or six-sector/60° cell configuration. The antenna panel can be assembled for horizontal or vertical polarization for reduced interference.

Conclusion

Fixed wireless data is a good option for networks in locations where DSL and cable modem access are not available. Small and midsize companies might also benefit from wireless data Internet in larger cities because of cost savings.

With the availability of the solid IEEE 802.11b products and the upcoming IEEE 802.11a and IEEE 802.16 products, network executives can count on having performance that exceeds DSL and cable modem access. However, network executives should strongly consider provisions in contracts for specific performance and availability. Because of potential interference in the 2.4-GHz band, the contract should be checked for provisions to recover investments if the system doesn't deliver what it's stated to do.

Finally, growing demand for fast information exchange to support business activities requires the implementation of low-cost, easily deployable communications networks. Fixed low-frequency BWDA radio systems at 3.5 and 10.5 GHz were presented as an attractive solution in this chapter. System architecture was presented from a signal processing and radio-frequency perspective. Architecture compromises were discussed, enabling the use of cost-effective solutions that meet quality and performance requirements.

References

1. Peter Scott, "The Value of Local Latin American Internet Infrastructure," Diveo Broadband Networks, Inc., 3201 New Mexico Ave. NW, Ste. 320, Washington, DC 20016, 2002.
2. Mina Danesh, Juan-Carlos Zuniga, and Fabio Concilio, "Fixed Low-Frequency Broadband Wireless Access Radio Systems," *IEEE Communications Magazine,* 445 Hoes Lane, Piscataway, NJ 08855, 2002.
3. John R. Vacca, *The Cabling Handbook,* 2d ed., Prentice Hall, 2001.
4. John R. Vacca, *Wireless Broadband Networks Handbook,* McGraw-Hill, 2001.
5. John R. Vacca, *Satellite Encryption,* Academic Press, 1999.
6. John R. Vacca, *Electronic Commerce,* 3d ed., Charles River Media, 2001.
7. John R. Vacca, *High-Speed Cisco Networks: Planning, Design, and Implementation,* CRC Press, 2002.

CHAPTER 9

Wireless Data Access Design

It is an exciting time for broadband[5] fixed wireless data access design, with key developments in frequency bands from 1 to 60 GHz and a range of new technologies being developed. While working on these new technologies, it is easy for us to forget that fixed wireless data access will form part of an integrated communications environment of the future, where users will have one communications device working in the home, at the office, and outdoors. This chapter predicts the communications environment of the next 30 years and looks at the role of fixed access within that environment. This involves assessing how fixed access systems will interface and integrate with in-home wireless data networks, how their architecture will enable multiservice operators to utilize the same core network across a range of different access technologies, and how they will act as a channel to carry mobile traffic originating within the building. On the basis of the requirements this vision and architecture imply, this chapter critically assesses the different fixed wireless data technologies available to date and compares their capabilities to provide future-proof broadband fixed wireless data platforms. (The Glossary defines many technical terms, abbreviations, and acronyms used in the book.)

Today's Communications

Communications today is a mixed and rather disorganized environment. The typical office worker in a developed country currently has a wide range of ways to communicate, including:

- The office telephone, used mostly for voice communications complete with mailbox system
- The office fax machine, now being used less as e-mail takes over
- The office LAN, providing high-data-rate communications such as e-mail and file transfer
- Dial-up networking for workers out of the office, providing the same capabilities as the LAN but at a much slower rate
- Mobile telephones providing voice communications, a mailbox, and in some cases low-speed data access
- A pager providing one- or two-way messaging
- A home telephone providing voice communications and dial-up access along with a home answering machine
- A computer at home linked to a different e-mail system, perhaps using high-speed connections such as asynchronous digital subscriber line (ADSL) or cable modems[1]

Managing all these different communication devices is complex and time-consuming. The worker who has all of these (and many do) will have five phone numbers, three voice-mail systems, and two e-mail addresses. There is no interconnection between any of these devices, so all the different mailboxes have to be checked separately, using different protocols and passwords. Contacting such an individual is problematic because of the choice of numbers to call, and many default to calling the mobile number as the one most likely to be answered. Although many are working on systems such as unified messaging, designed to allow all types of communications (voice, fax, e-mail) to be sent to one number, the wireless data industry is still some way from the ideal situation where individuals have only one "address" and all communications are unified. Effectively, there is little convergence, at least as far as the user is concerned, between all these different fixed and mobile systems. How this will change, and more detail on what the future will look like, especially for fixed wireless data design, is the subject of this chapter.

How You Will Communicate in the Next 20 to 30 Years

Based on an understanding of possible types of communications and the shortcomings of current communications systems, the following are advances predicted over the next 20 to 30 years:

- Video communications wherever possible
- Complete unification of all messaging
- Intelligent filtering and redirection
- Freedom to communicate anywhere
- Simplicity
- Context-sensitive information[1]

Video Communications Wherever Possible

When people are talking from the home or office, all communications should have the option of video links and hands-free talking to make communications as natural as possible. This may not always be appropriate, especially when users are mobile, but the option should be available.

Complete Unification of All Messaging

Each individual should have a single "address." This will typically be of the form "john.mulder@my-isp.com" (some further detail may be required to overcome the problem of multiple John Mulders) to which all communications will be directed.

Intelligent Filtering and Redirection

Upon receiving a message, the network, on the basis of preferences and past actions, will determine what to do with it, knowing the current status of John Mulder (whether mobile, at home, etc.). Work calls might be forwarded during the weekend only if they are from certain individuals, otherwise stored and replayed on return to the office, and so on.

Freedom to Communicate Anywhere

It should be possible to have almost any type of communications anywhere. However, the higher the bandwidth and the more "difficult" the environment, the higher the cost.

Simplicity

For example, upon walking into a hotel room, communications devices should automatically network with the hotel communications system. They should also be able to determine whether the tariff charged by the hotel is within bounds set by the user, and automatically start downloading information and presenting it to the user in accordance with his or her preferences.

Context-Sensitive Information

Besides being able to get information from the Internet on request, the user should be able to obtain the information he or she needs. This of course depends on the user's location, plans, and circumstances.

Technically, all this is relatively straightforward. No fundamental breakthroughs in communication theory, device design, or computing power are necessary to realize this vision. The key issues preventing realization of this vision today are:

- *Lack of bandwidth.* Most homes and mobile phones do not have access to sufficient bandwidth to realize video transmissions of good or high quality.
- *Multiplicity of disparate systems.* As discussed earlier, there are many different communication systems, which, to date, are rarely linked in an intelligent fashion, partly because they utilize different protocols, technologies, and paradigms.
- *Multiplicity of different operators.* Different systems are often run by different operators who do not always perceive commercial justification for tightly integrating with other systems that may be run by competitors, particularly since many operators are now involved in a complex web of partnerships.
- *Economics.* Provision of some systems, such as a radio transmission node in each hotel room, is generally not economically viable today and must await lower cost realizations.
- *Lack of standardization.* For a user to enter a hotel and automatically download e-mails to a laptop, there must be an agreed-on radio standards and infrastructure in place so that the hotel and the laptop can communicate. In many areas, standards are being developed but are far from ubiquitous.[1]

Now, let's look at the developments under way today that might form the basis of realizing the vision and extrapolate these forward. Key for the fixed wireless data arena is the requirement for ubiquitous and high-speed wireless data access to the home. The wireless data industry is still some way from realizing this vision. In the rest of this part of the chapter, let's consider some of the constraints and technologies that might be adopted.

The Future Architecture: A Truly Converged Communications Environment

A summary of the network of the future that would deliver the requirements discussed earlier is shown in Fig. 9-1.[1] Much is missing from this figure, and much has been simplified in order to show all the key elements in one picture. This figure demonstrates just how fixed and mobile systems will converge: Both will be linked back to the same postmaster by common protocols and possibly a common core network (when

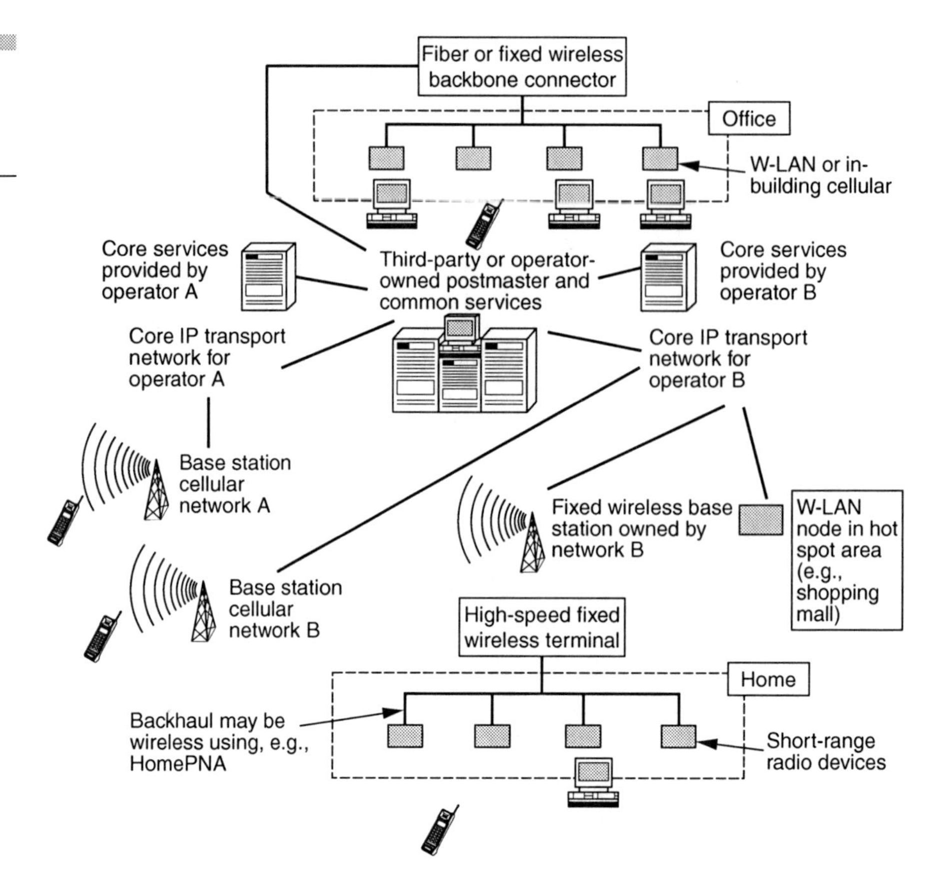

Figure 9-1
A possible network architecture of the future.

both are owned by the same operator), and mobile devices will also utilize in-home and office radio networks connected back through fixed networks into the postmaster, which coordinates their use. In summary, the key elements of realizing the network of the future are:

- Ubiquitous broadband access to the home delivered by using a range of different technologies, including fixed wireless data, based on technologies discussed in later in this chapter.
- Standardized in-home networks consisting of simple radio devices in each room connected to a home LAN. It is likely these will be enhanced developments of standards like Bluetooth.
- Standardized radio devices in most home and office appliances using the same short-range radio standard.
- The provision of an "intelligent postmaster" function, probably provided by third-party entities.

- A standard protocol for all networks to communicate with the postmaster and each other, probably using IP as the underlying transport mechanism and building on the protocols developed for the third generation (3G).
- Widespread cellular architecture using a single 3G standard or, alternatively, multimode phones operating over a multiplicity of standards with high-speed access delivered by wireless data LAN (WD-LAN) solutions in certain important areas.
- A standard approach for office wireless data networks, most likely based on wireless data LANs, common to all offices.
- Communicator devices able to work on the cellular, home, and office networks in a seamless manner.
- An environment that enables the development of innovative services by third parties, probably delivered through the Internet, and can be downloaded and run by all communicator devices using languages such as Java.[1]

Now, let's consider the ability of broadband fixed wireless data to play its envisaged role in this future vision.

Technical Constraints on Broadband Fixed Wireless Data Systems

If fixed wireless data is to play a key role in this network of the future, it must be able to deliver high data rates to most homes. Being more specific about high data rates is difficult because it depends on the user.

> ***NOTE*** High-definition video transmission requires around 8 Mbps, and if you allow for multiple simultaneous transmissions to or from the home, data rates in excess of 10 Mbps will be needed.

To date, fixed wireless data has been unable to deliver data rates in excess of 10 Mbps to a high percentage of homes in a given area cost-effectively. In this part of the chapter, let's examine the theoretical and economic constraints on fixed wireless data to assess whether this might change in the future.

It is possible, by making some assumptions, to calculate the theoretical capacity that can be provided by fixed wireless data solutions. This approach is described in detail and summarized next. The approach starts with Shannon's law, setting out the maximum information that can be transmitted per second per hertz of spectrum, and adds equations to model operation in a clustered cellular environment. Key to modeling capacity in a fixed wireless data environment is an understanding that propagation conditions are different from those in a mobile environment. Because many systems use directional antennas with line-of-sight (LOS) or near-LOS paths, the path loss exponent is often closer to that of free space, namely 2, than the mobile case, where it tends to fall between 3.5 and 4. However, for interfering signals, there is often no LOS, and there may be isolation created by the directional antenna. As a result, the interfering path loss may be closer to that for mobile.

The mathematical analysis shows that a solution can be derived, indicating that the capacity is inversely proportional to the cell radius and the bit rate required per user. One of the key parameters is the modulation scheme that is adopted. Figure 9-2 shows the variation of capacity M with the signal-to-interference ratio (SIR).[1] The curve clearly shows that the highest efficiencies can be obtained at the lowest SIRs. This result is in line with earlier work where it was reported that the best results were obtained with single-level modulation as opposed to multilevel modulation. Hence, you can assume the use of quadrature phase-shift keying (QPSK) modulation.

Analysis With a range of assumptions, it is possible to determine the viability of broadband fixed wireless data systems that technically approach the Shannon limit and economically fall in line with current revenue expectations. The end result is shown in Fig. 9-3.[1]

Figure 9-3 shows that below a spectrum allocation in megahertz of 5 times the user data rate in megabits per second (if the user data rate were 10 Mbps, the spectrum allocation would be 50 MHz), profitable operation seems unlikely. Spectrum allocations above around 10 times the user data rate result in little extra increase in profitability; hence, a 10 times allocation is probably most appropriate, minimizing use of the scarce spectrum resource.

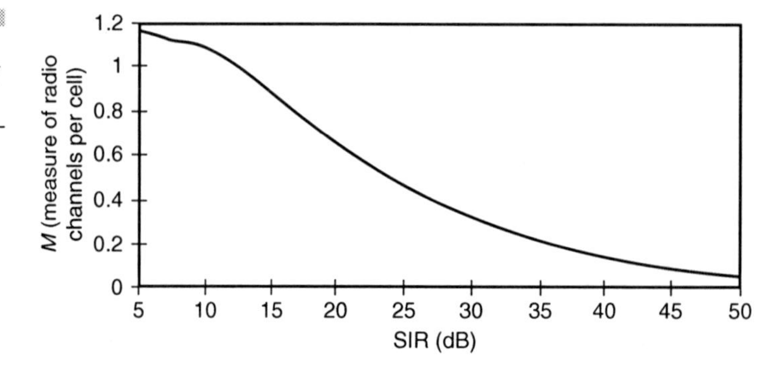

Figure 9-2 The relationship of M to SIR.

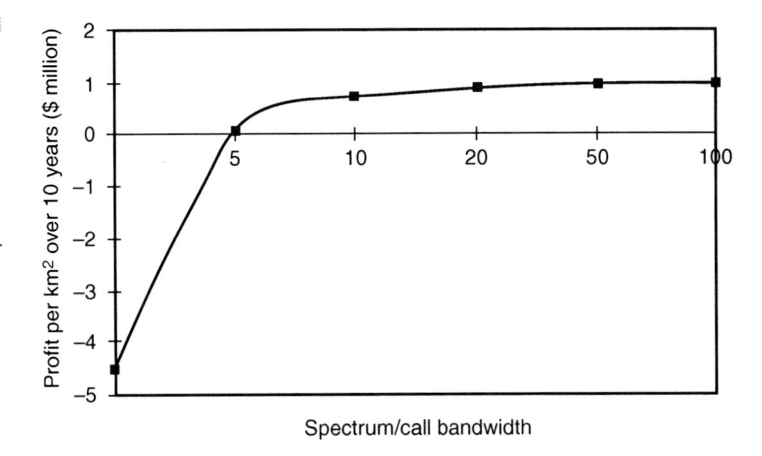

Figure 9-3 Variation of profit with spectrum (in megahertz) per call bandwidth (in megabits per second).

NOTE The assignment would need to be twice this size in practice to allow duplex communications; an uplink and a downlink assignment, both equal to 10 times the user rate, would be required. Hence, theoretically, with a 2 × 100 MHz spectrum assignment, a 10 Mbps duplex service can be profitably offered to residential users.

Another way of looking at this result is that it would appear to be profitable to operate a broadband fixed wireless data system in the region where the maximum data rate is around 10 percent of the spectrum assignment. The data rate per subscriber is then dependent only on the spectrum assigned by the regulator. For typical assignments in the frequency bands of 10 GHz and above, and some assignments at 2 GHz, bandwidths of 10 Mbps per subscriber using fixed wireless data would seem both technically and economically viable. In the next part of the chapter, specific technologies that might meet or exceed this performance are described.

Technologies for Broadband Fixed Access

There is a wide range of different technologies proposed for fixed access. Here, the key technologies you might consider in future systems are listed and briefly evaluated.

At the System Level

Here there appear to be two basic concepts available: a conventional point-to-multipoint (PMP) solution where each subscriber unit communicates directly with a base station, and a mesh approach where subscriber units communicate with the nearest neighbors and information is passed back through the mesh in a manner analogous to Internet traffic.

Examples of these two concepts are shown in Fig. 9-4.[1] The status quo is represented by the first option. Here, let's consider the merit of the mesh approach compared to the conventional PMP structure.

The mesh approach effectively changes the "rules" used for capacity and link budget calculations by turning each link into a point-to-point link. Arguably this has an effect similar to that of adaptive base station antennas in a conventional system, which can provide a narrow beam to each subscriber unit. The potential advantages of mesh solutions are:

- An increase in capacity as a result of frequencies being reused on a very localized level. Effectively, this is the equivalent of

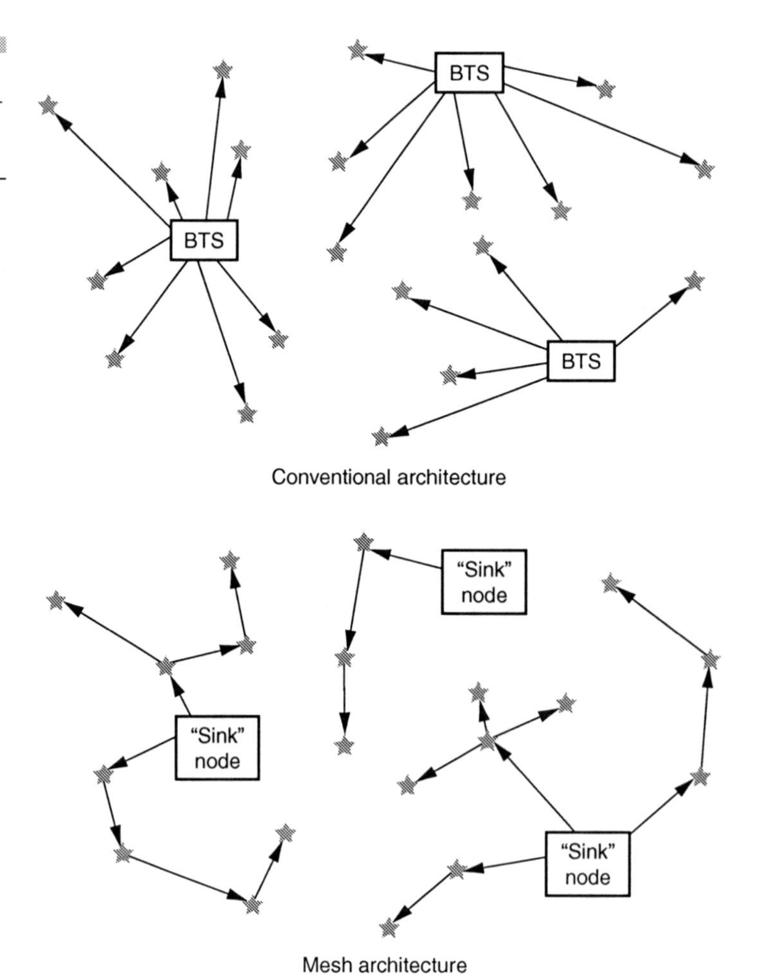

Figure 9-4 A comparison of conventional PMP and mesh deployments.

a microcellular approach on a conventional design, although these capacity gains could be offset by the need for each node to relay traffic.

- An improvement in quality as a result of each link being short and hence having a high link budget.
- A possible cost reduction in the subscriber unit as a result of the less demanding link budget. However, this may be offset by the additional complexity required to provide the repeater element needed within the mesh architecture.
- An ability to replan the system without repointing subscriber antennas (in cases where subscriber numbers grow more quickly than anticipated).
- A potentially nearly "infrastructureless" deployment.[1]

These need to be balanced against the potential disadvantages, which are:

- Highly complex algorithms are required to manage the system and avoid "hot spots," which may be unstable and result in poor availability.
- Different and novel medium access control (MAC) mechanisms may be required, which will need development and add to the complexity.
- The initial investment is relatively high since "seed nodes" have to be placed so that the mesh can form as soon as the first subscriber is brought onto the system.
- Marketing issues may be problematic in that customers may not want to rely on nodes not in their control and not on their premises for their connectivity, and may not want their equipment to be relaying messages for others.[1]

It is difficult to draw definitive conclusions at this point since many of the preceding variables are unknown. If the complexity and risk can be overcome, it seems highly likely that mesh systems will provide greater capacity than conventional systems for a given cost.

Layer One/Two

There are a number of discrete technologies here, which are mostly independent, so each can be considered separately. The issues are:

- Time-division duplex (TDD) versus frequency-division duplex (FDD)
- Adaptive versus fixed-rate modulation

- Orthogonal frequency-division multiplexed (OFDM) versus single carrier
- Code-division multiple access (CDMA) versus time-division multiple access (TDMA)
- Adaptive versus conventional antennas[1]

TDD versus FDD FDD represents the status quo. The question is whether TDD brings substantial benefits to the operator. Key to this question are:

1. Whether the spectrum is simplex or duplex. If it is simplex, TDD overcomes the need for a guard band, which can be wasteful of spectrum; however, TDD needs a guard time, which may be as large percentage-wise.
2. Whether the data are both asymmetric and the asymmetry is time-variable. If the asymmetry is known beforehand, unbalanced FDD assignments can be used; if not, TDD can bring some efficiency gains.

Determining the gains is then an issue of understanding the variability of the asymmetry and the simplex or duplex nature of the spectrum. It seems likely that if the asymmetry is highly time-variable, TDD will bring definite advantages, in principle up to a maximum of a 100 percent capacity gain (where traffic flows only in one direction—100 percent asymmetry). It also seems certain that in simplex bands TDD will bring advantages unless complex frequency assignment procedures are adopted for FDD, whereby the guard band is different in different cells, requiring complex planning and possibly greater expense for the subscriber unit. Hence, assuming that the cost of implementing TDD is not great, it is likely to bring significant benefits. Thus, forward predictions of asymmetry time variance are uncertain, so TDD gains cannot be definitively quantified, but it seems possible that there may be some worthwhile gains in certain situations.

Adaptive versus Fixed-Rate Modulation With adaptive modulation, instantaneous carrier-to-interference (*C/I*) and signal-to-noise ratio (*S/N*) measurements are made, and the number of modulation levels are modified dynamically. Hence, if the subscriber is experiencing relatively good *S/N*, perhaps because the subscriber's unit is not in a fade, more modulation levels can be used without a greater power requirement without adding interference to the system. Adaptive modulation is the technique proposed for EDGE, an enhancement to existing cellular systems. Adaptive modulation provides the greatest gains in fading channels where the number of modulation levels can be instantaneously matched to the channel conditions. In gaussian channels, more modulation levels bring no advantages. Fixed wireless data channels tend to fall somewhere between these two, so the advantages are likely to be less than in the mobile case, but they are still likely to bring some gains.

OFDM versus Single-Carrier Here single carrier is assumed to mean existing schemes (sometimes erroneously considered quadrature amplitude modulation, QAM), while OFDM is considered in the broadest sense to include vector OFDM (VOFDM) and other variants. OFDM brings two benefits: It effectively removes the need for an equalizer by turning a wideband signal into a multitude of narrowband signals, and it can overcome some specific types of narrowband interference more simply than other schemes. However, it also brings some disadvantages:

- It requires an overhead of around 12 percent for training sequences and cyclical redundancy; however, equalizers in non-OFDM solutions also require training sequences, so depending on the size of the cyclical redundancy, this may not be an issue.
- Because it transforms intersymbol interference (ISI) into narrowband Rayleigh fading, it foregoes the opportunity to make use of the effective diversity in multiple paths: An equalizer will actually increase the performance by combining the multiple signals, whereas within OFDM these signals are nonresolvable and appear as Rayleigh fading.
- The peak/average ratio of OFDM is perhaps 3 to 5 dB higher than, say, QPSK, putting more stress on power amplifier design. This is an issue especially for subscriber units.[1]

As a result, it is clear that, compared to a single-carrier solution with an equalizer able to accommodate the channel ISI, OFDM will result in inferior performance. However, equally, in the case where the equalizer is unable to accommodate the channel ISI, the single-carrier solution will typically fail, whereas the OFDM solution will mostly continue to work. The key unknown is to what extent the channel will exhibit ISI beyond the range of a commercially viable equalizer or whether there will be narrowband interference. It is generally agreed that, to date, there is insufficient information about the channel to be able to definitively answer this problem. Given the lack of information, OFDM represents the "more conservative" solution, guaranteeing operation in most environments while not necessarily maximizing performance.

CDMA versus TDMA CDMA is generally agreed to be the most efficient multiple-access scheme for mobile applications. However, there are different constraints within the fixed access environment related to the desire of operators to be able to instantaneously give all, or a substantial part of, the available bandwidth to an individual subscriber. Because of the manner in which CDMA is configured, within a sector in a clustered environment, typically only around 200 kbps of throughput is available per megahertz of spectrum. This compares with a QPSK TDMA solution where up to around 1.8 Mbps might be available. Of course, CDMA allows single frequency reuse, so overall efficiency is high, but this example shows that in order to provide a subscriber with, say, 10 Mbps of data, CDMA would require either a carrier of 50-MHz bandwidth or a multicarrier receiver, while TDMA would require only a carrier of about 6 MHz.

Thus, it seems clear that for constant-bit-rate narrowband services, CDMA is the most spectrally efficient solution by some distance. For broadband applications above circa 2 Mbps per user, CDMA solutions will probably not be economically viable. Below 2 Mbps, subject to there being sufficient spectrum and the cost of the CDMA system being competitive, CDMA is probably the optimal multiple-access scheme.

Adaptive versus Conventional Antennas The conventional approach is to use sectored antennas at the base station, possibly with diversity, and a directional antenna at the subscriber unit, again possibly with diversity. Adaptive antennas bring potential gains as follows:

- At the base station they can result in a narrow beam to an individual subscriber, limiting interference to other sectors and thus increasing capacity.
- At the base station they can be used to null interferers, enhancing the *C/I* of the received signal.
- At the subscriber unit they can be used to null interference, again increasing the *C/I*.[1]

It is not clear how great these gains will be. Because of the directionality already present on fixed wireless data links, it is likely that the gains would be less than those for the mobile case. However, deployment is also simpler than the mobile case since there is little need to track subscribers. In the mobile case, adaptive antennas tend to enhance the uplink rather than the downlink; for fixed wireless data, the most constrained link is typically the downlink because of the asymmetry of usage, so different techniques will need to be used.

NOTE Adaptive antennas will be simpler for TDMA transmission where one array can be steered to each subscriber, rather than CDMA transmissions where multiple arrays would be required to steer the different codes constituting a single carrier to different subscribers.

It would appear that probably the most useful deployment will be of antennas that can illuminate a subscriber using a narrow beam. But, substantial work is required to understand whether the capacity gains this would bring would be offset by the additional cost of the solution.

Recently, there has been considerable interest in the idea of cross-layer design of wireless data networks. This is motivated by the need to provide a greater level of adaptivity to variations of wireless data channels. This next part of the chapter examines one aspect of the interaction between the physical and medium access control layers. In particular, the impact of signal processing techniques that enable multipacket reception on the throughput and design of random access protocols is considered.

Random Access Wireless Data Networks: Multipacket Reception

Traditionally, the Medium Access Control (MAC) layer is designed with minimum input from the Physical layer and by using simple collision models. Most conventional random access protocols assume that the channel is noiseless and the failure of reception is caused by collisions among users; packets transmitted at the same time are destroyed, and retransmissions must be made later. The basic approach to improving performance has been "resolving" collisions by limiting the transmissions of users. One way is to randomize retransmissions as in Aloha; another is to split successively the set of users until collisions are resolved.

The advent of sophisticated signal processing has changed many of the underlying assumptions made by conventional MAC techniques. In code-division multiple access (CDMA), for example, one of the basic premises of multiuser detection is that signals from different users should be estimated jointly, which makes it possible for the node to receive multiple packets simultaneously. The use of antenna arrays also makes it possible to have multipacket receptions.

What are the impacts of these advances at the Physical layer on the performance and design of MAC protocols? If there is a high probability that simultaneously transmitted packets can be received correctly, should the MAC encourage, rather than limit, transmissions of users? Let's first consider receiver multipacket reception (MPR) capability at the Physical layer. Possibilities of obtaining receiver MPR at the modulation level through space-time processing are discussed. Impacts of receiver MPR on the network throughput are considered. The design of MAC protocols that can take advantage of the MPR property of the network is also a topic.

MPR Nodes

Users in a wireless data network share a common medium, and their transmissions may interfere with one another. An objective of receiver design is to extract, in some optimal way, signals of interest from interference and noise. If a node of the network is capable of correctly receiving signals from multiple transmitters, the node is an MPR node.

Signal Processing for MPR A general model of a multiuser system that includes spatial, temporal, and code diversities is the multiple-input, multiple-output (MIMO) channel shown in Fig. 9-5.[2] Here $s_i(t)$ are transmitted signals from M users, and $x_i(t)$ are received signals from antenna array elements or virtual receivers of temporal processing. The channel impulse response $H(z)$ depends on the form of modulation, the transmission protocol, and the configuration of transceiver antenna arrays.

The basic signal separation problem is to design an estimator such that multiple sources are extracted in some optimal fashion. Although optimal estimators are nonlinear in general, to reduce implementation cost, one is often restricted to an MIMO linear filter with finite impulse response $F(z)$.The design of $F(z)$ depends on knowledge of the channel $H(z)$ and the format of transmission.

It is unrealistic to assume that the receiver knows the channel response $H(z)$ in a wireless data mobile network. It is then necessary to train the receiver by introducing pilot or training symbols in the data stream. With known training symbols from user i, a linear estimator design for that user can be based on, for example, the least squares criterion. The least squares optimization, when there is a sufficient amount of training, can be implemented adaptively, offering the ability to track one or a group of users. Furthermore, the receiver needs to know only the training symbols from node i in order to design the optimal receiver for that node, and there is no requirement of synchronization among users. The performance of the receiver, however, does depend on the presence of other nodes and the level of interference. As users

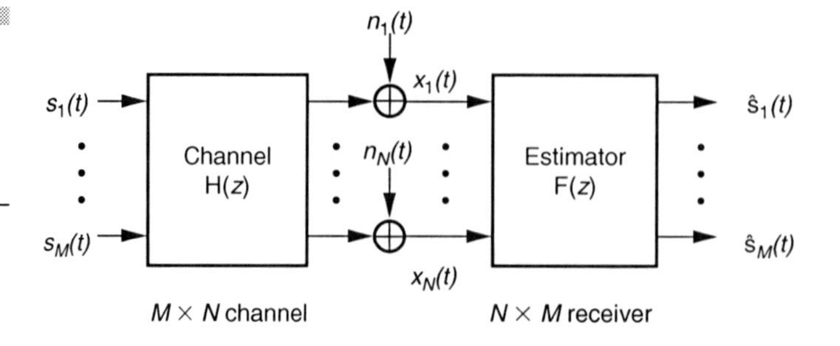

Figure 9-5 A general model for multiuser communications and receiver MPR.

drop in and out of the network and the channel varies with time, training needs to be done repeatedly.

The use of training significantly simplifies the problem of receiver design for MPR. However, it has several practical and theoretical drawbacks. For example, the training symbols and their locations in a data packet must be known to the receiver. This may be possible for networks with scheduled transmissions, but may not be practical for random access networks. The overhead associated with training may also be too excessive. It is therefore desirable to develop self-adaptive algorithms that are able to track users without relying on training.

There has been considerable research in blind and semiblind signal separation in recent years. Without a sufficient number of training symbols, the key to signal separation is to utilize the structure of the channel and characteristics of the input sources. For example, communication signals often have the constant modulus property, which enables the separation of multiple sources by minimizing the signal dispersion using the constant modulus algorithm (CMA). The finite alphabet property of communication signals may also be exploited for signal separation. The statistical dependency among sources is another condition that leads to a number of effective source-separation algorithms. In a transmitter-oriented CDMA system, code information can be exploited for signal separation. The diversity of the propagation channel from each transmitter to the receiver provides yet another possibility for packet separation. Simultaneously transmitted packets are separated according to the duration of their channel impulse responses.

Networks with MPR Nodes

To characterize the performance of a network with MPR nodes, MPR needs to be modeled at the node level. In between, the MPR channel matrix can take various forms as a function of the channel conditions and signal separation algorithms.

End-to-End Throughput For cellular systems, there is a one-to-one correspondence between receiver MPR at the base station and network MPR because all traffic goes through the base station, and all packets received by the base station are intended for it. This, however, is not true in general for ad hoc networks. Even if a node successfully receives multiple packets, some of these packets may not be intended for that node. To evaluate network throughput, one must convert the receiver MPR to the network MPR.

Unfortunately, the MPR model at the node level cannot accurately describe multihop ad hoc networks. Issues beyond MAC (routing) must be considered in the throughput evaluation. However, insights can be gained by examining networks with regular structures. An informative example is the rectangular grid, the so-called Manhattan network, shown in Fig. 9-6, where each node has four neighbors.[2] Although the receiver MPR at the node level is well defined, the network MPR cannot be defined by a single matrix.

MAC Protocols for MPR

MPR offers the potential of improving network performance. At the same time, it presents several new challenges. The outcome of a particular slot in the conventional collision channel can be a success, collision, or no transmission. In contrast, there is a higher level of uncertainty (hence a greater amount of information) associated with the outcome of a particular slot for networks with MPR. Specifically, the successful reception of a packet at an MPR node does not imply that only one neighbor transmitted. To improve the network throughput, you are no longer restricted to splitting users in order to resolve collisions. The two approaches that exploit the MPR capability of cellular networks are outlined next.

An Optimal MAC for MPR Channels The key to maximizing throughput is to grant an appropriate subset of users access to the MPR channel. For the conventional collision channel, this can be accomplished by splitting users in the event of collision. A more flexible approach is necessary for MPR channels because the protocol should allow the optimal

Figure 9-6
The Manhattan network.

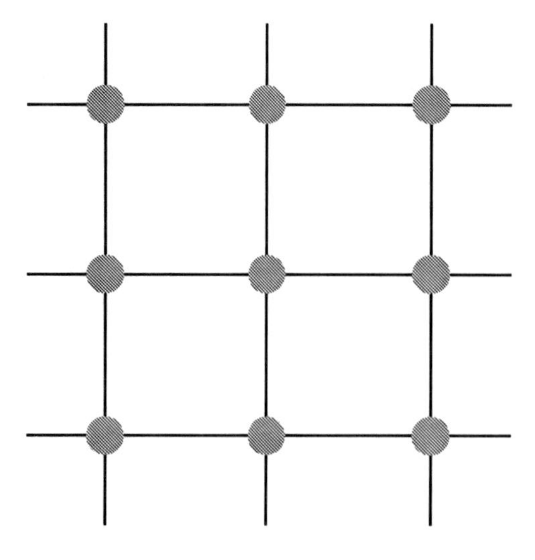

number of users to transmit. This implies that the set of users to access the channel should be enlarged if there were not enough users holding packets in the previous slot, and shrunk if too many users attempted to transmit. Ideally, N_o (the number that maximizes η_i) users should be allowed to transmit in order to achieve the maximum throughput. Unfortunately, this is not always possible because the number of users holding packets is a random variable not known to the receiver. One should extract information from the joint distribution of the states of all users.

The Multi-Queue Service Room (MQSR) protocol is designed explicitly for general MPR channels. The protocol is designed to accommodate groups of users with different delay requirements. Here, let's consider the case when there is only one group of users with the same delay requirement. As shown in Fig. 9-7, users are queued, waiting to enter a service room where transmissions are allowed.[2] The division of users into those inside and those outside the service room allows decomposition of the joint distribution of the user states so that this joint distribution can be updated effectively. To allow the flexibility to enlarge and shrink the set of users accessing the channel, the service room is divided into the access and waiting rooms. Only users in the access room are allowed to transmit. If there are too many users in the access room, the last users entering the access room are pushed back into the waiting room. If there are too few users in the access room, on the other hand, users in the waiting room, and users outside the service room if necessary, are allowed to enter the access room. The design of the optimal number of users entering the access room is based on the maximization of the network throughput for each slot.

Signal Processing versus MAC To gain insights into the roles of signal processing and MAC, as an illustration, let's compare first the performance of the optimal protocol with those of the URN and slotted Aloha protocols for a fully connected network with 10 users and a central controller. This example intentionally favors the two conventional protocols since the URN protocol assumes the knowledge of the number of nodes with packets, and the Aloha used in the comparison was implemented by using the optimal retransmission probability. Figure 9-8 shows that, for the conventional collision channel without MPR, three

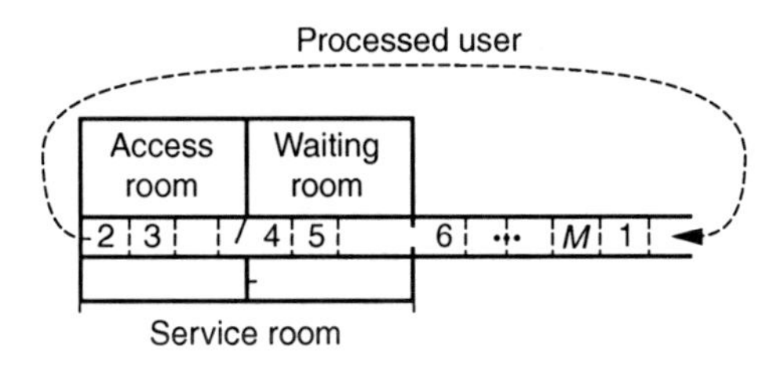

Figure 9-7 The basic structure of the service room protocol.

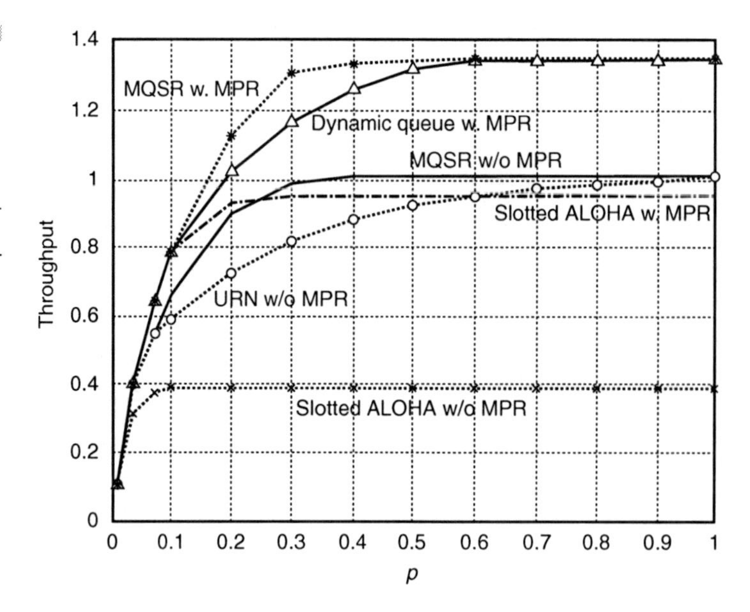

Figure 9-8 Throughput comparison. Ten users are present, each with probability p of generating a packet within one slot.

protocols behave similarly when the traffic is light.[2] As the traffic load increases, the throughput of the optimal protocol quickly reaches the maximum achievable throughput of 1, whereas the slotted Aloha remains at around 0.4. The URN protocol has the same performance in both light and heavy traffic, but lags in the midrange of the traffic load.

NOTE The gain of throughput from around 0.4 to 1 in Fig. 9-8 is due to the optimal MAC protocol without MPR.

If the receiver MPR is introduced using, in this example, the signal-processing-based collision resolution technique, another 30 percent gain can be achieved by the optimal protocol. This gain comes from the receiver MPR. The throughput of the Aloha protocol with MPR is twice that of the conventional collision channel.

Finally, the last part of the chapter presents the Terminal Independent Mobility for IP (TIMIP), which is a new architecture for IP mobility in the design of wireless data access networks. TIMIP is based on principles similar to those in the CIP and HAWAII architectures proposed at IETF and equally suited for micromobility scenarios. With TIMIP, terminals with legacy IP stacks have the same degree of mobility as terminals with mobility-aware IP stacks. Nevertheless, it still uses MIP for macromobility scenarios. In order to support seamless handoff, TIMIP uses context-transfer mechanisms compatible with those currently in discussion at the IETF SeaMoby group.

Mobility for IP

Increasing demand for user mobility throughout the global Internet has launched a successful wireless data LAN market and created the need for a new Internet architecture. While Layer 2 mobility is easy to accomplish and is already supported in most commercial WDLAN cards, it does not allow terminals to roam between different LANs and to cross between router domains. Layer 3 mobility allows Internet-wide mobility at the cost of more complex management. Several reference models for IP micromobility have already been proposed by the IETF, each with different advantages and disadvantages, the main proposals being MIP, HAWAII, and CIP.

NOTE These three proposals require the mobile terminal to be mobility-aware, which requires the replacement of the legacy IP protocol stacks (a hard task if you consider the variety of mobile terminal operating systems and versions).

This part of the chapter presents the specification of Terminal Independent Mobility for IP (TIMIP), which is a new proposal for IP mobility in wireless data access networks. Unlike the existent IETF proposals, TIMIP can be totally implemented in the network nodes and work transparently to the IP layer of the terminals. The proposed architecture is depicted in Fig. 9-9.[3]

A TIMIP domain is an IP subnet organized as a logical tree of access routers whose root is the access network gateway. The latter interfaces with the IP core network, which in turn connects to other access networks. The different elements of the wireless data access network have the following roles and capabilities:

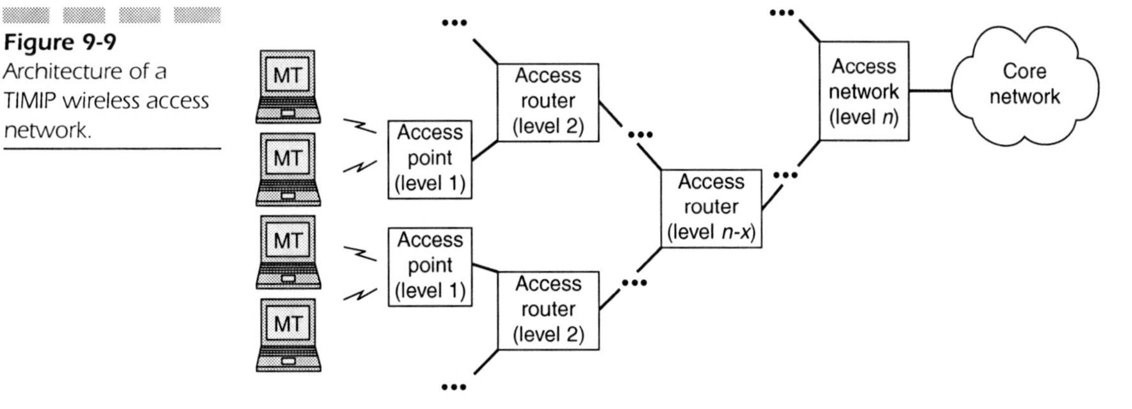

Figure 9-9 Architecture of a TIMIP wireless access network.

- Access router (AR)
- Access point (AP)
- Access network gateway (ANG)
- Mobile terminal (MT)

Access Router (AR)

The access network is formed by a number of routers organized in a logical tree topology. Each router incorporates mobility management functions.

Access Point (AP)

The AP is an AR that directly communicates with the mobile terminals at the radio interface. It is designed with the IP functionality of an AR because in this way IP mobility and QoS can be integrated at the radio interface. The AP sends/receives IP packets with application data to/from the mobile terminals. The AP is also responsible for detecting handoff and triggering mobility management procedures on behalf of the mobile terminal.

Access Network Gateway (ANG)

The ANG is the root AR of the wireless data access network, interfacing with the core IP network. The ANG also performs special mobility management functions related to the support of MIP-based macromobility.

Mobile Terminal (MT)

The MT runs the user applications. Roaming between different APs is performed by Layer 2 in a way that is transparent to the IP layer of the MT.

An overview of the IP Mobility reference models in discussion at the IETF is provided next. This is followed by the description of TIMIP.

IP Mobility in IETF

Of the IP mobility protocols already proposed at the IETF, MIP could be used in both micromobility and macromobility scenarios, though its use for micromobility presents some efficiency problems that can affect IP QoS. For this reason CIP and HAWAII were proposed as means to optimize micromobility, while they still rely on MIP to implement macromobility.

Mobile IP

The main framework for IP mobility in the IETF is Mobile IP (MIP), specified in RFC 2002. Its architecture and message flow are depicted in Fig. 9-10.[3] In the MIP model, a mobile terminal has two addresses: the home address (HAddr) and the care-of address (CoAddr). The HAddr is the address that the terminal retains independent of its location. This address belongs to the home network of the terminal, which is the IP subnetwork to which the terminal primarily belongs. The CoAddr is a temporary address assigned to the terminal within a foreign network.

When the mobile terminal is located within its home network, it receives data addressed to the HAddr through the home agent (HA). When the mobile terminal moves to a foreign network, it obtains a CoAddr broadcast by the foreign agent (FA) in router advertisement messages as defined in RFC 1256. This CoAddr is then registered with the HA with a registration request message. Whenever a packet arrives at the HA addressed to the HAddr of the mobile terminal, the HA checks if the mobile terminal is currently located on a foreign network. In this case, the HA tunnels the packet within an IP packet addressed to the FA. When the FA receives the packet, it de-encapsulates it and forwards it to the mobile terminal. Packets sent by the mobile terminal are routed normally, even if the terminal is located in a foreign network.

As MIP relies on normal routing, it presents several problems, namely the need for triangulation through the HA when the terminal is located on a foreign network. Triangulation and IP tunneling are difficult to integrate with RSVP. Besides, triangulation may cause a significant increase

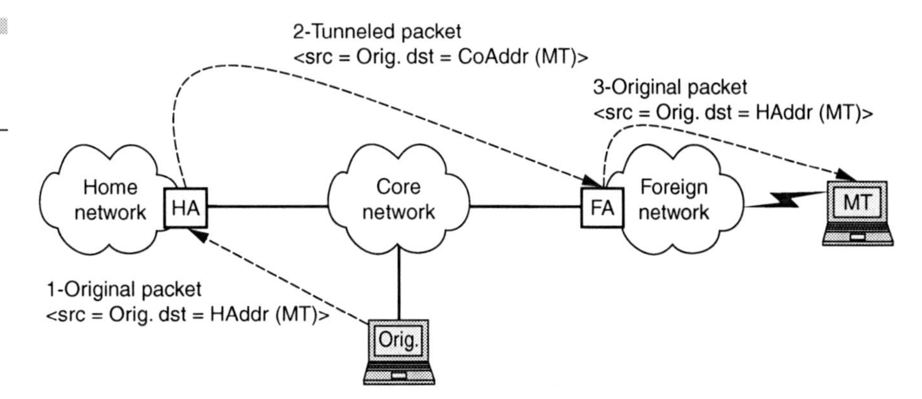

Figure 9-10 MIP architecture and message flow.

in end-to-end transmission delay, being especially inefficient when the mobile terminal is receiving data originated from the foreign network where it is currently located. This model can be optimized if the originator of the packets is a MIP terminal. In this case, the HA sends the originator a binding update, containing the CoAddr of the destination. Further packets are sent directly to the CoAddr instead of the HAddr.

HAWAII

The Handoff-Aware Wireless Access Internet Infrastructure (HAWAII) was proposed in order to solve the QoS and efficiency issues of MIP. In this model, the terminals implement MIP as before, while special forwarding entries are installed on specific routers, making them aware of the location of specific terminals. As such, routing outside a domain is performed as in MIP (per IP subnet); within a domain routing is performed per terminal by using direct routes (the terminal keeps its HAddr as before without any triangulation or IP tunneling). The HAWAII network architecture is depicted in Fig. 9-11.[3]

In HAWAII each domain is structured according to a hierarchy of nodes, forming a logical tree. Each domain owns a root gateway called the domain root router, which takes the role of HA. Each terminal has an IP address and a home domain. Whenever the terminal moves within its domain, its IP address is retained. Packets destined to the mobile

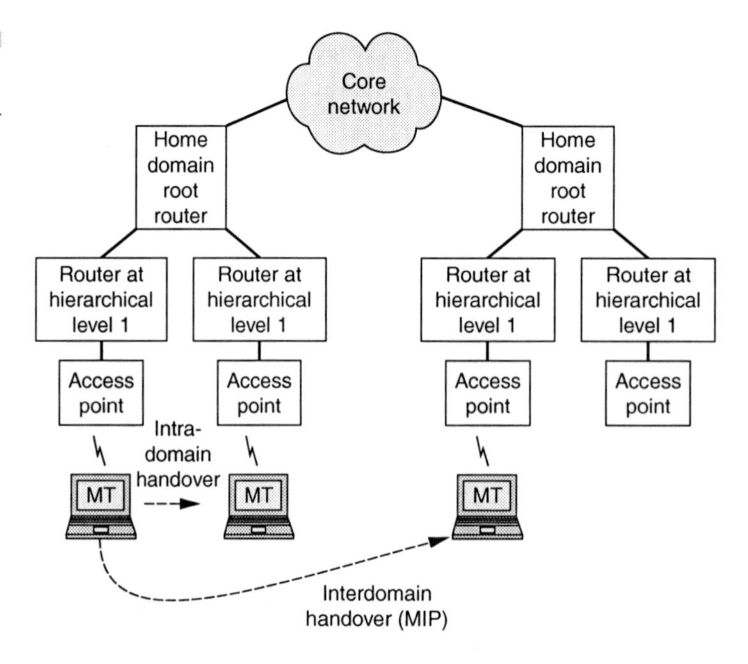

Figure 9-11
HAWAII architecture.

terminal are routed to the home domain root router in the normal way according to the IP subnet address of the domain. The received packets are then forwarded to the terminal by using special dynamically established paths. The establishment of these paths is triggered by the mobile terminal by means of the usual MIP registration messages whenever it moves between two APs, as each AP behaves as a different FA. Within the home domain, these messages create direct routing entries in the intermediate nodes they cross. When the terminal moves to a foreign domain, the usual MIP procedure is used where the foreign domain root router is now the FA, responsible for assigning a CoAddr and forwarding the packets to/from the mobile terminal.

Cellular IP

In both MIP and HAWAII, Layer 3 handover procedures are triggered by MIP signaling such as RFC 1256 when the terminal is already using the new access point. In this way the latency of Layer 3 handover may be high, originating significant packet losses. Cellular IP (CIP) makes use of Layer 2 information regarding access point signal strength in order to predict handover, allowing the terminal to trigger Layer 3 procedures earlier. Unlike HAWAII, in which the terminals run MIP, in CIP they must implement specific CIP procedures. The architecture of CIP is depicted in Fig. 9-12.[3]

Figure 9-12
CIP architecture.

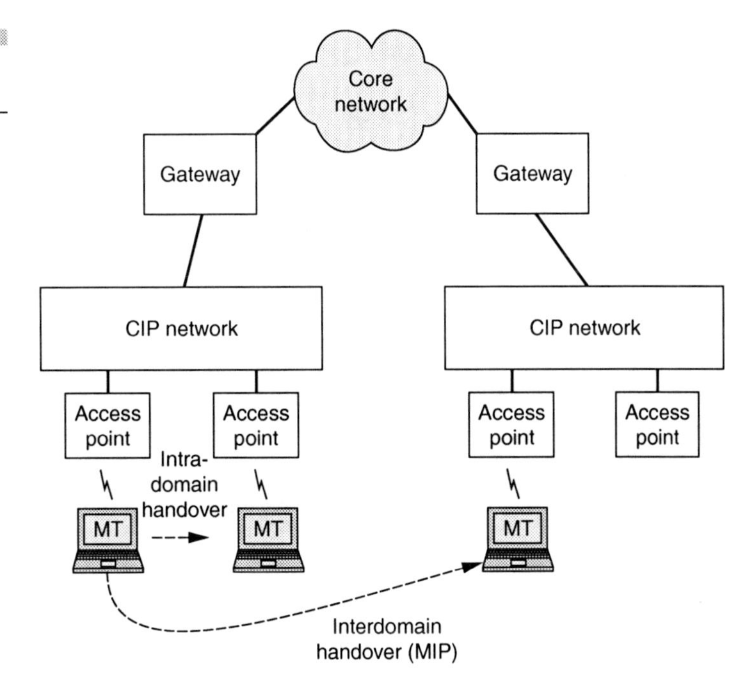

Each CIP domain is composed of a number of CIP nodes structured in a tree topology, having a MIP gateway as the root node. CIP nodes can route IP packets inside the CIP network and communicate with mobile terminals through the wireless data interface.

The CIP nodes maintain routing and paging caches. The routing caches are used to locate roaming mobile terminals, being updated by the IP packets transmitted by the mobile terminal. Throughout the CIP nodes, a chain of temporary cached records is created to provide information on a downlink path of packets destined to the terminal. After a successful roaming procedure, a CIP node can temporarily have several mappings for the same mobile terminal, leading to different interfaces. Whenever a packet arrives at the CIP node destined to the mobile terminal, that packet is sent to all interfaces mapped on the routing cache. Cached mappings must be refreshed periodically by the terminal; otherwise they expire and are deleted.

The paging caches are maintained by paging-update packets sent to the nearest access point each time the mobile terminal moves. These records are created by mobile terminals that do not send or receive packets frequently.

Within the CIP domain, when the terminal approaches a new access point, it redirects its outgoing packets from the old access point to the new access point, updating the routing caches all the way up to the gateway. All packets destined to the mobile terminal are forwarded to both access points during a time interval equal to the routing cache timeout. After the old path expires, the packets destined to the mobile terminal are forwarded only to the new path. Because of this, when the terminal has no packets to send during handover, it has to generate route-update messages in order to allow correct updating of the routing caches. Between CIP domains, normal MIP procedures are used for macromobility.

NOTE In CIP, all packets generated within the CIP domain must be routed by the gateway, even if the destination is located in a position adjacent to the source.

Terminal Independent Mobility for IP (TIMIP)

All IETF proposals for IP mobility require the mobile terminals to use a mobility-aware protocol stack, as it is the mobile terminal that notifies the access network about handoff by means of special IP layer signaling. This prevents terminals with legacy IP protocol stacks from taking advantage of mobility even when they are attached to a mobile access network. Replacing the protocol stacks of all legacy terminals can be a hard task if you consider the variety of mobile terminal operating systems and versions. Coupling the IP layer with Layer 2 handoff mechanisms at the APs by means of a suitable interface avoids the need for special IP layer signaling between the terminal and the AP. Such is the approach followed by Terminal Independent Mobility for IP (TIMIP).

In order for a terminal to be recognized by the TIMIP network, it has to be registered. This is accomplished off line through management procedures. The ANG keeps information on all mobile terminals recognized by the mobile network. For each terminal, this information consists of the following:

- MAC address
- IP address
- MIP capability
- IP address of the MIP home agent
- Authentication key
- Authentication option[3]

The MIP capability parameter specifies that if MIP is required, it's either implemented at the ANG on behalf of the legacy terminal (surrogate MIP) or implemented at the terminal itself. If the terminal has a legacy IP protocol stack, the next two parameters specify, respectively, the IP address of its home agent and the authentication key to be used between the terminal and the ANG when the authentication option is turned on.

NOTE TIMIP authentication is mandatory for macromobility scenarios for both MIP and legacy terminals.

The IP address of the home agent is not used when the terminal implements MIP, as the terminal itself is responsible for registering with the home agent, bypassing the ANG. Once this group of data is configured at the ANG, it is forwarded to the APs (except the authentication key) so that they are able to know the IP address of newly associated terminals based on their MAC address provided by Layer 2, as explained next.

Power-up

When an MT first appears in a TIMIP domain, a routing path is created along the hierarchy of ARs, as shown in Fig. 9-13.[3] The creation of the routing path takes the following steps:

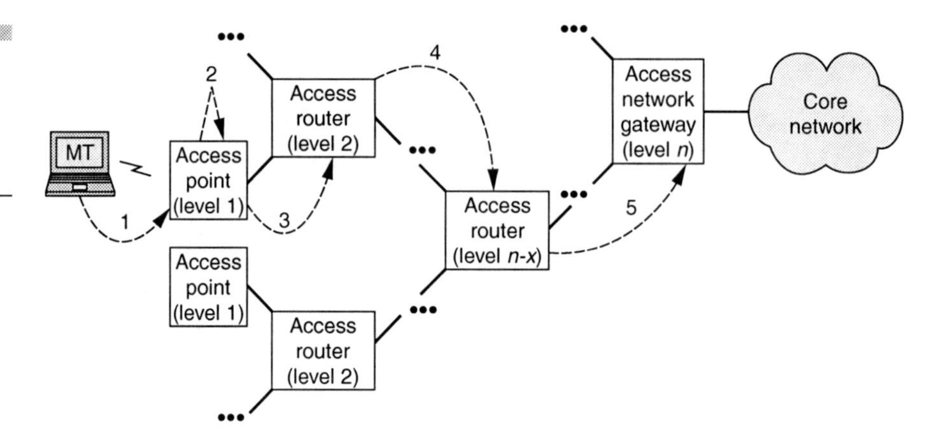

Figure 9-13 Establishment of routing path after power-up in a TIMIP domain.

1. The MT performs a Layer 2 association with an AP that belongs to the local TIMIP domain.

2. At the AP, Layer 2 notifies the IP layer about the presence of the MT in its wireless data interface, triggering the routing reconfiguration procedure. Layer 2 sends the MAC address of the terminal to the IP layer. The MAC address is matched against the terminal registration information broadcast by the ANG and the respective IP address is found. As the new AP currently has no routing table entry for the MT, the routing table is updated with the addition of this new entry.

3. The new AP sends a RoutingUpdate message up to the AR at hierarchical level 2. This AR acknowledges with a RoutingUpdateAck message, and updates its routing table accordingly with the addition of a new entry relative to the MT. This entry points to the source of the RoutingUpdate message (in this case the AP) in order to specify the path through which the terminal can be reached.

4. Exchange of RoutingUpdate/RoutingUpdateAck messages climbs up the hierarchy levels. At each level the routing table is updated with the creation of a new entry relative to the MT. This entry always points to the source of the RoutingUpdate message in order to specify the path through which the MT can be reached.

5. Exchange of RoutingUpdate/RoutingUpdateAck messages reaches the ANG, completing the creation of the new routing path.3

The MT is now reachable through the routing path established by the preceding procedures. The ARs that do not belong to this path have no routing entry for the MT. At these ARs, all packets destined to the MT are forwarded up the hierarchy of routers by default. All packets that arrive at an AR whose routing table has an entry to the destination are forwarded down the hierarchy of routers until they reach the radio interface in which the MT is located. Packets destined for a terminal located in the same TIMIP domain as the source reach the ANG only in the worst case.

The RoutingUpdate and RoutingUpdateAck messages include a timestamp generated at the new AP. As in TIMIP, all APs are synchronized by means of the Network Time Protocol (NTP); this guarantees consistency even when the MT moves faster than the route reconfiguration.

NOTE The routing path is soft-state, and after its establishment it is refreshed by the data packets sent by the MT. Nevertheless, as the packets are routed within the TIMIP domain, some of the ARs may not be refreshed. When this occurs, the routing entry for the MT becomes invalid after a predefined timeout (10 s). The AR where the timer expired starts to send ICMP EchoRequest messages to the terminal, filling the source address field of the IP header with the IP address of the ANG. This forces the MT to reply with EchoReply messages destined to the ANG, which will refresh the routing path within the TIMIP domain. If the MT does not reply within a predefined timeout (60 s), the routing entry for the MT is removed.

This basic TIMIP configuration is adequate to have micromobility in wireless data access networks where security is not an issue. Nevertheless, as in other unprotected IP networks, it allows MTs to power-on with false MAC and IP addresses. In order to avoid this, a minimal security functionality must be implemented at the MT itself. However, this can be done in the Application layer with no need to change the IP protocol stack. When the authentication option is turned on, it is assumed that the MT runs a special security application, which uses a database of authentication keys for the different TIMIP domains in which the MT is allowed to power up. This database is indexed by the IP addresses of the ANGs that are the root of the respective networks. The authentication takes place in step 2 of the power-on procedure, immediately after Layer 2 notifies the IP layer of the AP about the association of the MT. The AP sends a SignatureRequest message to a well-known UDP port in the MT. This message carries <IP address of the MT, IP address of the ANG, rand, timestamp>, where rand is a random value and the timestamp is an NTP-formatted 64-bit value. The same message is sent to the ANG. Both the MT and the ANG answer the AP with a SignatureReply message containing the same fields present in the SignatureRequest message, plus its 128-bit MD5 message digest calculated with the authentication key of the MT for this network. The latter is known only by the MT (based on the authentication key database and the IP address of the ANG) and the ANG (based on the registration information). The AP compares the signatures of the two SignatureReply messages, and proceeds with the routing reconfiguration procedures in case there is a match.

Micromobility

Handoff between two APs that belong to the same TIMIP domain is depicted in Fig. 9-14.[3] The first four steps of the handoff procedure are the same as those of the power-up procedure. The remaining steps are as follows:

5. Exchange of RoutingUpdate/RoutingUpdateAck messages climbs up the hierarchy levels, until the crossover AR (the AR that belongs simultaneously to the old path and to the new path) is reached. Now that the new routing path is completely created, the old path must be deleted. This procedure starts when the crossover AR sends a RoutingUpdate message addressed to the MT through the old routing path. The AR that receives the message realizes that the MT is no longer accessible through it, updates its routing path by deleting the entry that corresponds to the MT, and replies with a RoutingUpdateAck message.

6. Exchange of RoutingUpdate/RoutingUpdateAck messages goes down the AR tree following the old path, until the old AP is reached. At each level, the routing table is updated by deleting the entry relative to the MT.[3]

A problem might arise because a TIMIP domain consists of a single IP subnet. In a normal shared-media LAN, when a terminal has a packet destined to an address within the same IP subnet (which is known through the analysis of the IP address prefix), it tries to obtain the MAC address of the destination through an ARP request before sending the packet directly to it. In TIMIP, as the APs have the functionality of routers, if the destination is associated with a different AP (and hence a separate wireless data interface, though in the same IP subnet), the ARP request will not reach its destination. In order to prevent this situation, the MT must be forced to address all MAC frames to the local AP, which in turn will route them properly to their destination. A simple implementation is to have the APs answer to the ARP requests on behalf of the target MTs with their own MAC address. Nevertheless, this is a complex

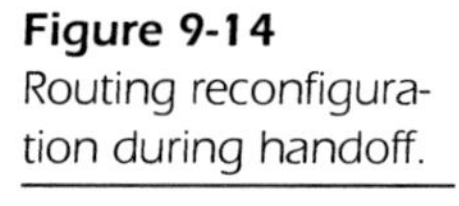

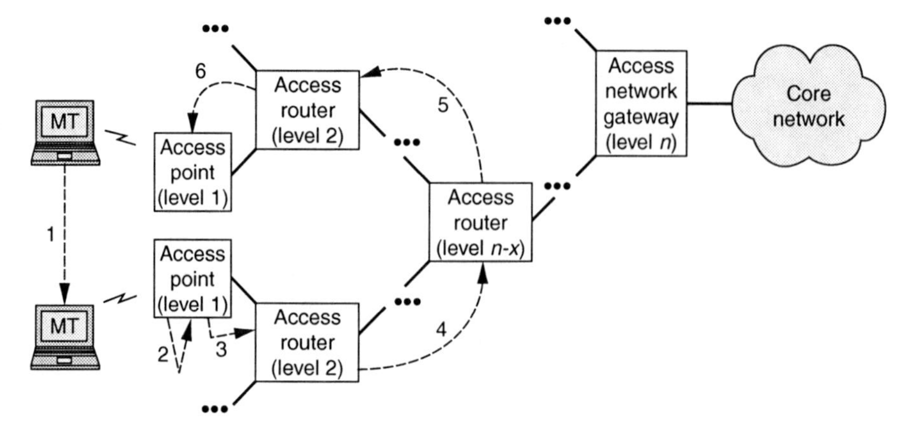

Figure 9-14 Routing reconfiguration during handoff.

task and can lead to an increase of the traffic broadcast within the radio interfaces due to the ARP messages. It is preferable to configure the MTs with a special subnet mask of 255.255.255.255 and the ANG as the default router to force the MT to send all IP data to the ANG, and to have the APs performing proxy ARP of the ANG with their own MAC address.

Macromobility

Finally, like HAWAII and CIP, TIMIP relies on MIP to support macromobility. The ANG implements the home agent (HA) for MTs whose home network is its TIMIP domain. The ANG implements surrogate MIP on behalf of foreign legacy terminals and the role of the foreign agent (FA) for foreign terminals that support MIP.

Conclusion

This chapter demonstrates that fixed wireless data technology has a significant role to play in the future of broadband communications, being used in areas where the copper or cable[4] infrastructure is not appropriate or by new operators who do not have access to these legacy resources. It also demonstrates that operators can economically and technically offer broadband services to users of 10 Mbps or more provided that they have a spectrum allocation of 100 MHz or more. Finally, it demonstrates that there is a plethora of technical options that can be used to provide fixed wireless data solutions, including a number of promising new ideas that could overcome problems of poor coverage at higher frequencies.

Furthermore, this chapter has also considered potential impacts of receiver MPR at the Physical layer on the performance and design of MAC protocols. Cross-layer design is a methodology that requires further investigation, and issues involved are broad and deep. Is it simple enough to implement? Does it scale? Is it robust? A critical element in cross-layer design is choosing an appropriate set of parameters that serve as agents carrying information between layers, parameters that are simple, but not too simple, so that the network can be designed to be adaptive to channel variations, but at the same time robust to modeling errors.

Finally, this chapter has presented a new proposal for mobility in IP networks called Terminal Independent Mobility for IP (TIMIP). In TIMIP, power-on and handover are inferred from Layer 2 notification at the wireless data access points. Consequently, IP mobility signaling is completely implemented in the network nodes and thus transparent to the IP layer of the terminals. Although authentication still requires some functionality to be performed at the terminals, it can be implemented as an independent application with no impact on the IP protocol stack. This contrasts with the CIP and HAWAII solutions proposed to the IETF that require the IP protocol stack of the mobile terminals to be changed to support special mobility signaling, which can be a hard task if you consider the variety of mobile terminals, operating systems, and versions available.

TIMIP combines some advantages from CIP and HAWAII. Like CIP, refreshing of routing paths is performed by data packets sent by the mobile terminals, with signaling being employed only when no traffic is detected at the routers for a certain time interval. Like HAWAII, routing reconfiguration during handoff within a TIMIP domain only needs to change the routing tables of the access routers located in the shortest path between the new AP and the old AP. Another feature similar to HAWAII is that routing of data packets within a TIMIP domain does not need to reach the access network gateway, involving only the access routers located in the shortest path between the sender and the receiver.

Finally, preliminary tests in a simple configuration with two APs and one ANG have shown that handoff latency due to TIMIP is not higher than 4 ms, which is satisfactory given the fact that the APs and the ANG used in the tests were based on PCs running LINUX with a TIMIP user-space implementation. Test scenarios with more network nodes will be performed in the near future.

References

1. William Webb, "Broadband Fixed Wireless Access as a Key Component of the Future Integrated Communications Environment," *IEEE Communications Magazine,* 445 Hoes Lane, Piscataway, NJ 08855, 2002.
2. Lang Tong, Qing Zhao, and Gokhan Mergen, "Multipacket Reception in Random Access Wireless Networks: From Signal Processing to Optimal Medium Access Control," *IEEE Communications Magazine,* 445 Hoes Lane, Piscataway, NJ 08855, 2001.
3. Antonio Grilo, Pedro Estrela, and Mario Nunes, "Terminal Independent Mobility for IP (TIMIP)," *IEEE Communications Magazine,* 445 Hoes Lane, Piscataway, NJ 08855, 2002.
4. John R. Vacca, *The Cabling Handbook,* 2d ed., Prentice Hall, 2001.
5. John R. Vacca, *Wireless Broadband Networks Handbook,* McGraw-Hill, 2001.
6. John R. Vacca, *Satellite Encryption,* Academic Press, 1999.
7. John R. Vacca, *High-Speed Cisco Networks: Planning, Design, and Implementation,* CRC Press, 2002.
8. John R. Vacca, *i-mode Crash Course,* McGraw-Hill, 2001.

CHAPTER 10

Designing Millimeter-Wave Devices

Gigabit data transport and processing technologies are required to respond to present and future information distribution and high-speed Internet application needs. Moreover, for broadband[2] communication channels to reach individual users in any geographical environment demands integration of network segments in such media as metallic wire, fiber, radio frequency (RF), and free-space optical wireless (FSOW). Fiber-optic technology has already matured to terabit-per-second data transport. However, for places lacking fiber infrastructure, wireless data technologies (RF and/or FSOW) are emerging as the transport media of choice in response to the daily increased demand for broadband networking. On the other hand, the maximum communication channel speed/data rates, link availability, and performance are limited in the wireless data domain (microwave and millimeter wave in particular) by wireless data range, propagation effects, atmospheric turbulence, and environmental factors. Typical bit rates for an RF wireless data system are in the lower megabits-per-second range for mobile,[4] and a few hundred megabits per second for fixed wireless data links. In addition, even at these low data rates, the link error performance and service quality are many orders of magnitude below those of fiber-optic transmission systems.

In response to these needs, this chapter proposes and demonstrates several new broadband network architecture and interface technology solutions based on the combined and complementary aspects of RF/microwave/millimeter wave, as well as FSOW links for integrated network operation. The combined scheme and architecture have extended the fiber-optic reach and bandwidth utilization closer to the end user and, more important, into the wireless data domain.

In this chapter, millimeter-wave devices' design and implementation scenarios for a gigabit-capacity and high-data-rate fixed wireless data access technology demonstrator are discussed. The system is based on a broadband wireless data access concept and implementation techniques utilizing millimeter-wave and newly introduced free-space optical wireless data high-speed links. The demonstration platform is to provide broadband "last mile" access and networking solutions to Internet users in densely populated areas with homes and businesses (building-centric and inner city environments) in need of high bandwidth not served by fiber infrastructure. The investigation focuses on the radio link design, network architecture, system integration, and compatible interface to the existing ATM fiber and satellite[3] core networks in support of the next-generation Internet (NGI) reach network extension by the wireless data technology. (The Glossary defines many technical terms, abbreviations, and acronyms used in the book.)

System Description

The fixed broadband wireless access (BWA) system trial is a short-range cellular-based point-to-point and/or point-to-multipoint distribution system that resembles the traditional local multipoint distribution services (LMDS) network architecture to enable gigabit capacity and high-speed data link capabilities. The system utilizes microwave, millimeter-wave, and FSOW technologies for access and distribution. Special emphasis is given to wideband wireless data local loop (W-WLL) applications, versatile service access, rapid system deployment, and dynamic network reconfiguration. A segmented functional subnetwork topology, shown in Fig. 10-1, is adapted as described next.[1]

Short-Range Micro/Picocell Architecture

In contrast to the conventional LMDS standard-size cell 2 to 5 miles in diameter, micro/picocells of less than 500 m radius were selected for high-density populated regions. Figure 10-2 illustrates cell options and scenarios for customers concentrated in small urban areas such as inner city environments, college campuses, business parks, multistory/high-rise buildings, or planned housing complexes and development in small communities.[1] An access point (AP) and a hub are established utilizing a remote antenna. Direct line-of-sight "illumination," say, the multistory building faces and windows, is achieved by either rooftop or sidewall-mounted shower-type antennas. The campus and small community access can be provided by projecting the signal from antennas mounted on street lampposts or neighboring buildings, as shown in Fig. 10-2.

Hybrid Fiber-Radio Backbone Interconnection

Hybrid fiber radio (HFR), RF photonics, and radio on fiber technologies are adapted to interconnect the APs to the backbone fiber network. The links are capable of transporting both high-speed digital and analog signals as well as multiple wireless data services based on subcarrier modulation (SCM) and wavelength-division multiplexing (WDM) technologies.

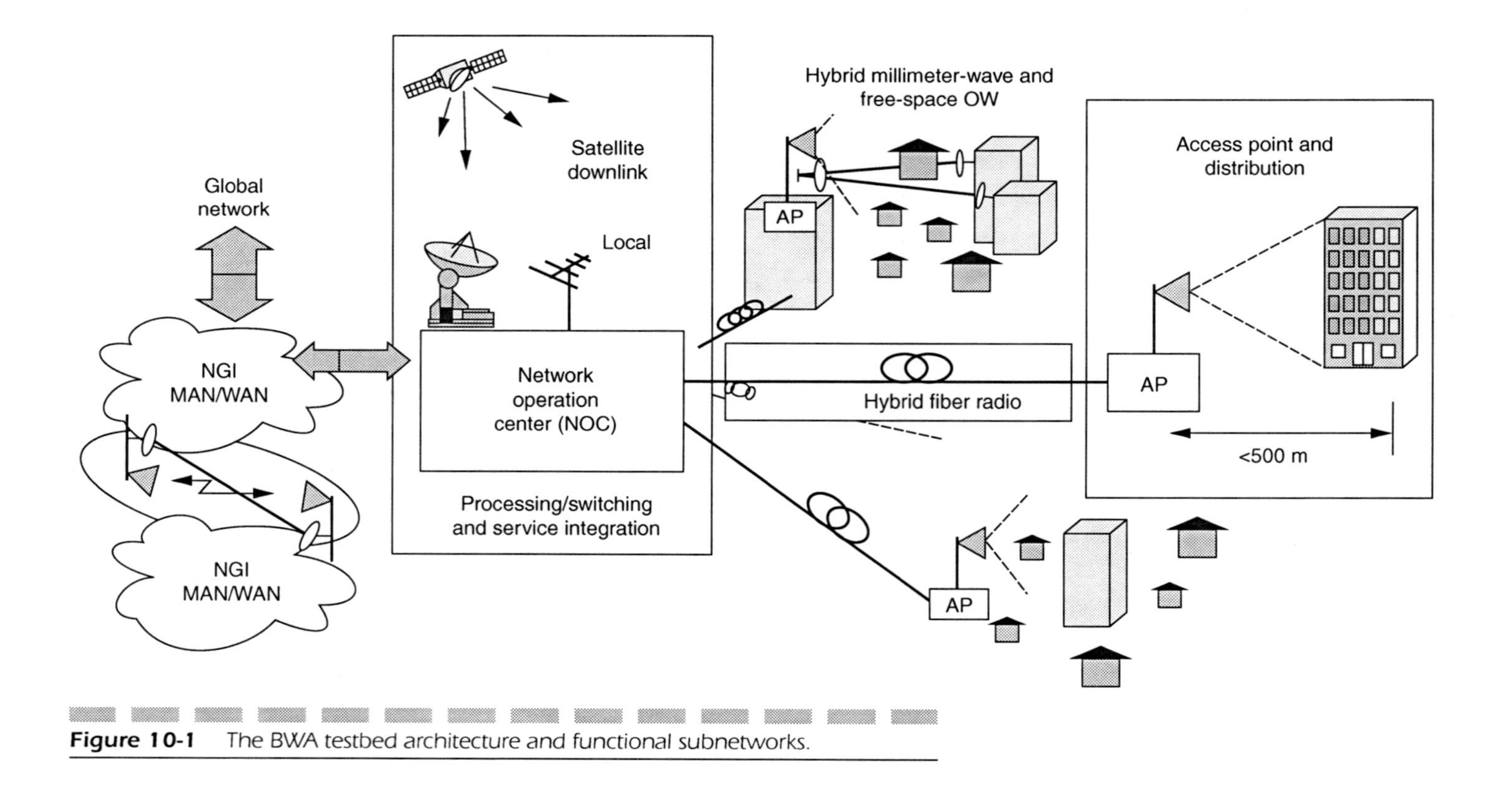

Figure 10-1 The BWA testbed architecture and functional subnetworks.

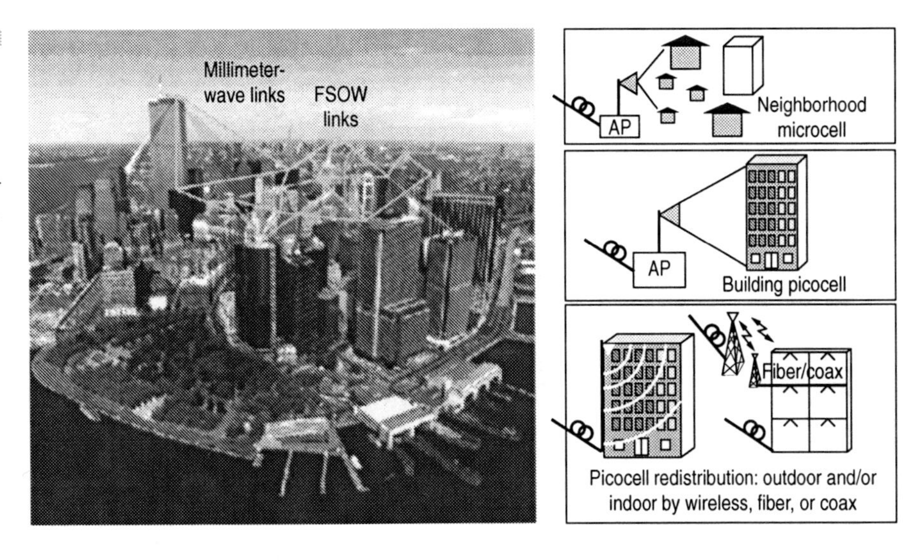

Figure 10-2 Integrated hybrid millimeter-wave, fiber, and optical wireless data access and distribution system scenarios. Implementation options for integrated HFR for picocell access and distribution systems for inner city environments and interconnection options. (*Note:* The World Trade Center towers in New York City are shown in this figure to remember those who died in the terrorist attack of September 11, 2001.)

The possibility of using the existing embedded fibers to the curb and neighborhood as well as FSOW tandem links permits broadband backbone network integration and combined services through a single shared infrastructure, leading to faster deployment and lower system cost for service providers.

Network Operation Center

A consolidated network operation center (NOC) for end-to-end network management and control is implemented to relocate the conventional base station control and switching facilities into the NOC to perform the required switching, routing, and service-mixing-function operations. The integration and merging of multiband HFR, FSOW, and digital fiber-optic technologies at the NOC with fixed BWA has provided flexible and unified network operation as well as the possibility of end-to-end network management and control. The consolidation will benefit through lower infrastructure complexity and cost, resulting in a more reliable and centralized database and operations.

Portable Broadband Wireless Data Bridge and Access Node

This chapter will now discuss the concept and realization of a portable wireless data access node for a bidirectional ATM-based connection to reach a fixed broadband fiber network. The goal of this effort is to demonstrate the feasibility of a rapidly deployed access node and backbone interconnection to the NOC for application in specialized scenarios, such as military theaters, emergency response, and disaster relief operations. Two portable nodes could also serve as a point-to-point wireless bridge to connect two or more isolated networks in places not served by fibers, as depicted in the lower left corner of Fig. 10-1.

Free-Space Optical Wireless Data Access and High-Speed Backbone Reach Extension

This is an emerging advanced technology providing many new approaches and platforms for high-bandwidth wireless data access and distribution networks. The technology, in combination with the millimeter-wave network topology, has created potential for increased capacity and extended the fiber-based bandwidth and services to users via wireless data. In the demonstrator, an FSOW point-to-point link is employed to complement and extend the NGI wireless data access capabilities for true gigabit-per-second data transport. The combined and side-by-side millimeter-wave/FSOW hybrid network topology shown in Fig. 10-1 provides direct performance comparison with the millimeter-wave links in various environmental conditions (multipath, rain fade) required for the design and implementation of high-reliability networks. Moreover, this topology ensures a higher degree of link availability when the millimeter wave fails during the rain or the FSOW power budget falls below the specified threshold during foggy weather. It has been shown that the hybrid technology can increase the current millimeter-wave network capacity and high-speed data transport capabilities.

A Measurement-Based Channel Model

To investigate millimeter-wave propagation issues, a high-resolution channel sounder at the 38-GHz LMDS band to model the channel on the basis of the measurements and simulation results is used. The model addresses the performance limits for broadband point-to-multipoint wireless data access in terms of data transport capability under realistic commercial deployment conditions. The model is used to examine a broadband channel-adaptive radio modem for dynamic selection of channel quality, channel switching, and bandwidth allocations. Propagation characterization, modeling, and simulation were performed for a short-range BWA system to provide sight selection design rules and solutions for adaptive channel configuration and operation mechanisms. A set of comprehensive data processing tools has been developed that, in combination with the channel sounder, can be used to develop statistical models for the broadband millimeter-wave channels.

System Architecture Advantages

Compared to the traditional LMDS system, the system technology and heterogeneous network topology previously described possess many technological and operational advantages:

- Increased coverage and user penetration percentage in each individual cell due to densely positioned users in the service area. This relaxes the tedious effort of cell frequency and polarization reuse planning.
- This in turn leads to a simpler design of overlapping cells for higher coverage and permits more efficient utilization of the spectrum.
- The required AP hub and customer transmitting power (at millimeter wave) are immediately scaled down (15 dB minimum) because of the relatively short cell radius. The result is a low-power, low-cost system solution and less complex MMIC hardware design.
- A major reduction in system interference (adjacent channel and adjacent cell) comes from constraints and limitations imposed by the power amplifiers' nonlinearities in high-power systems, due to spectral regrowth.
- As a result, possible reduction in the required radio channel spacing can be achieved, leading to increased system capacity due to higher spectrum utilization and efficiency.
- The near-short-range directly projected line-of-sight (LOS) propagation path becomes free from "major" multipath interference, intercell interference, and obstructions (buildings, moving objects, trees, and foliage). Consequently, the propagation path loss approaches that of square law, leading to a power-efficient system.

- An additional improvement in the system gain margin (7 to 10 dB) and link availability comes from the short LOS distance that removes the signal reception limitation due to excessive rain attenuation and system downtime experienced in higher-power, longer-range LMDS systems.
- The utilization of a hybrid millimeter-wave/FSOW network topology extends the broadband network reach without utilizing the radio spectrum. It can also provide high-capacity links, increased frequency reuse of millimeter waves, and greatly enhanced network reliability and availability.[1]

Implementation and Test Results

Now, let's look at the implementation of experimental BWA links and an asynchronous transfer mode (ATM)–based networked testbed infrastructure for experimentation toward high-speed Internet applications and W-WLL performance evaluation. The testbed comprises a single AP and three user nodes (two fixed and one portable), as shown in Figs. 10-3 and 10-4, operating in the 5.8/28/38-GHz bands.[1] A side-by-side high-speed point-to-point FSOW link (see Fig. 10-1), in parallel or tandem, was also implemented to extend the backbone fiber bandwidth to the AP operating up to 622-Mbps rates. On all the links, network demonstrations have been carried out for mixed services: broadcast 80-channel

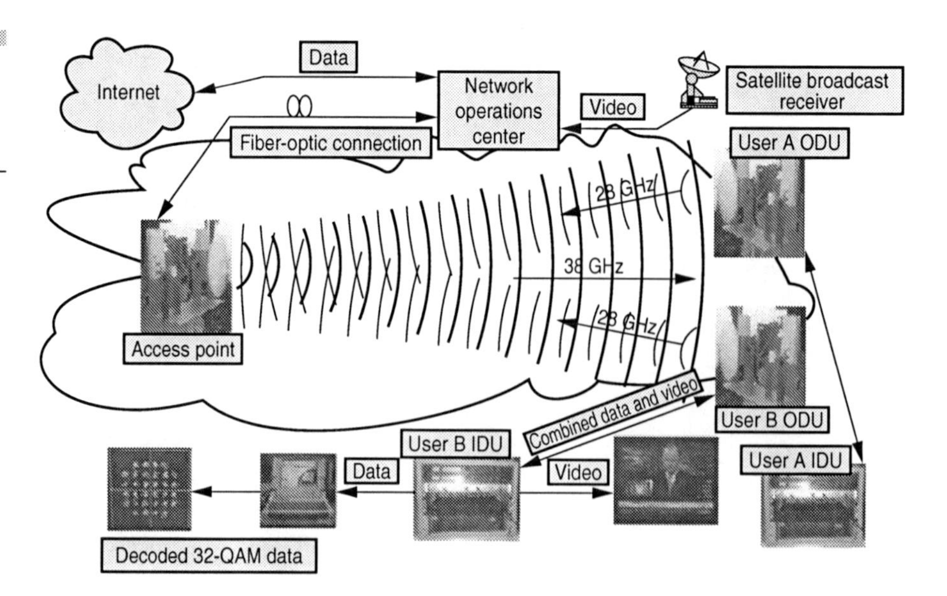

Figure 10-3 Multiband multiuser BWA testbed configurations.

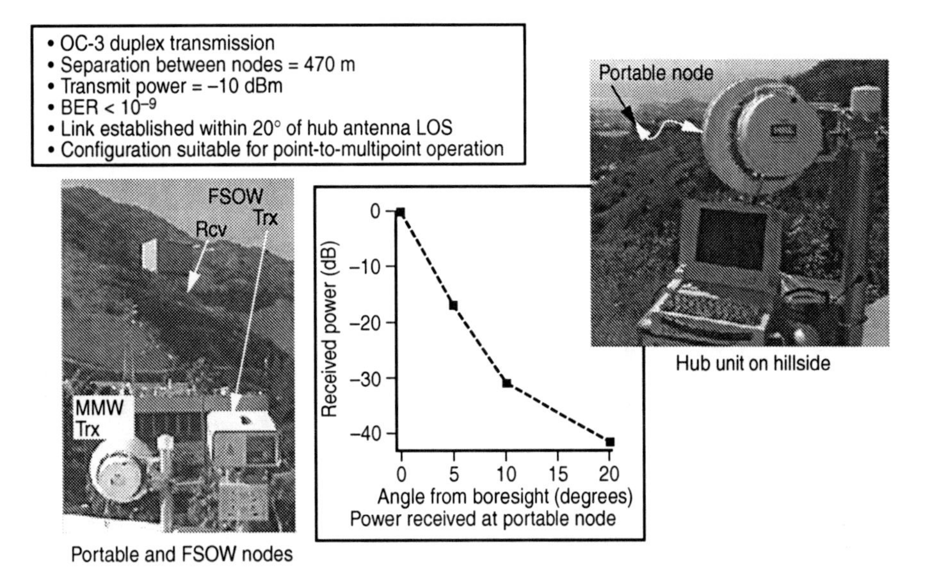

Figure 10-4 Portable node experimentation and measured BER.

video and RF wireless data channels with speeds at 1.5-, 25-, 45-, and 155 (OC-3)–Mbps rates in 4-, 16-, 32-, or 64-quadrature amplitude modulation (QAM) formats. The key issue in the topology described here is that the AP transmitter has the low power practical for mass deployments.

The implemented portable node of Fig. 10-4 is equipped with an OC-3 connection that occupies 50 MHz of bandwidth for 16 QAM. The performance of the OC-3 portable node was also field-tested using a data stream supplied by either a bit error test set or an Internet advisor ATM analyzer. Error-free operation was achieved in a 20° sector of a 470-m microcell environment.

Figure 10-5 depicts the functional elements and interconnection in the ATM-based BWA and distribution network in the NOC.[1] The ATM switch is programmed to combine and distribute traffic, integrate mixed services, and create dynamic user interconnection paths. The combined ATM wireless data/fiber network operation, as well as service integration, has been evaluated and tested using an Internet advisor ATM analyzer. Error-free millimeter-wave/optical transmission and network operation were achieved for 155-Mbps data channels switched between three users in cells up to 470 m in radius.

Figure 10-6 illustrates several examples of integrated HFR and RF photonics for wireless data/fiber internetworking and interface options.[1] The advantage of microwave and RF photonics is that it not only expands and merges broadband distribution and access, but it also incorporates "networked" functionality and control into the wireless data links. The top figure indicates integration of several different wireless

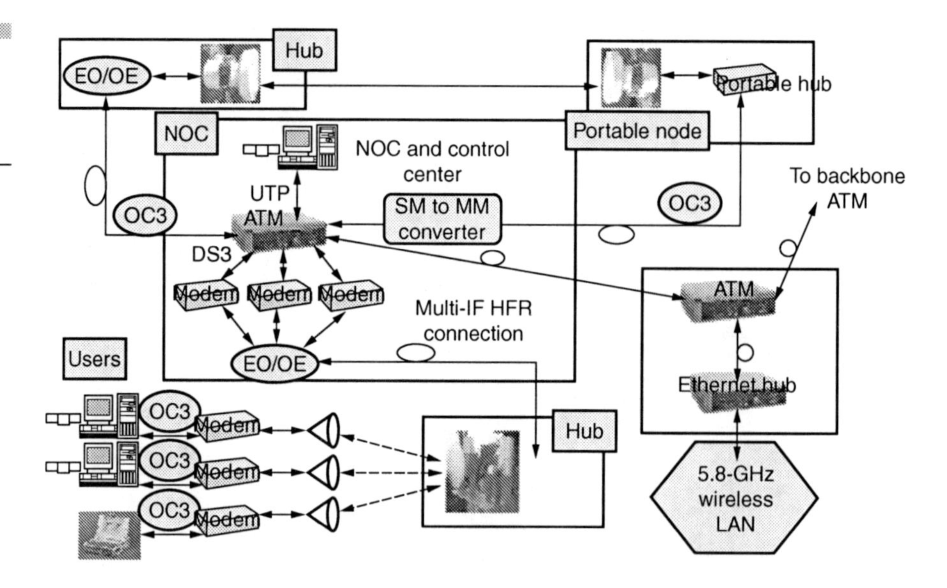

Figure 10-5 A three-user testbed and ATM network topology.

data bands (PCS, NII, millimeter-wave, FSOP) into a single HFR using WDM technology. The system integration has also been demonstrated for a single optical wavelength and synchronized multicarrier millimeter-wave radios with modular IF stages. The millimeter-wave subcarriers are selected with one-to-one fiber/wireless data channel mapping to provide unified end-to-end network operation and continuity.

The lower left part of Fig. 10-6 depicts the role of HFR for multiple AP signal distribution, centralized control of individual antenna beam and phases, and frequency band selections. Here, the otherwise traditional "antenna remoting" function has been replaced by a multiple service access link with centralized network management and control.

The lower right part of Fig. 10-6 depicts yet another example—utilizing the HFR technology to distribute high-stability, low-phase-noise local oscillator (LO) and sync signals to the millimeter-wave up/downconverters in the APs and base terminals. The experimentally deployed LO distribution demonstrated lower harmonics and superior phase quality in millimeter-wave systems, as well as lowered electrical intermediate frequency (IF)/RF terminal design complexity, component counts, and overall cost compared to pure all-electrical solutions. A two-channel (12- and 16-GHz) photonic unit was demonstrated for evaluating the performance of a switched dual-band photonic link in distributing LO/sync signals. The scheme provides the flexibility of frequency tuning, channel selection, and dynamic bandwidth allocations for wireless data access systems.

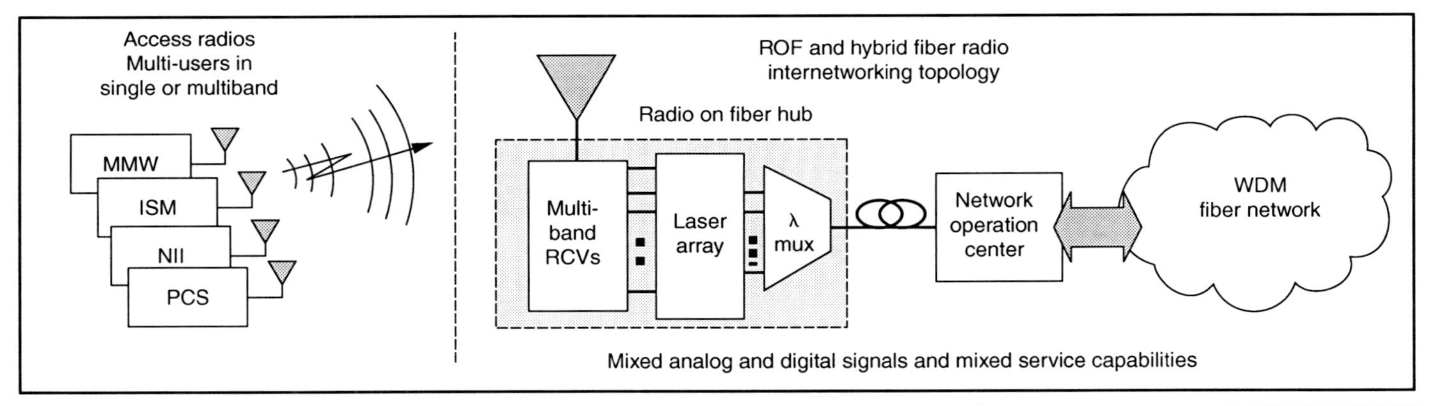

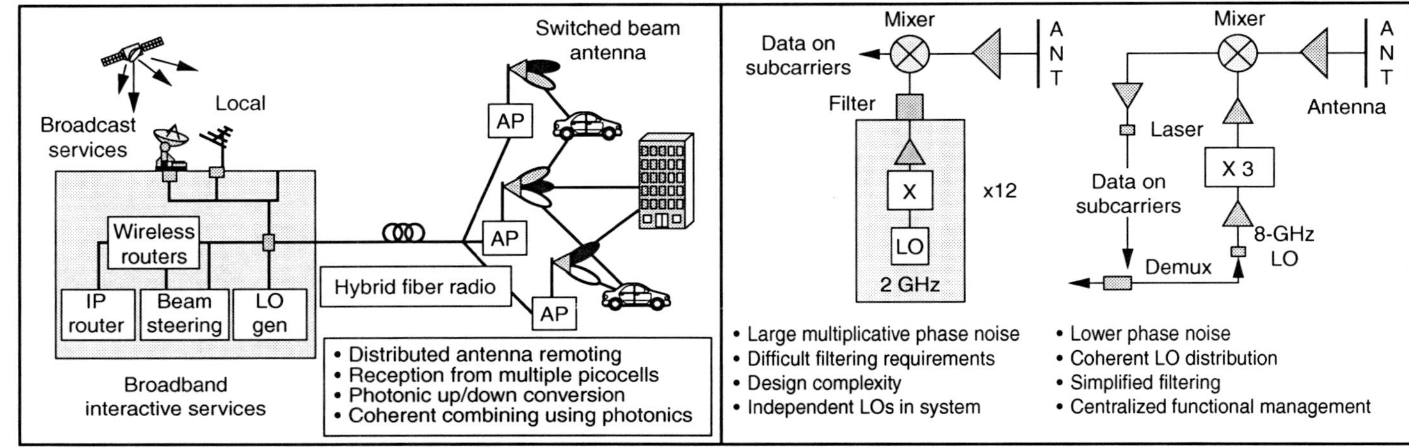

Figure 10-6 *Multiband ROF and HFR interconnection examples for a unified end-to-end network. Top: the role of WDM and RF photonics in a wireless data/fiber network interface. Lower left: multiple AP signal distribution and control. Lower right: centralized high-stability low-phase-noise LO distributed to the APs and base terminals.*

Conclusion

This chapter has introduced and demonstrated a short-range LOS LMDS-like millimeter-wave and FSOW architecture for a BWA system that possesses many technological and operational advantages. These include ease of installation and alignment; low radiation power; and, effectively, a link free from major multipath, obstructions (trees, buildings, and moving objects), and adjacent cell interference. The chapter also presented several system architecture and implementation scenarios for a complementary millimeter-wave/FSOW system highly suitable for integration of a BWA network with the existing backbone fiber network. The proposed system architecture is suitable for deployment in a highly developed, densely populated, urban inner city environment where large-capacity broadband services are in great demand, but lacking wired broadband access infrastructure.

References

1. Hossein Izadpanah, "A Millimeter-Wave Broadband Wireless Access Technology Demonstrator for the Next-Generation Internet Network Reach Extension," *IEEE Communications Magazine,* 445 Hoes Lane, Piscataway, NJ 08855, 2002.
2. John R. Vacca, *Wireless Broadband Networks Handbook,* McGraw-Hill, 2001.
3. John R. Vacca, *Satellite Encryption,* Academic Press, 1999.
4. John R. Vacca, *i-mode Crash Course,* McGraw-Hill, 2001.

CHAPTER 11

Wireless Data Services: The Designing of the Broadband Era

Loose coalitions of tech geeks, amateur radio hobbyists, and social activists worldwide have begun to design free broadband wireless data networks.[3] Sit in a park or cafe near one of these networks with your laptop and modem, and you can access files on your home or office computer, or access the Web without a hard-wired connection.

While some of these broadband wireless data networks are designed to extend free Internet access to people who otherwise couldn't afford the service, others are building what amounts to a community intranet. It's not about Internet access. It's about building up a broadband wireless data network, connecting people through their computers in the community.

The broadband wireless data networks are based on the 802.11b wireless data networking standard. Participants purchase access points, then create or buy antennas and place them on the roofs of their houses or apartment buildings and become nodes on a broadband wireless data network that links members' computers together. Many members with antennas already have high-speed data lines, such as DSL or cable modems, and they can share that Internet access for free with anyone who has an 802.11b modem and is within range of an access point. (The Glossary defines many technical terms, abbreviations, and acronyms used in the book.)

A growing number of local businesses will raise antennas and join the broadband wireless data network as a way to establish a presence among the other users of the network. A couple of coffee shops in Seattle are already part of SeattleWireless' data network, which so far has nine nodes.

As more people join the broadband wireless data network, the community grows and gives more impetus for businesses, for example, to maintain sites on the community network for free. Instead of paying a recurring monthly fee for a Web site, members incur only the one-time cost of putting up an antenna and linking to the broadband wireless data network.

Other businesses may want to add nodes on the broadband wireless data network so workers can access the corporate network from home or nearby cafes or restaurants. The broadband wireless data network doesn't have to hit the public Internet, and can use virtual private network technology to tunnel securely into the corporate intranet.

The independent way the broadband wireless data networks grow, however, may be one of the drawbacks.

Word Spreads

These volunteer projects seem to grow in fits and starts, yet the momentum in Seattle has spread quickly outside the city. Seattle is the pioneer in doing this in the world.

The idea is to have an independent broadband wireless data network. If the Internet backbone goes down, this will act as a network that would still be up in an emergency.

These groups run the risk of angering ISPs that might not like the fact that some of their network users are accessing the Internet without paying. So far, leaders of the free wireless data groups believe that they are just a blip on the ISPs' radar and not worth worrying about.

That may be true among the more open-minded ISPs. If some people are experimenting with cool stuff, there won't be a problem.

Most ISPs aren't happy to learn that customers are sharing connections for free, but the practice isn't expected to blossom to a threatening size. The problem with grass-roots local-area networks (LANs) is that someone has to pay for that service, and the reliability and performance of the link will be limited because no one has the incentive to invest additional dollars.

That fact may slow the growth of the free broadband wireless data networks and affect the networks' quality, but it also preserves the market for customers that might be willing to pay for the assurance of quality service. For example, MobileStar Network is one well-known company using 802.11b in places such as Starbucks coffee shops to offer high-speed wireless data Internet access to paying subscribers. The company has backup measures in place to ensure that customers receive high-quality service, and indicates that assurance will continue to attract customers.

However, some DSL and cable modem service providers may have reason to complain. High-speed data providers oversubscribe on the basis of projections of how much bandwidth customers will use. An unexpected number of users on their networks could affect their business plans. The network providers are concerned about maintaining the bandwidth they have.

Now, let's look at how typical image compression algorithms produce data streams that require a very reliable communication—they are not designed for transmission in an environment in which data may be lost or delayed, as provided by current and next-generation broadband wireless data communication networks. Compression and transmission provisions that avoid catastrophic failure caused by lost, delayed, or errant packets are therefore imperative in order to provide reliable visual communication over such systems. This robustness is obtained by modifying the source coding and/or adding channel coding. This part of the chapter presents an overview of both lossy and lossless source coding techniques and combined source/channel techniques providing robustness, examples of successful techniques.

Wireless Data Channel Image Communications

Images contain a great deal of redundancy, from both signal processing and psychological perspectives, which effective compression attempts to remove. Typical image compression algorithms produce data streams that require a very reliable and in fact perfect communication channel—they are not designed for transmission in an environment in which data may be lost or delayed (in real-time imaging, delay is equivalent to loss). Broadband wireless data systems are characterized by their limited bandwidths and high bit error rates, and cannot provide the necessary quality of service guarantees for compressed image data; therefore, compression and transmission provisions that avoid catastrophic failure caused by lost, delayed, or errant packets are imperative. Robustness is obtained by modifying the source coding and/or adding channel coding. Source coding can be modified by increasing redundancy in the image representation and making the encoded bit stream itself more robust to errors (while the former typically increases the source data rate, the latter can often be obtained with minimal or no increase in source data rate). Channel coding adds controlled redundancy in exchange for source coding rate. When combined, the required robustness can be provided for many broadband wireless data environments.

To appropriately understand the image transmission issue, first consider two extremes of image transmission over unreliable channels that allow lost or errant data to be recovered from received data. The first extreme is an information-theory result given by Shannon's well-known joint source/channel coding theorem: A stochastic process can be optimally transmitted over a channel if the source coding and channel coding are performed independently and optimally. Zero redundancy is placed in the source coding, and maximum redundancy is placed in the channel coding. Recovery from transmission errors is possible, provided that restrictions placed by the channel coding on the errors are not exceeded.

NOTE Knowledge of the channel is required to select an appropriate channel code.

A second hypothetical extreme exists in which knowledge of the channel is not required to ensure reliable image transmission. The uncoded image is simply transmitted, and the redundancy present in the image is used to compensate for lost data. In this case, raw data can be corrupted, but an uncoded image has sufficient redundancy to allow successful concealment of the errors using the received data at the decoder, which is now perhaps more appropriately called a reconstructor. The reconstructed image will not be pixel-for-pixel equivalent to the original, but visually equivalent, which is as well as the first extreme performed anyway, because in the first extreme, the data was first source-coded via lossy compression to achieve visual but not exact equivalence. In general, the first extreme is far more efficient with respect to the total bandwidth required on the channel, so the second is only of hypothetical interest. But, the second extreme suggests the existence of a continuum between the two. This part of the chapter examines various points along this continuum to provide robust image transmission over broadband wireless data channels.

Following a brief review of image compression and a discussion of commonly used models for broadband wireless data channels, source coding techniques that increase robustness are described. Separate and combined source/channel coding techniques are then considered. Representative successful techniques in each category are discussed.

A Brief Overview of Image Compression

Image compression is essentially redundancy reduction and is performed in one of two regimes: lossless or lossy compression. Lossless compression permits exact recovery of the original signal, and permits compression ratios for images of not more than approximately 4:1, although in practice 2:1 is more common. In lossy compression, the original signal cannot be recovered from the compressed representation. Lossy compression can provide images that are visually equivalent to the original at compression ratios in the range of 8:1 to 20:1, depending on content. Higher compression ratios are possible, but produce a visual difference between the original and compressed images.

An image compression system consists of three operations: pixel-level redundancy reduction, data discarding, and bit-level redundancy reduction, as shown in Fig. 11-1.[1] A lossless image compression system omits data discarding. A lossy algorithm uses all three operations, although extremely efficient techniques can produce excellent results even without

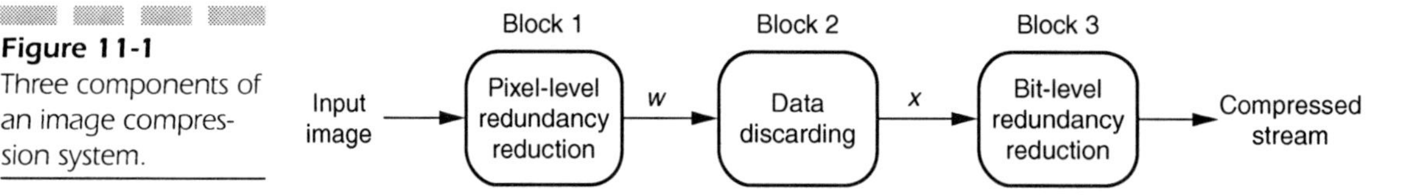

Figure 11-1 Three components of an image compression system.

bit-level redundancy reduction. While compression can be achieved using fewer operations, all three are required to produce state-of-the-art lossy image compression.

Pixel-level redundancy reduction performs an invertible mapping of the input image into a different domain in which the output data are less correlated than the original pixels. The most efficient and widely used mapping is a frequency transformation (also called a *transform code*), which maps the spatial information contained in the pixels into a frequency space. Such a representation is efficient because images exhibit high correlation, and it is also better matched to how the human visual system (HVS) processes visual information. Data discarding provides the "loss" in lossy compression and is achieved through quantization of w to form x. Both statistical properties of images and HVS characteristics are used to determine a quantization strategy that minimally impacts image fidelity. Finally, bit-level redundancy reduction removes or reduces dependencies in the data and is often called *lossless coding.* Lossless coding is often entropy-based, such as Huffman or arithmetic coding, but can also be dictionary-based, such as Lempel-Ziv-Welch coding. In this part of the chapter, such codes will be generically referred to as variable-length codes (VLCs). Each of these three operations can be adjusted to produce data that have increased robustness to errors and loss.

JPEG is the only current standard in existence for still gray scale and color image coding. Baseline JPEG image compression is a three-step operation consisting of applying a discrete cosine transform (DCT) to 8×8 pixel blocks, quantization of the resulting coefficients, and variable-length coding. The resulting JPEG data stream contains both header and image data. An error in the header renders the entire stream undecodable, while an error in the image data causes errors of varying seriousness, depending on location in the bit stream. JPEG permits periodic resynchronization flags known as *restart markers* at user-defined intervals in the compressed bit stream that reset the decoder in the event of a decoding error caused by transmission problems. A shorter period improves robustness, but decreases compression efficiency, since the restart markers represent no image data. Even with the use of restart markers, decoding errors are usually obvious in JPEG images, so some sort of error detection and concealment following decoding is often implemented.

Wavelet-transform-based image compression techniques have gained popularity in the last decade over DCT-based techniques such as baseline JPEG because these transforms operate on the entire image rather than individual blocks, and therefore eliminate blocking artifacts at high compression ratios. The wavelet transform is also argued to be better matched to the HVS frequency response than the DCT. The simplest wavelet coders are implemented as three-operation systems, previously described, with a wavelet transform followed by separate quantization of each band and variable-length coding. However, more efficient compression is possible with so-called zero-tree-based embedded wavelet coders, which produce a single embedded bit stream from which the best reconstructed images in the mean squared error sense can be extracted at any bit rate. An excellent representative of such a technique is the SPIHT algorithm. JPEG-2000 is wavelet-based, but does not use such an embedded bit stream.

Commonly Used Models for Broadband Wireless Data Channels

Two models are prevalent in developing robust image transmission techniques for broadband wireless data channels: bit error models and packet loss models. Bit error models assume random bit errors, occurring at some specified bit error rate (BER). They may also include burst errors, in which the instantaneous BER increases substantially for a fixed amount of time. The channel is assumed to be always available, although possibly with severely degraded conditions.

Packet loss models assume that the data are segmented into either fixed- or variable-length packets. Commonly it is assumed that lost packets are detected, and a lost packet does not disrupt reception of subsequent packets. Such a model is valid for a broadband wireless data channel when forward error correction (FEC) within packets is used to deal with any random bit errors in the stream; when the capabilities of FEC are exceeded, the packet is considered lost. A channel with packet loss is modeled as having a bandwidth and a packet loss probability (sometimes also called a packet error probability). It may also have an average burst length of packet losses, and an average frequency of burst losses.

More generally, a packet loss model can be applied when a data stream is segmented into and transmitted to the receiver in well-defined self-contained segments. Inserting resynchronization flags strategically in the compressed data stream allows periodic resynchronization at the receiver, and can transform transmission of a bit stream over a broadband wireless data link with deep signal fades into transmission of a packetized stream over a link exhibiting both packet loss and individual bit errors. If the receiver loses synchronization with the bit stream, data are lost only until reception of the next flag. Upon recognition of the flag, the receiver can again begin decoding. In this way, data between any two flags can be considered a packet, and inclusion of sequence numbers with the flag permits identification of lost packets. Adding FEC to each packet allows correction of errors within received packets.

Source Coding Techniques

The source coder performs frequency transformation, quantization, and lossless coding, and each of these operations provides an opportunity to improve robustness. Modified frequency transforms increase correlation in the transformed data above that provided by common transforms such as DCT or traditional wavelet transforms. Increased redundancy in the transmitted data facilitates error concealment, and these techniques allow reconstructed data of higher quality than is possible with traditional transforms. The increased redundancy incurs overhead, which is selectable during the design process and typically ranges from 30 percent to over 100 percent. In exchange for these high overhead rates, no hard limit is placed on packet loss rates. Rather, the quality of the received, reconstructed image degrades gracefully as loss increases, and loss rates of up to 30 percent are easily handled. Figure 11-2 shows an image coded by using a reconstruction-optimized lapped orthogonal transform and suffering 10 percent packet loss in known locations, both without and with reconstruction using averaging.[1]

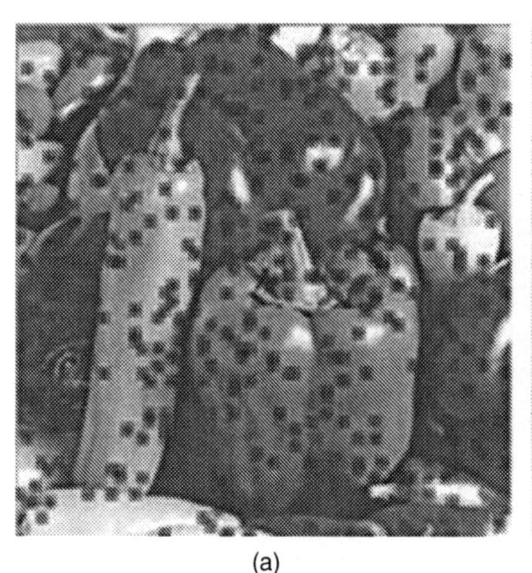

(a)

(b)

Figure 11-2 Peppers coded by using a reconstruction-optimized lapped orthogonal transform and suffering 10 percent random packet loss: (*a*) no reconstruction, PSNR = 17.0 dB; (*b*) reconstructed, PSNR = 29.6 dB.

NOTE The additional redundancy (90 percent over JPEG for this transform) in the representation is evident even when no reconstruction is performed.

Robustness can be incorporated into the quantization strategy through the use of multiple description (MD) quantizers. Such quantizers produce multiple indices describing samples; reception of all indices provides the most exact reconstruction, while reception of fewer indices allows reconstruction, but at reduced fidelity. MD quantization and more general complete MD compression algorithms are typically presented in the context of having multiple channels, and are inherently better suited to such transmission situations than to a single channel; however, the resulting data can be time-shared over a single channel.

The transform coding and quantization techniques previously described rely on the decodability of the source data. Transmission errors can cause catastrophic decoder errors when data have been encoded with a variable-length code (VLC). Even a single bit error left uncorrected by the channel code can render the remainder of the bit stream useless. One way to ensure that random bit or burst errors will not catastrophically affect decoding of the VLC through loss of synchronization is to use fixed-length rather than variable-length codes, but this is often at the expense of compression efficiency. Perhaps the simplest technique to deal with errors in VLC streams is to employ resynchronization flags, which are assigned to a source symbol that serves as a positional marker and whose reception ensures the correct placement of subsequently decoded data. Such flags are called *restart markers* in JPEG or *synchronizing codewords* in other work, and can be combined with error detection and correction techniques. They can be inserted at user-defined intervals; a shorter interval improves robustness, but decreases compression efficiency since the restart markers represent no image data.

More sophisticated techniques to provide robustness for VLC-coded data include both packetization strategies and specially designed VLCs. A packetization strategy to provide robustness is the error-resilient entropy code (EREC), which is applicable to block coding strategies (JPEG), in which the input signal is split into blocks that are coded as variable-length blocks of data; EREC produces negligible overhead. Reversible variable-length codes are uniquely decodable both forward and backward and are useful for both error location and maximizing the amount of decoded data; they also incur negligible overhead. Resynchronizing variable-length codes allow rapid resynchronization following bit or burst errors and are formed by designing a resynchronizing Huffman code and then including a restart marker at the expense of slight nonoptimality of the resulting codes; overhead is negligible at bit rates over approximately 0.35 b/pixel. The resulting codes are extremely tolerant of burst errors; if the burst length is less than the time to resynchronize,

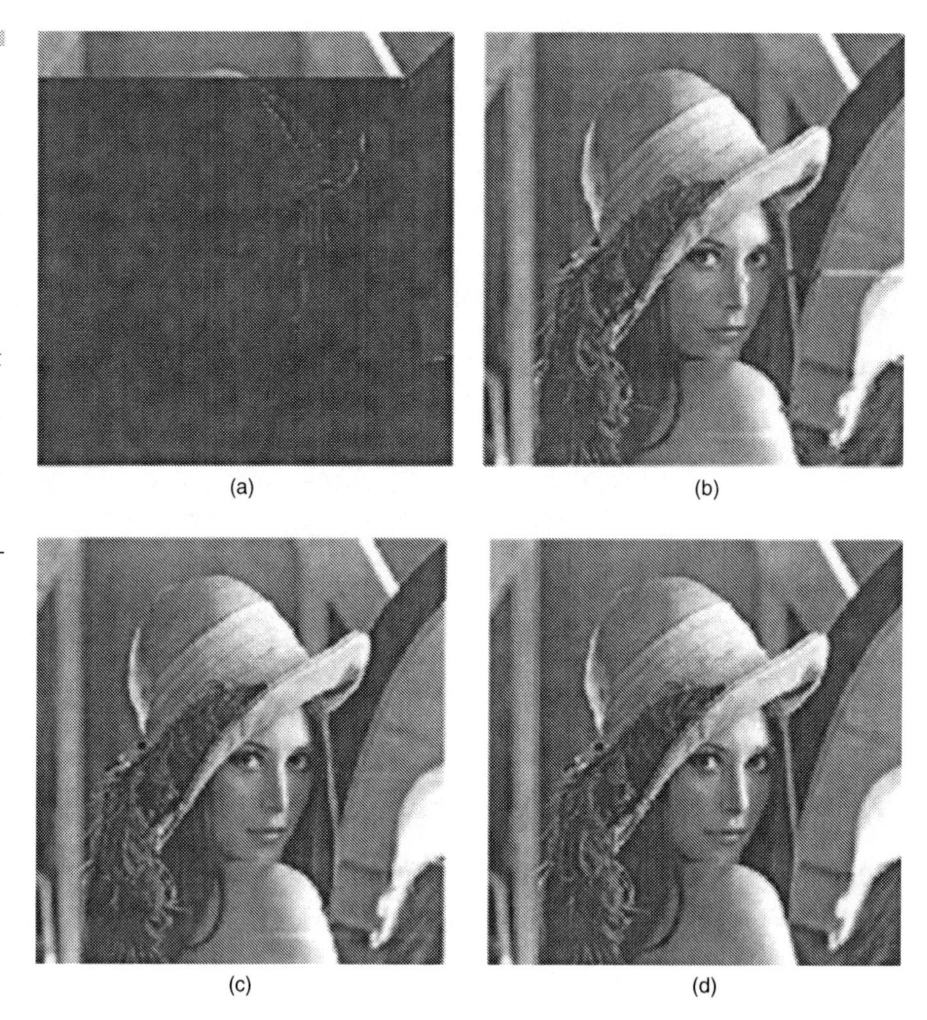

Figure 11-3
Lena at 0.38 b/pixel. (*a*) JPEG using standard Huffman coding: BER 2×10^{-4}. (*b*) JPEG using resynchronizing variable-length-coding: BER 2×10^{-4}, no error concealment. (*c*) Error concealment performed on (*b*). (*d*) JPEG using resynchronizing VLC: six burst errors of length 20 with error concealment.

the burst error is equivalent to a bit error. Figure 11-3 shows an image compressed to 0.38 b/pixel and compares JPEG using standard Huffman coding, and JPEG using resynchronizing variable-length codes at a BER of 2×10^{-4}, with error concealment on the latter.[1] An error-concealed image suffering six burst errors of length 20 clearly demonstrates the robustness of this technique to burst errors.

Separate and Combined Source and Channel Coding

The previous part of this chapter described modifications to source coding to increase robustness to transmission errors. This part of the chapter discusses adding controlled redundancy through FEC, with no or little modification to the source coding algorithm. Knowing the channel characteristics beforehand is necessary to select an appropriate FEC code. Interleaving can be, and often is, used to lessen the effect of burst errors. Additionally, the use of the source coding techniques previously described, along with channel coding, can further improve robustness and minimize such failures. Techniques for source and channel coding for robust image transmission can be classified in many ways: those that deal with bit errors only, packet loss only, or a combination of both; those that simply concatenate (separate) source and channel coding; those that jointly optimize the bit distribution between source coding bits and channel coding bits; those that apply equal error protection (EEP); and those that apply unequal error protection (UEP).

Bit errors only are typically dealt with by using a convolutional code or other appropriate channel code. The packet loss transmission model is addressed by applying FEC at a packet level: Data are segmented into packets and an FEC (usually systematic) is applied vertically to a block of packets. When an (n, k) code is applied vertically to a block of k packets, $(n - k)$ additional packets are created and represent the additional redundancy. Because the locations of lost packets are known, reconstructing them is treated as erasure correction, and up to $(n - k)$ erasures (lost packets) can be reconstructed. The capability to deal with random bit errors within packets (errors within packets no longer produce a packet that is labeled as lost) is provided by applying FEC within each packet. Such an application can be considered a product code, with FEC applied both across and within packets.

An appropriate source coding rate and channel coding rate can be selected in a jointly optimal fashion or simply sequentially. Joint optimization involves selecting the number of bits assigned to both source and channel coding together to satisfy an overall rate constraint while minimizing a distortion metric or achieving a throughput measure. This often involves dynamic programming or simplified solutions that run quickly, but may provide nonoptimal solutions. Alternatively, a source coding rate can be selected, and appropriate channel coding then added to achieve reliable transmission over a given channel.

Use of a single FEC code treats all source coding bits as equally important, providing EEP. However, since the SPIHT data stream can be decoded at any point to produce a full-resolution, but lower-rate image, UEP can easily be applied by increasing the strength of the ECC for earlier portions of the bit stream. For JPEG-encoded images, a stronger ECC is often applied to the header information. In the remainder of this part of the chapter, several example systems are provided that include various combinations of the previously described techniques.

A joint optimization of source bit rate, FEC selection, and assignment of unequal loss protection to the source data suggests an unequal loss protection framework applied to SPIHT-encoded image data, in which the FEC is selected to maximize the expected received quality for a given packet loss rate, subject to an overall bit rate constraint. This technique provides graceful degradation with increasing packet loss. Packet loss is approached by selecting a source coding algorithm in conjunction with a packetization scheme that facilitates reconstruction for wavelet-coded images; this produces a less efficient source coder that is, however, much more robust to packet loss.

The previously mentioned solutions are for packet loss, but cannot deal with individual errors within packets. Product codes successfully solve this problem. A concatenated channel coder is applied within packets, while a systematic Reed-Solomon code is applied across packets. The technique allows tuning of error protection, decoding delay, and complexity through the choice of particular codes. Unequal error protection can be achieved by including additional codes in the channel coder. A target overall bit rate is selected, appropriate codes are selected, and the remaining bits are filled with the SPIHT-encoded data. As such, no joint optimization is performed. The benefits of this technique stem from the efficiency of the product code, so more source coding bits can be included and hence produce a higher-quality image for the same overall bit rate. Unequal error protection, using rate-compatible punctured convolutional codes (RCPCs), is advocated. A key feature of this work is the assumption that the source bit stream is decodable only up to the first error, and that the optimization criterion should therefore be maximizing the length of the useful source bit stream. This results in a different choice of codes for different source bit rates, and therefore is not as easily applicable as previously mentioned techniques, but is perhaps more realistic.

Now, let's look at how hardware-based multipath fading simulators have traditionally been used to generate up to two simultaneous fading channels. Mobile network testing[5] and future wireless data applications like geolocation, smart antennas, and multiple-input, multiple-output (MIMO) systems, however, require more channels.

Wideband Wireless Data Systems: Hardware Multichannel Simulator

With the advancement of mobile multimedia systems, required data rates and system bandwidths are increasing, and the development of such systems puts demands on the associated test equipment to have increased features and performance. Future radio channel simulators will have to have multiple channels, wide bandwidth, high dynamic range, a sufficient number of fading paths, advanced channel modeling, and very high RF performance. Offering eight fading channels, 70-MHz RF bandwidth, and spatial channel modeling, the PROPSim C8 wideband multichannel simulator has been designed to meet these requirements.

PROPSim C8 is a hardware multichannel simulator, where a maximum of eight independent channels are run in one simulator unit with more channels possible if multiple simulators are synchronized together. Applications for testing antenna diversity include carrier-to-interference ratio (*C/I*), adaptive antennas, geolocation systems, handover, repeaters, and other multiantenna or multiterminal systems. Also, MIMO systems use multiple antennas both in transmission and reception.

An important feature of the simulator is that because it is independent of the incoming signal, it provides a very versatile platform for different tests. Any signal can be connected to the input when the RF bandwidth is 70 MHz or less, center frequency is between 350 MHz and 6 GHz, and the RF power is below 0 dBm. The hardware performs simulation in real time with digital path and digital channel combining, providing accurate and realistic radio channel simulations.

The multichannel simulator provides three simulation interfaces: RF, analog baseband (ABB), and digital baseband (DBB). Regardless of the selected interface, however, multipath fading simulation and signal combining and splitting are done in the digital domain to achieve the best possible accuracy, flexibility, and repeatability. A block diagram of the function is shown in Fig. 11-4.[2] The signal is downconverted from RF to analog baseband (I and Q branches), transferred into the digital domain by an analog-to-digital converter (ADC), and vice versa via a digital-to-analog converter (DAC). The digital baseband processing performs very high speed multipath fading simulation, and the faded analog baseband signal is upconverted back to RF.

In addition, a hardware simulator is implemented by removable plug-in units (see Fig. 11-5), utilizing a simulator controller unit (SCU), where an internal PC is installed.[2] The baseband unit (BBU) consists of the ADC and DAC, along with digital baseband processing, and multipath fading is implemented in the digital domain. The BBU has two interfaces, ABB and DBB, while the RF unit (RFU) makes quadrature down- and upconversions and provides the RF interface.

As mentioned earlier, the system architecture supports three different simulation interfaces, as shown in Fig. 11-6.[2] Represented are the transmitter (TX) under test or the test signal generator, the radio channel

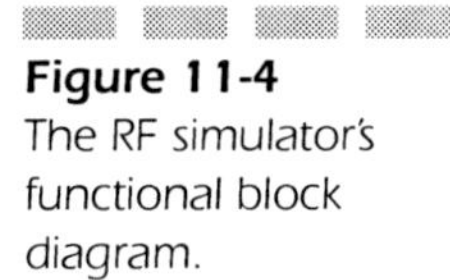

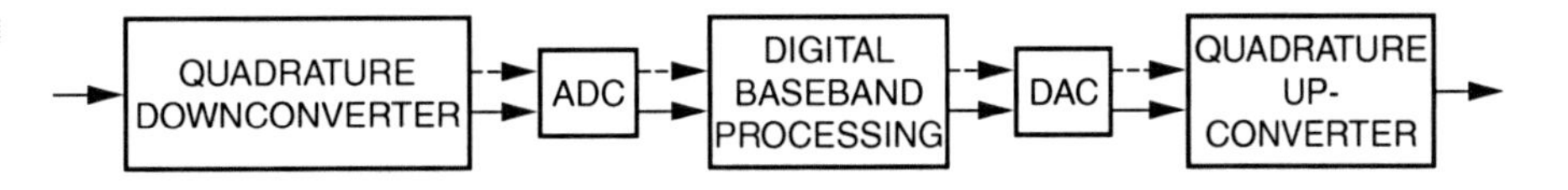

Figure 11-4 The RF simulator's functional block diagram.

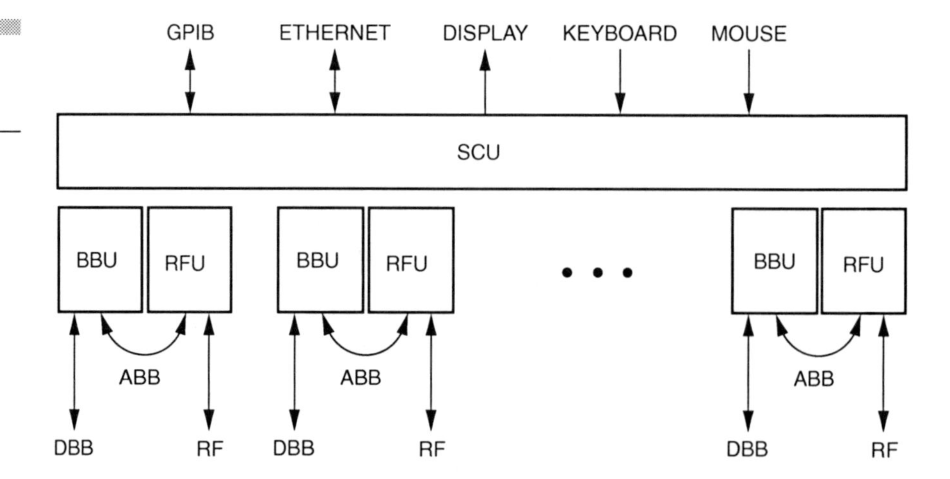

Figure 11-5 Simulator plug-in units.

simulator (RCS), and the receiver (RX) under test. A typical transmitter has DBB components, a DAC, and an upconverter (UC). In the receiver, there may be a downconverter (DC), ADC, and DBB. Similar parts can be found in the RCS, which facilitates the use of different interfaces. The use of the three interfaces brings the same radio channel used in the lab through the whole development cycle.

Also, the DBB interface extends the use of fading simulation to a very early phase of product development when analog parts are not available. It can be used as an accelerator of software simulation, non-real-time field-programmable gate array (FPGA) testing, or testing different parts, such as application-specific integrated circuits (ASICs). Consequently, it helps to improve the quality of product design and reduces the time-to-market and cost of product development. When all three interfaces are in the same product, similar channel models are available in all phases from early algorithm design to final product tests. These phases can be non-real-time macro model, ASIC, analog baseband, and RF performance tests, together with system verification and type approval.

Versatile channel modeling is required to ensure that the performance of the system is adequate in all situations. Testing only with models defined in various standards is often not sufficient to guarantee that the terminal actually operates in difficult fading environments. The testing of wireless data products with scenarios that stretch requirements beyond type approval models is important in all phases of the product development cycle. Existing standards do not model the spatial dimension of the radio channel.

Advanced channel modeling software will help to design realistic scenarios, and the use of the spatial dimension sets new requirements for radio channel simulators. For example, it can be utilized to improve sys-

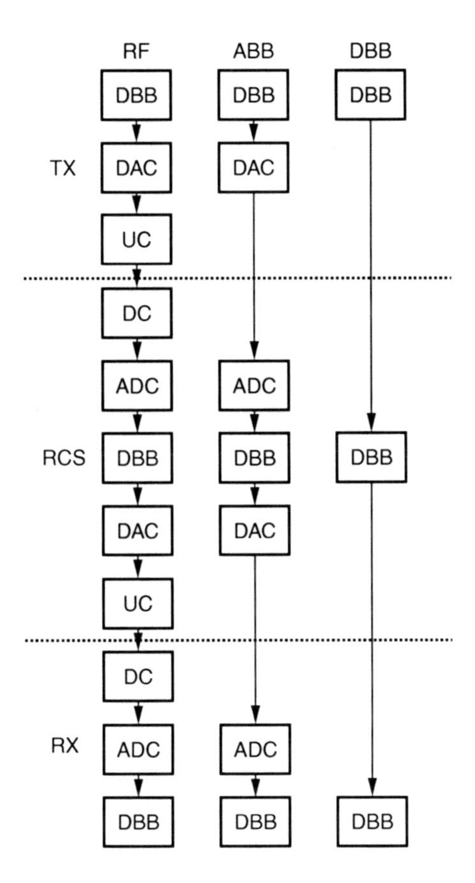

Figure 11-6 RF, ABB, and DBB interfaces.

tem capacity. Multiple frequency-selective fading channels with controlled correlation must be produced, and the PROPSim C8 uses two alternative methods to implement these spatial requirements: correlation matrix and geometric constellation. The first method uses an operator-selected set of mutual correlation values between the channels, while the second utilizes antenna array and direction of arrival (DoA) information to determine the correlation between channels.

This geometric constellation–based method is shown in Fig. 11-7.[2] The modeling method assumes that the source is so far away that the received wave is a plane wave and the angular spread of the incoming wave follows a laplacian distribution. Zero angular spread will lead to a situation where phase shifts between antenna elements stay constant during simulation.

Finally, typical applications for this multichannel simulator are different antenna array systems, mobile networks, MIMO systems, and geolocation applications. Figure 11-8 shows a typical test setup for antenna

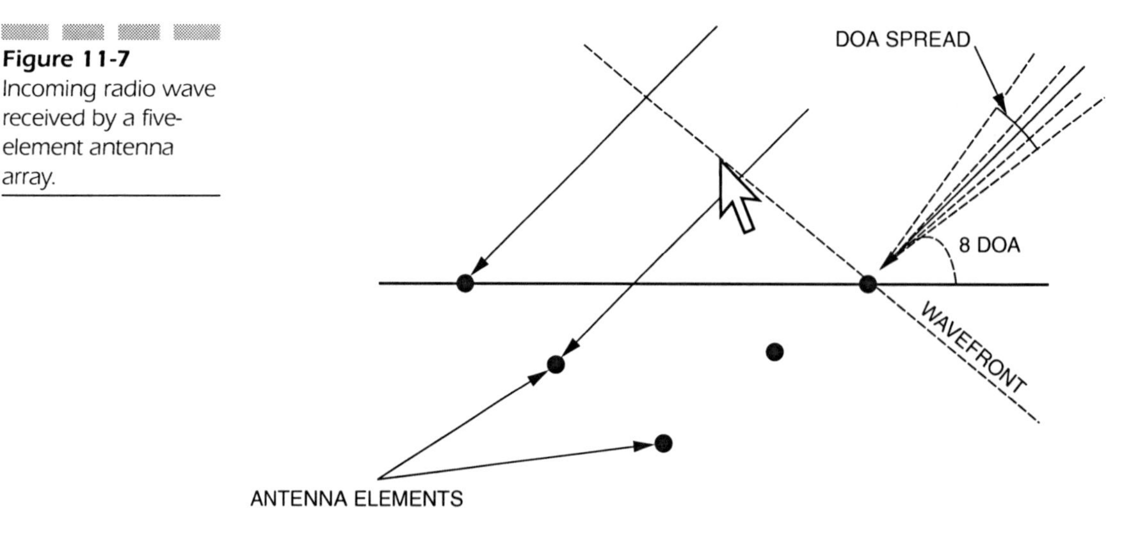

Figure 11-7 Incoming radio wave received by a five-element antenna array.

array tests where signal combining and splitting is done digitally.[2] Applications illustrated in Figs. 11-9 and 11-10 are multiple terminals and base station, and a MIMO system, respectively, with the latter being planned for use in upcoming third- and fourth-generation units.[2] Another feature of the simulator is that each channel has an integrated digital noise source, whereby additive white gaussian noise is generated internally and added to the faded signal. Typical wireless data test systems require transmitter, channel, noise, and receiver, so a combined noise source and fading channel simplifies the test setup.

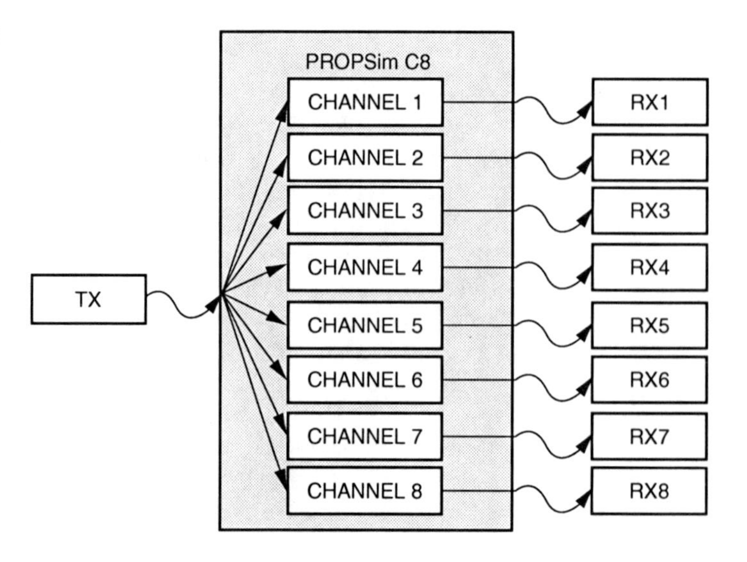

Figure 11-8 Antenna array application.

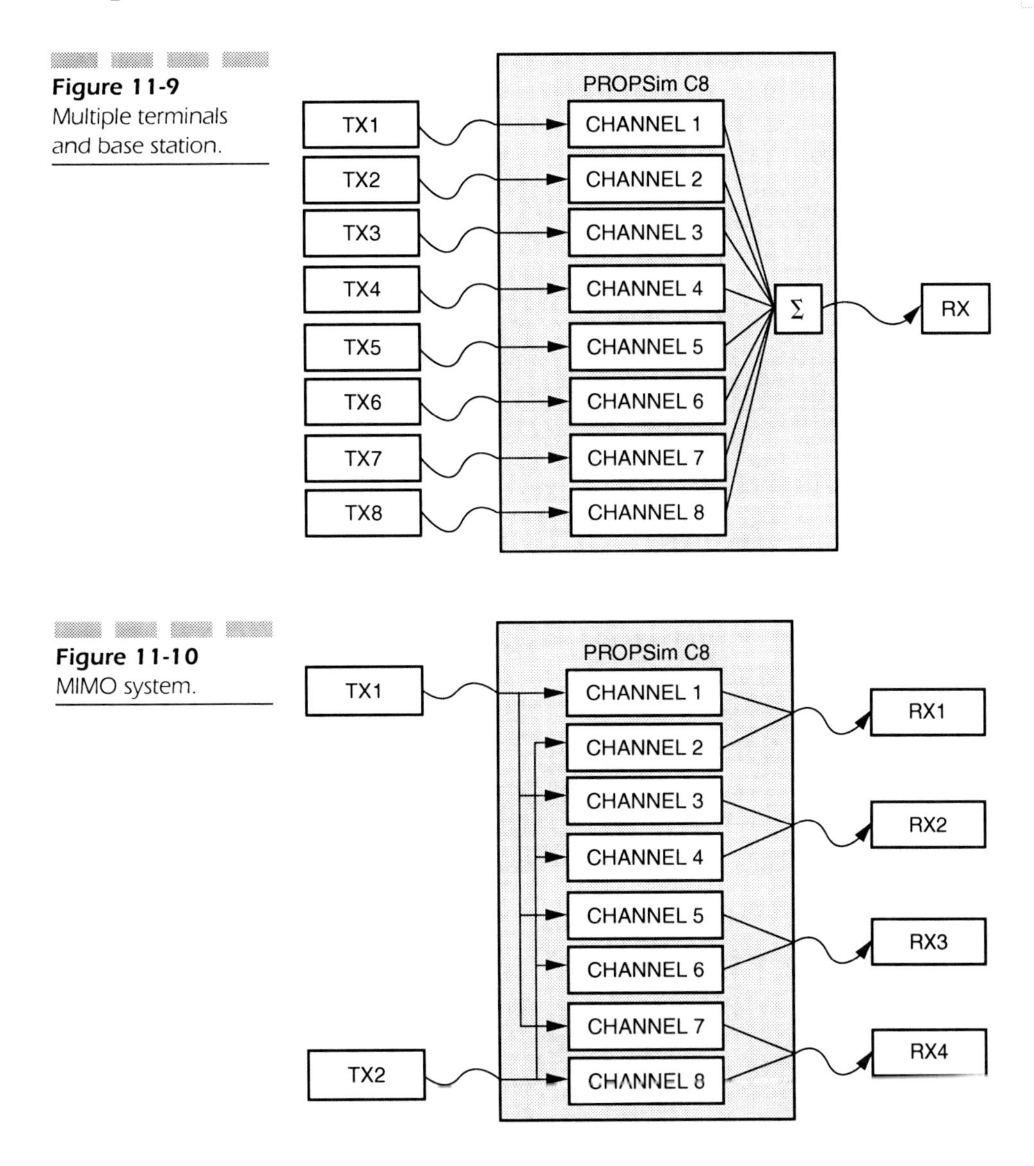

Figure 11-9 Multiple terminals and base station.

Figure 11-10 MIMO system.

Conclusion

This chapter provides an introduction to a variety of techniques used to provide robust image transmission over wireless data channels. Controlled redundancy can be added in the source coding and/or channel coding, and lossless compression techniques can be made more robust to transmission errors with little or no sacrifice in efficiency. While many example solutions are given, these solutions and in fact the techniques presented are not exhaustive; modified signaling to improve performance and the many variants of multiple-description source coding were not discussed, but can also improve the performance of image transmission systems. Whether the discussed techniques or others are used, received image quality can be greatly improved in transmission over imperfect channels.

Finally, with its multichannel capability, wide RF bandwidth, and three simulation interfaces, the PROPSim C8 fading simulator offers enhanced features and performance to produce accurate and realistic simulations. It has also been developed to provide the flexibility and adaptability needed to meet the requirements of future wideband wireless data systems.

References

1. Sheila S. Hemami, "Robust Image Communication over Wireless Channels," *IEEE Communications Magazine,* 445 Hoes Lane, Piscataway, NJ 08855, 2002.
2. "A Hardware Multichannel Simulator for Wideband Wireless Systems," *Microwave Journal,* 685 Canton St., Norwood, MA 02062, 2002.
3. John R. Vacca, *Wireless Broadband Networks Handbook,* McGraw-Hill, 2001.
4. John R. Vacca, *Satellite Encryption,* Academic Press, 1999.
5. John R. Vacca, *i-mode Crash Course,* McGraw-Hill, 2001.
6. John R. Vacca, *The Cabling Handbook,* 2d ed., Prentice Hall, 2001.

CHAPTER 12

U.S.-Specific Wireless Data Design

The General Packet Radio Service (GPRS) is a next-generation packet data service that provides wireless data connectivity support across the Global System for Mobile Communication (GSM)[2] and IS-136 time-division multiple-access (TDMA) wireless data networks. It also complements existing services such as circuit-switched data and short message service (SMS).

With over 500 million subscribers today, the GSM mobile communication standard is the leading digital wireless data communication standard in the world. The size of the current subscriber base indicates that there is a very large potential U.S.-specific marketplace for GPRS design and services. GPRS service deployment is already beginning in Europe and in the United States. The primary features of GPRS networks include:

- Faster data transfer rates
- Always-on connectivity
- Robust application support
- Dynamic IP addressing
- Prioritized service
- Migration path to 3G networks[1]

With the preceding in mind, this chapter presents an overview of U.S.-specific GPRS, as well as U.S.-specific wireless data design considerations for mobile applications being developed for GPRS deployment. (The Glossary defines many technical terms, abbreviations, and acronyms used in the book.)

Faster Data Transfer Rates

GPRS services support data transfer rates that are much higher than can be supported by circuit-switched data services on GSM networks. In theory, GPRS can support a maximum data transfer rate of 171.2 kbps when using the full capacity of the service. However, the physical radio interface consists of a carrier-configurable number of time slots. The theoretical GPRS maximum speed is not achievable unless all eight of the available time slots are allocated for GPRS packet data. Figure 12-1 depicts typical data transfer rates that compare GRPS with currently deployed wireless data networks.[1]

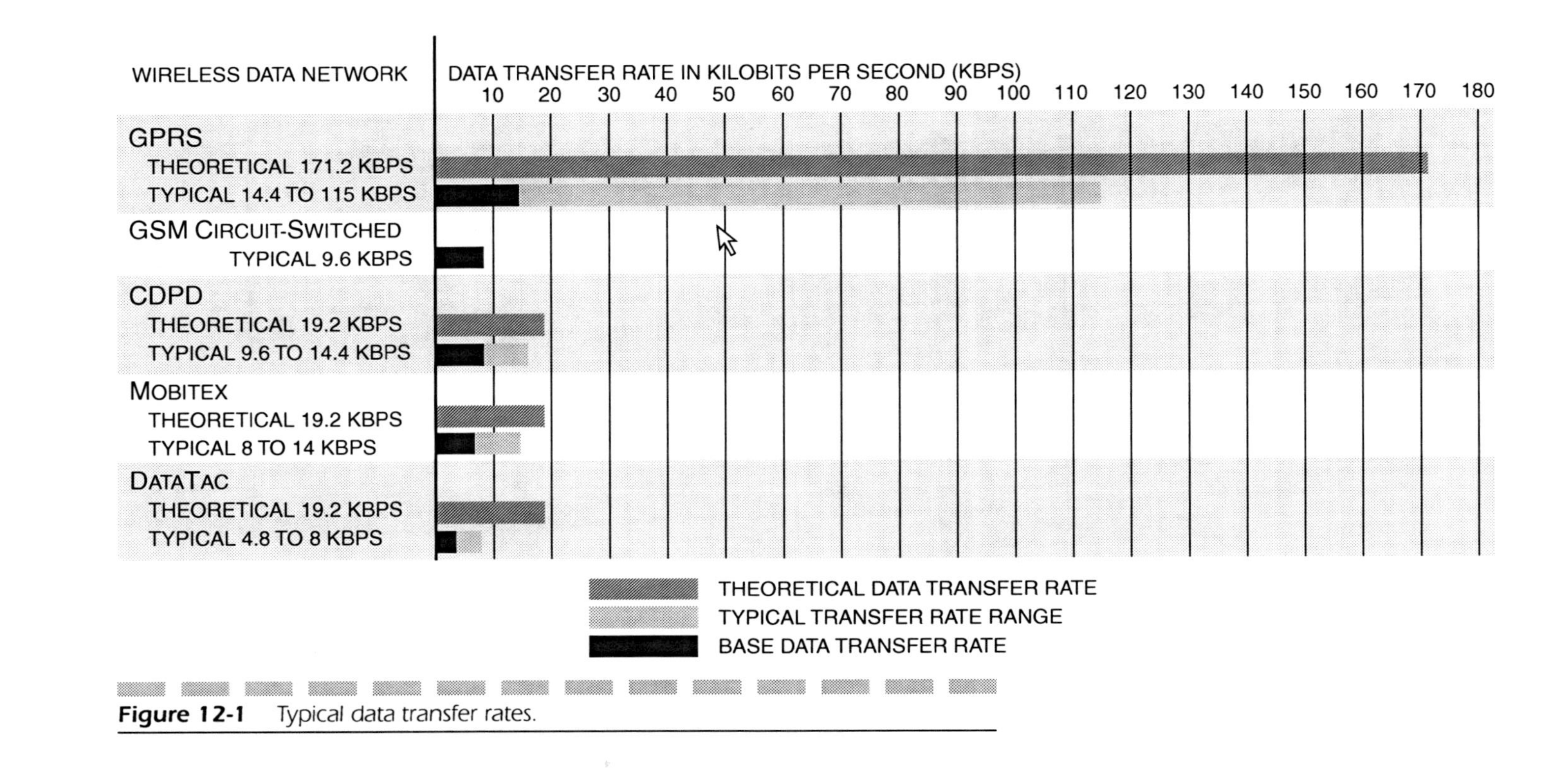

Figure 12-1 *Typical data transfer rates.*

Currently deployed wireless data networks such as CDPD, Mobitex, and DataTac operate at speeds of up to 19.2 kbps. Current circuit-switched data services on GSM networks operate at 9.6 kbps, while strapped-channel high-speed circuit-switched data (HSCSD) trials conducted in Europe have shown data rates of up to 64 kbps.

It is expected that typical GPRS networks will support data transfer rates that range from 19.2 to 115 kbps. The data transfer rate will depend on the individual carrier's allocation of resources.

This means that a typical GPRS network could have transfer rates over 5 times faster than those possible over the wireless data networks in use today, and could also be over 10 times faster than the circuit-switched data services available on GSM networks today.

Always-on Connectivity

Unlike GSM circuit-switched data services, access to the GPRS network does not require a lengthy network connection setup procedure to access the network. This feature allows data to be immediately sent or received on demand.

The perception to the end user is that the service is always available and that the device is always connected. This feature also allows data to be pushed to the subscriber device. The user can define a profile that causes data to be automatically delivered (pushed) on the occurrence of an event or on a specified time-driven basis.

Robust Application Support

An important feature of using GPRS is that the increase in speed is directly coupled to the types of applications that can be supported. The slow speed of a circuit-switched data network combined with its lengthy connection start-up time, and the limited message length (160 characters) of the SMS solution, results in wireless data applications that are limited in functionality.

Because the GPRS network is IP-based, it allows subscribers to access the full range of Internet applications. Many services that are currently available over the Internet, including wireless data Web access, e-mail, instant messaging, and file transfer, will be available over the GPRS network.

Other new applications that will be available to GPRS subscribers include peer file transfer support and home automation system support.

Subscribers will use a home automation system to remotely access and control in-house appliances and security systems.

GPRS enables mobile Internet functionality by allowing interworking between the existing Internet and the GPRS network. In fact, many network operators are evaluating use of GPRS to become wireless data Internet service providers. The combination of these characteristics provides a wide spectrum of robust applications that can be offered to mobile subscribers.

Dynamic IP Addressing

Use of the Internet Protocol (IP) addressing scheme is commonplace on wireline telecommunications networks that power the Internet. However, there is currently a shortage of IP addresses available. This problem could potentially limit the number of users on the Internet and the future 3G wireless data networks.

One method to mitigate the shortage of IP addresses is to dynamically allocate and assign IP addresses to a mobile device on a temporary basis by using the Dynamic Host Configuration Protocol (DHCP). The GPRS architecture recognizes this requirement and enables the wireless data carriers or third parties to dynamically allocate and assign IP addresses to a mobile device on a temporary basis. The architecture also allows for the use of statically assigned IP addresses.

Allowing wireless data carriers or third parties to dynamically allocate IP addresses provides additional addressing flexibility. Given its worldwide deployment, IP-based addressing is an obvious choice for the underlying GPRS network technology.

Optimizing implementations at the IP level benefits traditional Internet application protocols as well as emerging protocols that are layered on top of IP. IP is such a fundamental element of the Internet that non-IP alternatives face substantial obstacles to deployment and market acceptance.

Prioritized Service

Each time a device connects to a GPRS network, a Quality of Service (QoS) profile is associated with the connection. The QoS profile is a single parameter with multiple data transfer attributes. The QoS is defined by these attributes:

- Precedence

- Delay
- Reliability
- Throughput[1]

Precedence

Under normal conditions, the GPRS network attempts to meet the service requirements for all devices without regard to the associated QoS profile. However, under abnormal network conditions such as extensive congestion or limited network resources, this attribute is used to prioritize how service is to be allocated. Table 12-1 shows how the service precedence class attribute is used to determine the relative importance of maintaining service commitments under abnormal conditions.[1]

Delay

The delay attribute is used to determine the maximum amount of delay associated with packet transfers. There are four delay classes. Table 12-2 shows that the first three classes are predictive and class 4 is characterized as best effort.[1]

TABLE 12-1

Precedence Classes

Precedence	Interpretation
1—high	Service commitments are to be maintained ahead of precedence classes 2 and 3.
2—normal	Service commitments are to be maintained ahead of precedence class 3.
3—low	Service commitments are to be maintained after precedence classes 1 and 2.

TABLE 12-2

Delay Classes

	Packet Size			
	128 Octets		1024 Octets	
Class	Mean Delay, s	95th-Percentile Delay, s	Mean Delay, s	95th-Percentile Delay, s
1	0.5	1.5	2	7
2	5	25	15	75
3	50	250	75	375
4	Unspecified			

The predictive classes define the maximum delay values for small (128-octet) and large (1024-octet) data packet transfers. The maximum amount of delay is predicted for the mathematical mean and the 95th percentile of all data packet transfers. For example, the 95th percentile delay is the maximum delay guaranteed in 95 percent of all data packet transfers.

Reliability

The reliability attribute is used to specify the transmission characteristics and error rate tolerance required by an application. The GPRS architecture defines five reliability classes that guarantee certain maximum values for lost, duplicated, out-of-sequence, or corrupted data packets.

Table 12-3 summarizes the type of traffic that is associated with each of the reliability classes.[1] For real-time traffic, the QoS profile would require the appropriate settings for the delay and throughput attributes.

Throughput

The throughput attributes specify the rate at which data are transferred across the network. Throughput is measured in octets per second. Two classes are specified: The peak throughput values are shown in Table 12-4, and the mean throughput values are shown in Table 12-5.[1]

The peak throughput attribute is independent of the delay class. There is no guarantee that the specified peak rate can be achieved or sustained for any period of time. The eventual data exchange rate is network resource–dependent.

The network may also limit the subscriber to a negotiated peak rate even if additional capacity is available. The mean throughput class is used to specify the average rate at which data are expected to be transferred

TABLE 12-3

Reliability Classes

Class	Traffic Type
1	Non-real-time, error-sensitive application that cannot tolerate data loss.
2	Non-real-time, error-sensitive application that can tolerate infrequent data loss.
3	Non-real-time, error-sensitive application that can tolerate data loss. This is the classification that is associated with SMS and GMM/SM.
4	Real-time traffic, error-sensitive application that can tolerate some data loss.
5	Real-time traffic, error-insensitive application that can tolerate data loss.

TABLE 12-4

Peak Throughput Classes

Class	Octets per Second	kbps
1	Up to 1000	8
2	Up to 2000	16
3	Up to 4000	32
4	Up to 8000	64
5	Up to 16,000	128
6	Up to 32,000	256
7	Up to 64,000	512
8	Up to 128,000	1024
9	Up to 256,000	2048

across the GPRS network during the remaining lifetime of an activated communication context.

Once again, the network may limit the subscriber to the mean data rate even if additional transmission capacity is available. The best-effort mean throughput class can be characterized as on a per-need and as-available basis.

Support of different QoS classes allows the carrier to define performance profiles using precedence, reliability, delay, and throughput parameters. It also allows carriers to create billing stratification schemes by which they can charge different rates for different classes of service. However, in early phases of development, it does not appear that application developers will be able to dynamically change the QoS attributes through program control.

GPRS System Architecture

Figure 12-2 shows a representation of the GSM/GPRS system architecture.[1] To integrate GPRS into the existing GSM architecture, a new class of network nodes, called GPRS support nodes (GSNs), was created. The GSNs are responsible for routing data packets between the mobile stations and external packet data networks (PDNs). Two GSNs were created: the gateway GPRS support node (GGSN) and the serving GPRS support node (SGSN).

TABLE 12-5

Mean Throughput Classes

Class	Octets per Second	Approximate Bit Rate
1	Best effort	Best effort
2	100	0.22 kbps
3	200	0.44 bps
4	500	1.11 bps
5	1000	2.2 bps
6	2000	4.4 bps
7	5000	11.1 bps
8	10,000	22 bps
9	20,000	44 bps
10	50,000	111 bps
11	100,000	0.22 kbps
12	200,000	44 kbps
13	500,000	1.11 kbps
14	1,000,000	2.2 kbps
15	2,000,000	4.4 kbps
16	5,000,000	11.1 kbps
17	10,000,000	22 kbps
18	20,000,000	44 kbps
19	50,000,000	111 kbps

Gateway GPRS Support Node

The GGSN acts as an interface between the GPRS backbone network and inbound external packet data networks such as the Internet and corporate networks. It converts the GPRS packets coming from the SGSN into the appropriate packet data protocol (PDP) format (for example, IP or X.25) and sends them out on the corresponding packet data network.

The GGSN can also be used to connect to other GPRS networks to allow roaming. For data traversing from the PDP network to the GSM network, the PDP addresses of incoming data packets are converted to the GSM address of the destination.

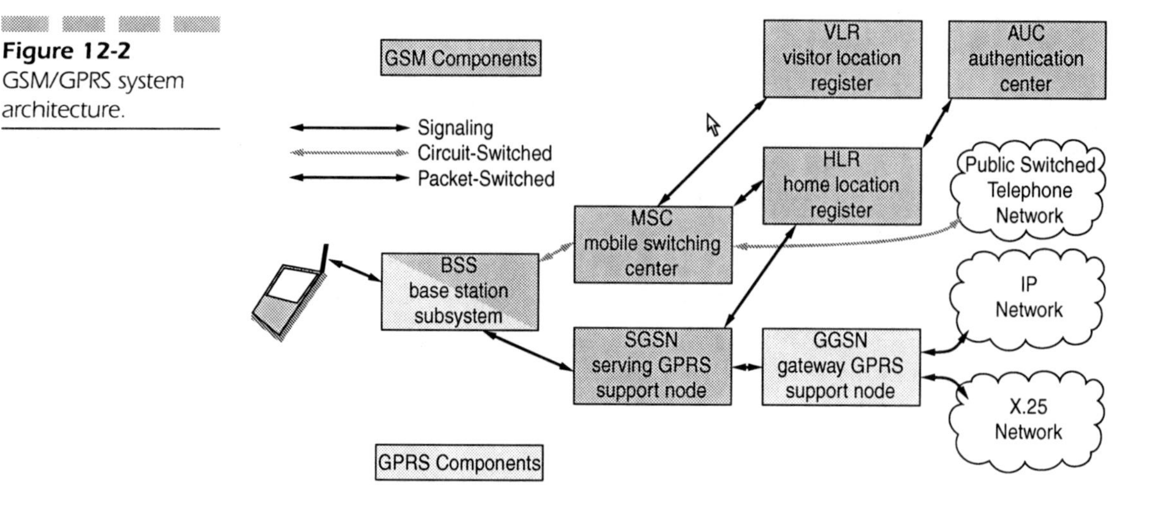

Figure 12-2 GSM/GPRS system architecture.

These readdressed packets are sent from the GGSN to the responsible SGSN. To support this routing function for each user, the GGSN stores the current SGSN address in its location register. In addition, the GGSN also performs authentication and charging functions.

In general, there is a many-to-many relationship between the SGSNs and the GGSNs. A GGSN may function as the interface to external packet data networks for several SGSNs. Likewise, an SGSN may route inbound packets over different GGSNs to reach different packet data networks.

The GGSN tracks the required routing information to the SGSNs that service particular mobile devices. One or more GGSNs can access multiple SGSNs.

Serving GPRS Support Node

The SGSN forwards packets to mobile devices within its service area. The SGSN interfaces to the home location registers (HLRs) to obtain subscriber profile information. SGSNs are responsible for detecting a new mobile device for a given service area and maintaining location information inside a specific service area.

In addition, the SGSN:

- Is a peer with the MSC
- Provides mobility management
- Performs authentication
- Performs encryption
- Provides access control[1]

The HLR stores location information (for example, the current cell and current visited location register) and user profiles of all GPRS users registered with this SGSN.

Mobile Application U.S.-Specific Design Considerations

Based upon the GPRS field trial experiences that occurred as a direct result of the recent Mobile Application Initiative (MAI) participation (see sidebar, "Aether Participation in the Ericsson Mobile Application Initiative"), the following are important U.S.-specific design points that developers should consider when creating wireless data client/server GPRS solutions: (1) design of the application protocol and (2) GPRS limitations and capabilities.

Aether Participation in the Ericsson Mobile Application Initiative

Aether Systems[1] recently participated in the Ericsson Mobile Application Initiative (MAI), a unique partner program that is dedicated to optimizing applications and services to function well over the next-generation wireless data networks, including GPRS and 3G networks. The purpose of Aether Systems' participation was to validate the functionality of the Aether middleware and to gain some indication of its performance while operating in the Ericsson GPRS acceptance test environment (GATE). The tests were performed at the MAI Berkeley Center in Berkeley, California.

This testing established that the Aether intelligent messaging (AIM) middleware was able to provide optimized support for GPRS applications. Additional information about the testing follows.

Application Overview

The AIM solution is wireless middleware that enables developers to create client/server applications that operate over a wide range of wireless data networks and devices without having to understand the complexities and nuances associated with the supporting wireless data protocols.

The AIM solution consists of three elements:

- AIM middleware that abstracts the wireless data networks and provides a client/server development environment.
- Aether server software development kits that are used to create the server-side interface to the AIM middleware.
- Aether client software development kits that are used to create the client-side interface to the AIM middleware.

Figure 12-3 depicts the AIM solution in relation to the GPRS network, the AIM middleware components, the client application interface provided by the associated Aether client API, and the server application interface provided by the associated Aether server API.[1]

Testing Success Criteria

The testing consisted of running an Aether-developed client/server suite over the GPRS application test environment (GATE) created by Ericsson. The GATE has provisions to simulate coverage, allocate uplink and downlink time slots, simulate continuous and non-continuous signal reception to the client, and simulate the number of simultaneous users.

The Aether client was configured to send four different message types of varying lengths at fixed and variable intervals. Some messages were compressed and some were not.

The Aether server was configured to echo the received messages back to the Aether client. A test was considered to pass if the number of acknowledgments, negative acknowledgments, and time-out acknowledgments added up to equal the expected results, and a sufficient explanation could be given for why an unexpected message was received. For example, a time-out acknowledgment message may have been received rather than an acknowledgment message because the GATE was modeling a lag time that was greater than the product of the number of retries and the retry interval.

Testing Environment

Figure 12-4 depicts the MAI test environment that is used to evaluate the AIM middleware.[1] The test environment consisted of the following components:

- A laptop PC that functioned as the client-side test environment for the Aether client application.
- A Linux PC running the GATE simulator.

- An Ethernet hub that was used to route the IP packets.
- A Windows NT server that ran the components that make up the Aether middleware and the Aether server application that provided the echo services.[1]

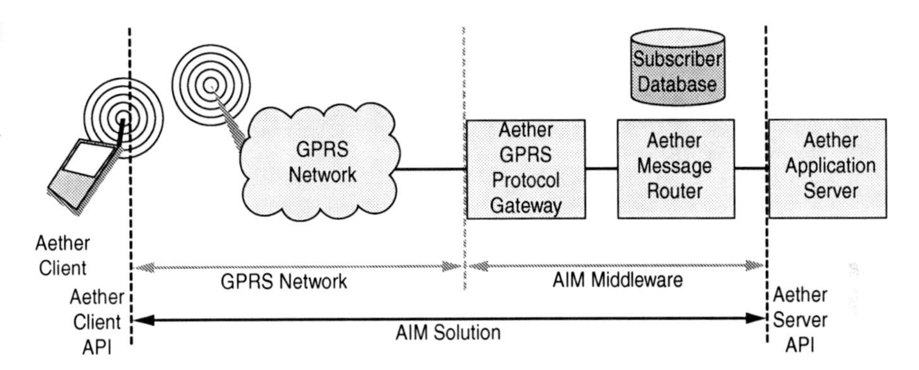

Figure 12-3 AIM solution for GPRS network.

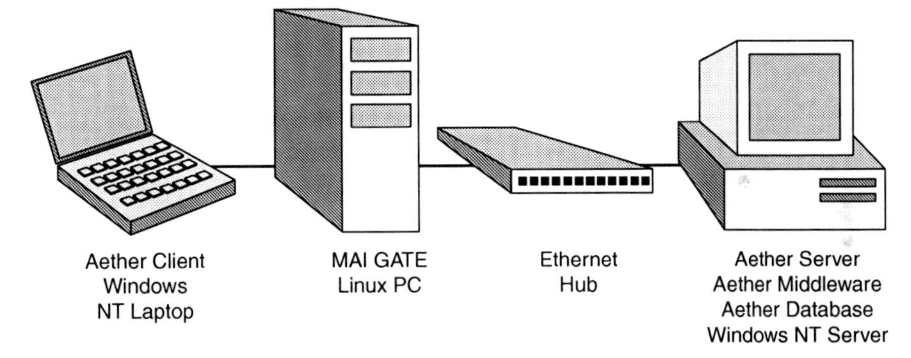

Figure 12-4 MAI test environment.

GPRS Limitations and Capabilities

It is clear that GPRS is an important new packet data technology that offers a major improvement in data transfer rate capability and functionality when compared with the other wireless packet data networks that are widely available today. However, GPRS has limitations. Each limitation and capability must be considered in designing and implementing a wireless data application to be deployed on a GPRS network:

- Typical data transfer rates much lower than the theoretical transfer rate

- Limited cell capacity
- No store-and-forward capability
- Limited modes of operation
- Uneven uplink and downlink message support
- Payload compression[1]

Typical Data Transfer Rates Much Lower Than the Theoretical Transfer Rate

To achieve the maximum theoretical GPRS transmission speed of 171.2 kbps, all eight time slots from a single base station must be allocated for use by a single GPRS terminal. Whether carriers will allocate all their resources for use by a single GPRS terminal is unclear.

NOTE Initially, GPRS handsets will be configured only to support a maximum of three or possibly four time slots.

Therefore, the bandwidth available to a GPRS user will be much less than the theoretical maximum. GPRS operators have not yet developed pricing levels for use of all eight time slots. Be advised to check the available GPRS network speeds and the wireless data device capacity against your development requirements.

You can also expect GPRS network operators to allocate more time slots for the downlink than for the uplink. This also has implications that you must consider when designing an application protocol.

Limited Cell Capacity

GPRS packet data and existing network circuit-switched voice and data services compete for the same network resources. Therefore, capacity for GPRS services cannot be increased without decreasing the capacity for these other circuit-switched network services.

There will probably be a strong migration of circuit-switched data users to GPRS data services, leaving the network resource tradeoff to be primarily between the GPRS data services and the circuit-switched voice services. The extent of the impact depends on the amount of resources that are reserved for exclusive use by GPRS.

To help mitigate this situation, the network uses dynamic channel allocation to allocate channels for GPRS and non-GPRS services. To support GPRS services, a cell must allocate physical channels, known as packet data channels (PDCHs), for GPRS data transfer.

PDCHs are allocated out of a common pool of physical channels that are available within the cell. There are only a limited number of these physical channels in the pool and they must support all GPRS and non-GPRS services offered by the cell.

A load supervision procedure monitors the load of the PDCHs in the cell. Depending on the current demand, the number of channels allocated for PDCHs can be dynamically changed.

Physical channels not currently in use by conventional GSM/IS-136 services can be allocated as PDCHs to increase the quality of service for GPRS. However, application designers must be aware that when there is a resource demand for services with higher priority, PDCHs can be reallocated. This results in a priority-load-based degradation of service.

No Store-and-Forward Capability

There is no store-and-forward mechanism built into the GPRS standard. However, in most GPRS networks, there will be an interconnection between the GPRS network and a short message service (SMS) facility.

Although the store-and-forward engine in the SMS is the heart of the SMS Center, and a key feature of the SMS, the store-and-forward capability may be limited by the relatively small SMS packet size. Also, application developers must be aware that the store-and-forward implementation for SMS is not consistently implemented among the carriers.

Interconnection links between SMS and GPRS may require different implementations for different carriers. Wireless data application developers must consider if store and forward is a requirement for their solution, and if so, they must implement this feature themselves or seek a third-party wireless data middleware solution.

Limited Modes of Operation

Table 12-6 shows the operating modes that are supported for GPRS wireless devices.[1] Initially, commercially deployed GPRS handsets will be configured as Class B GPRS wireless data devices. This eliminates the possibility of integrating voice and data services.

For example, an application server may want to push an unsolicited notification alert to a mobile user. A mobile user in voice mode may not be able to receive packet data.

To enable an application server to deliver the notification alert, a GPRS solution that requires the ability to push data might require

TABLE 12-6

Wireless Device Operating Modes

Class	Support
Class A	Simultaneous voice and data
Class B	Separate voice and data
Class C	Data only

a complementary bearer service (for example, SMS) that uses a different type of radio resource to provide this service.

Wireless data application developers must consider if pushing data is a requirement for their solution. If it is, they must implement an alternative delivery path feature themselves.

Uneven Uplink and Downlink Message Support

Figure 12-5 depicts how the testing bandwidth varied between 16 and 24 kbps on the downlink and 3 and 6 kbps on the uplink.[1] The typical network uplink (from the wireless data device to the carrier) will be small (one or two time slots), while the corresponding downlink (from the carrier to the wireless data device) will be larger (four or five time slots). For developers, uneven uplink and downlink message support will have an impact on how applications should be designed.

Small Uplink Messages All messages originating from the mobile device should be as small as possible. The rationale for this network configuration is based on the HTTP request/response transaction model. With HTTP, the client request is typically small, while the server response is typically larger.

During the GPRS field trial, the impact of sending small messages and large messages from the mobile device was tested with respect to the round-trip time (RTT). Table 12-7 summarizes the GPRS test net-

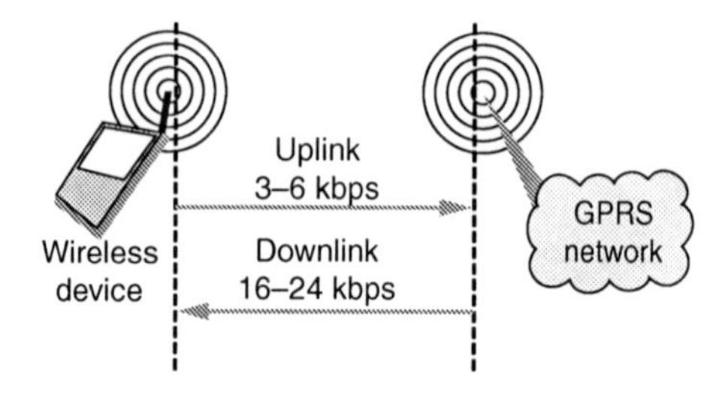

Figure 12-5
Uneven uplink and downlink bandwidth.

TABLE 12-7

RTT from Mobile Device

Message Size, bytes	Radio Shadow	Uplink Time Slots	Downlink Time Slots	Average Number of Background Users	Round-Trip Time, ms
Small, <540	Noncontinuous	1	4	15–35	503
Large, >540 and <8192	Noncontinuous	1	4	15–35	25,006

work configuration and the test results.[1] The noncontinuous radio shadow component of the testing is used to simulate a wireless data mobile device that randomly moves into and out of GPRS coverage.

As the results highlight, small messages ranging from 10 to 540 bytes from the mobile encountered no problem even with the limited uplink time slots and bandwidth. However, when the mobile device sent large messages ranging from 540 to 8192 bytes, the round-trip time increased significantly.

The longer RRT is attributed to the numerous retries that were encountered because of the uplink bandwidth not being available and the messages not reaching their target destination. In the second test, the messages from the mobile device eventually reached their target destination because the transport automatically retried sending the messages.

If the messages eventually reached their target destination, what is the concern? There are two important reasons why sending larger messages has an undesirable outcome: Because the messages had to be retried, an additional OTA byte count occurred, and the user experience in the second test case was diminished because of the slow RTT. If the mobile application must send large messages, consider segmenting the message into multiple segments and implement a pacing algorithm to minimize overrunning the small uplink bandwidth.

Larger Downlink Messages Server solutions should incorporate some form of message pacing on the downlink to allow the mobile device to acknowledge inbound messages, because of the limited bandwidth available on the mobile device uplink. If the solution does not effectively accommodate message pacing on the server side, the server will attempt to retry nonacknowledged messages, which will result in additional byte costs and slower RTT.

Because the downlink is configured for higher bandwidth, the solution cannot abuse it without increasing the OTA costs and decreasing the user experience. To illustrate a better approach, consider an e-mail application that is optimized for the wireless data environment.

Initially, the e-mail server should send only the subject lines associated with the new e-mail messages that are waiting to be read. A user who wants to read the body associated with a subject can request more information.

However, the server does not have to send the entire body in one message. Many times, the first paragraph of the message body contains enough information to allow the user to determine if additional information is required.

In this case, the server sends only what the mobile user is requesting. It does not send a 25K message when the read body request was sent to the server.

This results in extra byte count savings and effective message throughput. Consider implementing a feature in the wireless data application that lets the user decide when to get more messages.

During the GPRS field trial, the impact of message size over the downlink with respect to round-trip time was tested. Table 12-8 summarizes the GPRS network configuration and the test results.[1]

During these tests, bandwidth ranged from 16 to 24 kbps on the downlink and 3 to 6 kbps on the uplink. As the results highlight, sending large messages over the downlink impacted the total RTT.

Payload Compression

Because wireless data networks are bandwidth-constrained and have a long network latency, compression of the application message payload is an important design consideration for any wireless data solution. By applying compression mechanisms to the application message payload, the OTA byte count and RTT are reduced. This improves the interactive response time and user experience while reducing the air time costs associated with using the service.

TABLE 12-8

RTT from Server

Message Size, bytes	Radio Shadow	Uplink Time Slots	Downlink Time Slots	Average Number of Background Users	Round-Trip Time, ms
Small, <540	Noncontinuous	1	4	15–35	20
Large, >540 and <8192	Noncontinuous	1	4	15–35	120

During the GPRS field trial, the impact of compressing messages, with respect to OTA byte savings and round-trip time, was also tested. For the testing, the adaptive Huffman compression mechanism was applied to the application message payload.

Table 12-9 summarizes the GPRS network configuration and the test results.[1] As the results highlight, compressing large messages ranging from 540 to 8192 bytes from the mobile encountered no problem even with the limited number of uplink time slots and resulting bandwidth.

However, when the mobile sent large uncompressed messages ranging from 540 to 8192 bytes, the RRT increased significantly. The longer RRT can be attributed to two factors: numerous retries encountered because of the uplink bandwidth not being available and the messages not reaching their target destination, and the nature of moving in and out of coverage, which would again cause the message to be retried.

In the second test, the messages from the mobile device eventually reached their target destination, thanks to retry logic implemented at the client. Because the messages had to be retried, an additional byte count was incurred that has a direct relationship with the OTA costs.

The slower RTT has a negative impact on the user experience. Therefore, application designers must be aware that a client application that must initiate large data transfers will need to consider incorporating an effective compression mechanism that is optimized for wireless data messages.

Finally, from an end-to-end perspective, the developer must also be aware of the hidden costs associated with performing data compression. The major hidden costs are associated with the possible impact on the client-device battery life and the use of limited client-device memory and processing resources as a result of compressing and decompressing the message payload.

TABLE 12-9

RTT Using Compression

Message Size, bytes	Compressed	Radio Shadow	Uplink Time Slots	Downlink Time Slots	Average Number of Background Users	Round-Trip Time, ms
Large, <8192	Yes	Noncontinuous	1	4	15–35	3,075
Large, <8192	No	Noncontinuous	1	4	15–35	26,689

Conclusion

GPRS is an important evolutionary migration step as we all move toward 3G wireless data networks such as Enhanced Data GSM Environment (EDGE) and Universal Mobile Telephone Service (UMTS). Although GPRS introduces higher data transfer rates than existing wireless data networks, the real data rate will be much less than the theoretical maximum, which could be different for different carriers.

This means that the higher mobile data speeds might not be available to individual wireless data mobile users until EDGE or UMTS is introduced. The bandwidth required by wireless data applications will steadily increase over time. As this chapter has presented, an optimized OTA transport, intelligent application protocol design, and payload compression are some of the key factors to consider in designing a mobile application for GPRS. It is only after evaluating these factors and the resultant compression ratio that the developer will be able to make a value decision as to the most efficient method to implement the particular solution.

References

1. Dale R. Shelton, "General Packet Radio Service (GPRS)," Aether Systems, Inc., 11460 Cronridge Drive, Owings Mills, MD 21117, 2002.
2. John R. Vacca, *i-mode Crash Course,* McGraw-Hill, 2001.

PART 3

Installing and Deploying Wireless High-Speed Data Networks

CHAPTER 13

Deploying Mobile Wireless Data Networks

Hand-helds are a liberating technology for the mobile worker.[2] The same cannot always be said for IT administrators grappling with the swirl of issues surrounding enterprise hand-held deployment—or worse, turning the other way as hand-helds creep in their back door. The journey into hand-held deployment doesn't have to be a perilous one. Emerging enterprise solutions are empowering IT managers to take control of their mobile future.

Getting a Handle on Hand-Helds

The key to success is thorough planning. Begin with your existing infrastructure, and think about how you want to extend your information assets out to mobile devices. How will you integrate them into your existing processes and systems? Early development of a mobile wireless data network backbone, capable of supporting future growth and proliferation of new devices, is essential to the long-term success of your mobile program. The following is an introduction to enterprise hand-held wireless data deployment issues, from getting started to nurturing and protecting your investment. (The Glossary defines many technical terms, abbreviations, and acronyms used in the book.)

Getting a Plan in Place

Indeed, many IT shops have been reluctant to embrace and support enterprise hand-held wireless data deployments. But the time to move is now. Hand-helds have invaded the organization, with senior management being some of the early adopters. They know the benefits of having personal information available anytime, anywhere. They want to see this technology multiplied and adopted throughout the organization.

The Wave Is Coming

The mandate to mobilize business data is clear. Mobile commerce is the new benchmark. And just as companies have scrambled to do business on the Internet, the evolution of technology continues through hand-helds and smart phones. The benefits of pervasive information access and the ability to do business anywhere are evident in organizations that have seized mobile commerce opportunities.

How will your company measure up against the m-commerce benchmark? Are you supplying business data via hand-held devices? Are you supporting critical business functions with mobile technology? Is your IT shop providing support to end users?

There is a light at the end of the tunnel: It's mobile commerce coming your way. Meet it head on with a proactive mobile wireless data deployment strategy that includes a comprehensive, enterprise-endorsed platform and basic device synchronization. Kick off your learning curve by offering basic services to internal users. Those bedrock elements will enable you to meet the rapidly evolving demands of mobile commerce.

Delve into mobility with an easy-to-implement, rapid-return-on-investment mobile solution: Consider your sales people. Most of them carry PDA devices already. In just a few weeks, you can build a simplified version of your existing sales force automation tool, including the infrastructure to synchronize SFA data onto hand-helds. Synchronization eliminates time-consuming and cumbersome imports and exports and duplicate data entries.

Take a Step Back

The first step toward planning a mobile wireless data enterprise system is to take a step back. When it comes to mobility planning, many fail to see the forest for the trees. They consider devices ahead of applications, often at the behest of non-IT decision makers educated by the marketing schemes of hand-held manufacturers and the media. Start by looking at how and why the system is going to be created, not which devices will fit into that system. You should also get some perspective on the matter by asking the following questions:

- Why are you going mobile?
- Which mission-critical applications will flourish?
- Which business functions stand to benefit?
- What applications will you need for your mobile initiative?
- Will they be shrink-wrapped or fully customizable?
- Will you build custom applications?
- What kinds of devices will you need to run those applications?
- Who will use and support this technology?
- How are you going to train users? The support team?
- How will you manage the mobile applications and data?

- How will the hand-helds be brought into the fold of the existing IT environment?[1]

The very best hand-held wireless data deployment strategies are developed from the back end forward by answering the preceding questions. Yet in many cases, IT is not even consulted before hand-held devices are doled out to the masses. Start with the basics and work forward from there. Avoid purchasing the devices before planning for how they will be used.

Begin by capitalizing on quick-win opportunities. Build out your mobile network capability with easy, fast-turnaround projects that guarantee rapid returns. Use tools and technologies that will serve you well in the future. Comprehensiveness is key. Avoid getting stuck with a hodgepodge of administrative tools and integration issues.

Budgeting for Hand-Helds: Don't Underestimate

The typical hand-held device itself is relatively inexpensive when compared to traditional enterprise computing assets. But don't be fooled. The total cost of ownership of an enterprise hand-held device is much more than the initial cost of the hardware. Even so, the investment for hand-helds is still comparatively low and the return on investment for hand-held projects is often measured in months, versus years for traditional IT projects.

Nevertheless, it is important to budget realistically so that the project is not waylaid by budgetary constraints. When budgeting, keep in mind the following costs:

- Initial device purchase
- Communications and networking
- Synchronization software
- Replacement devices
- Systems management software
- Enterprise application purchase/build
- End user and support staff training[1]

Since this is emerging technology, seek out expert industry advice and explore how it applies to your organization and support structures. For example, Gartner Group has built an interesting TCO model around hand-held wireless data deployments. Likewise, the vendors you call on for assistance with various IT components should be prepared to help you understand the costs and benefits of each element.

This is new ground for most IT shops. Don't be afraid to leverage the experience of technology vendors and system integrators. Think about working with a service provider to build out your budget and project plan. Though hand-helds are relatively new to the enterprise IT world, there is a growing industry of systems consultants who are experienced in the space and can help you budget, justify, and plan your project.

Take Inventory

It's a virtual given that hand-helds have penetrated your enterprise. Most of your executives and sales team carry Palm devices. Pocket PCs are gaining in popularity. Your help desk takes support calls on devices they may not even know exist. These hand-helds walk out your doors everyday, carrying your company's valuable information assets. The question then becomes whether or not to bring those devices into the fold.

Survey Enterprise and Nonenterprise Assets

Survey your employees to find out who is currently using a hand-held device. What platforms are they using? What applications are they using? What types of information assets are stored on these devices? Then create a mobile business plan that addresses the integration of existing devices into the system or prohibition of nonenterprise hand-held assets. Consider the security implications of allowing nonenterprise assets into your system.

A first step is providing basic device synchronization capabilities linked to a systems management effort. Users will be mercifully spared from the task of duplicate data entry and rewarded with pervasive access to their day-to-day information. You'll be backing up devices to the server in preparation for the inevitable. And you'll emerge as a hero when a user's device fails or gets lost. At the same time, you can start gathering device inventory information upon which to build your mobile wireless data network plan.

The Reality of Multiple Devices

Just as in the early days of PCs, hand-helds of all different flavors are creeping in the back door. To get on top of the situation, you need tools that accommodate a range of devices types—Palm OS, Windows CE, Pocket PC, RIM Blackberry, EPOC phones, etc. And you don't want a different set of management tools for each platform or a different set to provide different types of synchronization support. Look for the vendor with the most comprehensive set of tools and a clear vision for providing the most comprehensive mobile management solution.

Why not just pick a single hand-held standard and enforce it? Unfortunately, two factors come into play. Because of the relatively low cost of the hand-held device itself, users have and will continue to purchase them on their own and begin storing corporate data on them. Within any organization, there will be different user classes or profiles. Each of these users stands to benefit from added mobility in different ways, and a variety of factors will dictate the type of hand-held device that is appropriate for any given user class. Applications must consider individual user patterns if the device is to achieve maximum assimilation into the enterprise.

Many of those who have purchased their own PDA also use a laptop. The laptop has not been replaced by the hand-held; rather, it serves as an adjunct device. This pattern will continue as new devices spring up to complement existing IT assets. You'll want a comprehensive suite of products to keep all these new and existing assets in synchronization and adequately supported.

Speaking of existing IT assets, you'll need to make sure the devices you set up with enterprise applications are able to smoothly integrate with the current environment. Nobody is going to trash a current CRM system running against an Oracle data warehouse to accommodate a new hand-held project. Rather, you should look for an infrastructure layer designed to link typical existing systems with the new crop of hand-helds. This layer should serve to arbitrate between the back-end systems you employ and the variables inherent to the different mobile devices you must support.

Device Selection

In choosing devices for mission-critical applications, consider scalability and robustness. If you purchase an underpowered hand-held simply because of the attractive hardware price, you may sacrifice the return on your much larger infrastructure investment. Many organizations will find that different departments and divisions require different devices. As stated earlier, start with the high-level business objectives, move to application requirements, and then consider the most appropriate hardware platform.

And still, there is the deployment of hand-helds as companions to laptop and notebook PCs. Hand-held devices are increasingly augmenting, not replacing, portable PCs. Certain functions such as serious document editing require mobile PCs. The advent of hand-held devices will further differentiate between luggable versus portable technology.

Hand-held devices are often application-specific devices. They may be extending a specific application, almost becoming an appliance or utility. The burgeoning consumer hand-held market has flourished largely as an extension of personal information management. Today, organizations are looking to these same devices to extend supply chain management, sales force automation, inventory management, facilities management, point-of-care applications, law enforcement, scientific data collection, and production data collection. Each application will influence device selection, which should consider these fundamental issues:

- Battery life
- Display size
- Data input
- Form factor
- Processing power
- Storage
- Communications options
- Security
- Application development tools[1]

Choose a Device That Meets Users' Needs

Form should follow function in device selection. Choose a device that will fit the primary application of your user group. Then look for ways to augment functionality and extend other applications and information to that device. Leverage the device's full potential and look for ways to deliver the most value to each user. Inexpensive add-ons can add significant value to the user's experience and multiply your productivity gains.

Also, consider the nature of the application and the application development tools that will be required to build the application. For Palm OS hand-helds, a variety of tools are well known in the marketplace, each with its own pros and cons. Some are incredibly easy to use, but may not support the full functionality you require. For Windows CE and Pocket PC hand-helds, the development tool options are more limited, though the tools themselves tend to be more powerful.

Each additional platform you support may require additional development effort to create a separate version of the application, and to test the application on that platform. Even HTML form-based applications often require different versions of the forms optimized for different display and input capabilities. The good news is that developing applications for hand-helds is almost always far simpler than doing so for PCs.

The Importance of Training

Early-stage hand-held wireless data deployments require training to ensure success. Hand-helds are new to most IT organizations. And, while users may be accustomed to the native PIM applications that ship with most hand-helds, running a hand-held enterprise application is likely a new experience. The good news is that hand-helds are generally easier to work with, easier to build applications for, and require a shorter learning curve that traditional IT technology.

Empower Users: Budget for Training

Consider the training components required for your project. In the beginning, you may need help learning about the issues and milestones inherent to hand-held wireless data deployment. Look to vendors and integrators for insight on how to budget and plan for an enterprise hand-held initiative.

Systems analysts charged with translating business objectives into a wireless data production deployment may need training on application development tools, supporting infrastructure technologies, typical integration points with existing systems, and the tools and techniques to bring it all together. Your support staff will need training on the device and application, frequently asked user questions, and how to address issues that arise regarding hand-helds.

Finally, end users will need training on device basics, as well as how to "manage" their devices—keeping them secure, synchronized, and integrated smoothly into their daily processes and activities. In order to realize a quick return on your investment, which can be surprisingly fast with hand-held applications (anywhere from 1 to 3 months is not uncommon), industry experts advise earmarking at least 12 to 15 percent of the total cost of ownership for training. Enterprise-wide success rests on a holistic approach to training and support that includes training on applications as well as devices, and attention to IT staff and applications analysts as well as users.

Synchronization Overview

Synchronization is an integral and necessary component of the mobile infrastructure. The notion of mobility in and of itself creates the need for synchronization. For a variety of reasons, synchronization has been sporadically embraced for laptops, perhaps because they are so often used in environments where a landline is readily available. User inconvenience was traded for network simplicity, often resulting in expensive shelfware.

Forward-thinking companies that did synchronize laptops have realized increased performance, greater information availability, and decreased communications costs when compared with running a constant network connection. With the accelerating adoption of supermobile hand-helds, synchronization became a recognizable necessity, and in turn, a hot market in and of itself.

The Many Faces of Synchronization

Recently, focus on wireless data access has grown. Yet, even with a wireless data network in place, a server connection is not always available. Until pervasive wireless data bandwidth becomes a reality in the coming years, applications and systems absolutely must be fashioned to deal with low bandwidth and spotty coverage. This generally requires offline applications and data access as an option. And, if critical business information is to be stored locally at the device level, there is an inherent need to synchronize changes and updates when a connection is possible and the adequate bandwidth is available.

Hand-helds are inherently more mobile and more accessible than any preceding device, so the need for synchronization has become obvious. In the hand-held world, synchronization is a loosely used term that can mean a variety of things.

PIM and E-Mail Synchronization

Perhaps the most obvious service desired is that of maintaining the individual calendar, contact, and to-do's across devices. This information is commonly referred to as PIM data, short for personal information manager. It has immediate value to users when offered on a portable device that is always accessible such as a Palm hand-held. Therefore, centralized management of this service is often the first requirement considered by corporate IT.

Covering the Basics

Most calendar and contact management software (Microsoft Outlook) already supports mobile PC clients; thus, this function is probably already supported by corporate IT. It follows that synchronization of this information would be immediately desired by PDA and smart phone users.

Like PIM synchronization, the corporate maintenance and support of e-mail is another obvious user demand. E-mail is widely considered to be the "most killer" of killer applications —the one that users are most addicted to. As users adopt more mobile devices, they generally want IT to provide support for e-mail on these devices. Despite the obvious limitations of screen size, text input methods, and capacity for attachments, access to e-mail will be an important and necessary service to offer across all mobile computing platforms.

More Tips for Application Selection

Look for an application that will synchronize native device applications with the existing Exchange or Notes server. It should provide administrators with the ability to authorize certain users and hand-helds, determine which PIM/e-mail information will synchronize, and set defaults or enforced profiles with a variety of synchronization options—such as maximum number of records, maximum record length, yes/no for text attachments, and what data to back up to server. These will help protect the limited storage of the device.

Stick with enterprise-specific products. Is it an individual, personal hand-held–to–PC synchronization utility that has been retroactively improvised into an "enterprise" product? These products typically lack the necessary administrative features and ease of administration inherent to enterprise-developed products.

Don't buy your PIM synchronization from a vendor that can't support other advanced synchronization needs, such as those being discussed next. The inevitable result is a hodgepodge of tools requiring redundant administrative work and continuous reconciliation and integration of different systems.

File Synchronization

At many corporations, a vast amount of information is stored in desktop application file formats created by word processing, spreadsheet, and presentation programs. These are typically distributed through e-mail or via an intranet site. Both approaches require proactive effort by mobile users to maintain the current version of the document on their mobile devices.

Distributing Personalized File Content by Using Profiles

Instead, a mobile computing infrastructure should provide the capability to automatically deliver personalized file content to users according to their profile. The current version should be maintained automatically on their device without any effort on their part. Typical examples of such files include sales call reports, benefit information, product pricing, HR forms, contracts, presentations, expense reports, company policy statements, sales literature, company positioning, and press releases.

In addition, a catalog of optional documents should be made available for users on request. Users should have control over their synchronization sessions, so that they can bypass optional activities to keep their session length down when preferable, and to manage the storage capacity on their devices.

This capability should not require any change to the way the information is maintained and stored on the corporate file server and network. Instead, it should offer convenient graphical tools to define the information to be published and to note which users receive specific information on each user's mobile device.

Delivery of key files should be guaranteed along with extensive logging to track who got what. File backup capability for harvesting important documents from the mobile devices should also be included.

Finally, the ability to translate common file formats for viewing and editing on non-native platforms is extremely important. A common example is the out-of-box inability of a Palm OS-based hand-held to view MS Word or Excel documents without special user intervention. When evaluating content synchronization and delivery technologies, use the following features as a starting point for building your criteria:

- Basic enterprise infrastructure platform features (see below)
- Publish and subscribe model for distribution logic
- A publication wizard for configuring the logic
- Publication availability scheduling
- Device-to-server file backup
- Overwrite versus rename configurations
- Automated scanner to detect new/updated files
- File translation/transcoding for multiple platforms
- File delivery logging
- File versioning capabilities

- Scripting wrappers for file delivery specs
- Part of a total mobile and wireless infrastructure solution[1]

Data Synchronization Options

Data synchronization requirements for hand-helds generally fall into one of two categories. Basic data synchronization requirements are common for relatively simple data access and collection applications that are usually new and replace paper-based systems. The basic data synchronization server acts as a communications server, managing the simultaneous connection of many users and passing information between the device's data store and a server database through a conduit. A variety of communications modes are supported. Thus, users may connect through desktop cradles, wireless data communications, dial-up, or direct network access. Basic synchronization capabilities include:

- Basic enterprise infrastructure platform features (see below)
- Integration with the standard device synchronization mechanism such as HotSync for Palm hand-helds
- Multiple communications modes
- Field-level change posting
- Full device refresh
- Refresh from an intermediate repository
- Support for standard IP
- Detailed logging and alerts
- Extensive administrative control and configurations
- Part of a total mobile and wireless data infrastructure solution[1]

Basic synchronization servers typically have a documented API. The IT organization writes the conduit previously mentioned. This conduit is custom code written to the API of the synchronization server to define the data mappings, data sharing, and synchronization logic and supporting functionality. For straightforward applications, the conduit coding effort is easily accomplished. With more complex requirements, it is more efficient to utilize an advanced data synchronization server than to try to write additional functionality and management tools into your conduit. Typical requirements that favor the robust functionality and rich management tools of an advanced data synchronization server include the following:

- Support for PC or laptop clients is also required
- Business processes evolve rapidly
- Connected users share enterprise data
- Information flows between mobile users
- Data refresh/realignment is required
- A complex data model is being synchronized
- Very large transaction volumes
- Processes run against the central database
- Mission-critical transactions are synchronized
- Robust administrative tools are required[1]

Basic data collection applications, such as inventory, warehouse, shipping, inspections, and meter reading, often find basic synchronization solutions adequate to meet their requirements. When the preceding criteria are present, purchasing an advanced data synchronization engine will provide significant cost savings when compared with the cost of writing, supporting, and debugging a large amount of custom conduit code written to define sharing rules and provide the features mentioned next. If your needs require an advanced data synchronization solution, look for the following features to be sure you are getting everything you'll need:

- Basic enterprise infrastructure platform features (see below)
- Support for heterogeneous databases
- Store-and-forward architecture
- Fast synchronization sessions
- Rich administrative tools
- Change capture including triggers, ODBC, and logs
- Transaction support—serialization and rollback
- Flexible conflict management and resolution
- Support for multiple communications layers
- Full bidirectional synchronization with sharing logic
- Parameterized sharing rules
- Field-level synchronization, partitioning, and data mapping
- Performance tuning utilities
- Support for complex database schema
- Nonintrusive architecture to easily integrate

- Sharing realignment and full client refresh capabilities
- An API for custom modifications
- SQL function trapping to preserve integrity
- Client-side error logging
- Part of a total mobile and wireless infrastructure solution[1]

Building in all these features via custom coding to augment the basic data synchronization communication server can drain significant resources. However, these capabilities are vital in managing your mission-critical enterprise data. As a result, organizations should look to buy, rather than build, an advanced data synchronization engine.

With basic PIM and e-mail synchronization, personalized file distribution, and data synchronization set up, you'll be providing users with most of the information they need to be productive, and you'll be effortlessly gathering and consolidating data from the field. Next, let's address some of the administrator's needs.

System Management and Inventory

Synchronization of corporate data often requires that related applications be deployed on the mobile devices. Management of application software and systems quickly also becomes a major requirement of a mobile infrastructure strategy.

Application Management

LAN-based approaches to system management of mobile devices are usually unworkable. Mobile users are intermittently connected through unreliable communications and require event-driven synchronization sessions. These factors introduce unique requirements. Support staff are unable to physically access the machine and must somehow keep track of overall inventory and the individual characteristics of each device.

Thus, mobile and wireless data devices demand a different type of system management solution, one that complements existing systems, but meets the unique needs of the occasionally connected user. Key features of a mobile system management solution should include:

- Basic enterprise infrastructure platform features (see below)
- Self-upgrade support and healing
- Delivery logging

- Publish and subscribe model
- Software package wizard
- Device history tracking
- Software inventory
- Scripts that are able to access inventory data
- Scanner utility
- Byte-level file differencing
- Offline installation
- Versioning
- Part of a total mobile and wireless infrastructure solution[1]

Utilizing software distribution capabilities makes the IT staff more efficient at managing the mobile devices, protects user productivity, lowers costly support incidents, and prevents user downtime.

Managing the Mobile Network

A comprehensive mobile infrastructure solution is necessary to achieve consistent and efficient management of your mobile network. Reactive and fragmented support is not an option. Deployment of one integrated mobile computing suite, complete with common administrative and user interfaces, will provide a host of benefits.

Demand Comprehensive and Fully Integrated Service

Working with a single vendor means a single contract and a single source for support. Integrated administration reduces duplicated efforts, provides for a simple and efficient end-user interface, and reduces training for IT staff by eliminating multiple tool sets. Cost savings are also realized through the elimination of application integration costs and reduced license, maintenance, and support expenses.

You should look for a strong underlying platform in the solutions you consider. Must-have features include the following:

- Support for all major hand-held devices
- Integrated management
- Multiple connection modes

- Remote server administration
- Administrative database
- Compression
- Scalable architecture
- Alerts and notifications
- Microsoft Management Console plug-in
- Directory services integration
- NT domain authentication
- Manage performance thresholds and load balancing
- Server-side process execution
- Encryption
- Open APIs
- Checkpoint restart in communications
- Standard Internet technologies
- Guaranteed delivery
- Logging and reporting
- Administrator-set scheduled connections

Administrative Console

The administrative console is the focal point for configuring and managing your mobile and wireless data infrastructure. It should offer a robust administrative interface for maintaining the settings and rules that drive the behavior of server-based engines.

Ideally, you can have one administrative console that manages all functions for all mobile devices you support. This console should allow system managers to:

- Define the user base
- Define activities (file, software, e-mail, and data distributions)
- Subscribe users to activities
- Prioritize the order of activity execution
- Review extensive system logs
- Review mobile device inventory
- Set alerts and notifications
- Troubleshoot and address problems[1]

The administrative interface will allow you to interact with the mobile wireless data network, provision new users, model changing business processes, track down problems, monitor the aggregate device inventory for planning, and complete all the other day-to-day tasks involved in supporting mobile users. It's important to have one central console, instead of learning a variety of administrative interfaces and duplicating tasks such as assigning a new user to a profile. Ideally, you can manage the entire mobile network from one GUI.

Communications Options

A variety of communications options are available for connecting hand-held devices to company servers for synchronization. It is very likely, and often advisable, for your organization to employ a mix of the following:

- Cradle to PC synchronization, which requires some sort of staging on PC
- Cradle through PC to network synchronization
- Network cradle synchronization
- Network dial-up (wired and wireless data)
- Wireless data direct to server over Internet synchronization (also wired)[1]

The specific options available will vary for different hand-helds, and for different communications/networking providers. A full review of these is beyond the scope of this chapter. The important thing to keep in mind is that the networking and connection options must serve the user. And mobile users need options to stay connected when they are out of range of a wireless data connection, or have only a low-bandwidth dial-up connection available from a hotel, for instance.

So, your hand-held infrastructure solution, including the synchronization and management tools, will need to support a range of communications protocols and transport mechanisms. The vendor should be committed to providing a total solution with the flexibility you need today—and tomorrow.

Security Concerns

IT is ultimately responsible for the integrity, confidentiality, and availability of the enterprise system. A comprehensive infrastructure plan will include the early integration of security solutions. The most basic and inherent security risk posed by a mobile device is that it is not bolted down. It walks out your doors every afternoon. It is left in your salesperson's car. It is carried on planes, trains, and automobiles, and left in hotel rooms around the world. And it contains information critical to your business.

Security on the New Frontier

Mobile enterprise security is founded in policy and supported by cutting-edge technology. A comprehensive mobile enterprise security solution will include:

- *Broad platform support.* From enterprise and e-commerce[3] servers, to desktops and Java, all the way down to PDAs, smart phones, and Internet appliances.
- *Standards compliance.* Support for all the current and de facto standards, and complete interoperability with past, present, and planned installations.
- *Network independence.* The ability to build a solution for Internet, intranet, wireless data, and even nonstandard networks.
- *Mobile, hand-held, and embedded specialization.* Toolkits optimized to provide full-strength security that is small, fast, and efficient even in the smallest devices.
- *Client authentication.* Provide the highest form of security on any platform, even hand-helds—a must-have feature for enterprise data access and financial transactions.[1]

Conclusion

In midst of challenges to launching an enterprise hand-held wireless data deployment, how are companies realizing success? The answer is something of an enterprise epiphany: whenever CEOs and CFOs say to themselves, "We are going to have a mobile wireless data network and we are not afraid to support it." The moment strikes when decision makers realize that to become more effective, more efficient, and more competitive, they don't have to develop big pieces of software. They discover that mobility is less about moving people and more about moving information. They discover that smaller pieces of software can be rapidly deployed to hand-helds which link back to their server and effectively push critical data out into the field where users interact with the market.

Now, the person you could have given a $4000 laptop is doing the same effective work with a $149 Palm OS hand-held because you have harnessed the information technology power of your existing back-end infrastructure. It's a small price to pay when you stand to gain a full return on your investment in as little as 2 months. And, having deployed a flexible mobile and wireless data infrastructure platform, no matter what new mobile device is coming, your wireless data network is prepared to meet it head on.

References

1. *The Handheld Applications Guidebook: Getting Started Deploying and Supporting Enterprise Applications on Handheld PDAs,* Synchrologic, Inc., 200 North Point Center East, Suite 600, Alpharetta, GA 30022, 2002.
2. John R. Vacca, *i-mode Crash Course,* McGraw-Hill, 2001.
3. John R. Vacca, *Electronic Commerce,* 3d ed., Charles River Media, 2001.

CHAPTER 14

Implementing Terrestrial Fixed Wireless Data Networks

Terrestrial wireless data communication may be mobile[4] or fixed. The first and second generations of wireless data technology emphasized mobility and lower frequencies, since they sought to provide only basic wireless data telephony and low-speed data communications. The third generation saw mobile technologies operate in higher frequencies (such as the 2-GHz PCS band) while coverage zones expanded, power needs fell, data rates rose, and more services were supported. Simultaneously, third-generation terrestrial fixed technologies made dramatic leaps into upper-band frequencies (24 to 39 GHz), attaining data rates as high as wireline technologies so as to support a range of hypercommunication services including voice, video, high-speed data, and Internet.

In this chapter, the implementation of terrestrial (nonsatellite) fixed wireless data technologies is discussed. As with wireline technologies, almost every specific service can be provided by terrestrial fixed wireless data technologies. The ability of terrestrial fixed wireless data technologies to serve as access paths for hypercommunication services depends mainly on signal frequency, user mobility, and the availability of appropriate antennas, DCE, and DTE. (The Glossary defines many technical terms, abbreviations, and acronyms used in the book.) Table 14-1 details terrestrial fixed wireless data technologies that are used to support mobile and nomadic services.[1]

Typically, mobile user devices include DTE, DCE, and an antenna in a single unit. Mobile services are provided by carriers in a series of overlapping coverage zones (cells), each of which is served by a tower attached to base stations. As subscribers travel in their carrier's local footprint, calls are passed from one cell to another. Subscribers may roam regionally or nationally and use their own (or another) carrier's network if compatible technologies are available in the roamed area.

Available Terrestrial Fixed Wireless Data Technologies

Table 14-2 lists several terrestrial fixed wireless data technologies that are available.[1] For each technology, the typical frequency and channel bandwidth, data rate, and services supported are shown. Since terrestrial fixed wireless data technologies are works in progress, the table cannot convey more than a broad general categorization. Hence, the specific technologies listed in the table are often imprecise terms, based on a melding of traditional FCC definitions, proposed frequencies, experimental tests, and implementations by carriers.

The first fixed terrestrial wireless technologies are WLAN (wireless LAN) technologies. The first type of WLAN is infrared WLAN. Infrared

TABLE 14-1

Mobile Terrestrial Fixed Wireless Data Technologies

Technology	Market Applications	Frequency (Channel Bandwidth)	Data Rate
CDMA (narrowband PCS)	Two-way paging, digital mobile enhanced telephony	900–941 MHz (25 MHz)	6.4–25.6 kbps
CSCD	Fax, file transfer, Internet	800-MHz analog cellular (25 MHz)	1.2–9.6 kbps, 14.4 kbps optimally
CDPD (analog and early digital cellular overlay)	Two-way paging, POS, data queries, dispatch, on analog cellular line	800 MHz (12.5–30 MHz)	4.8–12 kbps
Digital cellular–PCS (IS-136 TDMA, IS-95 CDMA)	Two-way paging, fax, e-mail, enhanced digital telephony	800, 900, 1900 MHz (25–200 MHz)	9.6–14.4 kbps (current), 1.2 Mbps (spread spectrum, future)
ARDIS (Motorola DataTAC)	Two-way paging, transportation data support, in-building wireless	800 MHz (25 MHz)	2.4 kbps (nationwide), 9.6 kbps (limited)
BellSouth wireless data (Mobiltex)	Two-way paging, dispatch, database	900 MHz (12.5 kHz)	4.8 kbps
AMPS cellular telephone	Mobile telephony, modem applications	400 MHz (25 MHz)	4.8–9.6 kbps; see also CDPD, CSCD
iDEN (SMR, ESMR)	Enhanced digital mobile telephony, Internet, e-mail, dispatch	800–960 MHz (15–30 MHz)	9.6–64 kbps (Nextel)
GSM-1900, TDMA (broadband PCS)	Enhanced digital mobile telephony, file transfer, Internet, e-mail	1.9–2.1 GHz (200 MHz)	300 kbps (extended range), up to 2 Mbps (local CZ)

adapters plug into token ring cards to allow localized transmissions (within 80 ft). While infrared can be used between buildings, it is so susceptible to fog, rain, and smog interference (since it is a form of light) that it is usually used for remote controls, wireless keyboards, and wireless computer mice.

For example, the WLAN hardware market (which is driven in part by wireless data collection) will grow at better than 36 percent annually

TABLE 14-2

Terrestrial Fixed Wireless Access Technologies

Technology	Market	Typical Frequency (Channel Bandwidth)	Data Rate (Range)
WLAN via infrared	Segments of LANs	3000–30,000 GHz	4–16 Mbps (with token ring card, 80 ft)
WLAN via laser	Local-area network in office or campus	30–150 THz (800-nm waves)	To 16 Mbps (3280 ft, line of sight)
WLAN via RF	Local-area network in office or campus	5.1–5.8 GHz (30–100 MHz), unlicensed spread spectrum	10–100 Mbps (3280 ft in open, 650 ft in building)
2.4-GHz WCS	WWAN, WLAN, and Internet access 2.402–2.48 GHz	(1–25 MHz per channel), unlicensed spread spectrum	156 kbps to 11 Mbps (up to 10–25 miles)
DEMS	Broadband microwave access: telephony, WAN, video, Internet	24 GHz (100 MHz)	Up to 30 Mbps (2–10 miles)
MMDS	Microwave access: telephony, data, video, Internet	2–2.6 GHz (190–200 MHz)	Up to 10 Mbps (30–35 miles, one-way; 6 miles, two-way)
WLL	Broadband microwave access and WAN: data, Internet, ATM, telephony, video	39 GHz (100 MHz–1.4 GHz)	45–155 Mbps (2–5 miles, possibly 9–10)

through 2006. Bar-code data collection is another fixed terrestrial wireless data infrastructure that's increasingly wireless, distributed, and based on common IT standards (see sidebar, "Terrestrial Fixed Wireless Bar-Code Data Collection").

Terrestrial Fixed Wireless Bar-Code Data Collection

Deploying an enterprise resources planning (ERP) system put Laufen USA's ability to manage transactions and information on a whole new level. But managers at this Tulsa, Oklahoma–based manufacturer of prefired tile needed to find a way to extend that level of information management to Laufen's warehouse operations.

The solution the company settled on was the deployment of a terrestrial fixed wireless automatic data collection system that it integrated with the warehouse management and data capture functions of its ERP system, from Herndon, Virginia–based The Baan Co. The solution eliminates the need to manually enter data about the movement of goods in the warehouse, while doing away with paper-based methods.

Laufen's Baan ERP system allowed the company to control the warehouse, but it was still a manual data entry process. Laufen needed a real-time solution that would feed information back to their ERP database as warehouse and inventory transactions occur.

The solution that Laufen chose uses standard terrestrial fixed wireless local-area network (WLAN) technology, and builds upon the ERP investment the company was making. In so doing, the deployment exemplifies a couple of trends. One is a continuation of the long-held value proposition behind data collection that it increases the value of existing systems by improving the timeliness and accuracy of data. The second trend (the deployment of data collection solutions that are increasingly wireless and built on ubiquitous IT standards) is evolving more rapidly, and involves a host of technologies.

But, for the time being, data collection's use remains closely tied to the health of the market for transactional systems such as ERP that rely on data collection as a means of recording transactions. With the economy inching its way back to health, some observers expect that data collection deployments will pick up. The next step in many existing ERP deployments is to automate that "last mile," down to where the data are generated.

As the economy picks up, a lot of manufacturers are going to be looking around at where they can gain the next set of efficiencies and save money. Many will realize that one way to do that is with better data.

Research from Venture Development Corp. (VDC), Natick, Massachusetts, bears out the dual nature of the data collection market. VDC estimates that growth for some long-established segments of the data collection market (such as hand-held scanners and barcode printers) will be relatively modest through 2006, with scanner sales increasing just 6 percent, and printers by 9.7 percent. On the other hand, VDC estimates that the WLAN hardware market (which is driven in part by wireless data collection) will grow at better than 36 percent annually through 2006. The result is a data collection market in transition—with data collection continuing as a valued extension to enterprise systems and other applications, while new mobile applications also drive growth.

Solid Payback

Laufen operates on a 24 × 7 basis, shipping about 280 truckloads (5600 pallets) of tile each month. Prior to 1998, the company's main warehouse was located 15 miles from its headquarters. With more orders to fill, it became clear that the warehouse should be on site. The move was an ideal time to implement new systems.

As part of the move to the new 193,000-ft^2 facility, Laufen began its search for systems integration firms capable of adding wireless data-capture functionality to the ERP system. Laufen selected Jump Technology Services, also of Tulsa, as its integration firm, and Jump recommended terrestrial fixed wireless data collection equipment from Intermec. Going with a terrestrial fixed wireless data system made the most sense, given the size of the facility.

Intermec T2425 Trakker Antares hand-held and T2455 vehicle-mounted computers replaced the handwritten log sheets in the Laufen warehouse, all tied together with an Intermec 2.4-GHz terrestrial fixed wireless data backbone. The units are used to scan pallets of tile as they move through Laufen's production and shipping processes.

A single scan of the pallet label at the production line prompts the system to post a quantity to the production batch, and closes the batch if complete, placing it in a finished goods receipt location. These transactions used to require multiple manual entries.

Information from the scanned bar code travels from the mobile scanner through the terrestrial fixed wireless data backbone. Intermec access points strategically placed in the rafters of the warehouse work with a repeater to boost the radio signal to the network controller, which links to the ERP system.

As pallets of boxed tiles arrive at the warehouse aboard a truck from a holding location near the production line, a worker scans the bar-code labels. This updates the data-capture system on the pallets' status and automatically provides an audit trail.

A forklift driver takes the pallets to randomly chosen storage slots within the warehouse aisles. Each of the slots is identified with a bar-code label. The driver scans the bar-code labels on both the pallet and the intended slot. This prompts the system to link the product with its location. It also notes the production date, allowing the system to track product by age.

After relocating to the new warehouse and going live with the data collection system, Laufen began to see results almost immediately. It has better visibility of the product. The system allows it to operate more efficiently, and make better use of its people.

More Mobility

While terrestrial fixed wireless data collection has a relatively long history in warehouse settings, some see broader-ranging mobile data collection as the future. Today, the acceptance of PDAs, terrestrial fixed wireless data and cellular networks, and mechanisms for data synchronization is such that there are fewer boundaries to data collection and to simplified, bidirectional interaction with systems.

In a bidirectional mode, devices function as full-fledged computers that can access applications over terrestrial fixed wireless data networks and receive alerts, not simply scan bar codes. Such devices are Web-enabled and WLAN-compatible, allowing them to communicate wirelessly while on the road or within a sprawling facility.

Symbol Technologies (Holtsville, New York) recently announced that London-based petrochemical giant BP is implementing Symbol's mobile data collection solutions at multiple sites worldwide. The systems, utilizing WLAN technology, will support asset management of thousands of pieces of equipment in BP's refineries.

New devices running on Windows CE or the Palm operating system also are part of the mobile data collection trend. If you look at some of the units today, such as the Intermec Model 700, it's a pocket PC with all the ability to run applications that the platform brings with it.[2]

The last WAN technology includes the 5.8-GHz spread-spectrum technologies that make use of unlicensed spectra to operate on a single premises. Spread-spectrum technologies allow low-power operation and reduce interference. In open spaces, 100-Mbps ranges for wireless data LANs (such as the Breeze Net Pro.11 product line) can range from 1000 m (3280 ft) in open areas to 60 to 200 m (200 to 650 ft) inside buildings.

Wireless Local-Area Networks

WLANs use all-wireless data Ethernet and specialized hybrid technologies. An all-wireless data Ethernet LAN replaces cabling[6] among computers in a local network with wireless data paths, so individual computers require antennas to transmit to the network host. With all-wireless data Ethernet technologies, laptop computers connect to the WLAN through antennas in their PCMIA slots and remain portable in the office. Hybrid WLAN technologies use wireline cabling to connect each machine in a particular area to a hub (or other intermediate DCE), but use wireless data paths from hub to central server.

The second WLAN technology, WCS, may be particularly useful for agribusinesses seeking to interconnect LANs inside a 10- to 25-mile radius of a central site [creating wireless data WANs (WDWANs)], or to obtain Internet access. The 2.4-GHz WCS band is also unlicensed spectrum that can be used on the local or access level. Since spectrum is unlicensed, the FCC requires that spread-spectrum technologies be used that make radio signals appear as background noise to unintended receivers. For example, numerous service providers in many parts of Florida are currently offering wireless data Internet access in the 2.4-GHz band. However, interference from garage door openers, baby monitors, and other wireless data equipment can occur in the unlicensed frequencies used by WCS.

DEMS is a two-way all-digital system that uses digital termination systems (DTSs) on each end as DCE. An important characteristic of DEMS is proper antenna placement. User stations require directional antennas over a 2- to 10-mile-long path length. In 1998, Teligent began to provide the first DEMS service in Florida to metro Jacksonville, Tampa, Orlando, Palm Beach County, and Miami–Dade County.

Upper-Band Technologies

Teligent's DEMS technologies use a two-step wireless data layout. In the first step, at the access level, when a customer makes a telephone call or accesses the Internet, the voice, data, or video signals travel over the building's internal wiring to the rooftop antenna. These signals are then digitized and transmitted to a base station antenna on another building, usually less than 3 miles away. The DEMS base station functions as a POP for that area, gathering signals from a cluster of surrounding customer buildings, aggregates the signals, and then routes them to a broadband switching[7] center (see sidebar, "Terrestrial Fixed Wireless Broadband Links").

Terrestrial Fixed Wireless Broadband Links

HTE8 began deployment of its domestic terrestrial fixed wireless data broadband network in Houston, providing connectivity to businesses that require reliable, easily deployed bandwidth solutions. HTE8's scalable networks allow for connections up to 44 Mbps, enabling enterprises to conduct business functions over the Internet in a high-throughput, secure environment.

HTE8's network design consists of eight nodes, providing coverage of over 90 percent of the Houston market. The transmission nodes utilize fiber-optic backhaul from several network partners to HTE8's point of presence at Level 3's Houston facility, assuring quality transmissions. Each node reaches up to 24 miles line-of-sight, and up to 15 miles non-line-of-sight. HTE8 intends to reach areas that have inadequate access to broadband connectivity, yet have high bandwidth requirements. With only 4 percent of buildings connected to fiber, and only 50 percent of any given market addressable by DSL, terrestrial fixed wireless data solutions bridge the gap between those businesses that have the advantage of speedy Internet access and those that don't.

Based on Cisco Systems'[5] terrestrial fixed wireless data broadband technology, HTE8's networks utilize hardware incorporating vector orthogonal frequency-division multiplexing (VOFDM), which enables non-line-of-sight implementations. HTE8's networks also implement point-to-multipoint transmissions, providing more scalable network deployments, thus allowing HTE8 to reach customers with diverse bandwidth requirements in a variety of configurations. Multitenant units in areas outside major metropolitan centers can receive state-of-the-art bandwidth solutions at lower costs and with faster provisioning times than historically available.

HTE8 works with service providers, such as ISPs, data centers, storage facilities, ASPs, and CLECs, to provide options for last-mile connectivity. Additionally, these companies utilize terrestrial fixed wireless data broadband for redundancy in addition to traditional terrestrial connectivity to insure mission-critical applications. Fiber-optic networks do not always reach the curb. HTE8 is able to reach customers and extend networks to locations for a fraction of the cost of fiber.[3]

MMDS is authorized at 190 MHz of spectrum near 2.5 GHz. MMDS architectures are designed for fairly large coverage zones, up to 50 km across. Typical MMDS reflector antennas are up to 0.6 m (2 ft) in diameter.

Designed as "wireless data cable TV," downstream MMDS signals can cover up to a 35-mile radius. Hence, MMDS technology is expected to have a broader market and more coverage of rural areas. However, until now, upstream (symmetric) MMDS has been limited to a 6-mile range. MMDS is often seen as a Small Office Home Office (SOHO) technology if the technology can become more adept at achieving symmetric data rates over distance and at sharing scarce frequency. Typical MMDS or customer premises equipment includes a roof-mounted transceiver and antenna, an up/downconverter to change signal frequency to frequencies usable by DTE, a network interface unit (NIU)—possibly a telephone interface—and an Ethernet hub or router.

WLL is still another upper-band technology christened as "fiber in the sky." Wireless POPs are centrally placed in urban areas with access accomplished via multitenant environment (MTE) office building rooftop transmitters to wireless data hub paths. Estimates are that only 3 percent of office buildings have fiber, but that they represent one-third of all business communication lines. WLL is targeted at this 3 percent (the power users), while MMDS, and to a lesser extent 2.4 GHz, is aimed at the 97 percent without fiber access—the smaller business customer.

CDMA technology is used to economize on spectrum, so individual clients may maximize data rates. The 39-GHz band in which WLL will operate can carry data rates of up to 155 Mbps over several miles. Winstar is a WLL provider in Florida, with the right to provide coverage from Jacksonville to Miami along the Atlantic Coast and from Citrus County south to Everglades City on the Gulf Coast.

The new upper-band technologies are no panacea for rural areas. According to the NTIA, most applications are for dense urban MTE locations.

New and proposed terrestrial fixed wireless data systems are being developed to exploit shorter-range, cellular deployments able to serve a much denser subscriber base, using multipurpose digital bit streams. These emerging systems will need to be much smarter and more complex than traditional systems, and they will demand extensive infrastructure development and integration into existing telecommunication infrastructures. Additionally, upper-band technologies may not hold great promise for a high-rainfall state such as Florida, since the Southeast is one of the worst areas for microwave signal propagation

In spite of predictions by others that terrestrial fixed wireless data bandwidth would be boundless in 2003, this author has different predictions for 2003: Wireless data users may hit a speed bump because of limitations on spectrum and failure to deploy equipment in many areas.

While terrestrial fixed wireless data technologies are developing slowly, inexpensive, symmetric satellite technologies[8] are coming even more slowly.

Conclusion

The best way to explain what terrestrial fixed wireless data technologies services and technologies are is by reviewing Fig. 14-1.[1] The seven layers of the OSI model are important in regard to technical issues of private data networking and the Internet.

The bottom of the figure represents the local, physical level that contains the local communications network and equipment. Most businesses have some computer equipment that may be used for private data networking or the Internet and telephone equipment to connect with the PSTN. The larger a business is, the more dependent on technology it is, and the more it has to communicate over long distances, the sooner it is likely to try to converge voice, Internet, and data into a single terrestrial fixed wireless data network. On a corporate premises, convergence means that it's currently separate voice, Internet, and data equipment and currently separate conduit will evolve into a unified whole.

However, even if terrestrial fixed wireless data services and technologies allowed smaller businesses to unify their networks, and even if the CPE needed to do the job was available, a high-speed connection to access the advanced terrestrial fixed wireless data network of the future would be needed. In many places, this last-mile connection is not yet able to handle convergence inexpensively if at all.

There are several reasons the promise of the future is hindered by the reality of the present. The infrastructure that connects a communications provider's POP to the business location varies from one location to another. Currently, wireline service may rely on the copper loop from the ILEC, a hybrid fiber-coax mix from the cable company, and the ability to connect directly to a fiber-optic network. Terrestrial wireless data access can be fixed, nomadic, or mobile depending on the movement of the user. Of the three kinds of terrestrial wireless data access, only fixed terrestrial is likely to compete sufficiently with the wireline infrastructure. However, most providers (wireline or wireless) have not worked out all the kinks in urban areas to provide a single link to businesses. In rural areas, the situation is even less developed.

In the future, a particular business may be able to obtain high-speed network access from competing providers and services for each of these four sources. At present, many businesses can obtain high-speed dedicated digital or circuit-switched digital access over copper from a single

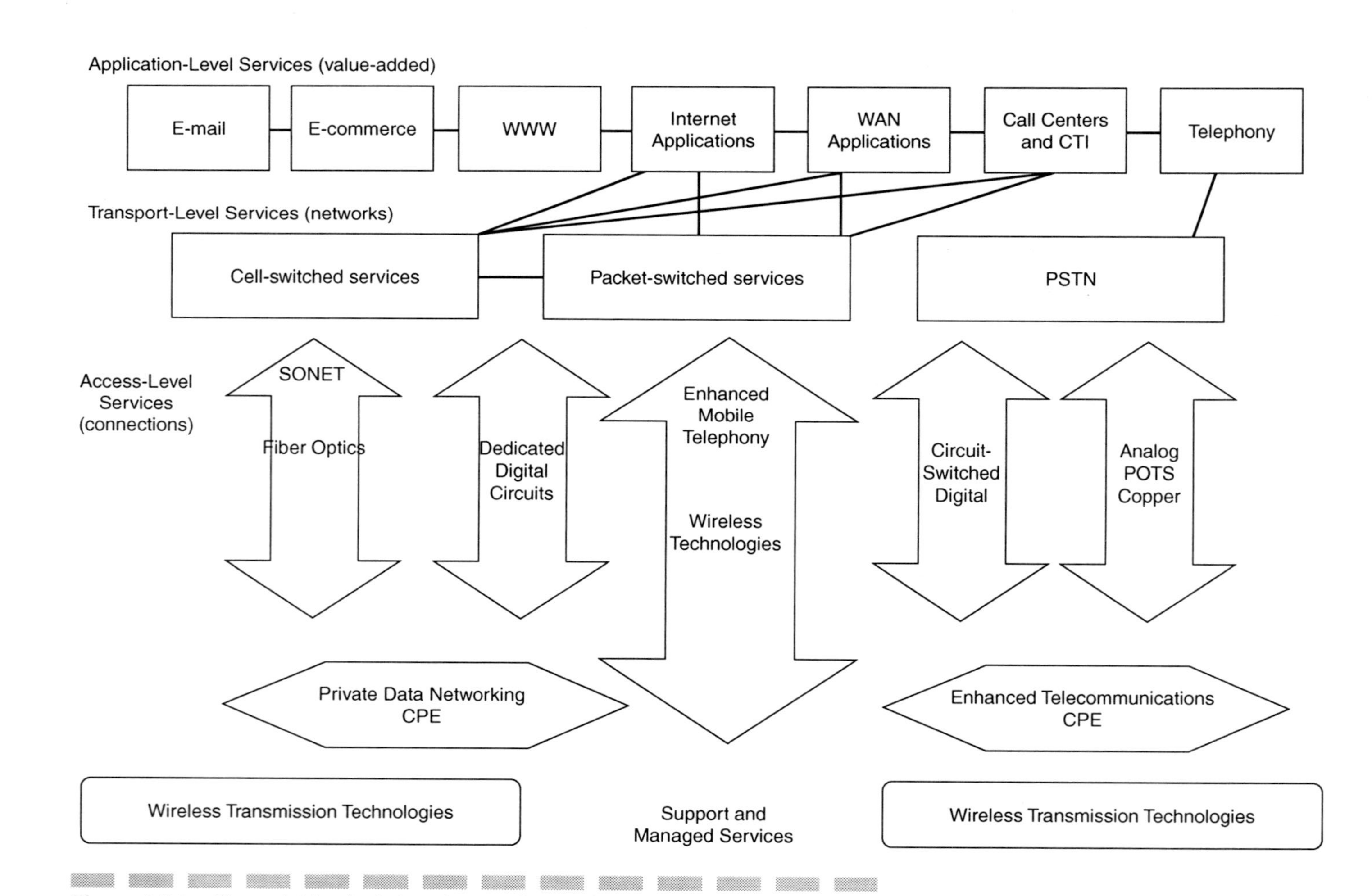

Figure 14-1 *Terrestrial fixed wireless data services and technologies.*

provider—the ILEC. Larger businesses with offices located in urbanized areas may have more providers (ILECs) willing to serve them by reselling the ILEC's copper loop or by using a fiber network that bypasses the local copper loop. Suburban locations may have cable or DSL access. However, most of these access level services are sold separately for Internet, data, and telephony.

Once businesses can obtain a single high-speed access-level connection to transport-level services (PSTN, ATM, and packet-switched networks), they will benefit from converged hypercommunication networks. Now, transport-level services are available and affordable only by large, strategically located businesses. As communication needs change, technology improves, and costs fall, transport-level services such as ATM should become more available and more demanded by businesses.

Application-level services or value-added services are currently available at least at low speeds to most businesses. New kinds of application-level services and value-added services will increase the benefits of high-speed terrestrial fixed wireless data networking. Even if high-speed access to intelligent networks was available to all businesses today, there would have to be demonstrable business reasons to adopt convergence technologies today. It may be expected that the earliest businesses to foresee innovative uses of terrestrial fixed wireless data will achieve supernormal profits from their use. This, in turn, will entice others to follow them. Just as with other technological changes, those who are slow to act may be left behind.

Now that a picture of what terrestrial fixed wireless data services and technologies are has been painted, and some of the most promising ones for businesses have been identified, the job is not over. Next, it is important to consider to what role location will play along with taxes and other government policies in allowing terrestrial fixed wireless data to reach rural locations.

References

1. Dr. Dean G. "Gordy" Fairchild, "Hypercommunications Convergence: Services and Technologies," 2645 S. 81st Lane, Phoenix, AZ 85043-5434, 2002.
2. Roberto Michel, "New Routes to Better Data," MSI, Reed Business Information, 2000 Clearwater Drive, Oak Brook, IL 60523, 2002.
3. "HTE8 Deploys Fixed Wireless Broadband Links throughout Houston," *Business Wire,* 40 E. 52 St., 14th Floor, New York, NY 10022, 2002.
4. John R. Vacca, *i-mode Crash Course,* McGraw-Hill, 2001.

5. John R. Vacca, *High-Speed Cisco Networks: Planning, Design, and Implementation,* CRC Press, 2002.
6. John R. Vacca, *The Cabling Handbook,* 2d ed., Prentice Hall, 2001.
7. John R. Vacca, *Wireless Broadband Networks Handbook,* McGraw-Hill, 2001.
8. John R. Vacca, *Satellite Encryption,* Academic Press, 1999.

CHAPTER 15

Implementing Wireless Data and Mobile Applications

The implementation of mobile wireless data applications[2] is becoming increasingly visible in virtually all industries. This interest has fueled explosive growth in the mobile wireless data device market—despite the recent slowdown in the tech sector of the worldwide economy. According to Gartner Group, the worldwide mobile PC market grew by 43.8 percent in the first quarter of 2002 compared with the first quarter of 2001.

Add hand-helds into the equation and today's enterprise is faced with a virtual flood of mobile wireless data devices. Meta Group has predicted that, by 2005, each corporate knowledge worker will have four to five different computing and information access devices that will be used to access various applications.

When mobilizing wireless data applications, the enterprise must answer several questions. Which devices support existing business practices? How will those devices connect to the network? How much functionality will reside on the devices? And how will the information on those devices be kept in sync with server information? This chapter answers the last question—too often the most overlooked component of going mobile. (The Glossary defines many technical terms, abbreviations, and acronyms used in the book.)

Why Synchronization?

In mobilizing wireless data enterprise applications, one of the most important questions your corporation must answer is what type of architecture will allow mobile users to connect to corporate servers. There are two options to consider here: synchronization and real-time access.

Real-Time Access

In real-time environments, users are perpetually connected to corporate servers and applications. Mobile wireless data devices are essentially terminals—storing no data or applications locally. Data are viewed as they are entered, with the mobile experience hinging on network performance and available bandwidth.

This environment, well-suited for LAN-connected users, faces challenges in the reality of wireless data instability. Occasional interruptions in network availability can be devastating to mobile users whose devices store no data or applications locally. Mobile users are at the mercy of the connection availability, unable to compensate for lapses in coverage.

Synchronization

In synchronization, or store-and-forward environments, users work with data and applications replicated from corporate servers. Some data and application components are stored on mobile wireless data devices, allowing mobile users to work effectively regardless of network status or available bandwidth. Data are updated locally and at the server only through synchronization sessions. In a wired or wireless world, synchronization offers the following benefits over real-time, always-connected environments:

- Reduced queries and network traffic
- Reduced user idle time
- Compression of staged data
- Reduced concurrent server processing loads
- Controlled communication costs[1]

What about Wireless?

The current wireless data landscape is marked by competing standards and relatively slow connection speeds. But this situation is only temporary. As wireless data protocols mature, the lines between real-time and store-and-forward architectures will begin to blur, and organizations will deploy both options in a complementary fashion. According to Gartner Group, the convergence of synchronization and real-time mechanisms is crucial in accommodating varying bandwidth and connection scenarios, and in graceful switching between modes of operation.

The current state of wireless data technology, coupled with the inconvenience of staying perpetually connected via wireline, has created the reality of the occasionally connected user. Synchronization is the best way to accommodate occasionally connected users.

Comprehensive Selection Criteria

Data synchronization is a deceptively complex technology. In evaluating vendors, it is important to select an organization with demonstrated experience in both the hand-held and mobile PC environment. Evaluating vendors based on the selection criteria that follow will ensure your project stays on track, protecting you from a variety of potential issues:

- Data corruption or data loss
- User frustration leading to system abandonment
- Long connection times and excessive bandwidth usage
- Prohibitively expensive communications
- Limited application functionality
- Extensive conduit coding and maintenance[1]

Don't settle on a vendor that cannot adequately address the aforementioned scenarios. Demand a robust synchronization platform that offers all of the features described next.

Multiple Device Support

Because most organizations are tasked with supporting multiple devices, your solution should be device-flexible, supporting a range of mobile hardware including:

- Laptops
- Remote desktop PCs
- Windows CE/Pocket PC devices
- Palm OS devices
- Industrial hand-held devices
- Point-of-sale systems
- Bar-code readers
- Portable data terminals
- EPOC smart phones and hand-helds
- RIM Blackberry e-mail pagers[1]

If the vendor does not support all of the devices today, it should have the stated intention of supplying, as a comprehensive solution, the additional devices as they become widespread.

Store-and-Forward Architecture

The wireless data synchronization solution you choose should feature a store-and-forward architecture. Store-and-forward environments zealously guard enterprise resources by minimizing connection times. Changes are accumulated, both on the client and server sides, with the connection time extending only as long as required to exchange compressed data files.

Offline Synchronization

Once the staged change files are passed from server to client, and vice versa, the connection to the server is closed. Only after the connection to the server is terminated are the change files actually applied. Alternative architectures feature online synchronization, where changes are applied during the connection session (resulting in longer, more costly communication sessions, while increasing the likelihood of concurrent synchronization sessions), dramatically reducing server performance.

Clustered Server Architecture

A clustered server architecture ensures scalability as the number of users or amount of wireless data transferred grows. Server performance can be continually monitored and new hardware added where appropriate.

Dynamic Load Balancing

Dynamic load balancing allows the system architecture to be more reliable and flexible, by assigning specific users to specific servers only at the time of connection. Servers request workloads instead of operating against a predefined set. Thus, if a particular server is not performing at optimal levels, the transactions are dynamically transferred to servers with smaller workloads. Dynamic load balancing reduces the stress on enterprise servers and aggregates available computing power for premier performance.

Server Failover and Recovery

Server failover and recovery elegantly allocate users to different servers in the event of server failure—a significant extension of the above architecture characteristics. Not only will users be dynamically allocated to the appropriate server according to workloads, but also, if that server fails at any point during synchronization, the processing and communications will instantly shift to another server. The failed transaction is recovered and applied correctly by the new server.

Failover and recovery ensure users can always synchronize, as there is always a server available to handle their synchronization session. Server failures are accommodated and transparently managed, so the end user is unaware of any problems.

Multiple Database Support

It is important that your wireless data synchronization solution features out-of-the-box support for a variety of industry standard databases including:

- Microsoft SQL Server
- Sybase SQL Anywhere (ASA)
- Sybase Adaptive Server Enterprise (ASE)
- IBM DB/2 UDB
- Oracle
- Microsoft Data Engine (MSDE)
- Microsoft Access[1]

Progressive solutions accommodate disparate client and server databases to allow the enterprise to leverage the most appropriate technologies for wireless data and mobile devices and applications alike. Selecting such a solution will protect your investment by not forcing you to choose devices or applications based on the restrictions of your platform.

Automated Data Type Conversion

In addition to supporting multiple client and server databases, your synchronization solution should include the ability to convert data types. This feature enables the synchronization engine to modify wireless data to fit the server or client database requirements. This feature is vital in organizations employing multiple database types. Typical wireless data type conversions include:

- Floats to numbers
- Decimals to numbers
- Bit fields to booleans
- Bit fields to character fields[1]

Open Application Development Support

In implementing wireless data and mobile applications, your software development process should not be taxed with accommodating a synchronization platform. Your wireless data synchronization solution should be independent of, and agnostic to, your application development environment. You should be able to develop your wireless data and mobile application with any tool that supports your target device. Common platforms that should be supported include Microsoft VB, VC++ for CE and PCs, Delphi, Java, CodeWarrior for Palm, AppForge, and Satellite Forms.[3]

Field-Level Synchronization

Field-level synchronization ensures that only field-level data changes are transmitted during synchronization sessions, a feature that results in dramatically reduced communication costs and transaction overhead. Synchronization session times are minimized and needless wireless data collisions are avoided, while multiple users can simultaneously update records, as data are locked only at the field level.

Flexible Change Capture

Flexible change capture at the server allows wireless data and mobile users to have access to updated enterprise data regardless of the source of modification, including centralized batched processes, LAN-connected user updates, and legacy systems. The benefit is that enterprise users, whether wireless data and mobile or LAN-based, all share access to the same updated information.

Many synchronization solutions available today rely on change flags or shadow tables that require changes to database schema. Your wireless data synchronization solution should feature a variety of change capture mechanisms including:

- Triggers
- ODBC drivers
- Transaction logs[1]

Triggers Using trigger capture allows the synchronization solution to capture changes from any wireless data and mobile application that makes modifications to the database. These possible sources include the client wireless data and mobile application, overnight feeds from other data sources, and LAN users connecting to the database. Triggers allow the application developers complete freedom in choosing how they access the database, capturing changes from native libraries, ODBC, ADO, etc.

ODBC Drivers All databases essentially provide an ODBC interface. Wireless data and mobile application developers that want to provide database-independent functionality can leverage the ubiquitous availability of ODBC drivers.

Transaction Logs Transaction logs are supported on IBM's DB/2 platform. Using transaction logs minimizes the run-time impact of any change capture mechanism, because it does not require any additional execution to log that a change has occurred.

NOTE Some synchronization solutions support a variety of database vendors by requiring the integrator to supply change capture and scripts for the target platform.

Database Schema Protection

Wireless data synchronization should be nonintrusive to existing enterprise applications and databases. Existing tables and the synchronization mechanism should not affect the wireless data and mobile applications' access to the database.

Light-Client API

A synchronization solution should have minimal impact on your wireless data and mobile application. The synchronization technology should be self-contained and not require complex interaction from the client application. Look for a solution that provides a simple and standard component mechanism that allows you to request that a synchronization session occur, and everything is handled appropriately from there.

Conflict Detection

The very reason synchronization exists is to govern information flow and protect wireless data integrity as it flows between multiple users. But, what happens when user A and user B update the same record in a database with conflicting data? Your data synchronization solution should include out-of-the-box support for detecting conflicts according to your business logic:

- Last in wins
- First in wins
- Manager wins
- Higher revenue impact wins[1]

It's important that your solution is flexible enough to allow you to define conflicts, not just detect them. Such a feature protects business logic and works within the context of your application rules. Wireless data synchronization solutions featuring an API for the development of more advanced detection mechanisms for shifting business needs should be considered premiere solutions.

Multiple Conflict Resolution Mechanisms

Detection is only half of the conflict management component of your wireless data synchronization solution. A complete suite of conflict resolution tools should accompany your data synchronization solution and allow for comprehensive transaction management:

- *Log-based resolution.* Wireless data conflicts are logged, providing the administrator with a snapshot of synchronization events to determine which data are correct.
- *Transaction rollback.* In the event of wireless data conflict, the user-initiated transaction is reset, ensuring database integrity. Rollback is useful in situations where duplicate order entry is a possibility.
- *Column-level data rejection.* In the event that one component of a wireless data transaction is conflicting, column-level data rejection ensures that only the most granular conflicts are discarded.
- *Operation rejection.* In the event of wireless data conflict surpassing acceptability thresholds, operations are rejected.[1]

Wireless Data Distribution Rules

Administrators should be able to graphically partition wireless data and allocate information to users and user groups according to need. This feature enables administrators to ensure communication costs are minimized by replicating only necessary wireless data to subscribers.

Graphical Rule Generator

Providing administrators with a graphical user interface to control the creation and maintenance of wireless data segmentation rules ensures that even the most complex of distribution rules is easy to create and maintain. A graphical rule generator guards your business logic and eliminates complex coding.

Multilevel Mapping

Often your server database may have a different schema from your client device's database—especially on your hand-held clients. Your synchronization solution should support mapping to eliminate dissemination of unneeded columns or tables and wireless data distribution to a database with schema differences.

Rule-Based Wireless Data Sharing

Allocating wireless data among users streamlines data flows and minimizes synchronization times through delivering only the most relevant information based on user and group profiling. A rule-based synchronization solution ensures your mobile users receive only wireless data that are relevant to them and yields the following benefits:

- *Device resource optimization.* By limiting replicated wireless data at the user or user group level, device resources such as RAM and disk space are zealously protected and optimized.
- *Reduced synchronization session times.* Connection times are minimized with transparent filtering mechanisms for relevant wireless data transmission.
- *Security.* Because access to enterprise databases is restricted to an as-needed level, the risk of misinformation corrupting the database is minimized.[1]

Parameterized Rules

Once consensus is reached regarding wireless data segmentation rules, administrators should be able to use parameterized rules to implement the distribution and partitioning logic. This feature eliminates the requirement to build and maintain rules for each unique user. Instead you can create flexible wireless-data-driven rules to model the segmentation, allowing simple changes to the database to ripple throughout the systems without requiring coding effort.

Wireless-Data-Driven Data Sharing

Your synchronization solution should support a dynamic business environment. People shift responsibilities, new employees are hired, new products are released. These daily business changes shouldn't require additional administration of your synchronization technology. Wireless-data-driven sharing automatically distributes information to your users according to data relationships in the database—all without the administrator having to update the sharing rules.

Localized Function Trapping

User-created SQL functions (MAX, MIN, COUNT, AVG) generate values for localized data stores—useful tools for customized data configurations. Be sure to select a synchronization solution that traps these functions at a local level so that only the value is synchronized with the server. Solutions without this feature return actual functions to server databases, a practice that jeopardizes data integrity.

Alert Conditions

The solution you select should provide a menu of alert conditions to notify administrators of typical system problems. Though your solution should provide myriad options for alerts, the most common reasons administrators wish to be notified include:

- Failed synchronization sessions
- Insufficient device space
- Elapsed time since last synchronization
- Slow synchronization sessions[1]

Alert Methods

It is important that your solution support a variety of alert methods that are most convenient, given the way you work. You do not want to have to adopt an additional communications device to receive alerts from your wireless data synchronization solution. A quality solution should support:

- E-mail messages

- Text messages to pagers
- Text messages to cell phones
- Network alerts[1]

MMC Administration

Administering a wireless data synchronization technology shouldn't require you to learn how to use a whole new set of tools. Your synchronization solution should utilize management standards to make administration more intuitive. Solutions that feature a Microsoft Management Console plug-in allow administrators to manage the core server attributes, the synchronization server software, and other common applications from the same administration console. In addition, it ensures a consistent and familiar look and feel to management tools.

Remote Administration

Remote administration lets administrators manage the system while not physically located at the server. As server hardware may be located in centralized data centers, remote administration allows administrators to observe server activity from their workstations.

Log Data Stored in Server Database

In addition to controlling the synchronization process, a sophisticated wireless data synchronization solution lets administrators track a variety of user behaviors and captures this in a server database. Storing log data in a server database enables hierarchical tree views of log data. It also allows administrators to define custom reports using their preferred reporting tool.

Comprehensive Logging Detail

Merely logging user connections and disconnections robs administrators of advanced troubleshooting and supporting functions. Detailed logging features let administrators monitor the following:

- Connection dates

- Connection duration
- Wireless data being processed
- When wireless data were created and delivered
- Errors on client devices
- Errors on server devices
- Performance information/results[1]

Flexible Log Viewing

Logs should be constructed in a hierarchical manner, with a variety of views and drill-down capabilities for optimized troubleshooting. Logs should showcase both aggregate trends and granular wireless data that should be viewable and sortable by:

- User
- Synchronization component
- SQL operation types
- Severity levels (errors, warning, debug, information)
- Machine[1]

Real-Time Logs

Logging should be viewable in a real-time format to facilitate rapid response times. Typically, this is used to monitor system-wide performance and investigate aggregate-level, multiuser problems. Beware solutions that offer limited reporting based on static points in time, as these make it more difficult to confirm and correct system-wide issues, leading to unnecessary downtime and increased support calls.

Console-Based Log Views

Console-based log views are preferred to predefined report generation. The console approach makes viewing logs an integral part of defining system behavior and managing users, simplifying the process. Console-based log views also allow rapid drill-down into specific log data to support troubleshooting.

User Disablement

Administrators should be able to turn off a specific user, user group, or device. For example, this would be useful in the event a specific division is sold to another company, rendering access inappropriate, or when a user loses a device and wants to make sure that no unauthorized persons are able to synchronize with company servers.

Encryption

The very notion of a wireless data mobile workforce can strike fear into the hearts of an IT staff charged with protecting network integrity. The solution you choose should incorporate industry standard security techniques to encrypt the connection between devices and servers. For example, Certicom's Elliptic Curve Cryptography (http://www.certicom.com/) is an excellent tool to protect wireless data during communications with certain hand-helds, while SSL works best for others and for PCs.

Encryption Toggling

Depending on the circumstances, encrypting wireless data flow may or may not be critical. When connected to the LAN inside the office, protected by the company firewall, encryption may not be as essential. Or certain user groups may deal with less sensitive information that does not require encryption—thus speeding their communications sessions. Providing toggle options for administrators ensures the network is secure and minimizes performance impacts.

Open APIs for Additional Security Measures

Open APIs should be available to enable more advanced security features if desired. Direct integration of other security-related products and processes is made available via published APIs to ensure maximum flexibility for this sensitive topic.

Pre- and Postprocessing

A pre- and postprocessing component allows administrators to define custom SQL functions to run on wireless data mobile devices both before and after data are synchronized. This feature results in minimized synchronization sessions and supports the ability to drop and rebuild indexes or create new tables on devices.

Automated Wireless Data Realignments

In situations where users' wireless data view changes because of changes to the database, this feature triggers automatic realignment of localized data sets. For example, when a sales organization restructures, account managers automatically get account information for customers in their new regions without any administrator intervention.

Wireless Data Refresh

When new users are added or existing users add new devices, it is important that the initial experience with synchronization is as painless as possible. Full wireless data refresh can be used to provide a complete set of data to initialize devices or repair corrupted data sets.

Flexible Transport Layer

Your wireless data synchronization solution requires a transport layer to transfer change files to and from application servers. A mature wireless data synchronization solution should not tie you to a specific transport mechanism and should support communicating with any device via a transport optimized for that device.

Multiple Connection Modes

As today's wireless data mobile worker logs more and more time away from office LAN connections, multiple device connection modes should be included in your enterprise synchronization solution, including:

- Ethernet cradles (hand-helds)
- Wireline PC pass-through or proxy (hand-helds)
- Dial-up
- Wireless data infrared[1]

NT Authentication

End users can authenticate to the server with existing NT user IDs and passwords. Adding new devices requires no additional IT intervention.

NT authentication eliminates requiring users to remember multiple passwords and prevents administrators from managing multiple password and username lists.

Transaction Integrity Support

A robust synchronization mechanism should maintain transactional integrity when wireless data are synchronized between servers and mobile devices. Transaction boundaries are used to guarantee that logical groups of wireless data are processed against a database consistently. These same logical boundaries need to be maintained when the wireless data are synchronized with other devices.

One Component of a Complete Wireless Data Mobile Infrastructure

An application wireless data synchronization solution is an important step in building a total mobile strategy. Ideally, your solution should be part of a large and more comprehensive mobile infrastructure (see Fig. 15-1).[1]

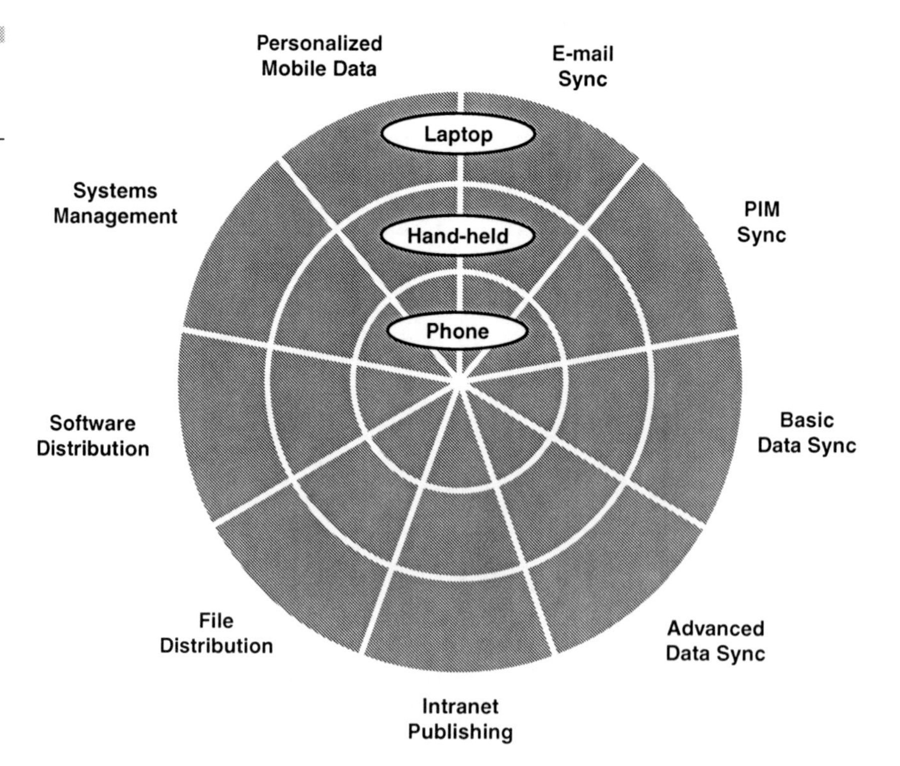

Figure 15-1 Wireless data mobile infrastructure.

In addition to application wireless data synchronization, components of a comprehensive mobile infrastructure solution include:

- *E-mail and PIM sync.* Groupware applications can be mobilized to enable pervasive hand-held access to e-mail, contacts, and calendars.
- *Software distribution/asset management.* The ability to deploy applications to wireless data mobile devices and remotely manage hardware and software inventories lowers cost of ownership of wireless data mobile devices and eases support burdens.
- *Content distribution*
  - *File distribution.* While data synchronization replicates structured data found in databases, file distribution helps move unstructured data found in spreadsheet, word processing, presentation, and graphics files.
  - *Intranet publishing.* Mobilizing your intranet site makes it available off line to mobile workers.
- *Third-party content.* Additional synchronization and third-party content aggregation capabilities can round out your wireless data mobile infrastructure with value-added Web content.[1]

Finally, by selecting an integrated and fully functional wireless data mobile infrastructure solution, you will minimize the total cost of ownership of mobile devices. Other benefits include:

- More rapid deployment of new solutions
- Minimized training for IT staff
- Elimination of middleware integration effort
- Improved ability to resolve support calls
- Ensured simplicity for end users
- Protection of existing mobile asset investments
- Lower license and maintenance fees

Conclusion

The aim of SUA is to remove the management burden of the SS7 network service part (MTP and SCCP) in a packet-switched network. In an SUA-only environment, network operations can do without signaling point code management, GTT to resolve the destination signaling point code and managing different national variants of SCCP and MTP Layer 3 (ETSI, ANSI, China, etc.) that are incompatible with each other. In an M3UA-only environment, the usage of point codes is inevitable, since M3UA requires SCCP above it to support MAP messages.

There is also the international version of both protocols (ITU-T) to ensure connectivity between national variants. SG is needed for both protocols whenever there is a need for connectivity between different countries or even different operators' networks.

In circuit-switched networks, it is the responsibility of the STP to act as a router while the routing is based on MTP Layer 3 (link by link) and SCCP (end to end). In IP networks, the IP router routes the signaling message from the originating point to the destination point. M3UA uses link-by-link signaling, while SUA uses end-to-end signaling. M3UA requires the GTT function in intermediate nodes since the point codes used are not globally unique. GTT is not required in SUA since the IP addresses involved have global significance. These greatly reduce network complexity. SUA uses its address mapping function (AMF) to resolve IP addresses from global titles, point codes, or host names.

From the examples shown, it is necessary to have SGs in hybrid networks for the migration of SS7-based signaling to IP-based signaling regardless of the adaptation layers (M3UA, SUA) used. While M3UA is an excellent adaptation layer for supporting MTP Layer 3 users such as ISDN User Part (ISUP), it presents an important limitation regarding the complexity of supporting different national variants. Therefore, SUA appears to have significant advantages over M3UA to transport MAP messages.

References

1. "Mobilizing Enterprise Applications with Data Synchronization," Synchrologic, Inc., 200 North Point Center East, Suite 600, Alpharetta, GA 30022, 2002.
2. John R. Vacca, *i-mode Crash Course,* McGraw-Hill, 2001.
3. John R. Vacca, *Satellite Encryption,* Academic Press, 1999.

CHAPTER 16

Packet-over-SONET/SDH Specification (POS-PHY Level 3): Deploying High-Speed Wireless Data Networking Applications

SONET/SDH systems have been the preferred transport technology over fiber optics for almost 2 decades now. Carriers have developed extensive expertise in implementing, operating, managing, and developing high-speed wireless data business models for these systems. Manufacturers' technical expertise in such systems has increased to a deep understanding of what high-speed wireless data transport over fiber is about. In short, SONET/SDH can be called a mature high-speed wireless data transport technology. Out of this mature expertise, new techniques for bettering high-speed wireless data transport over fiber services have recently appeared. These techniques are likely to considerably reshape the next generation of SONET/SDH systems in many aspects: new high-speed wireless data transport techniques, new high-speed wireless data transport services, new management systems and business models. In this chapter, several new high-speed wireless data transport techniques are described, and their impact on the creation of new high-speed wireless data transport services for next-generation SONET/SDH systems is discussed. (The Glossary defines many technical terms, abbreviations, and acronyms used in the book.)

High-Speed Wireless Data Transport Services for Next-Generation SONET/SDH Systems

For the last 16 years, synchronous optical network/synchronous digital hierarchy (SONET/SDH) has been the main high-speed wireless data transport technology over optical fibers. SONET/SDH systems allow the high-speed wireless data transport of constant-bit-rate clients, through synchronous transport modules (STMs) and virtual tributaries (VTs), as well as variable-rate packet-oriented clients, such as asynchronous transfer mode (ATM)/IP/frame relay, and others. These signals are transported over a synchronous frame, which is used to modulate a single-wavelength channel. Current SONET/SDH interface speeds range from 51 Mbps to 10 Gbps.

SONET/SDH legacy equipment was designed primarily for the high-speed wireless data transport of constant-bit-rate applications. This is evident in many characteristics of the transport technology; for instance, bandwidth is provisioned via a rigid hierarchy of bit-rate signals (STS-3, STS-12, STS-48, etc.).

However, with the explosion of datagram applications allowed by IP and other packet-switching technologies, solutions for the transport of high-speed wireless data over SONET/SDH systems were developed. For Internet traffic, for instance, IP packets are framed by using packet over SONET (POS) and placed into the synchronous payload envelope (SPE), the SONET/SDH frame payload area. Another example is multimedia traffic, with stringent quality of service (QoS) requirements. In this case, ATM is used as a way to access optical fiber, providing a predictable end-to-end transport service much needed by this type of application. ATM cells are placed into the SPE according to standardized optical interfaces.

Even though POS and ATM have been widely used as means of wireless data adaptation into SONET/SDH payload, neither of these is recognized to be the best for wireless data transmission purposes, as far as bandwidth usage and high-speed processing capability are concerned. POS uses HDLC framing that becomes difficult to implement in high-speed processing of 10 Gbps or even 40 Gbps. ATM has a well-known cell tax that consumes an extra 10 percent of bandwidth.

On the other hand, although SONET/SDH has been the single technology for Internet transport over fiber, it has limitations of its own. For instance, each transport path has a fixed bandwidth [time-division multiplexing (TDM) model], which is defined over a rigid rate hierarchy. Moreover, there is a lack of fine granularity to accommodate all potential clients' stream rates, especially data applications. Finally, because SONET/SDH nodes have limited network management functionalities, each transport path takes a long time to set up, typically weeks for U.S. coast-to-coast.

New techniques, however, are currently being developed to address many of these limitations. The generic framing procedure (GFP) has been developed as a new framing for wireless data accommodation into SONET/SDH and optical transport network (OTN). Virtual concatenation has been standardized for flexible bandwidth assignment of SONET/SDH paths. The link capacity adjustment scheme (LCAS) has been discussed for dynamic bandwidth allocation in support of virtual concatenation. One of the most important objectives of these new technologies is to enable flexible and reliable wireless data transport over SONET/SDH, which is referred to as wireless data over SONET/SDH (DoS). This chapter is aimed at describing these new technologies, elaborating on DoS architecture for new high-speed wireless data transport services, and considering implementation aspects.

The next part of the chapter introduces a DoS network architecture. A discussion ensues on upcoming high-speed wireless data transport services enabled by this architecture, including network scenarios and implementation aspects of these services.

Wireless-Data-over-SONET/SDH Network Architecture

DoS is a high-speed wireless data transport mechanism that provides a means to accommodate various data interfaces (Ethernet, Fibre Channel, ESCON/FICON) into SONET/SDH efficiently. In particular, DoS is effective to accommodate Gigabit Ethernet (GbE), which has been widely deployed for WAN interface application (see sidebar, "10G Ethernet").

10G Ethernet

It has beaten back a host of challengers in its more than 2 decades as a networking standard to the point where it is the undisputed ruler of local, campus, and metropolitan networks. Now the champion must transform itself to meet the future's demand for increased bandwidth.

Ethernet, the most venerable LAN standard and by far the most successful, will be getting yet another 10-fold boost in capacity in mid-2003. But rapid adoption of 10 Gigabit Ethernet products is not a sure thing, and the regularity of future speed boosts is uncertain as well. The result: 10 Gigabit Ethernet could be more of a plateau than a stepping stone.

The reason for interest in keeping the Ethernet topology alive and growing is obvious from an engineering point of view: End-to-end Ethernet connections simplify network management and thereby reduce costs. That's why it made sense to turbocharge the original Ethernet to create Fast Ethernet in the early 1990s, offering a boost in speed from 10 Mbps to 100 Mbps. Gigabit Ethernet products, which debuted in 1997, elevated Fast Ethernet's throughput by another factor of 10.

The 10 Gigabit Ethernet shares the same name as well as the same frame format as its lower-speed brethren, but that is where the similarities end. The emerging specification, which is expected to be ratified by the Institute of Electrical and Electronics Engineers in the second quarter of 2002, was designed for carrier networks as well as enterprise LANs, costs 20 to 90 times more than current Gigabit Ethernet links, and is expected to ramp up at a slower pace than its predecessors.

The business drivers and market potential for 10G Ethernet look much different today than they did a few years ago. At that time, Ethernet was riding the Internet boom. As corporations chewed up network bandwidth, their desire for higher-speed connections was insatiable. Ethernet emerged as the best option because of its low price: Traditionally, vendors have been able to deliver a 10-fold performance improvement at a price that was only three to four times as much as the previous technique. As a result, Ethernet emerged as the dominant networking option, first on the desktop, then in server farms with 100-Mbps links and recently in enterprise backbone networks with 1-Gbps connections.

With enterprises increasingly deploying Ethernet-only networks, vendors decided to push the technology out into the WAN with development of the 10-Gbps specifications. The IEEE formally began working on the standard, dubbed 802.3ae, in the spring of 1999.

The desire to push Ethernet into the WAN meant that the standard had to support a broader array of connectivity options. Consequently, the standard comes in eight flavors, works with four types of transceivers, and reaches distances from 300 m to 30 km.

Standards participants are heading into the homestretch and finishing up the final draft of the specification. There are no major technical issues that have to be addressed; the process revolves more around ensuring that the wording is correct.

In 2002, the Physical-layer chip sets (transceivers), which send and receive data signals over LAN and WAN connections, became available from companies such as Broadcom Corp., in Sunnyvale, California; Finisar Corp., also in Sunnyvale; and SwitchCore Group AB, in Lund, Sweden. Testing equipment designed to make sure there are no significant problems with these devices is emerging from companies such as Spirent.

With those building blocks in place, vendors such as Avaya Inc., of Basking Ridge, New Jersey; Cisco Systems Inc., of San Jose[6]; Extreme Networks Inc., of Sunnyvale; and Foundry announced plans for 10 Gigabit Ethernet products. And, the suppliers have tried to alleviate potential interoperability problems by forming the 10 Gigabit Ethernet Alliance, in Mountain View, California. Founded in February 2000, the group, which now has more than 200 members, conducts interoperability demonstrations.

What's more, compliance-testing mechanisms from organizations such as the University of New Hampshire are starting to arrive, so users should be able connect different vendors' products. Users can feel fairly safe deploying 10G products in their networks now. In fact, a handful of next-generation carriers and a few large enterprises are already tinkering with 10 Gigabit Ethernet devices.

Users might feel safe with the technology, but what would they use it for? In the carrier market, start-ups and established service providers are building out new IP networks and using Ethernet to deliver services, such as metropolitan-area networks (MANs). Demand for these services stems from enterprise deployments of bandwidth-intensive applications, such as video streaming and voice telephony, which need faster alternatives to low-bandwidth, dial-up, and even dedicated T-1 (1.5-Mbps) links. However, it has been difficult for carriers to deliver MAN services by Synchronous Optical Network (SONET), which is inflexible. One subset of the 802.3ae specification offers service providers the option of running 10 Gigabit Ethernet over dense wavelength-division multiplexing lines, a change that should make it simpler for them to deploy new high-speed services.

For example, Yipes Communications Inc., in San Francisco, a next-generation network service provider, is interested in high-speed Ethernet technology. The company prefers using Ethernet rather than SONET multiplexers to build out its network, and now offers customers bandwidth in increments from 1 Mbps to 1 Gbps. By adopting 10G Ethernet, you should be able to eliminate the conversions that now take place as you move wireless data onto your backbone links, and that change would improve network performance.

Another benefit of 10 Gigabit Ethernet is simpler network management. Network technicians can put their knowledge of managing Ethernet connections to use, employing one network management system to monitor a number of connections, rather than be forced to work with different systems for LAN connections, servers, backbones, and WANs.

Because of the potential benefits, Yipes has been experimenting with 10 Gigabit Ethernet switches and is now examining how to integrate them into its network. For them, the major issues are not the switch's speeds and feeds, but rather how the new system collects and passes information to their operation support system (OSS) applications. In the enterprise, 10 Gigabit Ethernet is expected to edge its way initially into a few niche applications. One use is in a large campus network where a company lays its own fiber and uses high-speed links to connect office buildings, in effect building a private MAN.

Storage-area networks (SANs) are another possibility. Traditionally, servers controlled storage devices, but this approach chewed up CPU processing cycles that could be focused on other activities, such as searching for information in a database management system. A SAN attaches to the back ends of a series of servers, usually high-end UNIX systems or mainframes, and collects storage information. Placing that data in one location simplifies management, and is more efficient than maintaining storage systems for each server. In addition, SAN connections support high-bandwidth links, so response is fast.

Some promising niches notwithstanding, there are plenty of hurdles in the way of 10 Gigabit Ethernet. Price is at the top of the list: Links are in the $20,000 to $90,000 per port range. For the first time in its history, a new version of Ethernet does not support a compelling financial model. The prices are high because the underlying technology is extremely complex and difficult to produce.

In addition, existing networks pose challenges. Corporations and service providers have deployed many different types of fiber. Suppliers may claim that all have the same performance characteristics, but that is not really the case, so corporations had to build products that work over all strands of fiber.

Still, service providers such as Yipes can build a sound business case for purchasing pricey 802.3ae links. A carrier will spend $250,000 to $300,000 to install a 10-Gbps link with SONET and optical equipment; a 10G Ethernet option costs significantly less.

In contrast, enterprises have not been spending as much on their high-speed links. In fact, Dell'Oro Group (http://www.delloro.com) pegged the price for a Gigabit Ethernet port, as deployed in many companies, at less than $1000.

Given the economic climate, companies have been reexamining their IT investments. Corporations are looking for an immediate return on any new IT investment; they don't want to wait 2 or 3 years for the payback.

There are also questions about the need for higher-speed links in the enterprise. The Genome Sequencing Research Center at Washington University in St. Louis seems to be a typical 10 Gigabit Ethernet customer. About 400 employees use Sun Microsystems Inc. workstations and servers to conduct DNA testing, and the applications have usually pushed up against network bandwidth ceilings.

But this is not the case at the moment for the Genome Sequencing Research Center, which relies on Cisco Catalyst 6500 switches to support 16 Gigabit Ethernet links. Right now, their network utilization is quite low, and the network has been able to handle any type of traffic burst. If the company hits 25 percent usage on any of its Gigabit Ethernet links, that is quite unusual.

Other technology building blocks may fall into place during that time and spur demand for 10 Gigabit Ethernet links, however. For instance, current SAN technology (SCSI and Fibre Channel) cannot be easily integrated with IP networks and does not have sufficient throughput to fill up a 10-Gbps line. But, standards to address these limitations, such as iSCSI and Fibre Channel/IP, are now making their way from vendor development laboratories into deliverable products.

To saturate a 10G connection, a company will need to install more powerful desktop and server systems. Just passing information from a computer onto a network is taxing: Moving 1 Gbps of data onto a LAN requires a microprocessor operating at a speed of at least 1 GHz.

For these reasons, companies such as Alcatel (http://www.alcatel.com) are waiting until the 802.3ae standard nears ratification in 2003 before announcing their 10 Gigabit Ethernet products. They don't expect to see significant demand for 10G Ethernet products until 2003.

Analysts do expect increases in desktop, server, and network power during the next few years. Recently, vendors began delivering network interface cards that support 10-Mbps, 100-Mbps, and 1-Gbps Ethernet transmissions to the desktop.

International Data Corp., in Framingham, Massachusetts, expects the worldwide installed base of Gigabit Ethernet ports to increase more than sixfold by 2005. As companies install multiple Gigabit Ethernet links, they will look for a way to consolidate them, and 10 Gigabit Ethernet could be the answer. There is no doubt that service providers and enterprises will adopt 10G Ethernet, but they just may make the transition a bit slower than everyone expected a few years ago.[1]

DoS utilizes three technologies: GFP, virtual concatenation, and LCAS. These technologies are being standardized in the International Telecommunication Union—Telecommunication Standardization Sector (ITU-T) and T1X1.5.

Generic Framing Procedure

GFP was first discussed in T1X1.5; now ITU-T is working to standardize it as G.7041. In G.7041, GFP is defined as a framing procedure to delineate octet-aligned variable-length payloads from higher-level client signals for subsequent mapping into octet-synchronous paths. The mapping procedure for the client signals into octet-synchronous paths is also defined in the standard document.

GFP is generic in terms of two stack directions: at the layer below it, with respect to transport services used by GFP, and at the layer above it, with respect to mapping services provided to applications by GFP. At the layer below it, GFP allows the use of almost any type of transport technology, although in the standards body the focus is mostly on SONET/SDH and OTN. The only requirement for the Transport layer is to provide an octet-synchronous path for GFP.1 At the layer above it, GFP supports various types of packets, including IP packets, Ethernet frames, and HDLC frames such as PPP. The only requirement for the upper layer is that packets should be octet aligned. This unique generic feature comes from the fact that GFP provides a simple packet delimitation scheme.

GFP has two mapping methods to accommodate client signals into SONET/SDH payload: frame-mapped GFP and transparent GFP. These two methods are designed to be suitable to several new applications and services coveted by wireless data carriers (see sidebar, "Wireless Data Carriers Address Packet Technology").

Wireless Data Carriers Address Packet Technology

The telecommunications market is experiencing unprecedented growth as voice and wireless data networks converge to satisfy subscribers that are increasingly mobile[5] and information hungry. Wireless data and IP technologies are working together to deliver content from the Internet and intranets over mobile phones. Third-generation (3G) wireless data services will change the way people connect and communicate. Packet-based wireless data communication will increase user mobility and enhance interactive communication by enabling personalized, media-rich services that can be delivered anywhere at any time.

Wireless data carriers face a number of issues as they consider the move to a packet-based network. They have made a very significant investment in the existing networks, and it is not economically feasible to abandon that infrastructure. The large geographic footprints of many global carriers may not be covered by the new 3G spectrum. Carriers will need to continue to support a large base of existing customers with 2G services as they transition to packet-based networks. And the compatibility and acceptance of new handsets must be considered.

The move from 2G to 3G will not take place quickly. 2G technology will be in place for a long while, with 3G developments taking place in parallel with the existing wireless data network. The shift from 2G to 3G likely will be a phased migration—one that will allow carriers to leverage their existing networks and maintain reliability and quality of existing services.

Getting to the Core

Mobile switching centers (MSCs) are the heart of the switching fabric in today's wireless data networks. The MSCs, which interface to the radio system and the public switched telephone network, perform critical functions including:

- Switching voice traffic from the wireless data network to the PSTN if the call is mobile-to-landline or to another MSC within the wireless data network if the call is mobile-to-mobile
- Delivering short message service (SMS)
- Providing subscriber mobility management

Typically, the circuit-switched wireless data network is a fully meshed architecture; each MSC is connected to every other MSC in the network by time-division-multiplexed (TDM) trunks. Mesh networks lack the flexibility to scale easily or economically to accommodate network expansion.

A Better Strategy

Wireless data carriers can accommodate increasing traffic loads today and create the foundation for a pure packet network by employing next-generation technology to create a common packet infrastructure to interconnect MSCs. This is accomplished by deploying packet tandems composed of centralized media gateway controllers (MGCs) or soft switches controlling an overlay of distributed media gateways (MGs) collocated with MSCs.

TDM trunks from each MSC are terminated on stand-alone MGs. The media gateways perform the IP or ATM conversion under the control of the soft switch, which can be located in a centralized server farm.

This architecture dramatically simplifies the network and reduces bandwidth requirements by eliminating the point-to-point connections between MSCs. As network traffic increases, new MSCs can be added to accommodate the growth with a single connection to the soft switch and without any interconnection to other MSCs. The use of media gateways also allows operators to deploy next-generation services without doing software upgrades at every

MSC in the network. And, by terminating TDM trunks from the PSTN at the media gateway rather than at the MSC, valuable mobile switching center resources are freed.

Packet-based technology also improves routing within the wireless data network. The soft switch can handle several of the MSC's functions including routing and home location register (HLR) lookups. In today's 2G network, calls entering the wireless data network can require extensive routing before they can be completed. Each call has to be routed back to the subscriber's home MSC or gateway MSC, which launches a lookup in an HLR to determine the subscriber's location. The call is then routed to the serving MSC, a trip that may traverse multiple switches. In a packet-based architecture, gateway intelligence can be added to the centrally located soft switch, which performs the lookup and then routes the call directly to the serving MSC.

On to 3G

With a common, core packet network in place, operators create a fabric that is optimized for advanced data services and can also carry wireless voice and data traffic. The evolution to full-blown 3G is greatly simplified with the major changes and additions taking place in the radio access network (RAN). A new interface defined by 3GPP, IU-CS, is required from the 3G RAN to the MSC server (the 3G equivalent of a MSC). While it's possible to upgrade the existing MSCs to support IU-CS and MSC server functions, it's an expensive and time-consuming proposition. The MSC server software can be added to the centralized soft switch and the physical interface can be terminated on the distributed MGs. With this approach, carriers do not have to undertake the expensive and cumbersome task of upgrading the hardware and software at each and every MSC in the network.[2]

Virtual Concatenation

Virtual concatenation is a mechanism that provides flexible and effective use of SONET/SDH payload. Historically, SONET/SDH was first defined as a (worldwide) unified digital hierarchy for the transport of 64-kbps-based TDM service. The capacity of payload was rigidly defined for plesiochronous digital hierarchy (PDH) service accommodation. However, the disadvantages of such a rigid SONET/SDH rate hierarchy, especially when data applications such as Ethernet are considered, were soon realized. Virtual concatenation breaks the limitation incurred by this rigidity via the definition of payloads with flexible bandwidth. It "virtually" concatenates several payloads to provide a payload with flexible bandwidth, appropriate for wireless data service accommodation.

Consider the case of Gigabit Ethernet (GbE) transport by SONET/SDH. According to the conventional SONET/SDH specifications, STS-48c SPE/VC-4-16c must be used to accommodate GbE signals at full speed. Since STS-48c SPE/VC-4-16c capacity is 2.4 Gbps, however, 1.4 Gbps of capacity is wasted. If STS-12c SPE/VC-4-4c is used to avoid bandwidth wastage, full-speed accommodation cannot be achieved. GbE could be suitably accommodated if a contiguously concatenated payload STS-21c SPE/VC-4-7c with 1.05 Gbps of capacity were defined. In such a case, however, every node in the network would need to handle this newly defined STS-21c SPE/VC-4-7c signal, which would not be practical because such a concatenated payload is not supported by legacy SONET/SDH equipment. Using the virtual concatenation technique, seven independent STS-3c SPE/VC-4 payloads are virtually concatenated to provide STS-3c-7v/VC-4-7v payload (suffix v stands for virtual) with 1.05 Gbps of bandwidth, which is perfectly suitable for GbE accommodation. This solution is viable because the implementation of virtual concatenation is limited to multiplexing nodes; there is no need to add virtual concatenation capability to every node of a SONET/SDH network. For instance, for GbE accommodation into STS-3c-7v/VC-4-7v, at the origination point, GbE is mapped into a STS-3c-7v/VC-4-7v payload, which is constructed with seven virtually concatenated STS-3c SPEs/VC-4s. There is no restriction on which OC-n/STM-N signal(s) should be used. The seven payloads may or may not reside in the same OC-n/STM-N contiguously, or may even reside at different OC-n/STM-N interfaces. Within the network, they are treated as seven separate and independent STS-3c SPE/VC-4 payloads.

At the destination, the seven payloads are combined to construct the original STS-3c-7v/VC-4-7v signal using inverse multiplexing, and GbE is subsequently demapped from it. This means that the intermediate nodes, through which each STS-3c SPE/VC-4 travels, do not need to handle STS-3c-7v/VC-4-7v at all, so that STS-3c-7v/VC-4-7v needs to be understood at both end nodes of the path only. Thus, carriers are free to introduce the virtual concatenation function without any serious impact on the existing network. Morever, an element management system/network management system (EMS/NMS) of today can easily support virtual concatenation. Figure 16-1*a* illustrates the virtual concatenation technique for a SONET/SDH system.[3]

Another valuable feature of virtual concatenation is that the bandwidth of a SONET/SDH interface can be divided into several subrates. In POS, the whole payload of OC-n/STM-N must be dedicated to IP packet accommodation. Therefore, it is not possible to accommodate IP services

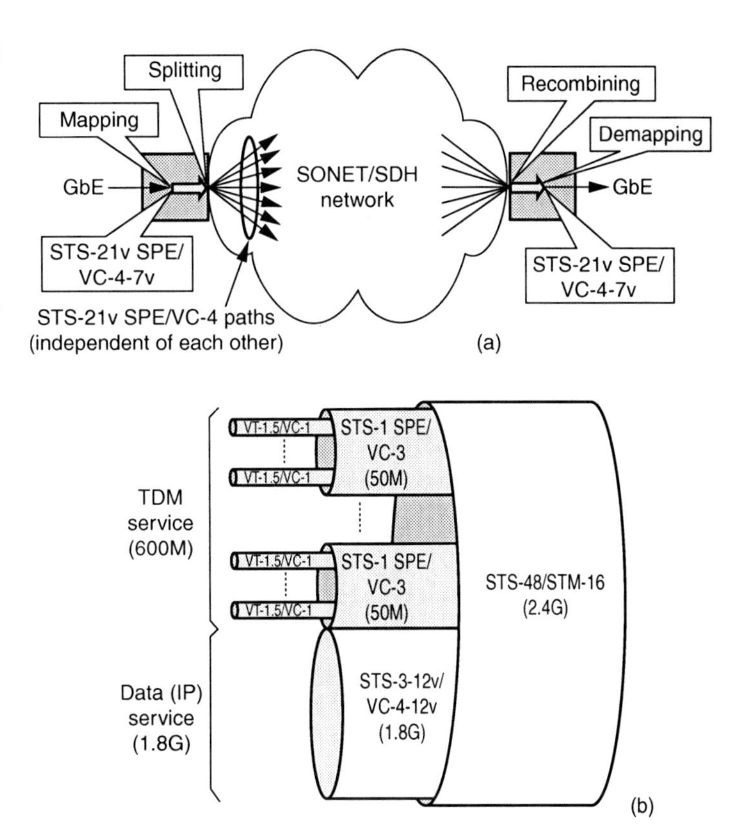

Figure 16-1 SONET/SDH virtual concatenation: (*a*) the virtual concatenation concept; (*b*) bandwidth partitioning with virtual concatenation.

into some portion of the SONET/SDH bandwidth. Virtual concatenation provides a way to partition SONET/SDH bandwidth into several subrates, each of which is capable of accommodating different services. Figure 16-1*b* illustrates the issue.[3] The figure shows an example of bandwidth partitioning over an STS-48/STM-16 signal. The 600 Mbps is dedicated to TDM (fixed-rate) services using VT1.5/VC-11 paths, and the rest of the 1.8-Gbps portion is virtually concatenated to construct a STS-3-12v/VC-4-12v payload assigned for data service. In such a manner, virtual concatenation can be used for partitioning an OC-n/STM-N bandwidth to accommodate various services within a single frame.

Data-over-SONET/SDH Architecture

GFP, virtual concatenation, and LCAS provide the fundamentals for the creation of a truly integrated wireless data services over a SONET/SDH transport system, or DoS. DoS has the following features:

- Flexible bandwidth assignment with 50-Mbps granularity
- No modification required for intermediate nodes
- Efficient framing scheme, with small overhead
- Accommodation of any type of data service, including IP packet, Ethernet Datagram, ESCON, and FICON
- Coexistence of legacy service and data service in a single SONET/SDH frame
- Dynamic bandwidth control
- Network management through an existing, quality-proven NMS[3]

Figure 16-2 illustrates the concept of integrated data accommodation of DoS.[3] Figure 16-2*a* shows current packet transport over SONET/SDH. Various framing methods are used, which segregate these applications from the transport service up to management level. Figure 16-2*b*, on the other hand, depicts an integrated data transport service over SONET/SDH based on DoS. Notice how GFP glues various applications into the same transport technology. This allows for the implementation of several new network level techniques, such as load balancing, multi-

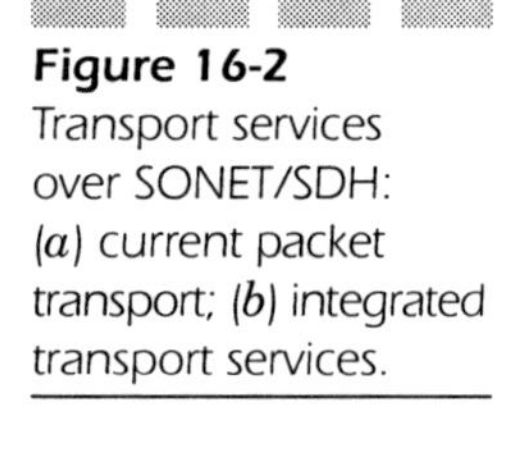

Figure 16-2 Transport services over SONET/SDH: (*a*) current packet transport; (*b*) integrated transport services.

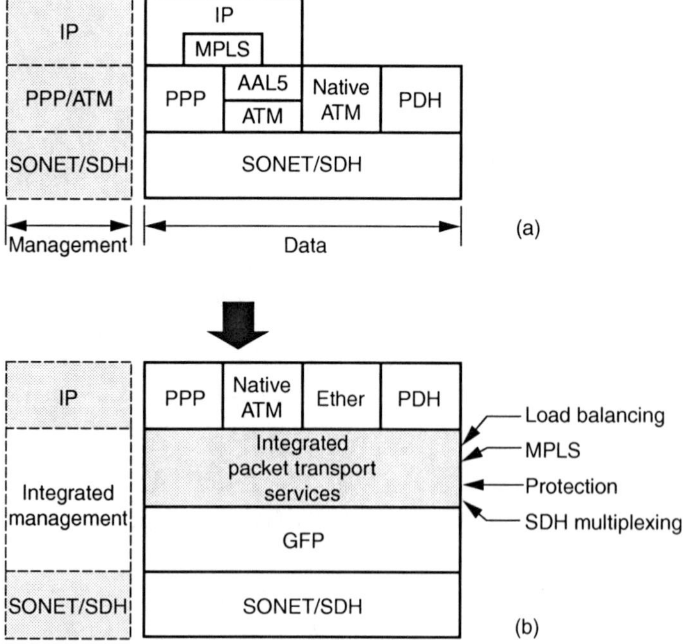

protocol label switching (MPLS), protection, and multiplexing, which can be developed for all applications seamlessly. These techniques are integrated into a powerful, efficient, and flexible NMS, bringing additional revenue to carriers.

Layer ½ Hybrid Network via DoS

One of the most important DoS applications is a Layer ½ hybrid network. DoS realizes coexistence of TDM and wireless data services in a single SONET/SDH frame. For TDM services, Layer 1 handling is required. For data services, packet handling is necessary in the GFP layer, which can be regarded as Layer 2 in some sense, because it is above the Physical layer and below the IP layer. Hence, if DoS is introduced in a transport network, the nodes should handle both Layer 1 (TDM) and Layer 2 (GFP) simultaneously. This means that the network element can be Layer ½ hybrid when DoS is applied.

Currently, SONET/SDH rings are widely used transport networks. If DoS is applied to one such ring, network nodes should perform GFP frame add/drop as well as conventional SONET/SDH path add/drop. Such a network is in fact a Layer ½ hybrid ring, realized by Layer ½ hybrid nodes. Figure 16-3 shows a Layer ½ hybrid add/drop function.[3] The bandwidth for data traffic is assigned in SONET/SDH section by section, according to the client bandwidth requirements.

NOTE A ring protection scheme, such as 2F-UPSR (SNCP ring) or 4F-BLSR (MS SPRING), can be applied to a Layer ½ hybrid ring, because the network is still based on SONET/SDH.

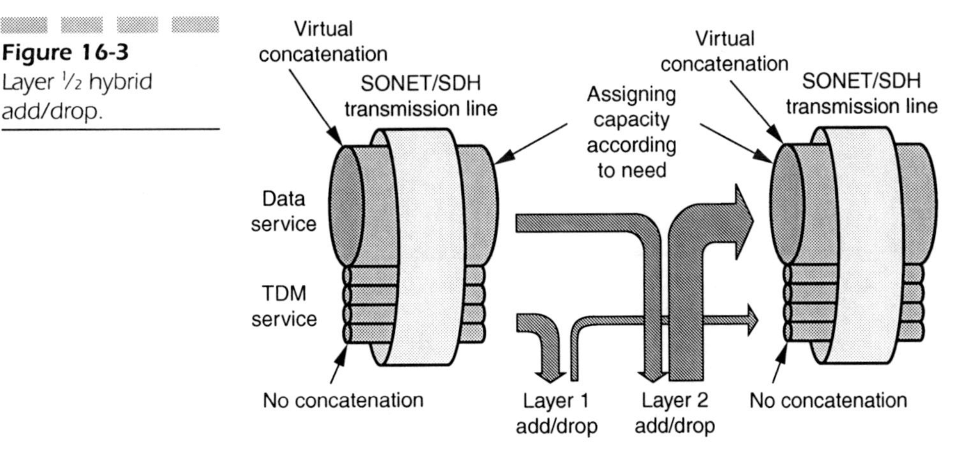

Figure 16-3 Layer ½ hybrid add/drop.

Novel SONET/SDH Transport Services

As described earlier, DoS allows coexistence of both TDM and wireless data traffic in a single OC-n/STM-N. In addition, it is also possible to configure the ratio between TDM and data traffic flexibly. These features lead to bandwidth on demand (BoD) services supported by Data over SONET/SDH architecture. BoD service has the following characteristics, according to current carriers' needs:

- Billing based on usage and SLA requirements, as well as length of contract
- Point-to-point, OC-n/STM-N bandwidth pipes, with flexible holding times
- Provisioning realized in near real time (seconds to minutes)
- Multiple classes of service based on protection and restoration[3]

A unique feature of BoD service based on DoS is best-effort service provisioning under network failure condition. In general, path failure leads to immediate service unavailability. In contrast, service can continue, albeit with some degradation, in BoD based on DoS, if the failed member of the virtual concatenation group is removed from the group.

BoD service can be deployed under both distributed and centralized control. In distributed control, user-network interface (UNI) and generalized MPLS (GMPLS)–based path setup is used. In centralized control, EMS/NMS-based network control is applied. An example of the distributed control scheme is depicted in Fig. 16-4*a*.[3] The end user requests bandwidth adjustment from the network provider (step 1, additional 50 Mbps, in Fig. 16-4*a*) through the UNI. Next, the network provider routes the additional member path and sets up the path based on GMPLS (step 2, additional VC-3, Fig. 16-4*a*). Every node in the network advertises its time slot usage using OSPF-TE or IS-IS, so the edge node can determine the end-to-end route for the additional member path using the advertised information. Next, the additional path is configured using RSVP-TE or CR-LDP signaling. After that, the LCAS protocol gets triggered and the additional member path is accommodated to the path group (steps 3 and 4 in Fig. 16-4*a*) in a hitless manner. The network provider is then able to start provisioning new bandwidth service, satisfying the current end user's bandwidth requirements.

The Optical Internetworking Forum (OIF) is currently working on the signaling specification between client and network—UNI 1.0 (end-user

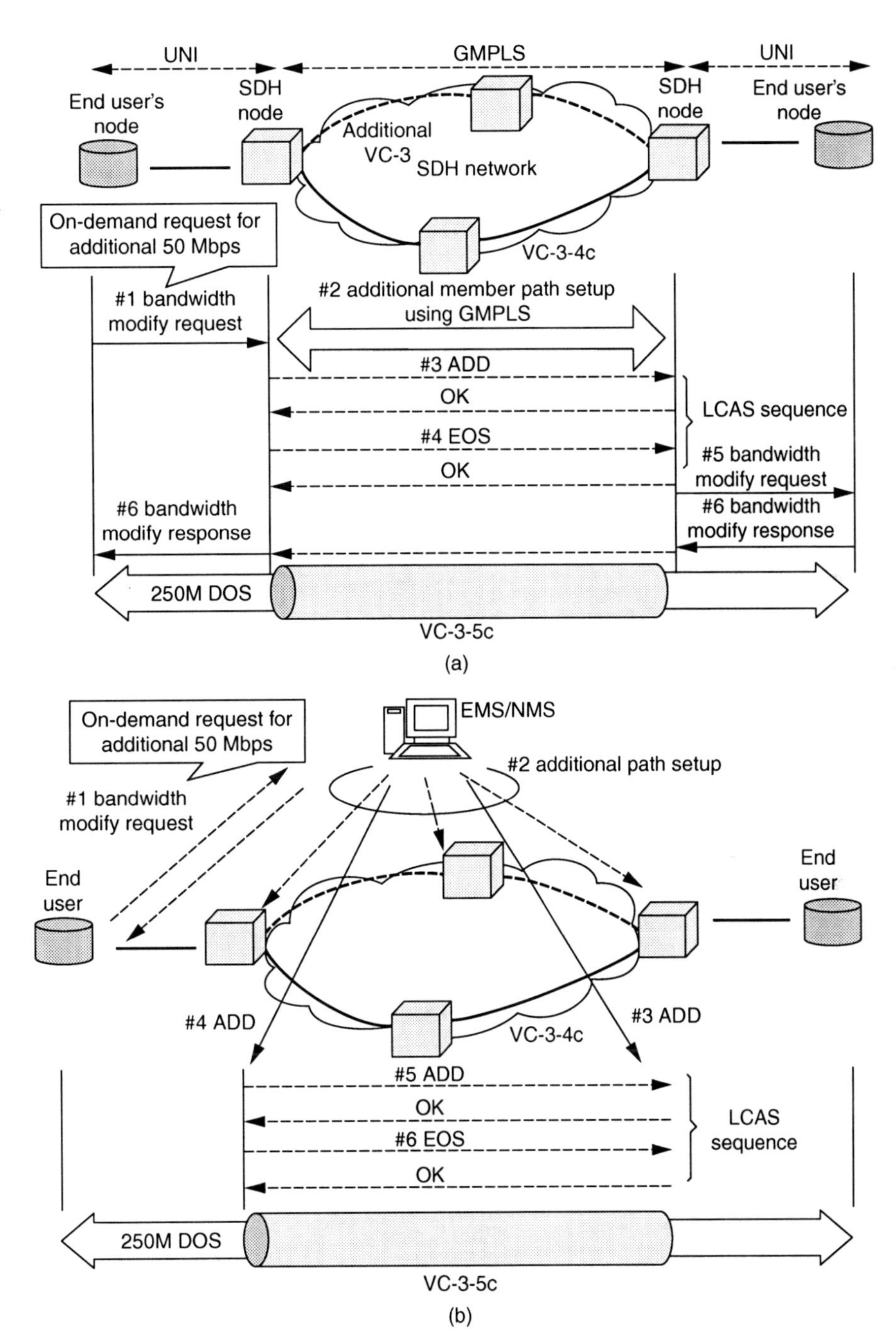

Figure 16-4 Examples of BoD setup: (*a*) UNI/GMPLS-based; (*b*) EMS/NMS-based.

and network provider interface in Fig. 16-4*a*). However, the bandwidth adjustment feature is out of the scope of that document, although expected to be present in a future UNI 2.0 document.

NOTE The mission of the Optical Internetworking Forum (OIF) is to foster the development and deployment of interoperable products and services for data switching and routing, using optical networking technologies. The OIF will encourage cooperation among telecom industry participants including equipment manufacturers, telecom service providers, and end users; promote global development of optical internetworking products; promote nationwide and worldwide compatibility and interoperability; encourage input to appropriate national and international standards bodies; and identify, select, and augment as appropriate and publish optical internetworking specifications drawn from appropriate national and international standards.

Figure 16-4*b* shows an example of EMS/NMS based on a centralized control scheme. In this case, the EMS/NMS receives a bandwidth modification request from the end user (step 1, additional 50 Mbps, in Fig. 16-4*b*). The EMS/NMS then routes the end-to-end path (additional VC-3 in Fig. 16-4*b*) in order to increase the end-to-end capacity and set up the path using a local command such as TL-1 (step 2 in Fig. 16-4*b*). After that, EMS/NMS starts the LCAS procedure at the edge nodes (steps 3 and 4 in Fig. 16-4*b*), so that hitless bandwidth increases are to be performed (steps 5 and 6 in Fig. 16-4*b*). Once these steps are completed, the network provider can supply the requested bandwidth to the user.

DoS Transport Node: Architecture and Applications

In this part of the chapter, DoS transport architecture is considered. Wireless data network applications are also considered.

DoS Node Architecture

Figure 16-5*a* illustrates an example of a novel transport network using DoS nodes.[3] The network is a ring that provides hybrid services: TDM and shared wireless data transport. In this example, TDM service is realized via dedicated bandwidth across the ring. In the same network, some shared bandwidth is used to provide efficient and reliable wireless data transport applications. Notice that this transport network differs from a recent resilient packet ring (RPR) initiative (IEEE 801.17), in the

Figure 16-5 A DOS node: (*a*) an example of network architecture using DOS; (*b*) functional architecture; (*c*) hardware architecture.

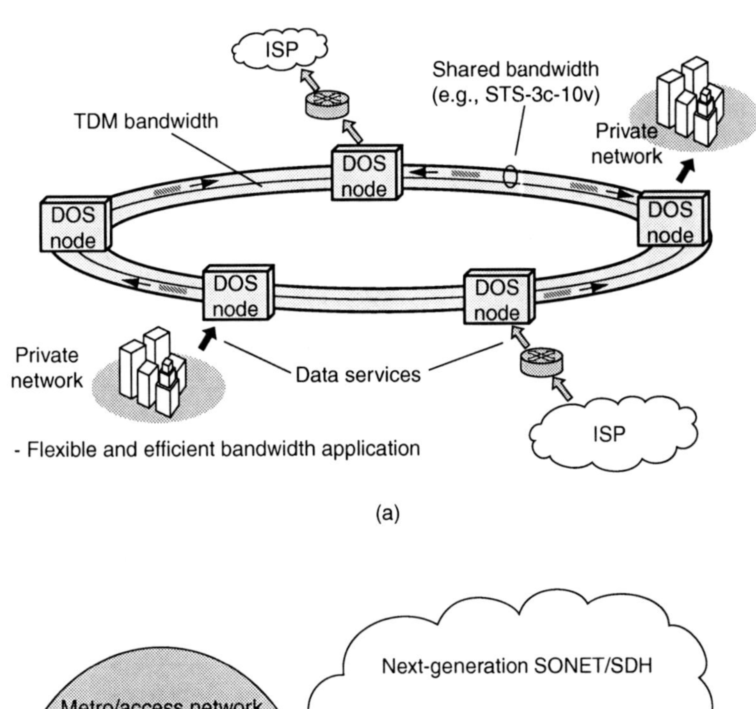

(a)

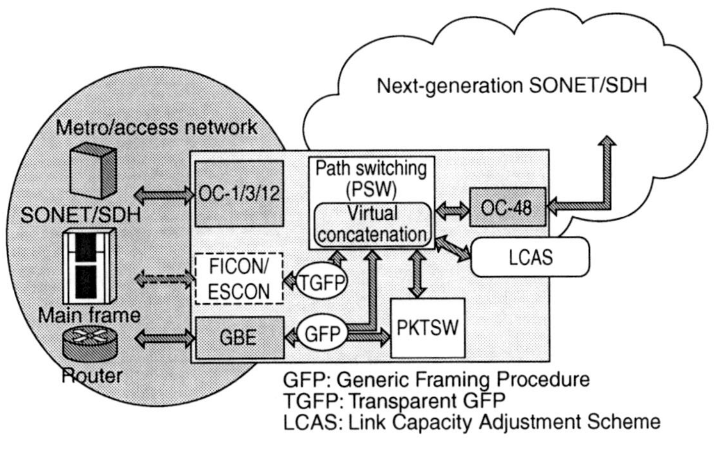

(b)

sense that the latter is focused on wireless data transport only (Ethernet directly over fiber is currently the interface of choice) for metro-area applications. In contrast, besides wireless data applications, the hybrid services support TDM applications as well, and can be used in both MAN and WAN scenarios.

NOTE The channel identifier field in GFP linear frame structure can be used to distinguish 256 data streams within a single SONET/SDH path (see Fig. 16-6).[3]

Figure 16-5 (Continued)

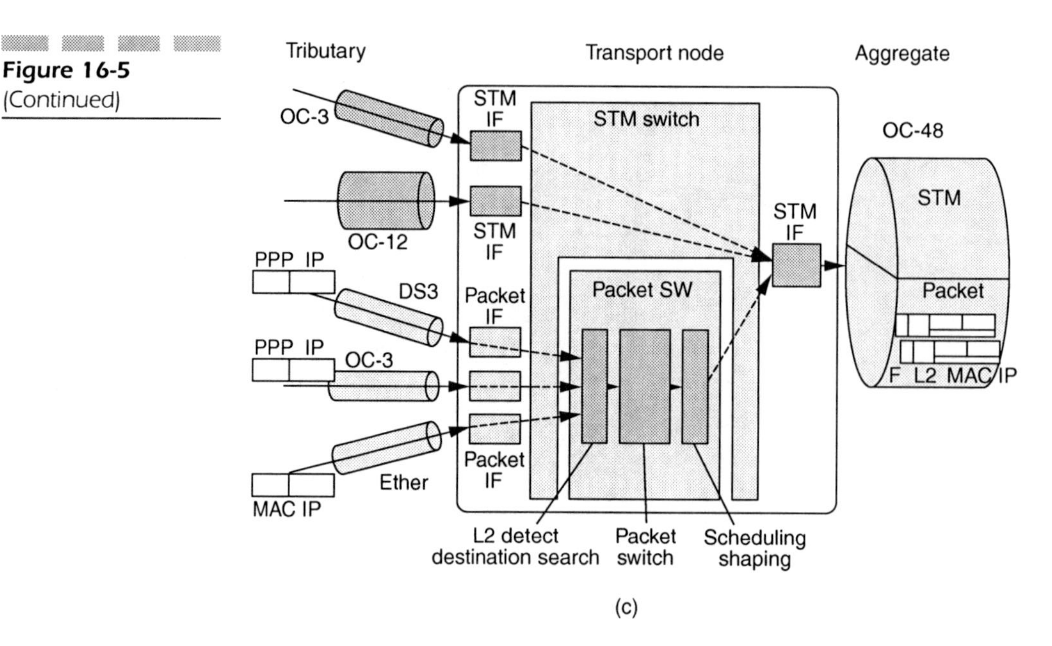

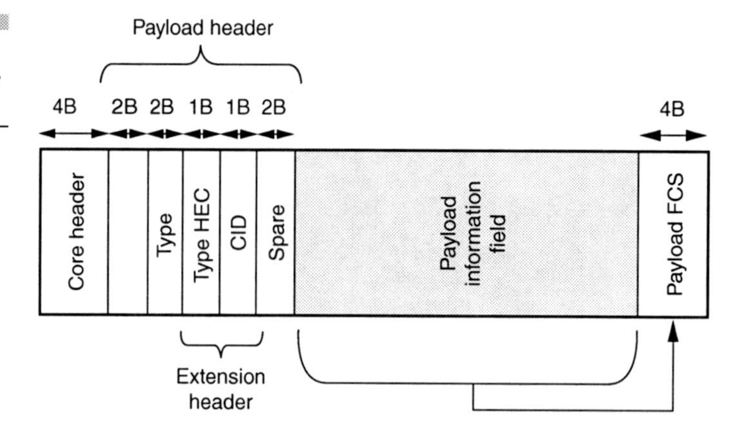

Figure 16-6 The GFP linear frame structure.

Figure 16-5*b* shows a typical functional architecture of a DoS node. The node provides transport interfaces, such as legacy SONET/SDH, ESCON/FICON, and GbE, for a wide range of applications. The DoS node uses virtual concatenation and GFP as enablers to efficiently pack wireless application data into SONET/SDH frames. It also uses LCAS to regulate the amount of bandwidth assigned to transport the client wireless data.

DoS nodes are designed to provide a wide variety of line interfaces so that new services can be launched without deployment of new nodes.

New line interface cards are installed as need arises. Interfaces for a data center (ESCON, FICON, Fibre Channel) and digital video (DVB-ASI) are also utilized. Figure 16-5*c* illustrates the hardware architecture of a DoS node with Layer ½ hybrid switch capability. The node is composed of the following modules.

- Switch modules:
  - STM switch
  - Packet switch

- Aggregate interface cards:
  - OC-48/STM-16
  - OC-192/STM-64
  - OC-768/STM-256

- Tributary interface cards:
  - Ethernet (10M/100M/1G)
  - Fibre Channel, ESCON/FICON
  - DVB-ASI (video interface)
  - POS (OC-3/STM-1, OC-12/STM-4, OC-48/STM-16)
  - ATM (OC-3/STM-1, OC-12/STM-4, OC-48/STM-16)
  - TDM (OC-3/STM-1, OC-12/STM-4, OC-48/STM-16, DS1, DS3, etc.)[3]

Node-to-node trunks are terminated on an aggregate interface card. On the receiver side of the aggregate interface, TDM traffic continues to be switched to either the tributary interface cards or aggregate interface cards, while the data traffic on virtually concatenated channels is routed to the packet switch. The packet switch performs termination of virtually concatenated payloads to produce GFP streams at the switch input ports. At the output ports of the packet switch, the virtual concatenation function maps the GFP streams into virtually concatenated payloads, which are sent to the STM switch. The packet switch performs the switching of GFP frames between ports, some connected to tributary interface cards and the rest to aggregate interface cards, through the STM switch. At the tributary interface card, GFP frames are terminated to extract the original data stream, which is then mapped to the appropriate Layer 1 and 2 protocols.

In the wireless data transmission direction, the incoming Layer 1 and 2 protocols are terminated, and wireless data streams are encapsulated into GFP frames at the tributary interface cards. If the line interface card happens to have several ports, the GFP frames from the various ports are aggregated and sent to the packet switch. The packet switch then switches GFP frames, maps the frames into virtually concatenated payloads, and sends them to the aggregate interface cards.

GFP Point-to-Point Frame Application

The structure of a GFP linear (point-to-point) frame is depicted in Fig. 16-6.[3] A typical application of a GFP linear frame is point-to-point connection and concentration. For example, data streams from multiple tributary interface cards can be aggregated into a same aggregate interface card. The 8-bit channel identifier (CID) in the GFP extension header is used to indicate one of 256 data streams. If the available bandwidth of the aggregate interface is below the sum of peak traffic of all data streams, statistical multiplexing is introduced to achieve concentration.

The optional payload FCS field in the GFP frame can be used for performance monitoring of an end-to-end GFP path. The area covered by FCS is the payload information field only, which contains the wireless user data. Therefore, at intermediate nodes, recalculation of FCS is not necessary, so that FCS is retained throughout the path. The end-to-end path monitoring can be used for path quality management as well as for triggering protection mechanisms.

SAN Interconnection by Transparent GFP

SAN deployment for disaster recovery applications has recently received a lot of attention. This application requires direct connection of SAN interfaces to a WAN in an efficient manner.

The conventional method for supporting this application is to simply assign one wavelength to each SAN interface. This method is inefficient in terms of bandwidth usage because the SAN bit rate is generally much less than the wavelength modulation rate. Better efficiency is achieved by multiplexing several SAN signals into a SONET/SDH-modulated wavelength. Transparent GFP (TGFP) allows transparent transport and multiplexing of 8B/10B clients such as Fibre Channel, ESCON, FICON, and DVB-ASI (digital video), as mentioned earlier. Transparency means that wireless data and clock rate received at the TGFP ingress node can be recovered at the egress node over a SONET/SDH network. TGFP can be seen as a kind of sublambda technique for 8B/10B interfaces over SONET/SDH (see Fig. 16-7).[3]

An additional benefit of this solution is that TGFP provides 6.25 to 16.25 percent bandwidth reduction from the original 10B rate. Table 16-1 shows typical VC path capacity required for SAN client transparent transmission.[3]

Finally, let's look at why generic framing procedure (GFP) is a new standard that has been developed to overcome wireless data transport inefficiencies or deficiencies with the existing ATM and packet over SONET/SDH protocols. Transparent GFP is an extension to GFP devel-

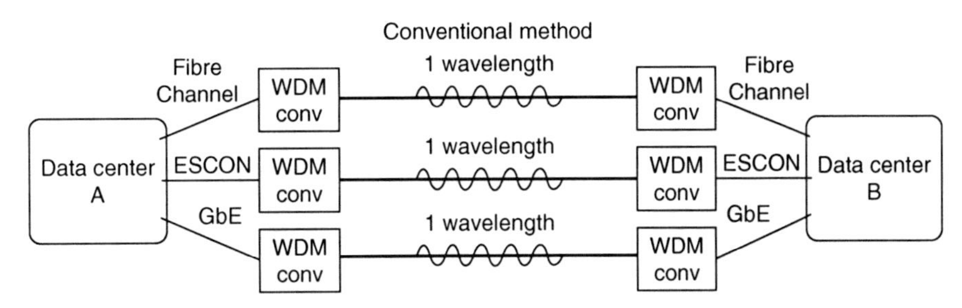

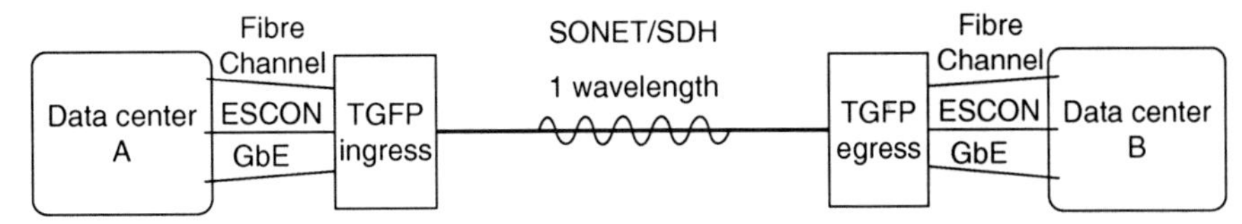

Figure 16-7 Application of TGFP.

TABLE 16-1 Bandwidth Reduction by Use of TGFP

Protocol	10B-Based Rate, Mbps	8B-Based Rate, Mbps	VC Path Size
ESCON	200	160	STS-1-4v
DVB-ASI	270	216	STS-3c-2v
Fibre Channel, FICON	1062.5	850	STS-3c-6v
GbE	1250	1000	STS-3c-7v
Infiniband	2500	2000	STS-3c-14v

oped to provide efficient low-latency support for high-speed WAN applications including storage-area networks. Rather than handle wireless data on a frame-by-frame (packet-by-packet) basis, TGFP handles block-coded (8B/10B) character streams. The next part of the chapter describes the GFP protocol along with technical considerations and applications for transparent GFP.

Transparent Generic Framing Procedure

Several important high-speed LAN protocols use a Layer 1 block code in order to communicate both wireless data and control information. The most common block code is the 8B/10B line code used for Gigabit Ethernet,

ESCON, SBCON, Fibre Channel, FICON, and Infiniband, which have become increasingly important with the growing popularity of storage-area networks (SANs). Since both client wireless data bytes and data source-to-sink control information are encoded into the 8B/10B codes, efficient transport of these protocols through a public transport network such as synchronous optical network/synchronous digital hierarchy (SONET/SDH) or the optical transport network (OTN) requires transporting both the wireless data and the 8B/10B control code information. The 8B/10B coding, however, adds a 25 percent wireless data bandwidth expansion that is undesirable in the transport network.

The previously available protocols for LAN transport through SONET/SDH networks were asynchronous transfer mode (ATM) and packet over SONET/SDH (POS). ATM is relatively inefficient from a bandwidth utilization standpoint and typically requires a much more complex adaptation process than GFP. POS requires terminating the client signal's Layer 2 protocol and remapping the signal into Point-to-Point Protocol (PPP) over HDLC, which suffers from a nondeterministic bandwidth expansion discussed previously on bandwidth considerations. Also, neither ATM nor POS supports the transparent transport of the 8B/10B control characters. In order to overcome the shortcomings of ATM and POS, GFP standardization began in the American National Standards Institute (ANSI) accredited T1X1 subcommittee, which chose to work with the International Telecommunication Union—Telecommunication Standardization Sector (ITU-T) on the final version of the standard, which has been published by the ITU-T. The transparent version of GFP has been optimized for transparently carrying block-coded client signals (both the data and the 8B/10B control codes) with minimal latency. This part of the chapter begins with a description of the transparent GFP protocol, followed by some special considerations such as bandwidth, error control, and client management. Potential extensions to the transparent GFP protocol are then also briefly discussed.

Transparent GFP Description: General GFP Overview

The basic GFP frame structure is shown in Fig. 16-8.[4] Protocols such as HDLC that rely on specific wireless data patterns for frame delimiting or control information require a nondeterministic amount of bandwidth because of the need for additional escape bits or characters adjacent to the payload strings or bytes that mimic these reserved characters. The amount of expansion is thus data pattern–dependent. In the extreme case, if the client payload data consist entirely of data emulating these reserved characters, byte-stuffed HDLC protocols like POS require nearly twice the

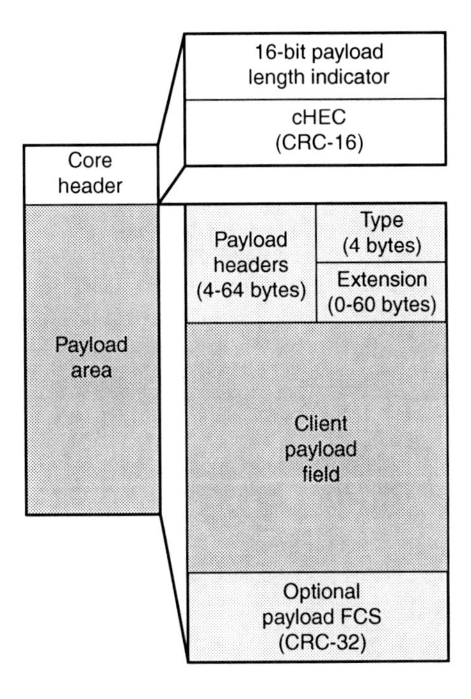

Figure 16-8
GFP frame format.

bandwidth to transmit the packet than if the payload did not contain such characters. GFP avoids this problem by using information in its core header for frame delimitation. Specifically, the GFP core header consists of a two-octet-long field that specifies the length of the GFP frame's payload area in octets, and a cyclic redundancy check (CRC-16) error check code over this length field. The framer looks for a 32-bit pattern that has the proper zero CRC remainder and then confirms that this is the correct frame alignment by verifying that another valid 32-bit sequence exists immediately after the current frame ends, as specified by the length field. Since no special characters are used for framing, there are no forbidden payload values that require escape characters.

NOTE CRC-16 also provides robustness by allowing single error correction on the core header once frame alignment has been acquired.

In frame-mapped GFP (GFP-F), a single client data frame [an IP packet or Ethernet medium access control (MAC) frame] is mapped into a single GFP frame. For transparent GFP, however, a fixed number of client characters are mapped into a GFP frame of predetermined length. Hence, the payload length is typically variable for frame-mapped GFP and static for transparent GFP. One of the primary advantages of TGFP over GFP-F is that TGFP supports the transparent transport of 8B/10B control characters as well as wireless data characters. In addition, GFP-F typically incurs the latency associated with buffering an entire client data frame at the ingress to the GFP mapper. As discussed next, TGFP requires only a few bytes of mapper/demapper latency. This lower latency is a critical issue for SAN protocols, which are very sensitive to transmission delay.

NOTE GFP-F is best suited to applications where latency is less important than bandwidth efficiency. For example, if the client signal is lightly loaded, GFP-F allows mapping the packets into a smaller transport channel or potentially frame multiplexing them into a shared channel with GFP frames from other client signals. Alternatively, GFP-F could make use of the link capacity adjustment scheme (G.7042) for handling client signals that experience temporary changes to their required bandwidth.

Transparent GFP 64B/65B Block Coding

The 8B/10B line code maps the $2^8 = 256$ possible data values into the $2^{10} = 1024$ value 10-bit code space such that the running number of ones and zeros transmitted on the line (the running disparity) remains balanced over very short intervals. Twelve of the 10-bit codes are reserved for use as control codes that may be used by the wireless data source to signal control information to the wireless data sink. The first step of TGFP encoding in the source adaptation process is to decode the client 8B/10B codes into control codes and 8-bit data values. Eight of these decoded characters are then mapped into the 8 payload bytes of a 64B/65B code. The leading (flag) bit of the 64B/65B code indicates whether there are any control codes present in that 64B/65B code (with flag = 1 indicating the presence of a control code). The 64B/65B block structure for various numbers of control codes is illustrated in Fig. 16-9.[4] Control codes are placed in the leading bytes of the 64B/65B block as illustrated in Fig. 16-9. A control code byte consists of a bit to indicate whether this byte contains the last control code in that 64B/65B block (= 0 if it is the last), a 3-bit address (aaa–hhh) indicating the original location of that control code in the wireless client data stream relative to the other characters mapped into that 64B/65B block, and a 4-bit code (Cn) representing the control code. Since there are only 12 defined 8B/10B control codes, 4 bits are adequate to represent them. One of the remaining 4-bit codes is used to communicate that an illegal 8B/10B character has been received by the GFP source adaptation process so that the GFP receiver can output an equivalent illegal 8B/10B character to the client signal sink. Figure 16-10 illustrates mapping of control and wireless data octets in the 64B/65B block.[4]

Aligning the 64B/65B payload bytes with the SONET/SDH/OTN payload bytes simplifies parallel wireless data path implementations, in

Input client characters	Flag bit	64-bit (8-octet) field							
		Octet 0	Octet 1	Octet 2	Octet 3	Octet 4	Octet 5	Octet 6	Octet 7
All data	0	D1	D2	D3	D4	D5	D6	D7	D8
7 data, 1 control	1	0 aaa C1	D1	D2	D3	D4	D5	D6	D7
6 data, 2 control	1	1 aaa C1	0 bbb C2	D1	D2	D3	D4	D5	D6
5 data, 3 control	1	1 aaa C1	1 bbb C2	0 ccc C3	D1	D2	D3	D4	D5
4 data, 4 control	1	1 aaa C1	1 bbb C2	1 ccc C3	0 ddd C4	D1	D2	D3	D4
3 data, 5 control	1	1 aaa C1	1 bbb C2	1 ccc C3	1 ddd C4	0 eee C5	D1	D2	D3
2 data, 6 control	1	1 aaa C1	1 bbb C2	1 ccc C3	1 ddd C4	1 eee C5	0 fff C6	D1	D2
1 data, 7 control	1	1 aaa C1	1 bbb C2	1 ccc C3	1 ddd C4	1 eee C5	1 fff C6	0 ggg C7	D1
8 data	1	1 aaa C1	1 bbb C2	1 ccc C3	1 ddd C4	1 eee C5	1 fff C6	1 ggg C7	0 hhh C8

Legend:

- Leading bit in a control octet (LCC) = 1 if there are more control octets and = 0 if this payload octet contains the last control octet in that block
- aaa = 3-bit representation of the first control code's original position (first control code locator)
- bbb = 3-bit representation of the second control code's original position (second control code locator)

. . .

- hhh = 3-bit representation of the eighth control codes original position (eighth control code locator)

- C*i* = 4-bit representation of the *i*th control code (control code indicator)
- D*i* = 8-bit representation of the *i*th data value in order of transmission

Figure 16-9 The 64B/65B block code structure.

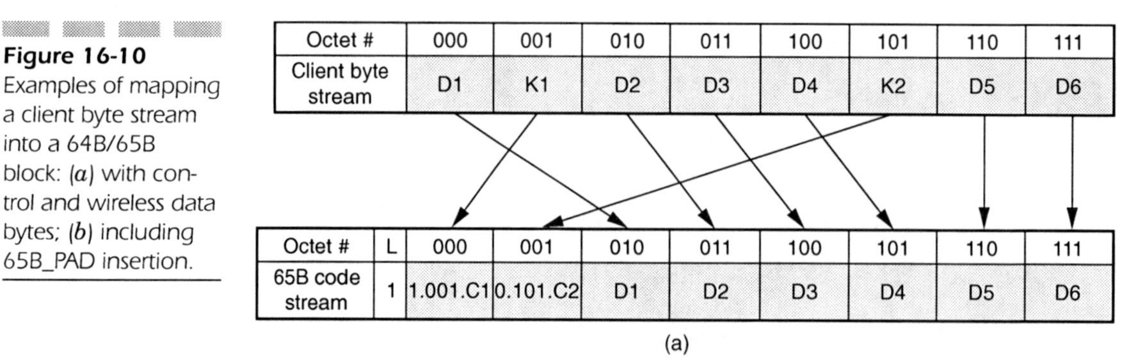

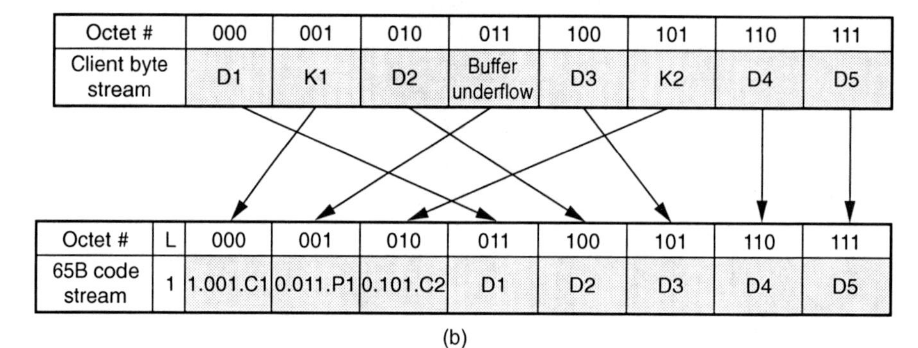

Figure 16-10 Examples of mapping a client byte stream into a 64B/65B block: (*a*) with control and wireless data bytes; (*b*) including 65B_PAD insertion.

addition to increasing the payload data observability within the SONET/SDH stream. In order to achieve this alignment, a group of eight 64B/65B codes are combined into a superblock. The superblock structure, as shown in Fig. 16-11, takes the leading flag bits of the eight constituent 64B/65B codes and groups them into a trailing byte followed by a CRC-16 over the bits of that superblock.[4] CRC-16 is discussed further in the section "Error Control Considerations," below.

Transport Bandwidth Considerations

TGFP channel sizes are chosen to accommodate the wireless client data stream under worst-case clock tolerance conditions (for the slowest end of the transport clock and fastest end of the client clock tolerance). In the case of SONET/SDH, while TGFP can be carried over contiguously concatenated channels, it will typically be carried over virtually concatenated signals. The concept of virtual concatenation is one in which multiple SONET synchronous payload envelopes (SPEs) are grouped together with SDH virtual containers (VCs) to form a higher-bandwidth pipe between the endpoints of the virtually concatenated path.

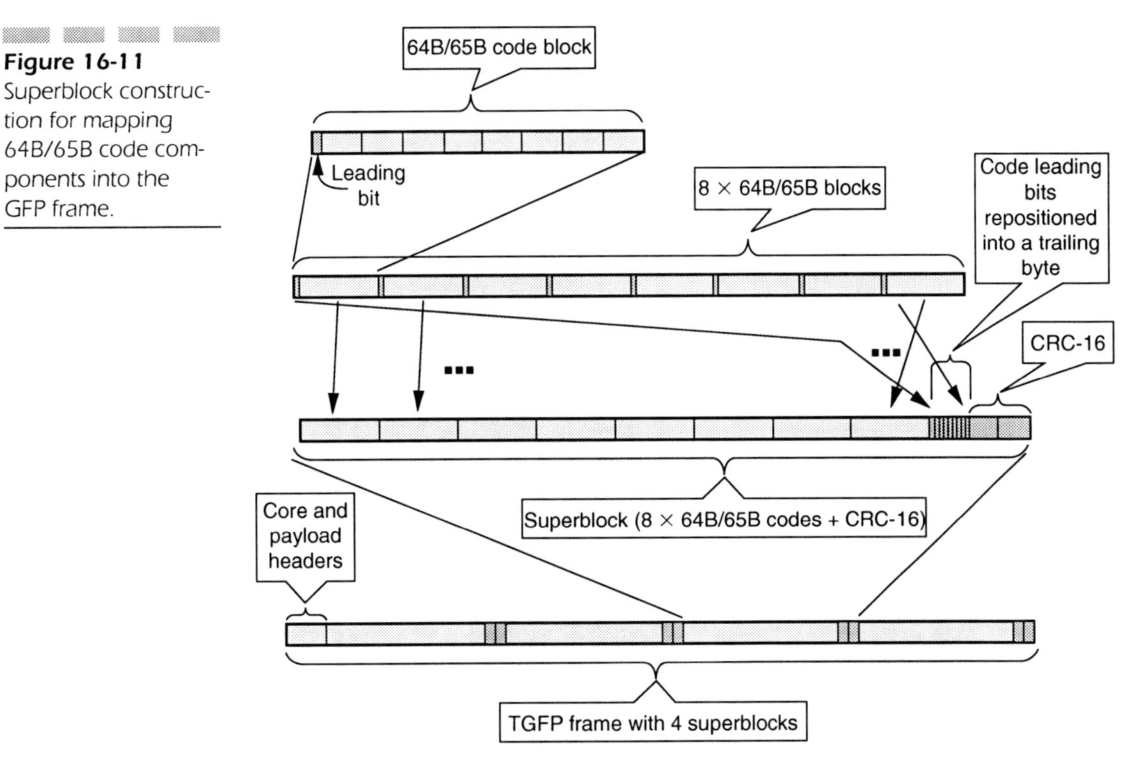

Figure 16-11 Superblock construction for mapping 64B/65B code components into the GFP frame.

NOTE The constituent SPEs/VCs do not need to be time-slot contiguous, which greatly simplifies the provisioning and increases the flexibility of virtual concatenation.

Another advantage of virtual concatenation is that it is transparent to intermediate nodes, with only the endpoints of the virtually concatenated path needing to be aware of its existence. The nomenclature for indicating a virtually concatenated signal is <SPE/VC type>-Xv, where X indicates the number of SPEs/VCs that are being concatenated. For example, STS-3c-7v is the virtual concatenation of seven STS-3c SPEs, which is equivalent to VC-4-7v for SDH. Virtual concatenation is specified in the ITU-T, the ANSI, and the European Telecommunications Standards Institute (ETSI). Table 16-2 shows the minimum virtually concatenated channel size that can be used for various TGFP clients.[4]

In practice, the SONET/SDH channel must be slightly larger than that needed to carry the GFP signal, a consequence of which is that the GFP mapper's client signal ingress buffer will underflow. There are two ways to handle this situation. One approach is to buffer an entire TGFP frame's worth of wireless client data characters prior to beginning the transmission of that GFP frame. This approach would increase the mapper

TABLE 16-2 *Virtually Concatenated Channel Sizes for Various Transparent GFP Clients*

Client Signal	Native (Unencoded) Client Signal Bandwidth	Minimum Virtually Concatenated Transport Channel Size	Nominal Transport Channel Bandwidth	Minimum Number of Superblocks per GFP Frame	Worst-Case (Best-Case) Residual Overhead Bandwidth (1)	Best-Case Client Management Payload (2)
ESCON	160 Mbps	STS-1-4v/VC-3-4v	193.536 Mbps	1	5.11 Mbps (24.8 Mbps)	6.76 Mbps
Fibre Channel	850 Mbps	STS-3c-6v /VC-4-6v	898.56 Mbps	13	4.12 kbps (85.82 Mbps)	2.415 Mbps
Gigabit Ethernet	1.0 Gbps	STS-3c-7v /VC-4-7v	1.04832 Gbps	95	281 kbps (1.138 Mbps)	376.5 kbps

NOTES: (1) The worst-case residual bandwidth occurs when the minimum number of superblocks is used per GFP frame. The best case occurs for the value of N that allows exactly one client management frame per GFP data frame. A 160-bit client management frame was assumed for the best case (with a CRC-32). For both cases, it was assumed that no extension headers were used. (2) The best-case client management payload bandwidth assumes 8 "payload" bytes per client management frame and the best-case residual overhead bandwidth conditions.

latency and buffer size. A second approach, which was adopted for the standard, is to use a dummy 64B/65B control code as a 65B_PAD character. Whenever there is no client character available in the ingress buffer, the mapper will treat the situation the same as if a client control character were present and will insert the 4-bit 65B_PAD character. Figure 16-10*b* illustrates the insertion of a 65B_PAD character. The demapper at the other end of the GFP link recognizes this character as a dummy pad and removes it from the wireless data stream. The result of using this 65B_PAD character is that the mapper ingress buffer size is reduced to effectively 8 bytes (the amount of data required to form a 64B/65B block) plus the number of bytes that can accumulate during the SONET/SDH overhead and the GFP frame overhead bytes. An 8-byte latency is always required since the mapper cannot complete the 64B/65B block coding until it knows whether there are any control codes present in the eight characters that will make up that block.

As discussed next, client management frames (CMFs) have been proposed for GFP that would make use of this "spare" bandwidth for client management applications. These CMFs would be up to 20 bytes long (including GFP encapsulation bytes) and, because they have lower priority than the wireless client data, would be allowed to be sent only when the ingress buffer is nearly empty. Support for these CMFs adds 20 bytes to the ingress buffer requirements to accommodate the wireless data arriving during the transmission of a CMF.

Demapping of the TGFP signal entails the removal of the 65B_PAD characters, and removal and interpretation of the interframe CMFs when present. Assuming that the egress of the client signal is done using a constant-rate local clock, if the egress buffer becomes empty as a result of reception of 65B_PAD characters and/or CMFs, interpacket fill words must be inserted according to the client signal type rules.

Error Control Considerations

The 8B/10B codes have built-in error detection capability since a single bit error will always result in an illegal code. The increased bandwidth efficiency gained by decoding the 8B/10B codes and remapping the data into 64B/65B codes comes at the expense of much of this error detection capability. There are four situations in which errors can cause significant problems with 64B/65B codes. The first and most serious problem results if the leading flag bit of the 64B/65B code is received in error. If the original block contained control codes, these codes will be interpreted as wireless data, and if the original block contained only data, some of these bytes may be interpreted as control codes. The number of data bytes that are erroneously interpreted as control codes depends on the value of the first bit (the last control code indicator bit position) of the bytes and whether the values of the location address bit positions contain increasing values (which would always be the case for a legal block). Wireless data erroneously converted into control codes could cause the truncation of a wireless client data frame, which in turn can cause error detection problems for the wireless client data, since there is a possibility of the truncated wireless client data frame appearing to have a correct CRC value. A similar situation occurs when control characters are present and the last control code indicator bit is affected by an error. Also, errors in the control code location address will cause it to be placed in the wrong sequence by the demapper, and errors in a 4-bit control code value will cause the demapper to generate an incorrect control code. Any error that results in a spurious or incorrect control code has potentially serious consequences.

It is these potential error problems that lead to the addition of a CRC-16 to each superblock. The most reliable mechanism for error control is for the demapper to discard all of the wireless data in a superblock in which an error is detected. The wireless data are discarded by having the demapper output 10B_ERROR 8B/10B codes for those clients that have defined such a code, or another illegal 8B/10B character for all of the characters in that superblock.

NOTE As discussed next, the CRC-16 optionally allows the possibility of single-error correction.

The payload area of the GFP frame is scrambled with a self-synchronous scrambler, and another error control issue concerns the interaction between the GFP payload scrambler and the superblock CRC-16. To understand the issue here, it is helpful to first understand the rationale and implementation behind the payload scrambler.

The reasons for using a self-synchronized payload scrambling process are related to the physical properties of the transport medium and the desire for robustness in public networks. The line code used for SONET/SDH and OTN is non-return-to-zero (NRZ) (after the data have been passed through a SONET/SDH/OTN frame-synchronous scrambler). For NRZ, the laser is turned on for the bit period to represent a 1 and off to represent a 0. The advantage of the NRZ line code is its simplicity and bandwidth efficiency. The disadvantage of NRZ, however, is that the receiver clock and data recovery circuits can lose synchronization after a long string of either 0s or 1s. The frame-synchronized scrambler, which is reset at regular intervals by the SONET/SDH/OTN frame, is adequate to defend against normally occurring user data patterns. It would be possible, however, for a malicious user to choose a packet payload that is the same as the frame-synchronized scrambler sequence. If this packet lines up in the correct position in the transport frame, an adequately long string of 0s or 1s can be generated to cause a loss of synchronization at the receiver. The resulting loss of synchronization will take down the transport link, while the receiver attempts to recover, thus denying the link to other users in the meantime. This problem was originally discovered in ATM networks and is exacerbated by the longer frames used in POS or GFP. In order to guard against such attacks ATM, POS, and GFP use a self-synchronous payload scrambler to further randomize the wireless payload data. This self-synchronous scrambler uses a polynomial of $x^{43} + 1$, which means that each bit of the ATM/POS/GFP payload area is exclusive ORed with the scrambler output bit that preceded it by 43 bit positions, as shown in Fig. 16-12.[4] The decoder's descrambler reverses this process.

NOTE The scrambler state is retained between successive GFP frames.

In order to use the same payload scrambling technique for both frame-mapped and transparent GFP, all of the GFP payload bits including the TGFP superblock CRCs must be scrambled. As a result, the superblock CRC has to be calculated over the superblock payload bits prior to scrambling and checked at the decoder after descrambling. The drawback to a self-synchronous scrambler, however, is that each transmission error results in a pair of errors (43 bits apart here) in the descrambled data, which means that the superblock CRC must cope with this error multiplication. It has been shown that a CRC will preserve its error detection capability in this situation as long as the scambler polynomial and the CRC generator polynomial have no common factors. Unfortunately, all of the standard CRC-16 polynomials contain $x + 1$ as a factor, which is also a factor in the $x^{43} + 1$ (or any $x^n + 1$) scrambler polynomial. Therefore, a new CRC generator polynomial was required that preserved the triple-error detecting capability (which is the maximum achievable over this block size) without having any common factors with the scrambler. In order to perform single-error correction, the syndromes for single errors and double errors spaced 43 bits apart must all be unique. The code selected for the superblock is $x^{16} + x^{15} + x^{12} + x^{10} + x^4 + x^3 + x^2 + x + 1$, which has both these desired properties, and hence retains its triple-error detection and optional single-error correction capabilities in the presence of the scrambler.

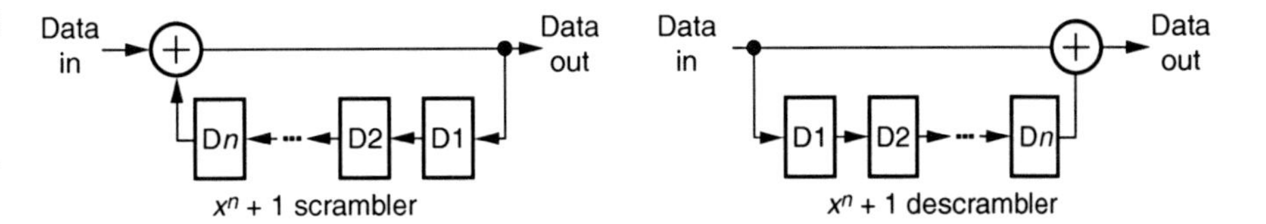

Figure 16-12 Payload self-synchronous scrambler.

Transparent GFP Client Management Frames

CMFs have the same structure as GFP wireless client data frames, but are denoted by the payload type code PTI = 100 in the GFP payload header. Like GFP wireless client data frames, CMFs have a core header, a payload header [both with 2-byte header error checking (HEC)], and an optional 32-bit FCS. The total CMF payload size in TGFP is recommended to be no greater than 8 bytes.

Assuming an 8-byte payload area along with the 8 total bytes for the mandatory core and type headers, the payload efficiency will be 50 percent. Use of FCS and especially extension headers will greatly reduce the efficiency of the CMFs.

As previously noted, there is some residual "spare" bandwidth in the SONET/SDH channel for each of the client signal mappings. As shown in Table 16-2, the amount of this spare bandwidth depends on the efficiency of the mapping, which in turn is partially a function of the number of superblocks used in each GFP frame. The residual bandwidth can be used as a client management overhead channel for client management functions, as described in this part of the chapter. CMFs are also used for downstream indication of client signal fail.

Remote Management If both ends of the GFP link are owned by the same carrier and the intervening SONET/SDH/OTN network is owned by another operator, the potential exists for sending GFP-specific provisioning commands using CMFs. It is not uncommon for interexchange carriers (IECs) to provide the customer premises equipment (CPE) and rely on a local exchange carrier (LEC) to provide the connection between the CPE and the IEC network (see Fig. 16-13).[4] Ideally, the IEC would like to manage the CPE as part of its own network, which frees the customer from having to manage the equipment and allows the IEC a potential revenue source from providing the management service. Normally, management information is communicated through a wireless SONET/SDH section data communication channel (SDCC). In order to prevent unwanted control access, however, carriers do not allow SDCC wireless data to cross the network interfaces into their networks; hence, there is currently no way for the IEC to exchange management communications with the CPE through the intervening LEC network. TGFP CMFs, however, provide a mechanism to tunnel the SDCC information through the intervening network. Table 16-2 shows the maximum amount of payload capacity that can be derived from the CMFs for SDCC tunneling or other operations, administration, and maintenance (OAM) applications, with the assumptions stated in the table notes and assuming a 20-byte CMF with an 8-byte payload field. For all client signal types, there is adequate bandwidth available to carry a 192-kbps SDCC channel.

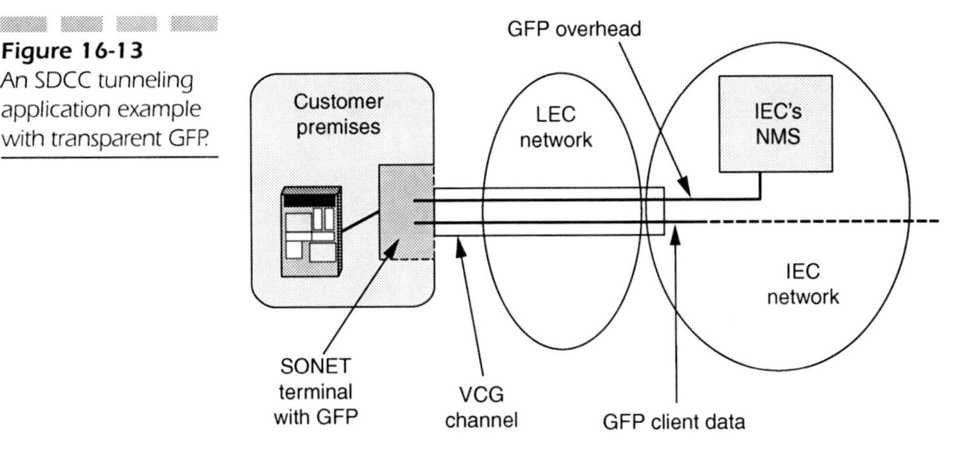

Figure 16-13 An SDCC tunneling application example with transparent GFP.

Conclusion

This chapter introduced several emerging techniques currently under development for next-generation SONET/SDH systems. On the basis of these new techniques, the chapter elaborated on new SONET/SDH transport services likely to become reality within a few years. Data over SONET/SDH, using GFP, virtual concatenation, and LCAS, is likely to become the dominant transport method over SONET/SDH transport networks.

Looking ahead, flexible transport services, combined with virtually unlimited bandwidth availability brought by WDM transport techniques, will ensure that sophisticated and bandwidth-hungry Internet applications of the future can be deployed. These yet to be seen applications will likely change computer and human communications in a revolutionary and unprecedented way.

Finally, transparent GFP provides an efficient mechanism for mapping constant-bit-rate block-coded wireless data signals across a SONET/SDH network or OTN. Performing the mapping on a client character basis rather than a client frame basis significantly reduces the transport latency to a fixed number of bytes rather than a whole client frame, which is a critical issue for SAN protocols including Gigabit Ethernet. The translation of client block codes into more efficient 64B/65B mapping provides significant bandwidth efficiency increase while the superblock structure provides robustness. Transparent GFP also allows increased performance monitoring capability for the Transport layer, and the ability to tunnel SDCC management information through an intervening network provides a powerful extension to network providers' capabilities.

References

1. Paul Korzeniowski, "10 Gigabits and Beyond," 12 Stagecoach Lane, Sudbury, MA 01776, 2002.
2. Ravi Ravishankar, "Wireless Carriers Address Network Evolution with Packet Technology," Technology Marketing Corporation, One Technology Plaza, Norwalk, CT 06854, 2002.
3. Dirceu Cavendish, Kurenai Murakami, Su-Hun Yun, Osamu Matsuda, and Motoo Nishihara, "New Transport Services for Next-Generation SONET/SDH Systems," *IEEE Communications Magazine,* 445 Hoes Lane, Piscataway, NJ 08855, 2002.
4. Steven S. Gorshe and Trevor Wilson, "Transparent Generic Framing Procedure (GFP): A Protocol for Efficient Transport of Block-Coded Data through SONET/SDH Networks," *IEEE Communications Magazine,* 445 Hoes Lane, Piscataway, NJ 08855, 2002.
5. John R. Vacca, *i-mode Crash Course,* McGraw-Hill, 2001.
6. John R. Vacca, *High-Speed Cisco Networks: Planning, Design, and Implementation,* CRC Press, 2002.

CHAPTER 17

Wireless Data Access Implementation Methods

The explosive growth of both the wireless industry and the Internet is creating a huge market opportunity for the implementation of wireless data access methods. Limited Internet access, at very low speeds, is already available as an enhancement to some existing cellular systems. However, those systems were designed with the purpose of providing voice services and (at most) short messaging, but not fast data transfers. In fact, as shown in this chapter, traditional wireless data technologies are not very well suited to meet the demanding requirements of providing very high data rates with the ubiquity, mobility, and portability characteristic of cellular systems. Increased use of antenna arrays appears to be the only means of enabling the types of data rates and capacities needed for wireless data Internet and multimedia services. While the deployment of base station arrays is becoming universal, it is really the simultaneous deployment of base station and terminal arrays that can unleash unprecedented levels of performance by opening up multiple spatial signaling dimensions.

Using Antenna Arrays: Lifting the Limits on High-Speed Wireless Data Access

As previously mentioned, the explosive growth of both the wireless data industry and the Internet is creating a huge market opportunity for wireless data access. Limited Internet access at low speeds (a few tens of kilobits per second at most) is already available as an enhancement to some second-generation (2G) cellular systems. However, those systems were originally designed with the sole purpose of providing voice services and, at most, short messaging, but not fast data transfers. Third-generation (3G) mobile wireless data systems,[3] currently under development, will offer true packet access at significantly higher speeds. Theoretically, user data rates as high as 2 Mbps will be supported in certain environments, although recent studies have shown that approaching those rates might be feasible only under extremely favorable conditions—in the vicinity of a base station and with no other users competing for bandwidth.

In fact, as will be argued in this part of the chapter, traditional wireless data technologies are not particularly well suited to meet the extremely demanding requirements of providing the very high data rates and low cost associated with wired access, and the ubiquity, mobility, and portability characteristic of cellular systems. Some fundamental barriers, related to the nature of the radio channel as well as to limited bandwidth availability at the frequencies of interest, stand in the way.

As a result, the cost per bit in wireless data is still high and not diminishing fast enough. In contrast, the wired world is already providing basically free bits, which has accustomed an entire generation of Internet users to accessing huge volumes of information at very high speeds and negligible cost. This part of the chapter establishes practical limits on the wireless data rates that can be supported by a wireless data access system with a typical range of parameters, and it shows how those limits can be lifted by using a combination of transmit and receive antenna arrays with powerful space-time processing techniques. (The Glossary defines many technical terms, abbreviations, and acronyms used in the book.)

Fundamental Limitations in Wireless Data Access

Ever since the dawn of the information age, capacity has been the principal metric used to assess the value of a communication system. However, several definitions of capacity exist. Link capacity or user capacity is used here to signify the highest data rate at which reliable communication is possible between a transmitter and a receiver. At the same time, system capacity is used here to indicate the total throughput (sum of user data rates) within a cell or sector. System capacity can be converted into area capacity simply via normalizing by the cell size. Since existing cellular systems were devised almost exclusively for telephony, user data rates were low and had minimal variability. In fact, source rates were purposefully reduced to the minimum level necessary to support a highly compressed voice call and implicitly traded for additional users. Therefore, systems were designed to accommodate a large number of low-data-rate users. With the emergence of wireless data services, many of these concepts are becoming obsolete. User wireless data rates are increasingly variable and heterogeneous. The value of a system is no longer defined only by how many users it can support, but also by its ability to provide high peak rates to individual users as needed—in other words, by its ability to concentrate large amounts of capacity at very localized spots. Thus, in the age of wireless data, user data rate surges again as an important metric.

Because of the logarithmic relationship between the capacity of a wireless data link and the signal-to-interference-and-noise ratio (SINR) at the receiver, trying to increase the wireless data rate by simply transmitting more power is extremely costly. Furthermore, it is futile in the context of a dense interference-limited cellular system, wherein an increase in everybody's transmit power scales up both the desired signals as well as their mutual interference, yielding no net benefit. Therefore, power increases are useless once a system has become limited in essence by its own interference. Furthermore, since mature systems designed for high capacity tend to be interference-limited, it is power itself (in the form of interference) that ultimately limits their performance. As a result, power must be carefully controlled and allocated to enable the coexistence of multiple, geographically dispersed users operating in various conditions. Hence, power control has been a topic of very active research for many years.

Increasing the signal bandwidth (along with the power) is a more effective way of augmenting the wireless data rate. However, radio spectrum is a scarce and very expensive resource at the frequencies of interest, where propagation conditions are favorable. Moreover, increasing the signal width beyond the coherence bandwidth of the wireless data channel results in frequency selectivity. Although well-established techniques such as equalization and orthogonal frequency-division multiplexing can address this issue, their complexity grows rapidly with the signal bandwidth. Altogether, it is imperative that every unit of bandwidth be utilized as efficiently as possible. Consequently, spectral efficiency (defined as the capacity per unit bandwidth) has become another key metric by which wireless data systems are measured. In order to improve it, multiple-access methods (originally rather conservative in their design) have evolved toward much more sophisticated schemes. In the context of frequency-division multiple access (FDMA) and time-division multiple access (TDMA), this evolutionary path has led to advanced forms of dynamic channel assignment that enable adaptive and much more aggressive frequency reuse. In the context of code-division multiple access (CDMA), it has led to a variety of multiuser detection and interference cancellation techniques.

Models and Assumptions

This analysis is conducted in the 2-GHz frequency range, which is where 3G systems will initially be deployed. This is a favorable band from a propagation standpoint. Also, and again in line with the 3G framework, the available bandwidth is assumed to be $B = 5$ MHz. However, for simplicity, frequency selectivity is ignored here, with the argument that it can be dealt with by using the techniques mentioned earlier and their extension to the realm of antenna arrays and BLAST.

The downlink has the most stringent demands for wireless data applications. However, a similar analysis could be applied to the uplink, although with much tighter transmit power constraints. A cellular system with fairly large cells is assumed, with every cell partitioned into 120° sectors.

The propagation scenario portrayed in Fig. 17-1 is based on the existence of an area of local scattering around each terminal.[1] Little or no local scattering is presumed around the base stations. From the perspective of a base station, the angular distribution of power that gets scattered to every terminal is characterized by its root-mean-square width, commonly referred to as angle spread. Typical values for the angle spread at a base station are in the range of 1 to 10°, depending on the environment and range. The antennas composing a base station array can be operated coherently if they are closely spaced, or decorrelated by spacing them sufficiently apart. At a terminal buried in clutter, on the other hand, angle spreads tend to be very large (possibly as large as 360°) and thus, large uncorrelation among its antennas is basically guaranteed.

M is used here to denote the number of antennas within every base station sector and N to indicate the number of antennas at every terminal. The channel responses from every sector antenna to every terminal antenna are assembled into an $N \times M$ channel matrix $\mathbf{H} = \{h_{nm}\}$. Ignoring frequency selectivity, the entries of $\mathbf{H}$ are complex gaussian scalars (Rayleigh distributed in amplitude), whose local average path gain has a range-dependent component and a shadow fading component. The range-dependent component is modeled here using the well-established COST231 model. The shadow fading is taken to be log-normally distributed with an 8-dB standard deviation. The correlation among the entries of $\mathbf{H}$ is determined by the antenna spacing and angle spread. Antennas within a terminal are assumed fully uncorrelated, whereas those within a base station sector are assumed to be either uncorrelated (for sufficiently large spacing) or fully correlated (for close spacing and coherent operation). Therefore, the rows of $\mathbf{H}$ are always independent, whereas the columns can be either linearly dependent or also independent.

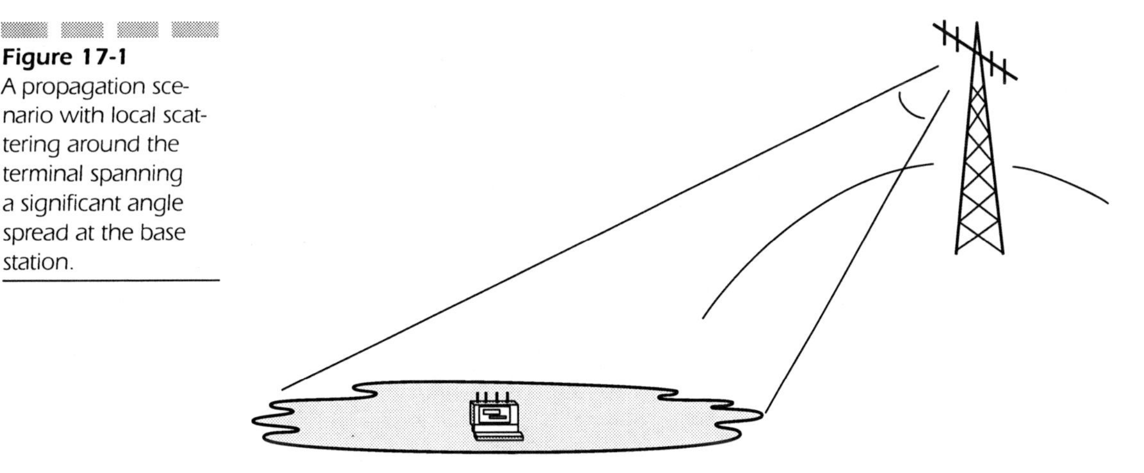

Figure 17-1 A propagation scenario with local scattering around the terminal spanning a significant angle spread at the base station.

While power control proved to be an essential ingredient in voice systems, where source rate variability was minimal, in wireless data systems rate adaptation becomes not only an attractive complement, but even a full alternative to power control. Hence, this part of the chapter restricts itself to the case where the total power per user is held constant while the wireless data rate is being adapted.

Furthermore, an open-loop architectures is used here—wherein transmitters do not have access to the instantaneous state of the channel. Only long-term information (information that varies slowly with respect to the fading rate) is available to the transmitters. In its original form, BLAST has no need for instantaneous channel information at the transmitter. However, more elaborate closed-loop forms of BLAST have been devised in order to exploit that information in those cases when it may be available, such as when a fast feedback link is available or when time-division duplexing is employed. It is assumed (in all cases) that the channel matrix **H** is known perfectly at the receiver.

NOTE This may require training overhead or, alternatively, blind acquisition algorithms.

Single-User Wireless-Data-Rate Limits

Let us first consider an isolated single-user link limited only by thermal noise. Within the context of a real system, this would correspond to an extreme case wherein the entire system bandwidth is allocated to an individual user. Furthermore, it would require that no other users be active anywhere in the system or that their interference be perfectly suppressed. Clearly, these are unrealistic conditions; thus, the single-user analysis provides simply an upper bound, only a fraction of which is attainable. Also, since for a system to be interference-limited, it is necessary that the signal-to-noise ratio (S/N) be large enough that the noise level is much lower than the interference, this analysis also determines what cell sizes can be supported in interference-limited conditions.

Steered Directive Array The use of an array is first considered here at the base station only, with M closely spaced antennas operating coherently. Since an estimate of the directional location of the terminal can be usually derived from the uplink, the use of directive array algorithms has been regarded as an attractive option for enhancing the performance of existing 2G systems.

Each individual base station antenna has a gain of 15 dBi. The terminal is equipped with a single omnidirectional antenna. Under these conditions, and assuming the beam synthesized at the base is properly steered toward the terminal, as M grows, the array becomes more directive; thus, more precise information about the directional location of the terminal is needed in order to fully illuminate its local scattering area. Also, since no further directional gain can be realized beyond the point at which the beamwidth becomes smaller than the angle spread, the size of a directive array has a fundamental bound imposed by the environment. Now, set $M = 8$, at which point the beamwidth falls below 10°, as the maximum number of 15-dBi antennas that can be aggregated.

The 90 percent single-user capacities corresponding to this scenario are presented in Fig. 17-2 with the transmit power set to $P_T = 10$ W.[1] Because of the logarithmic relationship between rate and power, the use of a base station directive array offers very limited improvement in terms of single-user data rate.

Transmit Diversity An alternative strategy, also based on the deployment of base station arrays only, which has already been incorporated into the 3G roadmap, is that of transmit diversity. In this case, the base station antennas must be spaced sufficiently far apart so that their signals are basically uncorrelated. The corresponding single-user results, as shown in Fig. 17-3, are nonetheless similar to their directive array counterparts.[1] However, in this case, there is no fundamental bound to the size of the array; there is little advantage in increasing it beyond $M = 3$ to 4 because of the diminishing returns associated with adding additional diversity branches to an already diverse link. Furthermore, in a true wideband system, frequency diversity would further reduce the benefits of transmit diversity.

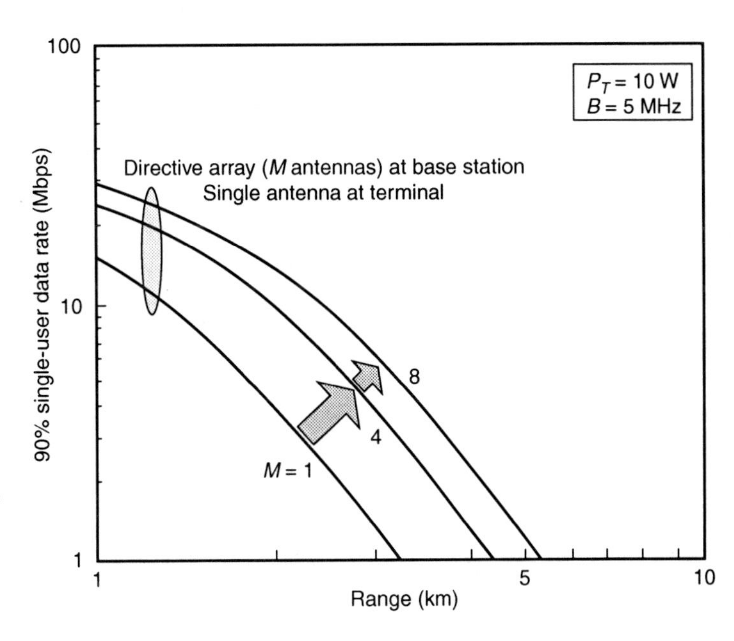

Figure 17-2 Single-user data rate supported in 90 percent of locations versus range with a directive array at the base station and a single omnidirectional antenna at the terminal. M is the number of 15-dBi antennas at the base; transmit power $P_T = 10$ W; bandwidth $B = 5$ MHz.

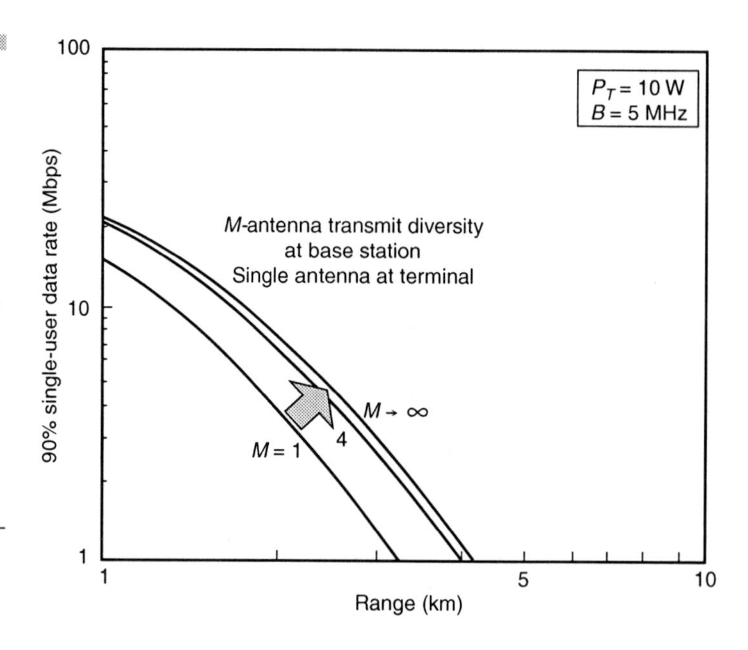

Figure 17-3 Single-user data rate supported in 90 percent of locations versus range with transmit diversity at the base station and a single omnidirectional antenna at the terminal. M is the number of 15-dBi uncorrelated antennas at the base; transmit power $P_T = 10$ W; bandwidth $B = 5$ MHz.

Multiple-Transmit Multiple-Receive Antenna Architectures Now, let's look at architectures with both transmit and receive arrays. As in the transmit diversity case, base station antennas must be spaced apart for proper decorrelation. In addition, the terminal must be equipped with its own array. Also, as in the diversity case, no information about the directional location of the terminal is required. In order to avoid cluttering the results with an excessive number of parameters, scale the size of both the base station and the terminal arrays simultaneously; that is, set $M = N$. The capacities are depicted in Fig. 17-4.[1] For completeness, the transmit diversity curves of Fig. 17-3 are also shown. Notice the extraordinary growth in attainable data rates unleashed by the additional signaling dimensions provided by the combined use of transmit and receive arrays. With only $M = N = 8$ antennas, the wireless single-user data rate can be increased by an order of magnitude. Furthermore, the growth does not saturate as long as additional uncorrelated antennas can be incorporated into the arrays.

Wireless-Data-Rate Limits within a Cellular System

In this part of the chapter, let's extend the analysis in order to reevaluate the wireless user data-rate limits in much more realistic conditions. To that end, let's incorporate an entire cellular system into the analysis.

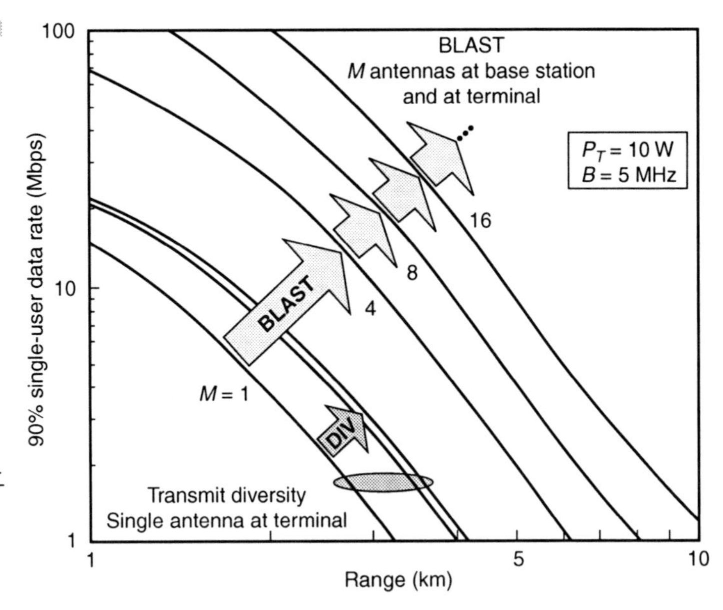

Figure 17-4 Single-user data rate supported in 90 percent of locations versus range with transmit and receive arrays. M is the number of 15-dBi antennas at the base station as well as the number of omnidirectional antennas at the terminal; transmit power $P_T = 10$ W; bandwidth $B = 5$ MHz.

Most emerging wireless data-oriented systems feature time-multiplexed downlink channels, certainly those evolving from TDMA, but also those evolving from CDMA. With that, same-cell users are ensured to be mutually orthogonal; thus, the interference arises exclusively from other cells. Accordingly, a time-multiplexed multicell system is considered with base stations placed on a hexagonal grid. Users are uniformly distributed and connected to the sector from which they receive the strongest signal. To further mimic actual 3G data systems, rate adaptation with no power control is assumed. Transmit signals are assumed gaussian, which maximizes capacity as long as no multiuser detection across cells is attempted. Altogether, the results presented in this part of the chapter can be considered upper bounds for a 5-MHz data-oriented 3G system.

The results correspond to Monte Carlo simulations conducted on a 19-cell universe: a central cell, wherein statistics are collected, surrounded by two rings of interfering cells. The cell size is scaled to ensure that the system is basically interference-limited. The simulation parameters are summarized, for convenience, in Table 17-1.[1]

Figure 17-5 displays cumulative distributions of system capacity (in megabits per second per sector) over all locations with transmit arrays only, as well as with transmit and receive arrays.[1] These curves can also be interpreted as user peak rates, that is, wireless user data rates (in megabits per second) when the entire capacity of every sector is allocated to an individual user. With transmit arrays only, the benefit appears

TABLE 17-1

System Parameters

Multiplexing	Time Division
Sectors per cell	3
Base station antennas	120° perfect sectorization
Terminal antennas	Omnidirectional
Frequency reuse	Universal
Propagation exponent	3.5
Log-normal shadowing	8 dB
Fading	Rayleigh (independent per antenna)
Power control	No
Rate adaptation	Yes
S/N in 90% of locations	≥25 dB

significant only in the lower tail of the distribution, corresponding to users in the most detrimental locations. The improvements in average and peak system capacities are negligible. Moreover, the gains saturate rapidly as additional transmit antennas are added. With frequency diversity taken into account, those gains would be reduced even further. The combined use of transmit and receive arrays, on the other hand, dramatically shifts the curves, offering multifold improvements in wireless data rate at all levels.

NOTE Without receive arrays, the peak wireless data rate that can be supported in 90 percent of the system locations (with a single user per sector) is only on the order of 500 kbps with no transmit diversity and just over 1 Mbps therewith. Moreover, these figures correspond to absolute upper bounds.

With modulation excess bandwidth, training overhead, imperfect channel estimation, realistic coding schemes, and other impairments, only a fraction of these bounds can be actually realized. Without receive arrays, user rates on the order of several megabits per second can be supported only within a restricted portion of the coverage area and when no other users compete for bandwidth within the same sector.

Finally, the broadband wireless access industry, which provides high-rate network connections to stationary sites, has matured to the point at which it now has a standard for second-generation wireless metropolitan-area networks. IEEE Standard 802.16, with its WirelessMAN air interface, sets the stage for widespread and effective deployments worldwide. The next part of the chapter is an overview of the technical medium access control and Physical layer features of this new standard.

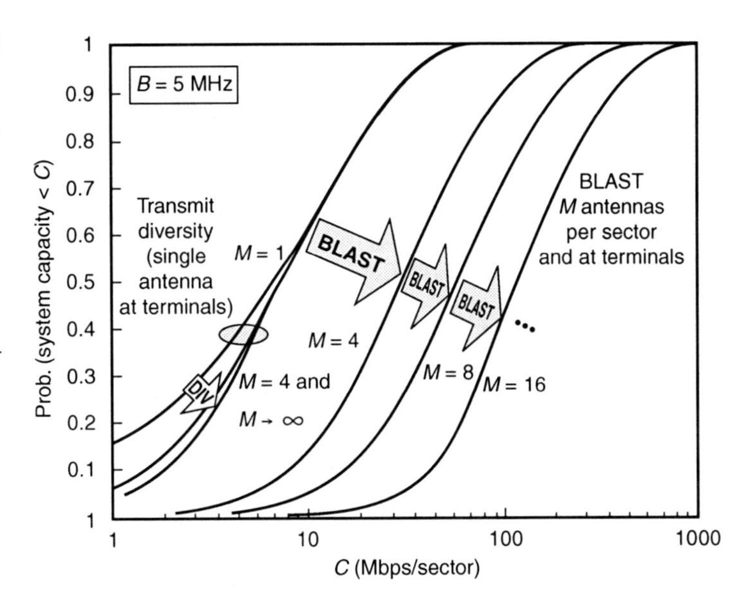

Figure 17-5 Cumulative distributions of system capacity with transmit arrays as well as with transmit and receive arrays. M is the number of antennas per array; system bandwidth B = 5 MHz.

WirelessMAN: Air Interface for Broadband Wireless Access

IEEE Standard 802.16-2001, completed in October 2001 and published on April 8, 2002, defines the WirelessMAN air interface specification for wireless data metropolitan-area networks (MANs). The completion of this standard heralds the entry of broadband wireless data access as a major new tool in the effort to link homes and businesses to core telecommunications networks worldwide.

As currently defined through IEEE Standard 802.16, a wireless data MAN provides network access to buildings through exterior antennas communicating with central radio base stations (BSs). The wireless data MAN offers an alternative to cabled access networks, such as fiber-optic links, coaxial systems using cable modems, and digital subscriber line (DSL) links. Because wireless data systems have the capacity to address broad geographic areas without the costly infrastructure development required in deploying cable links to individual sites, the technology may prove less expensive to deploy and may lead to more ubiquitous broadband access. Such systems have been in use for several years, but the development of the new standard marks the maturation of the industry and forms the basis of new industry success with second-generation equipment.

In this scenario, with WirelessMAN technology bringing the network to a building, users inside the building will connect to it with conventional in-building networks such as, for data, Ethernet (IEEE Standard

802.3) or wireless data LANs (IEEE Standard 802.11). However, the fundamental design of the standard may eventually allow for the efficient extension of the WirelessMAN networking protocols directly to the individual user. For instance, a central BS may someday exchange medium access control (MAC) protocol wireless data with an individual laptop computer in a home. The links from the BS to the home receiver and from the home receiver to the laptop would likely use quite different Physical layers, but design of the WirelessMAN MAC could accommodate such a connection with full quality of service (QoS). With the technology expanding in this direction, it is likely that the standard will evolve to support nomadic and increasingly mobile users. For example, it could be suitable for a stationary or slow-moving vehicle.

IEEE Standard 802.16 was designed to evolve as a set of air interfaces based on a common MAC protocol, but with Physical layer specifications dependent on the spectrum of use and the associated regulations. The standard, as approved in 2001, addresses frequencies from 10 to 66 GHz, where extensive spectrum is currently available worldwide, but at which the short wavelengths introduce significant deployment challenges. A new project, currently in the balloting stage, expects to complete an amendment denoted IEEE 802.16a before the end of 2003. This standards document will extend the air interface support to lower frequencies in the 2- to 11-GHz band, including both licensed and license-exempt spectra. Compared to the higher frequencies, such spectra offer the opportunity to reach many more customers less expensively, although at generally lower data rates. This suggests that such services will be oriented toward individual homes or small to medium-size enterprises.

The 802.16 Working Group

Development of IEEE Standard 802.16 and the included WirelessMAN air interface, along with associated standards and amendments, is the responsibility of IEEE Working Group 802.16 on Broadband Wireless Access (BWA) Standards (http://WirelessMAN.org). The Working Group's initial interest was the 10- to 66-GHz range. The 2- to 11-GHz amendment project that led to IEEE 802.16a was approved in March 2000. The 802.16a project primarily involves the development of new Physical layer specifications, with supporting enhancements to the basic MAC. In addition, the Working Group has completed IEEE Standard 802.16.2 ("Recommended Practice for Coexistence of Fixed Broadband Wireless Access Systems") to address 10- to 66-GHz coexistence and, through the amendment project 802.16.2a, is expanding its recommendations to include licensed bands from 2 to 11 GHz.

Historically, the 802.16 activities were initiated at an August 1998 meeting called by the National Wireless Electronics Systems Testbed

(N-WEST) of the U.S. National Institute of Standards and Technology. The effort was welcomed in IEEE 802, which opened a study group. The 802.16 Working Group has held week-long meetings at least bimonthly since July 1999. Over 800 individuals have attended a session. Membership, which is granted to individuals on the basis of their attendance and participation, currently stands at 140. The work has been closely followed; for example, the IEEE 802.16 Web site received over 3.9 million file requests in 2001.

Technology Design Issues

The IEEE 802.16 MAC protocol was designed for point-to-multipoint broadband wireless data access applications. It addresses the need for very high bit rates, both uplink (to the BS) and downlink (from the BS). Access and bandwidth allocation algorithms must accommodate hundreds of terminals per channel, with terminals that may be shared by multiple end users. The services required by these end users are varied in their nature and include legacy time-division multiplex (TDM) voice and data, Internet Protocol (IP) connectivity, and packetized voice over IP (VoIP). To support this variety of services, the 802.16 MAC must accommodate both continuous and bursty traffic. Additionally, these services expect to be assigned QoS in keeping with the traffic types. The 802.16 MAC provides a wide range of service types analogous to the classic asynchronous transfer mode (ATM) service categories, as well as newer categories such as guaranteed frame rate (GFR).

The 802.16 MAC protocol must also support a variety of backhaul requirements, including both asynchronous transfer mode (ATM) and packet-based protocols. Convergence sublayers are used to map the Transport-layer-specific traffic to a MAC that is flexible enough to efficiently carry any traffic type. Through such features as payload header suppression, packing, and fragmentation, the convergence sublayers and MAC work together to carry traffic in a form that is often more efficient than the original transport mechanism.

Issues of transport efficiency are also addressed at the interface between the MAC and the Physical layer (PHY). For example, the modulation and coding schemes are specified in a burst profile that may be adjusted adaptively for each burst to each subscriber station. The MAC can make use of bandwidth-efficient burst profiles under favorable link conditions, but shift to more reliable, although less efficient, alternatives as required to support the planned 99.999 percent link availability.

The request-grant mechanism is designed to be scalable, efficient, and self-correcting. The 802.16 access system does not lose efficiency when presented with multiple connections per terminal, multiple QoS levels per terminal, and a large number of statistically multiplexed users. It takes advantage of a wide variety of request mechanisms, balancing the stability of contentionless access with the efficiency of contention-oriented access. While extensive bandwidth allocation and QoS mechanisms are provided, the details of scheduling and reservation management are left unstandardized and provide an important mechanism for vendors to differentiate their equipment.

Along with the fundamental task of allocating bandwidth and transporting data, the MAC includes a privacy sublayer that provides authentication of network access and connection establishment to avoid theft of service, and it provides key exchange and encryption for data privacy.

To accommodate the more demanding physical environment and different service requirements of the frequencies between 2 and 11 GHz, the 802.16a project is upgrading the MAC to provide automatic repeat request (ARQ) and support for mesh, rather than only point-to-multipoint, network architectures.

The Physical Layer In the design of the PHY specification for 10 to 66 GHz, line-of-sight propagation was deemed a practical necessity. With this condition assumed, single-carrier modulation was easily selected; the air interface is designated WirelessMAN-SC. Many fundamental design challenges remained, however. Because of the point-to-multipoint architecture, the BS basically transmits a TDM signal, with individual subscriber stations allocated time slots serially. Access in the uplink direction is by time-division multiple access (TDMA). Following extensive discussions regarding duplexing, a burst design was selected that allows both time-division duplexing (TDD), in which the uplink and downlink share a channel but do not transmit simultaneously, and frequency-division duplexing (FDD), in which the uplink and downlink operate on separate channels, sometimes simultaneously. This burst design allows both TDD and FDD to be handled in a similar fashion. Support for half-duplex FDD subscriber stations, which may be less expensive since they do not simultaneously transmit and receive, was added at the expense of some slight complexity. Both TDD and FDD alternatives support adaptive burst profiles, in which modulation and coding options may be dynamically assigned on a burst-by-burst basis.

2 to 11 GHz The 2- to 11-GHz bands, both licensed and license-exempt, are addressed in IEEE Project 802.16a. The standard is in ballot, but is not yet complete. The draft currently specifies that compliant systems implement one of three air interface specifications, each of which provides for interoperability. Design of the 2- to 11-GHz Physical layer is driven by the need for non-line-of-sight (NLOS) operation. Because residential applications are expected, rooftops may be too low for a clear sight line to a BS antenna, possibly because of obstruction by trees. Therefore, significant multipath propagation must be expected. Furthermore, outdoor-mounted antennas are expensive in terms of both hardware and installation costs. The three 2- to 11-GHz air interface specifications in 802.16a Draft 3 are:

- *WirelessMAN-SC2.* This uses a single-carrier modulation format.
- *WirelessMAN-OFDM.* This uses orthogonal frequency-division multiplexing with a 256-point transform. Access is by TDMA. This air interface is mandatory for license-exempt bands.
- *WirelessMAN-OFDMA.* This uses orthogonal frequency-division multiple access with a 2048-point transform. In this system, multiple access is provided by addressing a subset of the multiple carriers to individual receivers.[2]

Because of the propagation requirements, the use of advanced antenna systems is supported. It is premature to speculate on further specifics of the 802.16a amendment prior to its completion. While the draft seems to have reached a level of maturity, the contents could change significantly in balloting. Modes could even be deleted or added.

Physical Layer Details

The PHY specification defined for 10 to 66 GHz uses burst single-carrier modulation with adaptive burst profiling in which transmission parameters, including the modulation and coding schemes, may be adjusted individually to each subscriber station (SS) on a frame-by-frame basis. Both TDD and burst FDD variants are defined. Channel bandwidths of 20 or 25 MHz (typical U.S. allocation) or 28 MHz (typical European allocation) are specified, along with Nyquist square-root raised-cosine pulse shaping with a rolloff factor of 0.25. Randomization is performed for spectral shaping and to ensure bit transitions for clock recovery.

The forward error correction (FEC) used is Reed-Solomon GF(256), with variable block size and error correction capabilities. This is paired with an inner block convolutional code to robustly transmit critical data, such as frame control and initial accesses. The FEC options are paired with quadrature phase-shift keying (QPSK), 16-state quadrature amplitude modulation (16 QAM), and 64-state QAM (64 QAM) to form burst profiles of varying robustness and efficiency. If the last FEC block is not filled, that block may be shortened. Shortening in both the uplink and downlink is controlled by the BS and is implicitly communicated in the uplink map (UL-MAP) and downlink map (DL-MAP).

The system uses a frame of 0.5, 1, or 2 ms. This frame is divided into physical slots for the purpose of bandwidth allocation and identification of PHY transitions. A physical slot is defined to be four QAM symbols. In the TDD variant of the PHY, the uplink subframe follows the downlink subframe on the same carrier frequency. In the FDD variant, the uplink and downlink subframes are coincident in time, but are carried on separate frequencies. The downlink subframe is shown in Fig. 17-6.[2]

The downlink subframe starts with a frame control section that contains the DL-MAP for the current downlink frame as well as the UL-MAP for a specified time in the future. The downlink map specifies when Physical layer transitions (modulation and FEC changes) occur within the downlink subframe. The downlink subframe typically contains a TDM portion immediately following the frame control section. Downlink wireless data are transmitted to each SS by using a negotiated burst profile. The wireless data are transmitted in order of decreasing robustness to allow SSs to receive their wireless data before being presented with a burst profile that could cause them to lose synchronization with the downlink.

In FDD systems, the TDM portion may be followed by a TDMA segment that includes an extra preamble at the start of each new burst profile. This feature allows better support of half-duplex SSs. In an efficiently scheduled FDD system with many half-duplex SSs, some may need to transmit earlier in the frame than they receive. Because of their half-duplex nature, these SSs lose synchronization with the downlink. The TDMA preamble allows them to regain synchronization.

Because of the dynamics of bandwidth demand for the variety of services that may be active, the mixture and duration of burst profiles and the presence or absence of a TDMA portion vary dynamically from frame

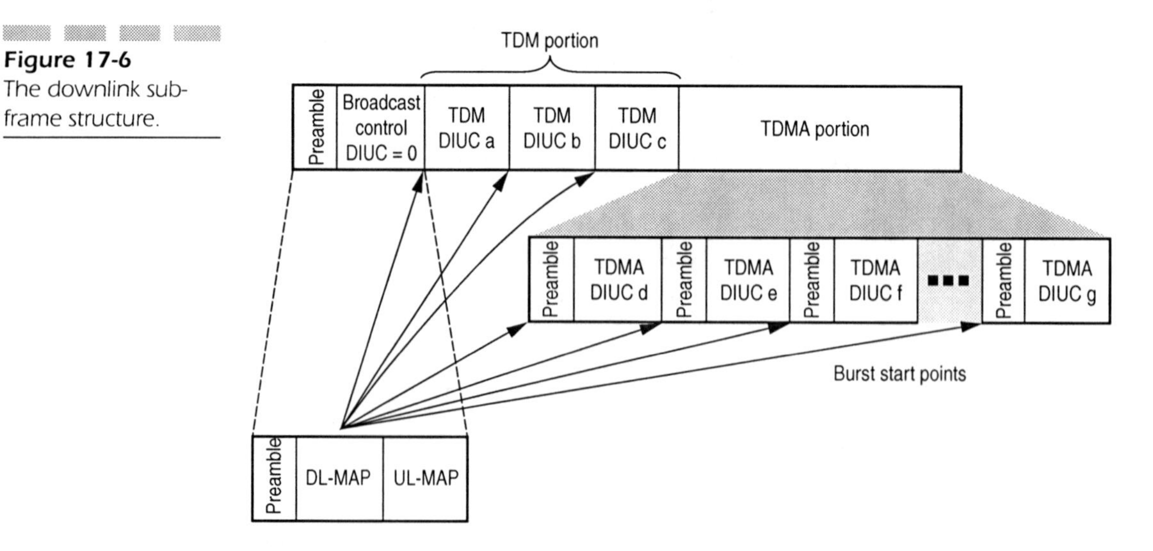

Figure 17-6 The downlink subframe structure.

to frame. Since the recipient SS is implicitly indicated in the MAC headers rather than in the DL-MAP, SSs listen to all portions of the downlink subframe they are capable of receiving. For full-duplex SSs, this means receiving all burst profiles of equal or greater robustness than they have negotiated with the BS.

A typical uplink subframe for the 10- to 66-GHz PHY is shown in Fig. 17-7.[2] Unlike the downlink, the UL-MAP grants bandwidth to specific SSs. The SSs transmit in their assigned allocation using the burst profile specified by the Uplink Interval Usage Code (UIUC) in the UL-MAP entry granting them bandwidth. The uplink subframe may also contain contention-based allocations for initial system access and broadcast or multicast bandwidth requests. The access opportunities for initial system access are sized to allow extra guard time for SSs that have not resolved the transmit time advance necessary to offset the round-trip delay to the BS.

Between the PHY and MAC is a transmission convergence (TC) sublayer. This layer performs the transformation of variable-length MAC protocol data units (PDUs) into the fixed-length FEC blocks (plus possibly a shortened block at the end) of each burst. The TC layer has a PDU sized to fit in the FEC block currently being filled. It starts with a pointer indicating where the next MAC PDU header starts within the FEC block. This is shown in Fig. 17-8.[2]

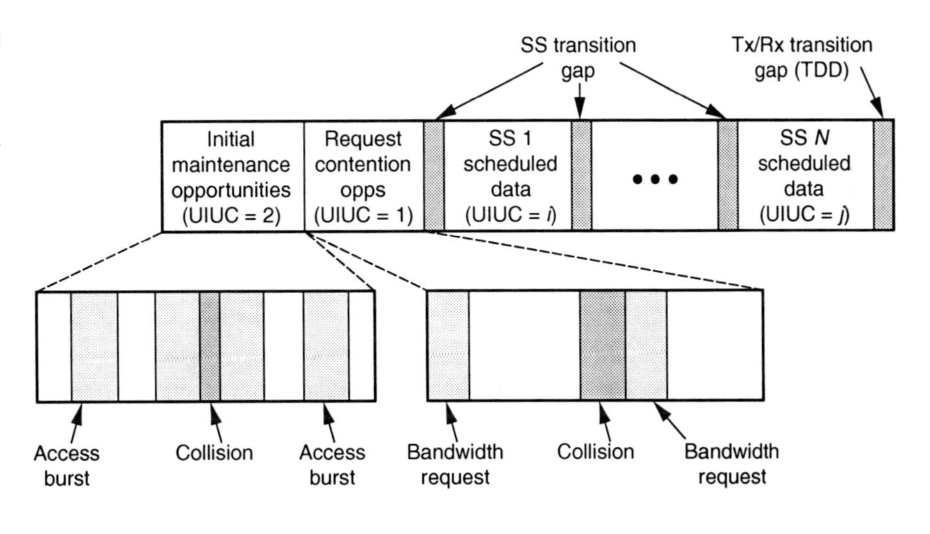

Figure 17-7 The uplink subframe structure.

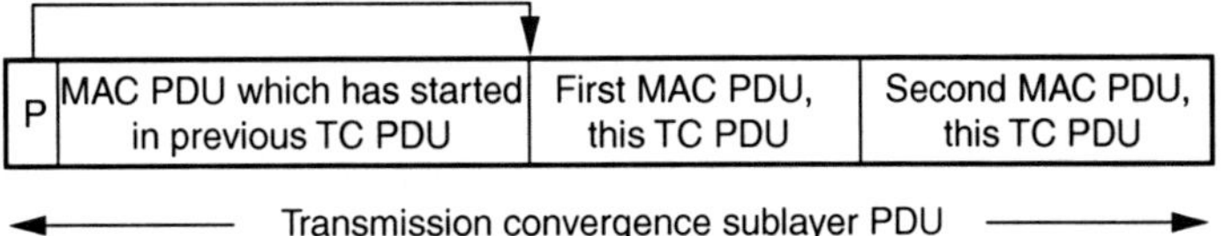

Figure 17-8 TC PDU format.

The TC PDU format allows resynchronization to the next MAC PDU in the event that the previous FEC block had irrecoverable errors. Without the TC layer, a receiving SS or BS would potentially lose the entire remainder of a burst when an irrecoverable bit error occurred.

Medium Access Control Details

The MAC includes service-specific convergence sublayers that interface to higher layers, and above the core MAC common part sublayer that carries out the key MAC functions. Below the common part sublayer is the privacy sublayer.

Service-Specific Convergence Sublayers IEEE Standard 802.16 defines two general service-specific convergence sublayers for mapping services to and from 802.16 MAC connections. The ATM convergence sublayer is defined for ATM services, and the packet convergence sublayer is defined for mapping packet services such as IPv4, IPv6, Ethernet, and virtual local-area network (VLAN). The primary task of the sublayer is to classify service wireless data units (SDUs) to the proper MAC connection, preserve or enable QoS, and enable bandwidth allocation. The mapping takes various forms depending on the type of service. In addition to these basic functions, the convergence sublayers can also perform more sophisticated functions such as payload header suppression and reconstruction to enhance airlink efficiency.

Common Part Sublayer In general, the 802.16 MAC is designed to support a point-to-multipoint architecture with a central BS handling multiple independent sectors simultaneously. On the downlink, data to SSs are multiplexed in TDM fashion. The uplink is shared between SSs in TDMA fashion.

The 802.16 MAC is connection-oriented. All services, including inherently connectionless services, are mapped to a connection. This provides a mechanism for requesting bandwidth, associating QoS and traffic parameters, transporting and routing data to the appropriate convergence sublayer, and all other actions associated with the contractual terms of the service. Connections are referenced with 16-bit connection identifiers (CIDs) and may require continuously granted bandwidth or bandwidth on demand. As will be described, both are accommodated.

Each SS has a standard 48-bit MAC address, but this serves mainly as an equipment identifier, since the primary addresses used during operation are the CIDs. Upon entering the network, the SS is assigned three management connections in each direction. These three connections reflect the three different QoS requirements used by different management levels. The first of these is the basic connection, which is used for the transfer of short, time-critical MAC and radio link control (RLC) messages. The primary management connection is used to transfer longer, more delay-tolerant messages such as those used for authentication and connection setup. The secondary management connection is used for the transfer of standards-based management messages such as Dynamic Host Configuration Protocol (DHCP), Trivial File Transfer Protocol (TFTP), and Simple Network Management Protocol (SNMP). In addition to these management connections, SSs are allocated transport connections for the contracted services. Transport connections are unidirectional to facilitate different uplink and downlink QoS and traffic parameters; they are typically assigned to services in pairs.

The MAC reserves additional connections for other purposes. One connection is reserved for contention-based initial access. Another is reserved for broadcast transmissions in the downlink as well as for signaling broadcast contention-based polling of SS bandwidth needs. Additional connections are reserved for multicast, rather than broadcast, contention-based polling. SSs may be instructed to join multicast polling groups associated with these multicast polling connections.

MAC PDU Formats The MAC PDU is the data unit exchanged between the MAC layers of the BS and its SSs. A MAC PDU consists of a fixed-length MAC header, a variable-length payload, and an optional cyclic redundancy check (CRC). Two header formats, distinguished by the HT field, are defined: the generic header (see Fig. 17-9) and the bandwidth request header.[2]

Except for bandwidth request MAC PDUs, which contain no payload, MAC PDUs contain either MAC management messages or convergence sublayer data. Three types of MAC subheader may be present. The grant management subheader is used by an SS to convey bandwidth

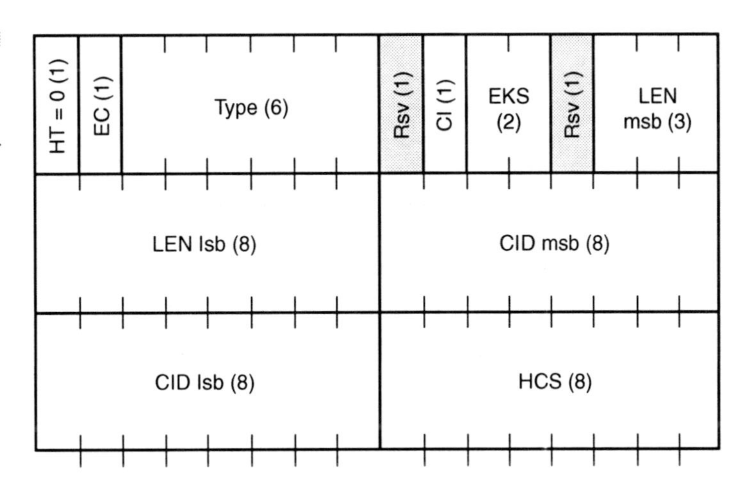

Figure 17-9 Format of generic header for MAC PDU.

management needs to its BS. The fragmentation subheader contains information that indicates the presence and orientation in the payload of any fragments of SDUs. The packing subheader is used to indicate the packing of multiple SDUs into a single PDU. The grant management and fragmentation subheaders may be inserted in MAC PDUs immediately following the generic header if so indicated by the Type field. The packing subheader may be inserted before each MAC SDU if so indicated by the Type field.

PHY Support and Frame Structure The IEEE 802.16 MAC supports both TDD and FDD. In FDD, both continuous and burst downlinks are supported. Continuous downlinks allow for certain robustness enhancement techniques, such as interleaving. Burst downlinks (either FDD or TDD) allow the use of more advanced robustness and capacity enhancement techniques, such as subscriber-level adaptive burst profiling and advanced antenna systems.

The MAC builds the downlink subframe starting with a frame control section containing the DL-MAP and UL-MAP messages. These indicate PHY transitions on the downlink as well as bandwidth allocations and burst profiles on the uplink.

The DL-MAP is always applicable to the current frame and is always at least two FEC blocks long. The first PHY transition is expressed in the first FEC block, to allow adequate processing time. In both TDD and FDD systems, the UL-MAP provides allocations starting no later than the next downlink frame. The UL-MAP can, however, allocate starting in the current frame as long as processing times and round-trip delays are observed. The minimum time between receipt and applicability of the UL-MAP for an FDD system is shown in Fig. 17-10.[2]

Radio Link Control The advanced technology of the 802.16 PHY requires equally advanced radio link control (RLC), particularly the

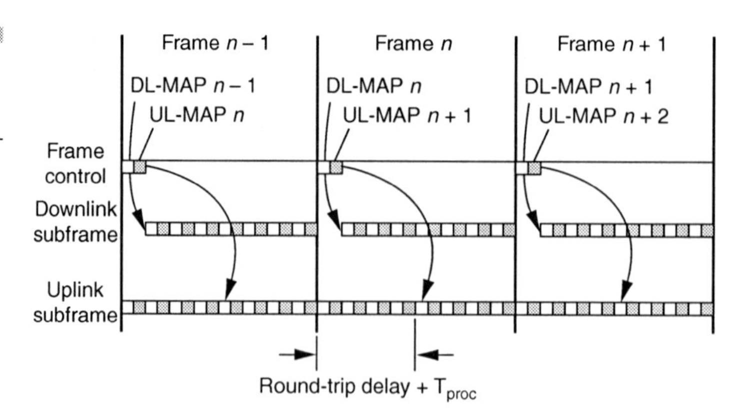

Figure 17-10 Minimum FDD map relevance.

capability of the PHY to transition from one burst profile to another. The RLC must control this capability as well as the traditional RLC functions of power control and ranging.

RLC begins with periodic BS broadcast of the burst profiles that have been chosen for the uplink and downlink. The choice of particular burst profiles used on a channel is based on a number of factors, such as rain region and equipment capabilities. Burst profiles for the downlink are each tagged with a downlink interval usage code (DIUC). Those for the uplink are each tagged with an uplink interval usage code (UIUC).

During initial access, the SS performs initial power leveling and ranging, using ranging request (RNG-REQ) messages transmitted in initial maintenance windows. The adjustments to the SS's transmit time advance, as well as power adjustments, are returned to the SS in ranging response (RNG-RSP) messages. For ongoing ranging and power adjustments, the BS may transmit unsolicited RNG-RSP messages commanding the SS to adjust its power or timing.

During initial ranging, the SS also requests to be served in the downlink via a particular burst profile by transmitting its choice of DIUC to the BS. The choice is based on received downlink signal quality measurements performed by the SS before and during initial ranging. The BS may confirm or reject the choice in the ranging response. Similarly, the BS monitors the quality of the uplink signal it receives from the SS. The BS commands the SS to use a particular uplink burst profile simply by including the appropriate burst profile UIUC with the SS's grants in UL-MAP messages.

After initial determination of uplink and downlink burst profiles between the BS and a particular SS, RLC continues to monitor and control the burst profiles. Harsher environmental conditions, such as rain fades, can force the SS to request a more robust burst profile. Alternatively, exceptionally good weather may allow an SS to temporarily operate with a more efficient burst profile. The RLC continues to adapt the SS's current UL and DL burst profiles, ever striving to achieve a balance between robustness and efficiency. Because the BS is in control and directly monitors the uplink signal quality, the protocol for changing the uplink burst profile for an SS is simple: The BS merely specifies the profile's associated UIUC whenever granting the SS bandwidth in a frame. This eliminates the need for an acknowledgment, since the SS will always receive either both the UIUC and the grant or neither. Hence, no chance of uplink burst profile mismatch between the BS and SS exists.

In the downlink, the SS is the entity that monitors the quality of the receive signal and therefore knows when its downlink burst profile should change. The BS, however, is the entity in control of the change. There are two methods available to the SS to request a change in downlink burst profile, depending on whether the SS operates in the grant per connection (GPC) or grant per SS (GPSS) mode (see "Bandwidth Requests and

Grants" next). The first method would typically apply (based on the discretion of the BS scheduling algorithm) only to GPC SSs. In this case, the BS may periodically allocate a station maintenance interval to the SS. The SS can use the RNG-REQ message to request a change in downlink burst profile. The preferred method is for the SS to transmit a downlink burst profile change request (DBPC-REQ). In this case, which is always an option for GPSS SSs and can be an option for GPC SSs, the BS responds with a downlink burst profile change response (DBPC-RSP) message confirming or denying the change.

Because messages may be lost as a result of irrecoverable bit errors, the protocols for changing an SS's downlink burst profile must be carefully structured. The order of the burst profile change actions is different when transitioning to a more robust burst profile than when transitioning to a less robust one. The standard takes advantage of the fact that an SS is always required to listen to more robust portions of the downlink as well as the profile that was negotiated. Figure 17-11 shows a transition to

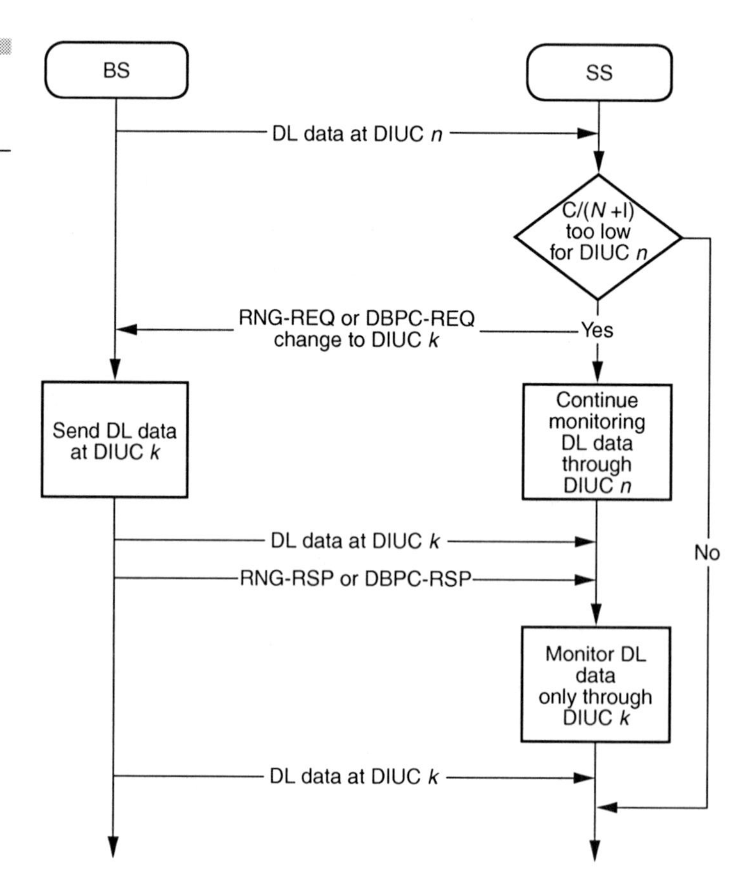

Figure 17-11 Transition to a more robust burst profile.

a more robust burst profile.[2] Figure 17-12 shows a transition to a less robust burst profile.[2]

Bandwidth Requests and Grants The IEEE 802.16 MAC accommodates two classes of SS, differentiated by their ability to accept bandwidth grants simply for a connection or for the SS as a whole. Both classes of SS request bandwidth per connection to allow the BS uplink scheduling algorithm to properly consider QoS when allocating bandwidth. With the grant per connection (GPC) class of SS, bandwidth is granted explicitly to a connection, and the SS uses the grant only for that connection. RLC and other management protocols use bandwidth explicitly allocated to the management connections.

With the grant per SS (GPSS) class, SSs are granted bandwidth aggregated into a single grant to the SS itself. The GPSS SS needs to be more intelligent in its handling of QoS. It will typically use the bandwidth for the connection that requested it, but need not. For instance, if the QoS situation at the SS has changed since the last request, the SS has the option of sending the higher QoS data along with a request to replace this

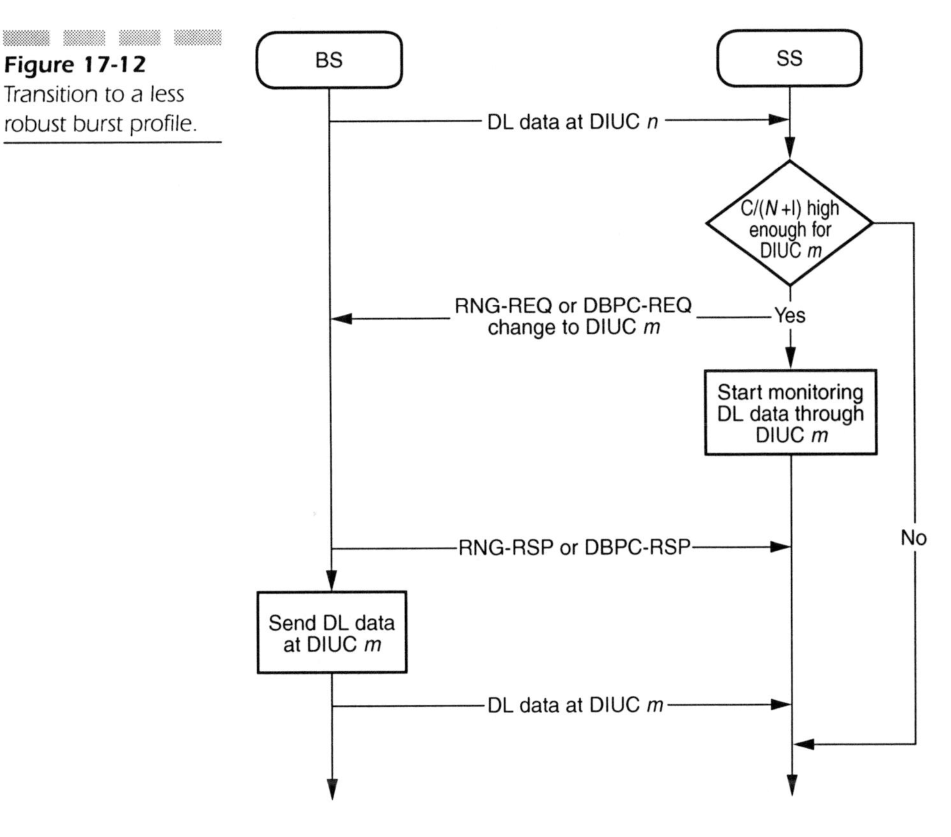

Figure 17-12 Transition to a less robust burst profile.

bandwidth stolen from a lower QoS connection. The SS could also use some of the bandwidth to react more quickly to changing environmental conditions by sending, for instance, a DBPC-REQ message.

The two classes of SS allow a tradeoff between simplicity and efficiency. The need to explicitly grant extra bandwidth for RLC and requests, coupled with the likelihood of more than one entry per SS, makes GPC less efficient and scalable than GPSS. Additionally, the ability of the GPSS SS to react more quickly to the needs of the PHY and those of connections enhances system performance. GPSS is the only class of SS allowed with the 10- to 66-GHz PHY.

With both classes of grants, the IEEE 802.16 MAC uses a self-correcting protocol rather than an acknowledged protocol. This method uses less bandwidth. Furthermore, acknowledged protocols can take additional time, potentially adding delay. There are a number of reasons the bandwidth requested by an SS for a connection may not be available:

- The BS did not see the request because of irrecoverable PHY errors or collision of a contention-based reservation.
- The SS did not see the grant because of irrecoverable PHY errors.
- The BS did not have sufficient bandwidth available.
- The GPSS SS used the bandwidth for another purpose.[2]

In the self-correcting protocol, all of these anomalies are treated the same. After a time-out appropriate for the QoS of the connection (or immediately, if the bandwidth was stolen by the SS for another purpose), the SS simply requests again. For efficiency, most bandwidth requests are incremental; that is, the SS asks for more bandwidth for a connection. However, for the self-correcting bandwidth request/grant mechanism to work correctly, the bandwidth requests must occasionally be aggregate; that is, the SS informs the BS of its total current bandwidth needs for a connection. This allows the BS to reset its perception of the SS's needs without a complicated protocol acknowledging the use of granted bandwidth.

The SS has a plethora of ways to request bandwidth, combining the determinism of unicast polling with the responsiveness of contention-based requests and the efficiency of unsolicited bandwidth. For continuous bandwidth demand, such as with CBR T1/E1 data, the SS need not request bandwidth; the BS grants it unsolicited.

To short-circuit the normal polling cycle, any SS with a connection running UGS can use the poll-me bit in the grant management subheader to let the BS know it needs to be polled for bandwidth needs on another connection. The BS may choose to save bandwidth by polling SSs that have unsolicited grant services only when they have set the poll-me bit.

A more conventional way to request bandwidth is to send a bandwidth request MAC PDU that consists of simply the bandwidth request header and no payload. GPSS SSs can send this in any bandwidth allocation they receive. GPC terminals can send it in either a request interval or a data grant interval allocated to their basic connection. A closely related method of requesting data is to use a grant management subheader to piggyback a request for additional bandwidth for the same connection within a MAC PDU.

In addition to polling individual SSs, the BS may issue a broadcast poll by allocating a request interval to the broadcast CID. Similarly, the standard provides a protocol for forming multicast groups to give finer control to contention-based polling. Because of the nondeterministic delay that can be caused by collisions and retries, contention-based requests are allowed only for certain lower QoS classes of service.

Conclusion

Traditional wireless data technologies are not very well suited to meet the demanding requirements of providing very high data rates with the ubiquity, mobility, and portability characteristic of cellular systems. Given the scarcity and exorbitant cost of radio spectrum, such data rates dictate the need for extremely high spectral efficiencies, which cannot be achieved with classical schemes in systems that are inherently self-interfering. Increased processing across the spatial dimension thus appears to be the only means of enabling the types of capacities and wireless data rates needed for ubiquitous wireless Internet and exciting multimedia services. While the most natural way of utilizing the space dimension may be to deploy additional base stations in order to allow for more frequent spectral reuse with smaller cells, economical and environmental considerations require that performance be enhanced on a per-base-station basis. That, in turn, calls for the use of antenna arrays. While the deployment of base station antenna arrays is becoming universal, it is really the simultaneous deployment of base station and terminal arrays that unleashes vast increases in capacity and data rates by opening up multiple signaling dimensions. Space-time processing techniques can exploit this dimensionality to concentrate large amounts of capacity in localized spots. Recognizing this potential, the 3G Partnership Project (3GPP) recently approved the use of transmit and receive arrays as a working item for the high-speed downlink packet access mode currently under development.

This chapter quantified the benefits of using antenna arrays (in the context of emerging mobile wireless data systems) as a function of the number of available antennas. Although absolute capacity and data-rate levels are very sensitive to the specifics of the propagation environment, the improvement factors are not. Hence, the relative scaling, rather than the absolute numbers themselves, is relevant.

Needless to say, a number of hurdles must be overcome before these new concepts can be widely implemented. First of all, it is necessary to assess the antenna arrangement and spacings that are required as well as the scattering richness of the environments of interest. In this respect, the BLAST prototype, operational for some time at Bell Labs' Crawford Hill facility, has yielded extremely encouraging results. Additional (and equally encouraging) results from other sources are also surfacing. Second, the historical opposition to installing multiple antennas on a terminal must be conquered. While the shrinking size of cellular phones was a powerful argument sustaining that thesis, terminals requiring higher data rates tend to naturally be larger in size, so they can take full advantage of the increased throughput. As a result, they also offer additional room for multiple closely spaced antennas. In fact, it is the cost of multiple separate radio chains that poses a limitation that might prove to be more stringent than the antennas themselves. Hence, the development of low-cost integrated multiple-chain radio solutions has become a research topic of the utmost importance.

Finally, the WirelessMAN air interface specified in IEEE Standard 802.16 provides a platform for the development and deployment of standards-based metropolitan-area networks providing broadband wireless data access in many regulatory environments. The standard is intended to allow multiple vendors to produce interoperable equipment. However, it also allows for extensive vendor differentiation. For instance, the standard provides the base station with a set of tools to implement efficient scheduling. However, the scheduling algorithms that determine the overall efficiency will differ from vendor to vendor and may be optimized for specific traffic patterns. Likewise, the adaptive burst profile feature allows great control to optimize the efficiency of the PHY transport. Innovative vendors will introduce clever schemes to maximize this opportunity while maintaining interoperability with compliant subscriber stations.

The publication of IEEE Standard 802.16 is a defining moment in which broadband wireless data access moves to its second generation and begins its establishment as a mainstream alternative for broadband access. Through the dedicated service of many volunteers, the IEEE 802.16 Working Group succeeded in quickly designing and forging a standard based on forward-looking technology. IEEE Standard 802.16 is the foundation of the wireless data metropolitan-area networks of the next few decades.

References

1. Angel Lozano, Farrokh R. Farrokhi, and Reinaldo A. Valenzuela, "Lifting the Limits on High-Speed Wireless Data Access Using Antenna Arrays," *IEEE Communications Magazine,* 445 Hoes Lane, Piscataway, NJ 08855, 2002.
2. Carl Eklund, Roger B. Marks, Kenneth L. Stanwood, and Stanley Wang, "IEEE Standard 802.16: A Technical Overview of the WirelessMAN™ Air Interface for Broadband Wireless Access," *IEEE Communications Magazine,* 445 Hoes Lane, Piscataway, NJ 08855, 2002.
3. John R. Vacca, *i-mode Crash Course,* McGraw-Hill, 2001.

PART 4

Configuring Wireless High-Speed Data Networks

CHAPTER 18

Configuring Wireless Data

Key technical aspects concerning the realization of a reconfigurable user terminal based on software-defined wireless data radio technology are discussed in this chapter. These include functionalities such as mode monitoring, mode switching, adaptive baseband, and software download. The implications of these functionalities and the need for related entities within the wireless data network are also discussed. The concepts presented here form part of the Fifth Framework IST European Research Project, Transparent Reconfigurable Ubiquitous Terminal (TRUST). It is the intention of this chapter to present the technical achievements of the TRUST project together with an overview of the potential complexity in the realization of a reconfigurable user terminal. (The Glossary defines many technical terms, abbreviations, and acronyms used in the book.)

Reconfigurable Terminals

Reconfigurable user terminals (RUTs) based on software-defined wireless data radio (SDWDR) are currently an active area of research in Europe. It is part of an overall drive toward the potential realization of reconfigurable radio systems and wireless data networks. Increasing user demand for flexibility, scalability, and multifunctional communication equipment has motivated the need for such futuristic systems. It is envisaged that SDWDR as a technology will help bring together the different forms of communications. The incorporation of mobile communications,[2] broadcast receivers, location services, Internet, multimedia, dedicated point-to-point communications, personal computing, and digital aids (PDAs) would be possible with the help of a mature and reliable SDWDR technology. This would eventually lead to the realization of reconfigurable radio systems and wireless data networks that would consist of self-organizing, self-evolutionary intelligent radio system infrastructures and user terminals for ubiquitous information interaction.

A RUT may be regarded as a programmable radio transceiver whereby user equipment (mobile terminals, hand-held computing devices, fixed terminals) is able to reconfigure itself, in terms of its capability, functionality, and behavior, in order to dynamically accommodate the needs of the user. In its purest form, the RUT will be able to reconfigure itself at any level of the radio protocol stack by implementing appropriate software within an adaptive hardware (analog and digital) environment. With the realization of a mature SDWDR technology, a terminal will be able to:

- *Adapt its behavior.* To change the applications, range, services, and functionality of the terminal to meet the demands of the user in accordance with the capability of the terminal.

- *Traverse across different communication standards.* To allow the terminal to switch between different modes, which would involve reconfiguration of the radio protocol stack, change in data rates, and different RF band and carriers.
- *Evolve with user demands.* To provide the users with the desired services, at the time of need, at a cost they can afford, while maximizing the quality of service (QoS) delivered.[1]

The CEC-sponsored Fifth Framework IST-TRUST Project is considering both the needs and design of a reconfigurable terminal in the context of a composite radio access environment. It is studying user requirements in order to identify the frameworks and systems needed to support software-reconfigurable radios. The primary goal of the TRUST project is to realize the user potential of reconfigurable radio systems, which will provide network connectivity and services when and where they are required.

This part of the chapter presents the basic architectural solutions derived within the TRUST project. These include functionalities such as mode monitoring, mode switching, software download, and reconfigurable baseband architecture. The implications of these functionalities and the need for related entities within the network are also discussed.

Mode Identification

One fundamental attribute of a RUT is its ability to adapt itself to a variety of different systems and standards (from GSM to UMTS). Among the many features required to support this intersystem roaming is the ability to detect, identify, and monitor alternative radio access technologies (RATs). Having located a RAT, the terminal then needs to connect to it and also monitor it in order to determine how viable the level of service offered by it is. The identification procedure can be broadly classified as either blind or assisted. *Blind methods* are where the terminal autonomously identifies the available RATs without any external (host network) support or advance knowledge. *Assisted methods* are where the terminal is presented with some information about the radio environment (a selection of available RATs).

In order to discover a RAT, the terminal has to perform some kind of energy detection. The amount of energy detected, its distribution across the spectrum, and its correlation with some predefined functions are sufficient to indicate the presence of a RAT. For TRUST, RAT identification requires a detector able to cope with both narrowband (GSM) and wideband spread-spectrum (UMTS) systems. It does not require prior knowledge of any air interface protocol. Monitoring, on the other hand, implies not only the ability to measure, but also to recognize specific parameters with values corresponding to predefined actions. Thus, monitoring does require prior knowledge of the observed RAT protocol or at least a part of it. For the terminal to store all possible protocols would require a large amount of memory. Thus, it seems that the combination of multimode detectors and flexible protocols is imperative for TRUST. Within TRUST, two methods of mode identification are considered.

Initial Mode Identification (IMI) The terminal searches all possible modes after power-up. The RUT will have to scan large areas of the radio spectrum in order to locate and use a mode. In the worst case, the RUT will have to do so without any help from the network (blind IMI). The IMI procedures for GSM and UMTS (and IS-95 and CDMA2000) follow the steps shown in Fig. 18-1.[1] First, the RUT selects (manually or automatically) the public land mobile network (PLMN), the provider. The next step is cell selection and reselection, in which the best cell is selected according to certain criteria. The RUT will synchronize with this cell in idle mode. Consequently, the terminal will be able to receive system information from the PLMN, and when registered (location registration, LR), it will also be able to initiate and receive a call. The cell selection step is vital for the RUT,

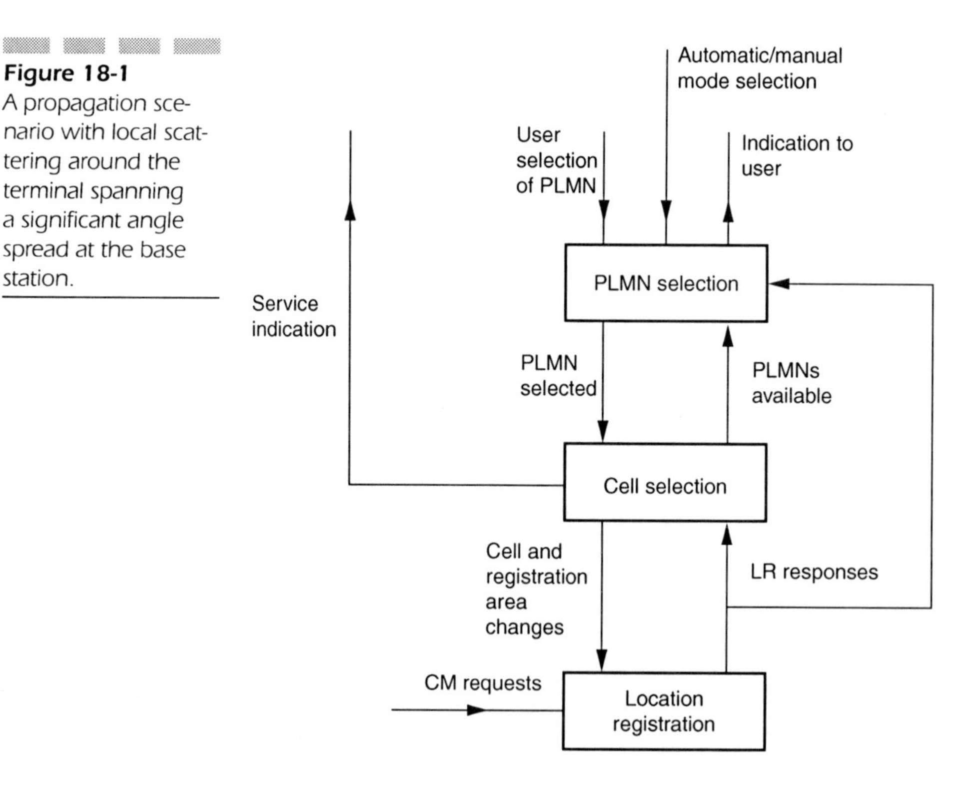

Figure 18-1 A propagation scenario with local scattering around the terminal spanning a significant angle spread at the base station.

since only after its completion will the terminal have the capability of connecting to a host network. After that the terminal is able to negotiate a channel for downloading software, if necessary, for reconfiguration.

Mode Switching

By definition, a RUT must be able to switch from one mode to another (perform a handoff to a different RAT) if the quality of the incumbent mode gets unacceptable or a new mode is preferred according to requirements specified by the user, service provider, or network operator. In order to minimize mode-switching negotiations across the radio link, it is necessary to implement an intelligent scheme that is able to combine information from the wireless data network, terminal, and user as effectively as possible.

The terminal performs a mode identification procedure (seen in the previous section) on a regular basis to detect if/when a new mode becomes available. After conducting a service negotiation to discover service availability and QoS trading with the provider, the terminal may perform a mode switch to a preferred mode while complying with the user profile, the terminal's static (hardware-related) and dynamic (instantaneous resource status) capabilities, service provider profiles, and the wireless data network operator. The mode switch itself may involve reconfiguration and/or allocation of resources on the terminal and in the wireless data network. This may involve downloading appropriate software if deemed necessary. Following the mode switch, the terminal must perform authentication and a location update before commencing any service over the new radio wireless data network.

The mode switching process could be triggered by a wireless data network entity in order to perform load balancing between different modes or to downgrade the quality of services used by some users in favor of other users. Wireless data network entities are involved in service negotiation occurring before and while deciding to switch from one RUT mode to another. Proxies in the wireless data network could assist in preparing and performing the mode switching procedure. On single-transceiver terminals, a mode switch may cause a temporary disruption of services. For dual-transceiver terminals, one radio transceiver may be configured for a target mode while applications and services operate on the other. In both cases, the use of proxies potentially reduces the amount of time required for service negotiation.

In order to minimize interactions between the terminal and the wireless data network entities, appropriate information could be generally obtained via the proxy reconfiguration manager (PRM), which is located in the radio access wireless data network (see Fig. 18-2).[1] It serves as a proxy for negotiations with other wireless data network entities, in particular the serving reconfiguration manager and home reconfiguration manager, respectively. Information about the currently available services and their customization parameters (QoS parameters) is contained in the wireless data network bearer service profile, which may also be retrieved via the PRM. Interaction with the authentication, authorization, and accounting (AAA) server, prior to service negotiation and prior to service invocation, is necessary in order to verify the mode switch and its compliance with the new network. The concept of the virtual home environment (VHE) currently being standardized by the Third Generation Partnership Project (3GPP) is employed to facilitate transparent QoS negotiation. It allows the user interface, applications, and their associated wireless data transfer characteristics to be separated from the underlying transmission networks. The VHE provides a standardized framework for the handling of user profiles and service customization information.

In short, the use of proxies allows inquiries for a new mode to be performed while connected to the old mode, minimizing the disruption interval. Thus, the terminal does not have to connect to the new wireless data network for service negotiation and resource reservation purposes. The required interactions between a mode switching decision unit and related entities are shown in Fig. 18-3.[1]

NOTE RUT connection with the new wireless data network can be prepared by a proxy obtaining instructions from another proxy located in the old network rather than from the terminal itself (Fig. 18-2).

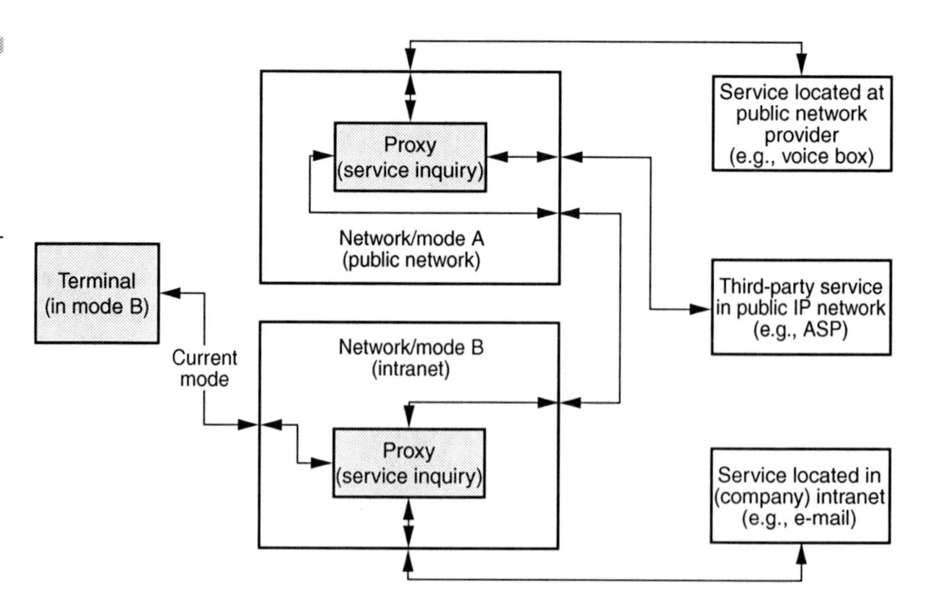

Figure 18-2 A proxy architecture (in the current mode) for service negotiation and resource reservation.

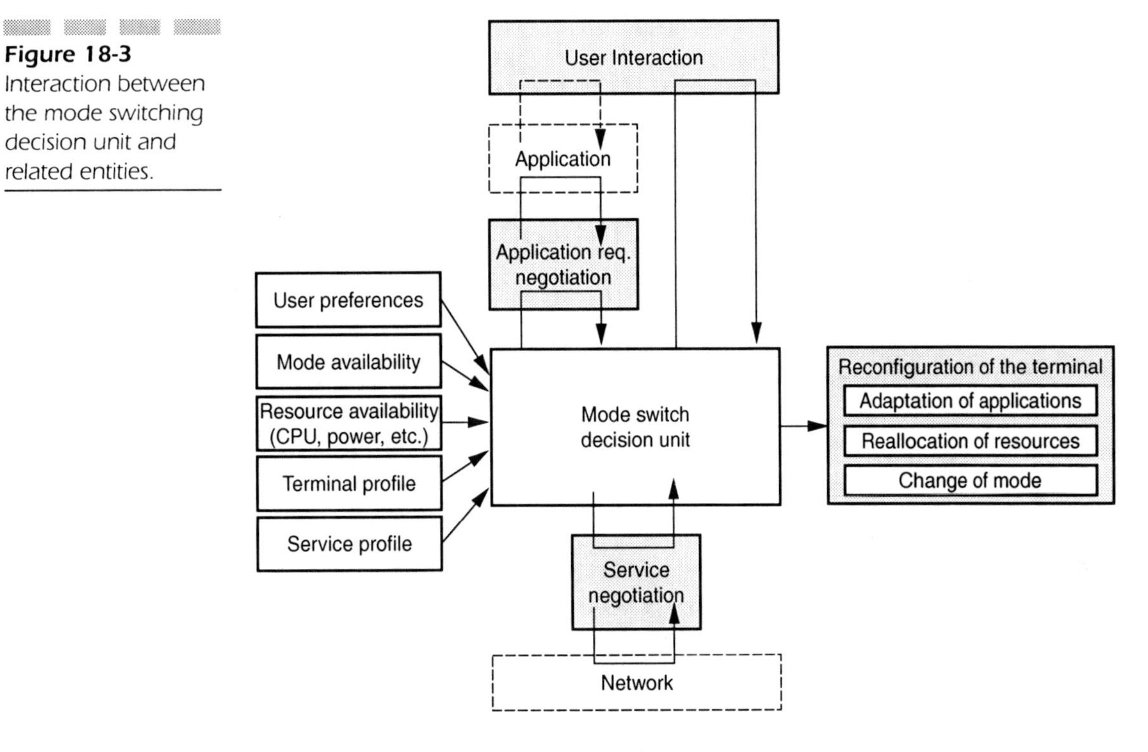

Figure 18-3 Interaction between the mode switching decision unit and related entities.

Software Download

Software download is a key enabler for ubiquitous RUTs. The process of software download involves the introduction of new functionality (defined in software) into the terminal, with the aim of modifying its configuration and/or content. Software download may consist of software patches, software upgrades, software licenses and keys, install scripts, validation test cases, and device configuration files. There are different categories of software that can be downloaded in a device, for example:

- Application and service provision software
- RAT protocol software for changing the air interface and/or bearer service
- Analog and digital signal processing software for reconfiguring the processing environment of the RUT[1]

Software download scenarios mainly fall in two categories: in-call or dynamic download, where software components are downloaded and installed during a call to support dynamic reconfiguration or distributed processing, requiring over-the-air download, and out-of-call or static download, where software components are downloaded into a secure store for installation at an appropriate time.

The TRUST project is considering over-the-air (OTA) software download, which may be either user-initiated or automatic. The latter can be initiated transparently by a reconfiguration agent (residing within the terminal or network), a service provider, and/or the operator within a distributed processing framework. On the other hand, user-initiated means the user explicitly triggers a software download process. The downloaded software may be received in different formats such as high-level software to be compiled or interpreted, hardware-specific binary code, software objects, reconfiguration parameters, and software agents. OTA software download offers great advantages in terms of cost-effective software deployment to a large number of devices. It enables manufacturers and service providers to introduce additional features in hardware and software components after shipment, to download software upgrades, bug fixes, local reference information, and new applications and services. It also facilitates configuration management of reconfigurable systems, distribution of software licenses and keys, certificates, and billing. Distributed application processing may also be supported, given an appropriate supporting framework and assuming the communications overhead can be justified. A generic software download process is shown in Fig. 18-4.[1] The security problem in software and data transactions can be summarized by the following four areas:

- *Privacy.* No one can see the content transferred on the wireless data network (applying data encryption).

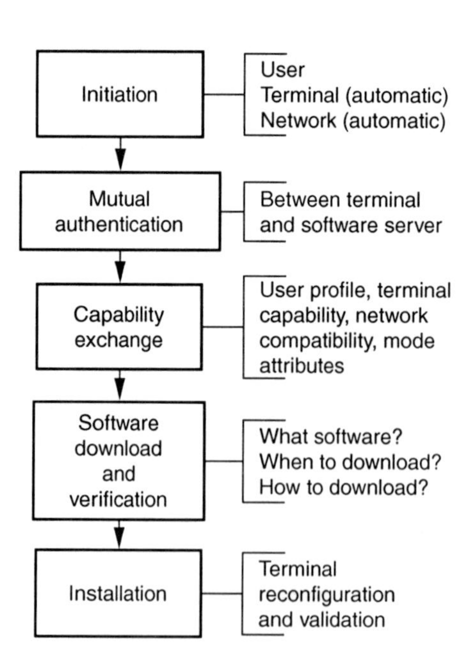

Figure 18-4 A generic software download flowchart.

- *Integrity.* No one can tamper with the content transferred on the wireless data network (signing transaction and content).
- *Authentication.* Both parties in a transaction are really who they say they are (applying simple password schemes, certificates, etc.).
- *Nonrepudiation.* A user or provider cannot deny having made a transaction. This can be ensured by the use of a digital signature mechanism.[1]

To create an environment where transactions can be made freely with minimal disruptions, those four items need to be addressed. In addition, a set of technical regulations is needed to minimize the potential of causing interference to other radio services due to inappropriate software being implemented on user terminals.

Adaptive Baseband Processing

The TRUST baseband is adaptive because of its ability to reconfigure itself. The software architecture of this reconfigurable baseband (R-BB) subsystem is based on object-oriented methodology. Each module of the baseband transceiver chain is reconfigurable by instantiation of an appropriate class and/or reinitialization of modules with new attributes. It is assumed that the software (a class) of each module (modulator, FEC decoder) is available, error-free, and compatible. The R-BB subsystem (see Fig. 18-5) consists of the following components[1]:

- *Reconfigurable baseband management module (RMM).* The overall authority of the R-BB subsystem. It is responsible for negotiating reconfiguration, creating active and shadow transceiver chains, and controlling the run-time behavior of each module. The RMM also controls the RF subsystem through appropriate signaling.
- *Active baseband transceiver chain.* The incumbent baseband transceiver chain.
- *Shadow baseband transceiver chain.* The target baseband transceiver chain.
- *Baseband software library.* Contains the active and shadow configuration maps of the baseband. These correspond to the active and shadow transceiver chains. A configuration map is the overall definition of the baseband processing subsystem, and is a piece of software in itself that is downloaded when a new standard is to be implemented. It includes the identity of the baseband modules, their interconnections, and the constituent module interface definitions. In addition, the library will also store all the baseband module leaf classes currently in use and those used previously.

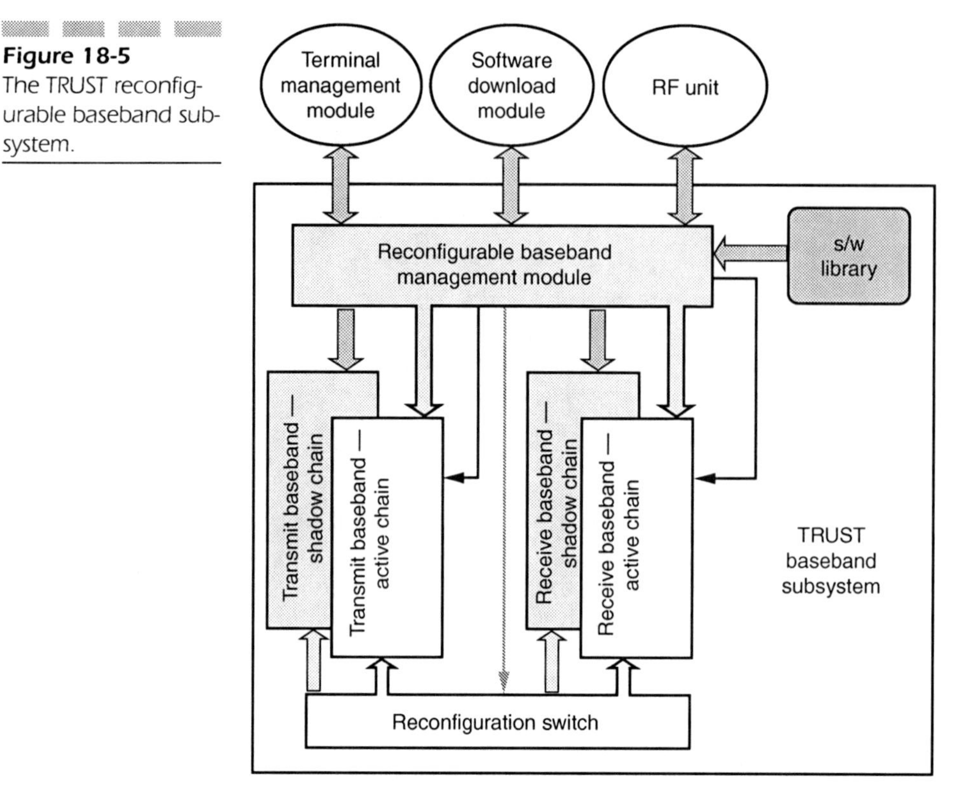

Figure 18-5 The TRUST reconfigurable baseband subsystem.

- *Reconfiguration switch.* A typical on/off switch. It implements the on/off signal from the RMM in order to switch the shadow chain on and the active chain off.[1]

Baseband Processing Cells The TRUST baseband transceiver chain is made up of several baseband processing cells (BPCs). These are the fundamental building block of any (active or shadow) transmit-receive baseband chain. The BPC class is a generic wireless data processing entity. The algorithmic functionality of a BPC object is defined by its process function, which in turn is defined by the downloaded leaf class. The software download mechanism ensures that the properties of the leaf classes are compliant with the terminal (type, hardware, user profile, etc.) and the considered configuration map. Once the leaf class is downloaded and available, the RMM will use its valid parameter lists to instantiate the required BPC object. Each BPC object has a life cycle, and is defined by its state transition diagram (STD) as shown in Fig. 18-6.[1]

Next, let's look at how a BPC object is created and then reconfigured. For the sake of clarity, psuedo-C++ code is given for each step. Assume that the BPC object in question is a convolutional encoder:

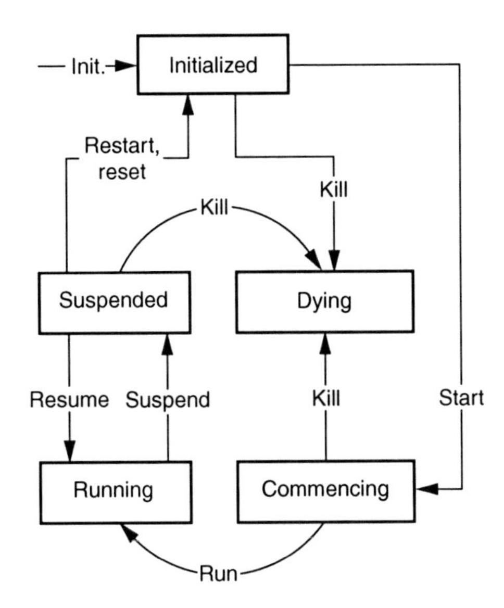

Figure 18-6 The state transition diagram of a BPC object.

1. The RMM creates a new BPC object, of identity _id, by instantiating the BPC class:

```
BPC_id = new BPC (Input Port
addresses, Output Port addresses)
```

2. Get ConvolutionalEncoder leaf class code from the software library. The RMM will create a virtual pointer in its memory and reserve some memory for the new object, which will be instantiated from the class.
3. The RMM will then create a new process function for the new BPC object by issuing the 'init' signal:

```
BPC_id.initialize (new Convolution
alEncoder(k1, G1(p)))
```

where k_1 is the constraint length of the code, and $G_1(p)$ is the galios field polynomial. These are the attributes of the ConvolutionalEncoder class.

NOTE The initialize() is a member method of the BPC class. Once the BPC object has been initialized, it is then placed by the RMM in its appropriate place, as defined by the configuration map.

4. The RMM will then put this new BPC object in the Commencing state. The Commencing state implies that the BPC object is

almost ready to operate. The RMM will determine the correct timing and synchronization in order to start running the newly created BPC object:

```
BPC_id.resume()
```

5. Once the RMM has established an appropriate time and synchronization, it will put the new BPC object in Running state, wherein wireless data is processed as intended:

```
BPC_id.run()
```

Now, consider that this BPC object is to be reconfigured by changing the process function within it. In other words, the functionality of the convolutional encoder needs changing without destroying the BPC object. Such reconfigurations are typical of the type that is called *partial reconfiguration* for such instances.

6. The RMM will make a pointer reference (copy) of the BPC object in the shadow chain (*BPC_id). It will then put *BPC_id in a Suspended state:

```
*BPC_id.suspend()
```

7. The RMM will reset the BPC object by instantiating a new process function. The new process function implies that there is no change in the input/output ports of *BPC_id, but simply a change in the resident wireless data processing entity process:

```
*BPC_id.reset(new convolutionalEncoder (k2, G2(p)))
```

where k_2, $G_2(p)$ are the new attributes of the process.

8. Once the BPC object has been reset, the RMM will then put it in Initialized state (step 4). The RMM will next issue the start signal to commence the newly configured BPC object in the shadow chain. Once that is done, it will then put the BPC object in Running state by issuing the run signal (step 5). Now consider that a new type of FEC encoder is required and that the incumbent BPC object is to be replaced by a new BPC object with new input/output ports and a new process within it. Such reconfigurations are typical of the type called *total reconfiguration* for such instances:
9. The RMM will suspend the BPC object in the shadow chain (step 6).
10. The RMM will then remove this shadow BPC object by issuing a kill signal. The kill signal to the BPC object will destroy only the

process function within it. Once that is successfully completed, the RMM will delete the BPC object completely:

```
BPC_id.KILL(), and then, delete
BPC_id
```

11. The RMM will replace the killed BPC object with a new one in the shadow chain, which it created in the background following steps 1 to 5.[1]

Reconfiguration Steps Finally, the following is a sequential list of steps that explain how baseband reconfiguration is managed and administered:

1. The RMM accepts a reconfiguration request from the terminal's management entity, that is, a terminal management module (TMM). The request includes information on:
   - Which BPC objects to reconfigure
   - How to reconfigure them
   - New configuration map
   - Run-time signaling changes
2. RMM then negotiates the reconfiguration request with the TMM. This includes details such as:
   - Complexity of reconfiguration
   - Processing and memory requirements
   - Time duration for reconfiguration
3. RMM will perform an RF capability check by referring to the RF property list.
4. Following a successful negotiation, the TMM will instruct the RMM with:
   - A list of BPCs to be reconfigured
   - How to reconfigure them
   - When to reconfigure them
5. As part of the successful negotiation, the RMM instructs the TMM if new software needs to be downloaded, or whether it intends to use the already present software from its local library store.
6. The TMM then instructs the software download module (SDM) if new software needs to be acquired and then instructs the RMM when it is available.
7. RMM reads the necessary software from the baseband software library. This could be either the newly downloaded code or that already present.

8. RMM then creates the shadow transceiver chain. The shadow chain contains the new baseband modules and pointer references of the unchanged modules, which are intended to remain from the current baseband chain.
9. RMM then validates the shadow chain such that it complies with the agreed configuration map in terms of interfaces between neighboring BPC objects and their input/output ports.
10. Once the RMM has successfully configured the shadow chain, it will then instruct the RF subsystem to retune its filters.
11. While the RF subsystem is being reconfigured, the RMM will reconfigure the chosen BPC objects in accordance with the STD.
12. The RF subsystem will send an acknowledgment back to the RMM after it has successfully reconfigured. Then the RMM is in a position to switch the shadow BPC object on and thus complete a given baseband reconfiguration.[1]

The switch-over between shadow and active chains needs to be authorized by the TMM in order to maintain network compliance.

Conclusion

The realization of a reconfigurable user terminal based on wireless data software-defined radio technology demands novel architectural solutions and conceptual designs, both from a terminal-centric viewpoint and also with regard to provisions in the host networks. Following investigations in the TRUST project, it is clear that in order to develop a terminal that is able to reconfigure itself across different radio access standards, there need to be some supporting mechanisms within the different wireless data networks. Considering these aspects together with the technical solutions needed, the TRUST project has proposed several entities needed to enable terminal reconfiguration. This chapter presented architectural solutions for the following aspects, identified in the TRUST project:

- Mode identification
- Mode switching
- Software download
- Adaptive baseband processing

Finally, these solutions provide insight into the type of entities necessary to develop a feasible RUT based on SDR technology. In addition, it also helps you to understand the framework (wireless data network and terminal entities and flexible processing environment) needed for adapting terminal functionality, behavior, and mode (radio access technology) in accordance with user requirements, terminal capability, and available services across detected modes. The added benefit of such a flexible solution will help yield improved QoS to the user, multimode capability, and adaptive pricing and service packaging.

References

1. Mehul Mehta, Nigel Drew, Georgios Vardoulias, Nicola Greco, and Christoph Niedermeier, "Reconfigurable Terminals: An Overview of Architectural Solutions," *IEEE Communications Magazine,* 445 Hoes Lane, Piscataway, NJ 08855, 2002.
2. John R. Vacca, *i-mode Crash Course,* McGraw-Hill, 2001.

CHAPTER 19

Configuring Broadband Wireless Data Networks

The wireless data communications industry is gaining momentum in both fixed and mobile applications.[2] The continued increase in demand for all types of wireless data services (voice, data, and multimedia) is fueling the need for higher capacity and data rates. Although improved compression technologies have cut the bandwidth needed for voice calls, data traffic will demand much more bandwidth as new services come on line. In this context, emerging technologies that improve wireless data systems' spectrum efficiency are becoming a necessity, especially in the configuration of wireless data broadband applications. Some popular examples include smart antennas, in particular multiple-input, multiple-output (MIMO) technology; coded multicarrier modulation; link-level retransmission; and adaptive modulation and coding techniques.

Popularized by cellular wireless data standards such as Enhanced Data GSM Evolution (EDGE), adaptive modulation and coding techniques that can track time-varying characteristics of wireless data channels carry the promise of significantly increasing data rates, reliability, and spectrum efficiency of future wireless data-centric networks. The set of algorithms and protocols governing adaptive modulation and coding is often referred to as *link adaptation* (LA).

While substantial progress has been accomplished in this area to understand the theoretical aspects of time adaptation in LA protocols, more challenges surface as dynamic transmission techniques must take into account the additional signaling dimensions explored in future broadband wireless data networks. More specifically, the growing popularity of both multiple transmit antenna systems [MIMO and multiple-input, single-output (MISO)] and multicarrier systems such as orthogonal frequency-division multiplexing (OFDM) creates the need for LA solutions that integrate temporal, spatial, and spectral components together. The key issue is the design of robust low-complexity and cost-effective solutions for these future wireless data networks.

This chapter is organized as follows. First, the traditional LA techniques are introduced. Then other emerging approaches for increasing the spectral efficiency in wireless data access systems with an emphasis on interactions with the LA layer design are discussed. Next, the chapter focuses on smart antenna techniques and coded multicarrier modulations. The chapter then continues with a short overview of space-time configuration broadband wireless data propagation characteristics. Then the chapter explores various ways of capturing channel information and provides some guidelines for the design of sensible solutions for LA. Finally, the chapter emphasizes the practical limitations involved in the application of LA algorithms and gives examples of practical performance. (The Glossary defines many technical terms, abbreviations, and acronyms used in the book.)

Link Adaptation Fundamentals

The basic idea behind LA techniques is to adapt the transmission parameters to take advantage of prevailing channel conditions. The fundamental parameters to be adapted include modulation and coding levels, but other quantities can be adjusted for the benefit of the systems such as power levels (as in power control), spreading factors, signaling bandwidth, and more. LA is now widely recognized as a key solution to increase the spectral efficiency of wireless data systems. An important indication of the popularity of such techniques is the current proposals for third-generation wireless packet data services, such as code-division multiple-access (CDMA) schemes like cdma2000 and wideband CDMA (W-CDMA) and General Packet Radio System (GPRS, GPRS-136), including LA as a means to provide a higher data rate.

The principle of LA is simple. It aims to exploit the variations of the wireless data channel (over time, frequency, and/or space) by dynamically adjusting certain key transmission parameters to the changing environmental and interference conditions observed between the base station and the subscriber. In practical implementations, the values for the transmission parameters are quantized and grouped together in what is referred to as a set of modes. An example of such a set of modes, where each mode is limited to a specific combination of modulation level and coding rate, is illustrated in Table 19-1.[1] Since each mode has a different wireless data rate (expressed in bits per second) and robustness level [minimum signal-to-noise ratio (*S/N*) needed to activate the mode], they are optimal for use

TABLE 19-1

EGPRS Modulation and Coding Schemes and Peak Data Rates

Scheme	Modulation	Maximum Rate, kbps	Code Rate
MCS-9	8 PSK	59.2	1
MCS-8		54.5	0.92
MCS-7		44.8	0.76
MCS-6		29.6	0.49
MCS-5		22.4	0.37
MCS-4	GMSK	17.6	1
MCS-3		14.8	0.80
MCS-2		11.2	0.66
MCS-1		8.8	0.53

in different channel/link quality regions. The goal of an LA algorithm is to ensure that the most efficient mode is always used, over varying channel conditions, based on a mode selection criterion (maximum data rate, minimum transmit power, etc). Making modes available that enable communication even in poor channel conditions renders the system robust. Under good channel conditions, spectrally efficient modes are alternatively used to increase throughput. In contrast, systems with no LA protocol are constrained to use a single mode that is often designed to maintain acceptable performance when the channel quality is poor to get maximum coverage. In other words, these systems are effectively designed for the worst-case channel conditions, resulting in insufficient utilization of the full channel capacity.

The capacity improvement offered by LA over nonadaptive systems can be remarkable, as illustrated by Fig. 19-1.[1] This figure represents the link-level spectral efficiency (SE) performance (bits per second per hertz) versus the short-term average *S/N* Y in decibels, for four different uncoded modulation levels referred to as binary phase-shift keying (BPSK), quaternary PSK (QPSK), 16 quadrature amplitude modulation (QAM), and 64 QAM. The SE was obtained for each modulation by taking into account the corresponding maximum data rate and packet error rate (PER), which is a function of the short-term average *S/N*. The SE curve of two systems is highlighted. The first system is nonadaptive and constrained to use the BPSK modulation only. Its corresponding SE versus *S/N* is represented by the BPSK modulation curve that extends from the intersection of SE 1*y* (bps) and *S/N* 10*x* (dB) straight across to the intersection of SE 1*y* and *S/N* 30*x*. The second system uses adaptive modulation. Its corresponding SE is given by the envelope formed by the BPSK, QPSK, 16 QAM, and 64 QAM curves that extend from the intersection of SE 0*y* and *S/N* 0*x* to the intersection of SE 1*y* and *S/N* 10*x*, to the intersection of SE 2*y* and *S/N* 17*x*, to the intersection of SE 4*y* and *S/N* 24*x*, and to the intersection of SE 6*y* and *S/N* 30*x*, respectively. It is seen that each modulation is optimal for use in different quality regions, and LA selects the modulation with the highest SE for each link. The performance of the two systems is equal for *S/N* up to 10 dB. However, in the range of higher *S/N*, the SE of the adaptive system is up to 6 times that of the nonadaptive system. When averaging the SE over the *S/N* range for a typical power-limited cellular scenario, the adaptive system is seen to provide a close to threefold gain over the nonadaptive system.

The example in Fig. 19-1 is ideal since it assumes that the modulation level is perfectly adapted to the short-term average *S/N*, and that the probability of error as a function of the *S/N* is exactly known; for example, here an additive white gaussian noise (AWGN) channel is considered, which corresponds to an instantaneous channel measurement. That assumption is true only for instantaneous feedback and is not practical because of delays in the feedback path. When there is delay, as

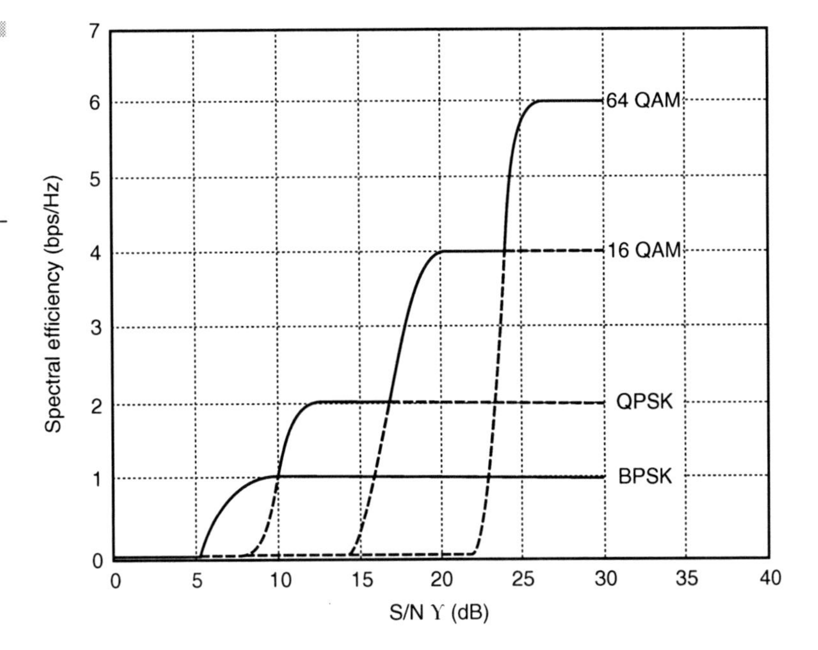

Figure 19-1 Spectral efficiency for various modulation levels as a function of short-term average S/N.

explained later, the first- and higher-order statistics of the fading channel should be incorporated to improve the adaptation. Furthermore, other dimensions such as frequency and space (where different transmission schemes may be adapted) may yield further gains simply by providing additional degrees of freedom exploitable by LA.

Expanding the Dimensions of Link Adaptation

"Smart antenna" technology is widely recognized as a promising technique to increase the spectrum efficiency of wireless data networks. Systems that exploit smart antennas usually have an array of multiple antennas at only one end of the communication link [at the transmit side, as in MISO systems, or at the receive side, in single-input, multiple-output (SIMO) systems]. A more recent idea, however, is multiarray or MIMO communication where an antenna array is used at both the transmitter and receiver. The potential of MIMO systems goes far beyond that of conventional smart antennas and can lead to dramatic increases in the capacity of certain wireless data links. In the so-called BLAST scenario, each antenna transmits an independently modulated signal simultaneously and on the same carrier frequency. Alternatively, the level of redundancy between the transmitting antennas can be increased to improve robustness by using so-called space-time codes.

Multiantenna-Element Systems and LA

In MIMO and MISO systems, the presence of multiple transmit antenna elements calls for an efficient way of mapping the bits of the messages to be sent to the various signals of individual antenna elements. The mapping can and must be done in different ways as a function of both the channel characteristics and the benefit desired from the smart antennas. For instance, in the MIMO case, the mapping in a multiplexing/BLAST scheme tends to minimize the redundancy between the various antenna signals in order to favor a maximum wireless data rate. In contrast, a typical space-time coding approach will introduce a lot of redundancy in an effort to maximize the diversity gain and achieve a minimum bit error rate (BER). The properties of the instantaneous or averaged space-time channel vector/matrix (the rank and condition number) play a critical role in the final selection of mapping strategy, just as the *S/N* does in picking a modulation or coding scheme for transmission. This is because independent signals transmitted over a rank-deficient MIMO channel cannot be recovered. In this respect, it is clearly understood that the antenna mapping strategy must be treated as one component in the joint optimization of the signaling by the LA layer. Practical examples of this are considered later in the chapter when performance simulation results are described.

Multicarrier Systems

Broadband transmission over multipath channels introduces frequency selective fading. Mechanisms that spread information bits over the entire signal band take advantage of frequency selectivity to improve reliability and spectrum efficiency. An example of such a multipath-friendly mechanism is frequency-coded multicarrier modulation (OFDM). Transmission over multiple carriers calls for a scheme to map the information bits efficiently over the various carriers. Ideally, the mapping associates an independent coding and modulation scheme (or mode) with each new carrier. The idea is to exclude (avoid transmitting over) faded subcarriers, while using high-level modulation on subcarriers offering good channel conditions. While this technique leads to high theoretical capacity gain, it is highly impractical since it requires significant knowledge about the channel at the transmitter, thereby relying on large signaling overhead and heavy computation load. Alternative solutions based on adapting the modes on a per-subband basis (as opposed to per-subcarrier basis) are less demanding in overhead and select a unique mode for the entire subband, while still profiting from the frequency selectivity.

Adaptive Space-Time-Frequency Signaling

Before presenting the possible approaches to designing and configuring LA in broadband multiantenna systems, let's look at a brief overview of wireless data propagation channel modeling aspects in space, time, and frequency dimensions relevant in LA design and configuration.

The Space-Time-Frequency Wireless Channel

An ideal link adaptation algorithm adjusts various signal transmission parameters according to current channel conditions in all of its relevant dimensions. It is well known that, unlike wired channels, radio channels are extremely random and the corresponding statistical models are very specific to the environment. Propagation models are usually categorized according to the scale of the variation behavior they describe:

- Large-scale variations include, for instance, the path loss (defined as the mean loss of signal strength for an arbitrary distance between transmitter and receiver) and its variance around the mean, captured as a log-normal variable, referred to as *shadow fading* and caused by large obstructions.
- Small-scale variations characterize the rapid fluctuations of the received signal strength over very short travel distances or short time durations due to multipath propagation. For broadband signals, these rapid variations result in fading channels that are frequency selective.[1]

A typical realization of a fading signal over time is represented in Fig. 19-2.[1] The superposition of component waves (or multipath) leads to either constructive (peaks) or destructive (deep fades) interference. Time-selective fading is characterized by the so-called channel coherence time, defined as the time separation during which the channel impulse responses remain strongly correlated. It is inversely proportional to the doppler spread and is a measure of how fast the channel changes: The larger the coherence time, the slower the change in the channel. Clearly, it is important for the update rate of LA to be less than the coherence time if one

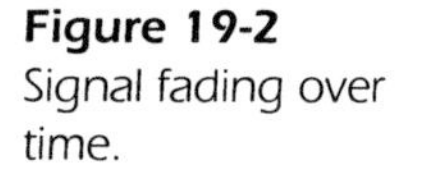

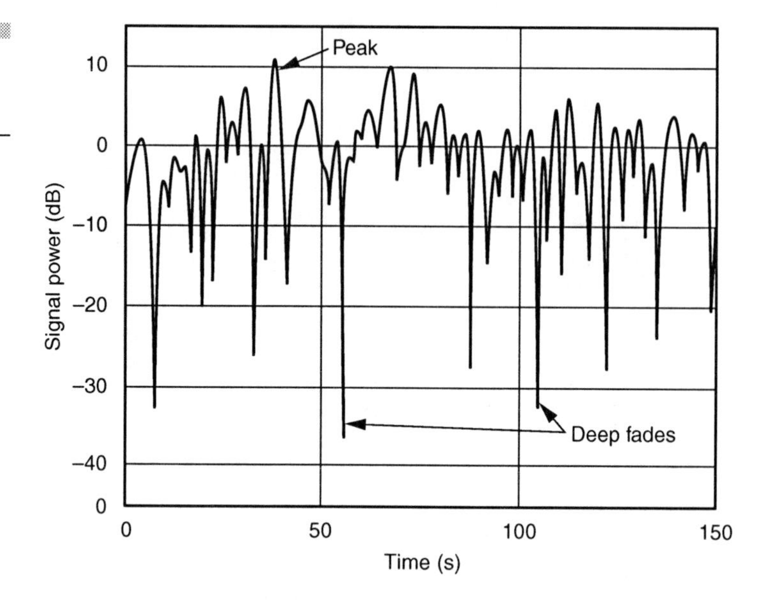

Figure 19-2 Signal fading over time.

wishes to track small-scale variations. However, adaptation gains can still be realized at much lower rates thanks to the large-scale variations.

In a multipath propagation environment, several time-shifted and scaled versions of the transmitted signal arrive at the receiver. When all delayed components arrive within a small fraction of the symbol duration, the fading channel is frequency nonselective, or flat. In wideband transmission, the multipath delay is often non-negligible relative to the symbol interval, and frequency-selective fading results. Figure 19-3 shows the time-varying frequency response of a channel model taken from over 2 MHz of bandwidth.[1] In this type of channel, the variation of the signal quality may be exploited in both the frequency and time domains.

The third dimension LA may exploit is fading selectivity over space, which will be observed in a system that employs multiple antennas. Space selectivity occurs when the received signal amplitude depends on the spatial location of the antenna and is a function of the spread of angles of departure of the multipaths from the transmitter and the spread of angles of arrival of the multipaths at the receiver.

Adaptation Based on Channel State Information

The general principle of LA is to (1) define a channel quality indicator, or so-called channel state information (CSI), that provides some knowledge on the channel and (2) adjust a number of signal transmission

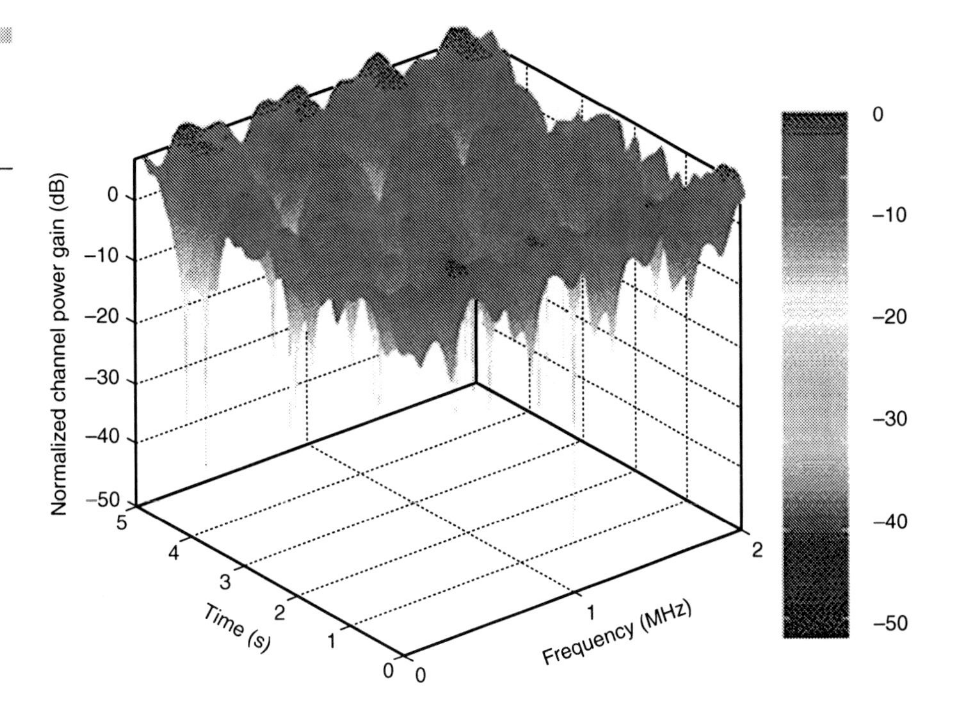

Figure 19-3 Frequency response versus time for a multipath channel.

parameters to the variations of that indicator over the signaling dimensions explored (time, frequency, space, or a combination thereof).

There are various metrics that may be used as CSI. Typically, *S/N* or signal-to-noise-plus-interference ratio (SINR) may be available from the Physical layer (by exploiting power measurements in slots without intended transmit data). At the Link layer, packet error rates (PERs) are normally extracted from the cyclic redundancy check (CRC) information. BERs are sometimes available. This part of the chapter reviews the respective use of this type of CSI in the design of the LA protocol, with emphasis on time adaptation and for an error-rate-constrained system. Let's first consider the traditional example of LA using *S/N* measurement with the perfect instantaneous feedback introduced earlier. This part of the chapter shows the limitations of this scheme, and moves on to more sophisticated types of adaptation.

Adaptation Based on Mean *S/N*

To implement adaptive transmission, the CSI must be available at either the transmitter or receiver. Often, such information consists of the *S/N* measured at the receiver. In this case, a possible solution for LA is as follows:

1. Measure the *S/N* at the receiver.
2. Convert the *S/N* information into BER information for each mode candidate.
3. Based on a target BER, select for each *S/N* measurement the mode that yields the largest throughput while remaining within the BER target bounds.
4. Feed back the selected mode to the transmitter.[1]

Step 1 corresponds to the assessment of the CSI. Step 2 refers to the computation of the adaptation (or switching) thresholds. In this case, a threshold is defined as the minimum required *S/N* for a given mode to operate at a given target BER. Step 3 refers to the selection of the optimal mode, based on a set of thresholds and *S/N* measurement. Step 4 is concerned with the feedback of information to the transmitter. Under ideal assumptions, the implementation of these steps is straightforward. For example, let's assume a channel that is fading over time only (left aside are the two other dimensions for simplicity). The conversion from mean *S/N* to BER can be made only if the mean *S/N* is measured in a very short time window, so each window effectively sees a constant nonfading channel. Let's therefore assume further that the *S/N* can be measured instantaneously, and the LA algorithm aims at adapting a family of uncoded *M* QAM modulations to each instantaneous realization of the *S/N*. Established closed-form expressions for the AWGN channel may be used to express the BER as a function of the *S/N* Υ assuming ideal coherent detection. Figure 19-4 represents a set of these BER curves for modulations BPSK, QPSK, 16 QAM,

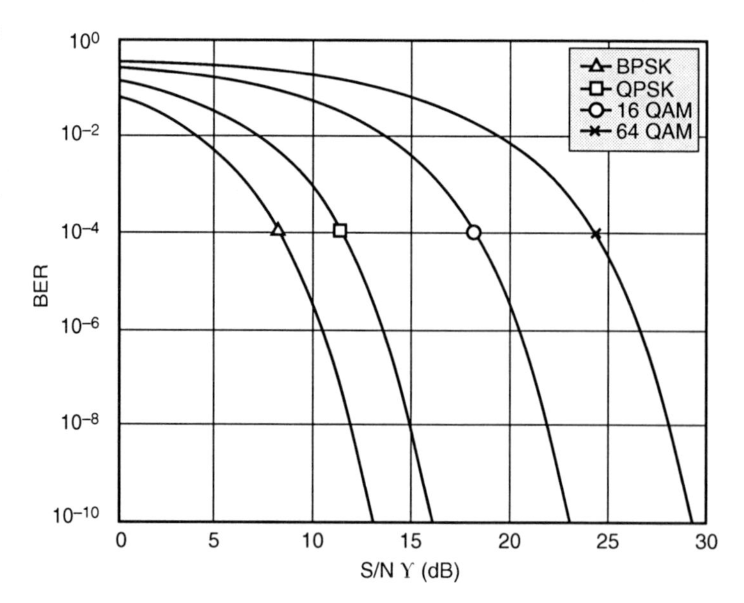

Figure 19-4 BER for various modulation levels as a function of short-term average *S/N*.

and 64 QAM.[1] The set of adaptation/switching thresholds is then obtained by reading the *S/N* points corresponding to a target BER. For example, if the target BER is 10^{-4}, the thresholds are 8.4, 11.4, 18.2, and 24.3 dB, respectively (as indicated by the markers on the figure).

Of course, the scenario presented in the preceding relies on ideal assumptions. In practice, the feedback delay and other implementation limitations will not allow for mode adaptation on an instantaneous basis, and the effective update rate may be much slower than the coherence time. In that case, the conversion of the *S/N* into BER information is not as simple as the formulation available for the AWGN model, because the real channel may exhibit some fading within the *S/N* averaging window. This calls for the use of second- and higher-order statistics of the *S/N* instead of just the mean.

Adaptation Based on Multiple Statistics of the Received *S/N*

Let's assume here that the CSI is measured over an arbitrary time window (flat fading case) set by the system-level constraints of the LA protocol. If multicarrier modulation is used, a two-dimensional time-frequency window may be used. The mapping between the *S/N* and the average BER is determined by using the probability density function (pdf) of the *S/N* over that window. Unfortunately, in real channels, this pdf cannot be obtained via simple analysis because it is a function of many parameters. It depends on:

- The channel fading statistics over frequency and time (which often have different distributions).
- The relationship between the length of the window in time and the channel coherence time.
- The relationship between the length of the window in frequency and the channel coherence bandwidth.
- In the case of multiantenna-based systems, the *S/N* is determined after antenna combining; therefore, the pdf also depends on such system parameters as the number of antennas used on the transmit and receive sides, antenna separation, antenna polarization, and transmission and reception schemes.[1]

Instead of trying to estimate the full pdf of the *S/N* over the adaptation window, one can simplify the problem by estimating limited statistical information from the pdf, such as the k-order moment over the adaptation window, in addition to the pure mean (first-order moment). These statistics provide only an approximation of the pdf of the received *S/N*. They are useful, however, when k can be kept low and yet yield sufficient information for a reasonably accurate mapping of the S/N into BER information. The first moment of the S/N captures how much power is measured at the receiver on average. The second moment of the S/N over the time (cf. frequency) dimension captures some information on the time (cf. frequency) selectivity of the channel within the adaptation window. Higher-order moments give further information on the pdf. However, they are also more computationally demanding, so there is a tradeoff between accuracy and computation efficiency.

With moment-based CSI, the adaptation thresholds are a function of multiple statistics of the received S/N. This introduces simplicity and flexibility to the LA algorithm, since the adaptation thresholds no longer rely on any particular channel conditions. They remain valid for any doppler spread, delay spread, and ricean K-factor. In the case of multi-antenna systems, they do not depend on any assumption made on the number of transmit and receive antennas, antenna polarization, and so on, since the effect of all these factors is captured by the low-order moments of the S/N ($k > 1$) and, to a large extent, by the first- and second-order moments alone.

Space-Time-Frequency Adaptation

In a system with multiple antennas at the receiver and/or transmitter, the S/N not only varies over time and frequency, but also depends on a number of parameters, including the way the transmitting signals are mapped and weighed onto the transmit antennas, the processing technique used at the receiver, and the antenna polarization and propagation-related parameters such as the pairwise antenna correlation. In a space-time-frequency adaptation scheme, it is desirable that the adaptation algorithm be able to select the best way of combining antennas at all times (choosing between a space-time coding approach, a beamforming approach, and a BLAST approach in a continuous way). Furthermore, it should do so in a systematic way that is transparent to the antenna setup itself. For example, ideally, the same version of the LA software is loaded in the modem regardless of the number of antennas of this particular device or their polarization. One possible solution to capture the effects of all these parameters in a transparent manner is to express the channel quality (and therefore the adaptation thresholds) of multiantenna-based systems in terms of postprocessing S/N as opposed to simple preprocessing S/N levels measured at the antennas. In this case, the variation of the postprocessing S/N is monitored over time, frequency, and space, thus enabling the LA algorithm to exploit all three dimensions. For CSI based purely on error statistics, the channel quality of multiantenna-based systems is directly expressed postprocessing, since errors are detected at the end of the communication chain.

Performance Evaluation

In this final part of the chapter, the performance attainable by an LA algorithm (where the adaptation is based on a combination of *S/N* and PER statistics) in a simulated scenario is illustrated. As an example, a broadband wireless data MIMO-OFDM-based system is used that provides wireless data access to stationary Internet users.

NOTE Despite the users' stationarity, the environment is still time-varying, and some gain may be obtained by tracking these variations.

The adjustable transmission parameters are the modulation level, coding rate, and transmission signaling scheme. One possible transmission signaling scheme is to demultiplex the user signal among the several transmit antennas [referred to as *spatial multiplexing* (SM)]; the other is to send the same copy of the signal out of each transmit antenna with a proprietary coding technique based on the concept of delay diversity. This latter scheme is referred to as *transmit diversity* (TD). A particular combination of modulation level, coding rate, and transmission signaling scheme is referred to as a *mode.* The system may use six different combinations of modulation level and coding rate, and two different transmission signaling schemes, resulting in a total of 12 candidate modes, indexed as modes SM i and modes TD i, where $i = 1,\ldots,6$.

Figure 19-5 presents the system spectral efficiency (SE) performance in bits per second per hertz versus the long-term average *S/N* Υ_0 for a frequency-flat and a frequency-selective (rms delay spread is 0.5 μs) environment.[1] The results highlight the great capacity gain achievable by LA over a system using a single modulation. The dashed line in the lower part of the figure shows the SE versus *S/N* of a system using mode TD 1 only. It is seen that the SE remains at 0.5 bps/Hz, regardless of the *S/N*. In contrast, the SE of the adaptive system increases with *S/N* as different modes are used in different *S/N* regions. It is also shown that exploiting the extra dimension (over frequency) provides additional gain in a frequency-selective channel, mostly for higher *S/N* where the higher mode levels (with larger SE) may be used.

Finally, Fig. 19-6 shows the system spectral efficiency (SE) performance in bits per second per hertz versus the normalized adaptation

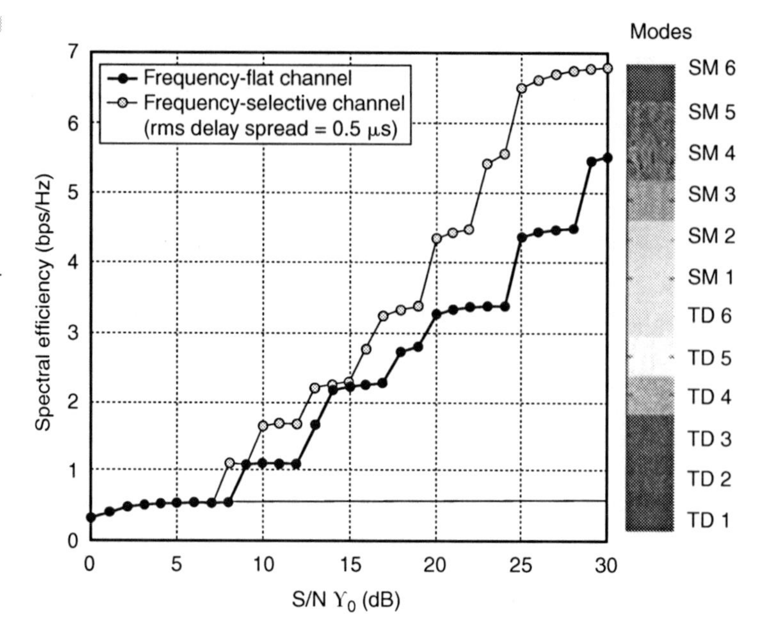

Figure 19-5 Spectral efficiency versus long-term average *S/N* with and without adaptive signaling with and without frequency selectivity.

window defined as T_a/T_c, where T_a and T_c denote the adaptation window and channel coherence time, respectively.[1] The channel is considered frequency flat. Thus, the results are independent of the channel doppler spread. Three curves are plotted for a fixed long-term average *S/N* of 10 dB. The upper curve represents the SE of instantaneous LA; that is, the mode is adapted for each instantaneous realization of the *S/N*. This scenario may not be achievable in practice because of practical limitations, but it is used here as an upper bound on the performance obtainable with a practical adaptation rate. For an average *S/N* at 10 dB, the upper bound is almost 2 bps/Hz. The lower curve represents the SE of provisioned LA; that is, the mode is adapted on the basis of long-term *S/N* statistics. The SE value is taken from Fig. 19-5 at *S/N* of 10 dB, where it is read to be equal to 1.1 bps/Hz. Since provisioned LA is the slowest way of adapting to the time-varying channel components, the corresponding SE is used as a lower bound on the performance obtainable with a practical adaptation rate. Finally, the middle curve shows the variation of the SE as a function of the normalized adaptation window. When the ratio T_a/T_c is small, the adaptation is fast and the performance approaches the theoretical upper bound. When the ratio T_a/T_c is large, the adaptation is slow and the performance converges toward the lower bound. In general, it is seen that a twofold capacity gain may be achieved between the slowest and fastest adaptation rates.

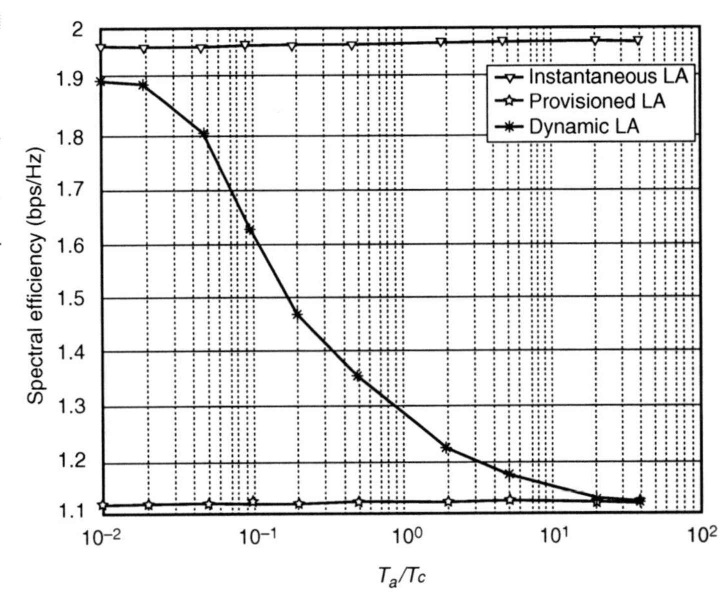

Figure 19-6 Spectral efficiency versus normalized adaptation window, for various adaptation rates at fixed long-term average S/N.

Conclusion

Link adaptation techniques, where the modulation, coding rate, and/or other signal transmission parameters are dynamically adapted to the changing channel conditions, have recently emerged as powerful tools for increasing the data rate and spectral efficiency of wireless data-centric networks. While there has been significant progress on understanding the theoretical aspects of time adaptation in LA protocols, new challenges surface when dynamic transmission techniques are employed in broadband wireless data networks with multiple signaling dimensions. Those additional dimensions are mainly frequency, especially in multicarrier systems, and space in multiple-antenna systems, particularly multiarray multiple-input, multiple-output communication systems. This chapter gave an overview of the challenges and promises of link adaptation in future broadband wireless data networks. It is suggested that guidelines be adapted here to help in the design and configuration of robust, complexity/cost-effective algorithms for these future wireless data networks.

Finally, this chapter also reviewed the fundamentals of adaptive modulation and coding techniques for MIMO broadband systems and illustrated their potential to provide significant capacity gains under ideal assumptions. Other emerging techniques for increasing spectral efficiency in wireless data broadband access systems were presented; smart antenna techniques and coded multicarrier modulations, and their interactions with the LA layer design and configuration, were emphasized. Following a short overview of space-time broadband wireless data propagation characteristics, the chapter explored various ways to capture channel information and provided some guidelines on the design of sensible solutions for LA. Implementing and configuring optimum LA is challenging because of practical limitations, but simulated performance of a realistic broadband wireless data MIMO-OFDM-based system using LA is very encouraging.

References

1. Severine Catreux, Vinko Erceg, David Gesbert, and Robert W. Heath, Jr., "Adaptive Modulation and MIMO Coding for Broadband Wireless Data Networks," *IEEE Communications Magazine,* 445 Hoes Lane, Piscataway, NJ 08855, 2002.
2. John R. Vacca, *i-mode Crash Course,* McGraw-Hill, 2001.

CHAPTER 20

Configuring Wireless Data Mobile Networks

Configuring wireless data connectivity has implications for the specific mobile computing hardware you choose.[2] While all types of devices support at least some types of wireless data connectivity, the specific type of wireless data network you configure will influence your hardware options, and vice versa. There are many devices you can use to support mobile computing initiatives, with multiple manufacturers for each. Table 20-1 shows some of the major manufacturers for each type of device.[1] (The Glossary defines many technical terms, abbreviations, and acronyms used in the book.)

Configuring Wireless Data Connectivity to Hand-Helds

Wireless data connectivity with laptops employs fairly standard technologies, typically using PCMCIA cards that are compatible with any machine. The situation with hand-helds is more complex, with proprietary hardware and resulting limited network choices for most hand-

TABLE 20-1

Mobile Device Manufacturers

Type of Device	Manufacturers
Laptop PCs	Dell, Gateway, IBM, NEC, Compaq, Toshiba, Sony, and many others
Tablet PCs	Fujitsu, ViewSonic, DT Research
Palm OS hand-helds	Palm, Handspring, Sony, Symbol, HandEra
Pocket PC hand-helds	HP, Compaq, Casio, URThere, Intermec
PC hand-helds	HP, Casio, NEC, Sharp
Other CE devices	Symbol, HP, NEC, Intermec
E-mail pagers	Motorola, RIM
SMS-enabled phones	Ericsson, Motorola, Samsung, Nokia, and many others
WAP-enabled phones	Ericsson, Motorola, Samsung, Nokia, and many others
Palm OS smart phones	Kyocera, Samsung, others
Stinger smart phones	Not yet available
EPOC devices	Psion, Nokia, Ericsson, Siemens

helds. Table 20-2 summarizes this information for the most popular hand-helds.[1]

Smart phones, WAP phones, and SMS phones all offer inherent network connectivity if they operate by digital technology. The wireless data service provider the phone is configured for must also offer wireless data Internet service.

TABLE 20-2

Device Options for Wireless

Hand-Helds	Networks	Modems	Available from
Compaq iPAQ H3600	CDPD	Sierra Wireless Aircard 300	GoAmerica, Compaq, Omnisky
	CDPD	Novatel Wireless Merlin	Verizon
	Ricochet	Sierra Wireless Aircard 400	GoAmerica
	Ricochet	Novatel Wireless Merlin for Ricochet	Compaq
HP Jornada 540	CDPD	Novatel Minstrel 540	GoAmerica, Omnisky
Casio Cassiopeia E125	CDPD	Enfora Pocket Spider	GoAmerica
Handspring Visor (multiple models)	CDPD	Novatel Minstrel S	Go America, Omnisky
	GSM	VisorPhone Module	Cingular, VoiceStream
Palm VII	Mobitex	(Built in)	Palm.net
Palm m500	CDPD	Novatel Minstrel m500	Verizon, GoAmerica, Omnisky
Palm V	CDPD	Novatel Minstrel V	Verizon, GoAmerica, Omnisky
Palm III	CDPD	Novatel Minstrel III	Verizon, GoAmerica, Omnisky
Palm m100	NA	NA	NA
All Palms	CDMA	Palm mobile Internet kit (requires data-enabled phone)	Verizon, Sprint
RIM Blackberry	Mobitex, Datatrac, GPRS	(Built in)	GoAmerica, Cingular, AOL, BT Cellnet, and many others

NA = not available.

The Device Wars

Laptop PCs have largely become commodity items, similar to the PC, with recent ultraslim, highly styled laptops being the exception. Tablet PCs have yet to become very popular, though many organizations that have adopted them are having very positive experiences. The hand-held wars for market share are widely reported on. The once-dominant Palm is seeing steady erosion of market share to its own operating system licensees, as well as to the Pocket PC coalition. Compaq and Hewlett-Packard in particular are leading the Pocket PC charge with very rapid increases in share, principally through enterprise sales. With the rate of innovation and new product releases, it is difficult to predict how the market will evolve. A case in point is Compaq—its iPAQ Pocket PCs have been on the market for less than a year. Yet in the second quarter of 2001, the iPAQ secured 16 percent of worldwide unit shipments for the quarter, doubling Compaq's market share in just 3 months! See Table 20-3 for more detail of the rapid shifts in public favor that hand-held manufacturers are experiencing.[1]

There is a lot of commentary on the plusses and minuses of the different hand-held platforms in the industry press. The major points seen repeated are that Palm's battery life, simplicity, small form factor, and consumer appeal are that platform's historical draws. Meanwhile Pocket PC devices are rapidly gaining ground because of better display quality, integration with MS Office, suitability for enterprise applications, and innovative functionality such as support for rich media files. And the RIM Blackberry devices are adored for the small form factor, integrated wireless data capabilities, and small keyboard that all contribute to making e-mail anywhere an enjoyable reality. In any initiative, carefully consider the business needs and investigate the appropriate choices against the preceding criteria before choosing a device.

TABLE 20-3

Hand-Held Market Share Based on Estimated New Unit Shipments

Manufacturer	Second Quarter 2001	First Quarter 2001
Palm	32.1%	50.4%
Handspring	10.7%	15.9%
Compaq	16.1%	7.8%
Hewlett-Packard	6.9%	3.7%
Research in Motion (RIM)	4.1%	4.6%
Others	30%	17.6%

Smart Phones and Futures

When doing long-term planning and configuration, don't forget about smart phones based on Windows CE, Palm OS, and EPOC. This is an emerging category that is still relatively immature but holds great promise. The new Microsoft "Stinger" standard for smart phones should kick-start this category. The emergence of smartphones has sparked a wide-ranging debate on the future of mobile devices—with two camps emerging. One believes in device convergence and sees smart phones as harbingers of the death of pure phones and pure PDAs. The other sees smart phones as proof of the ongoing proliferation of new device types and the trend toward users having more and more devices.

Choosing the Right Device

It is important for corporations looking to contain the costs of procuring and supporting devices to standardize on a small portfolio of devices. The selection of these devices should take into consideration these factors:

- Length of battery life
- Size of the display area
- Readability of the display
- Mechanisms for data input
- Cost of procurement and support
- Overall form factor
- Processing power
- Third-party application availability
- Amount of local storage
- Available connectivity options
- Security factors
- Supporting application development tools[1]

Different groups of mobile workers may be best served by different devices. For each user community, you need to consider their usage patterns and the business processes that are being facilitated and determine the appropriate devices to support them. Make sure to budget for a short lifespan, keep plenty of spares on hand, and train your help desk staff on how to support the devices. And don't rely on your existing LAN-based system management tools to do a good job of servicing these highly mobile assets. Like all assets, look to manage the total cost of ownership—in this case, by utilizing specialized mobile system management software.

Device Selection Examples

The following are some typical examples. They will give you a feel for how to relate business needs to the selection factors listed.

Warehouse Inventory One large shoe manufacturer, for example, uses simple ruggedized hand-helds for capturing inventory information—a basic data collection task. Symbol Corporation (http://www.symbol.com) manufactures the units, which include a built-in bar-code scanner. The manufacturer uses a mix of Symbol units based on both Palm OS and Windows CE. Hand-helds in general are very well suited to simple data collection tasks, and the decreased cost of hand-helds makes it possible to cost-effectively automate a wide variety of process that have been paper-based until now.

Document Authoring A large law firm in Southern California has a very mobile workforce of people who typically work from home or often at client sites. For these professionals, the laptop is still the device of choice for extensive document authoring—legal briefs, client memos, etc. Creating lengthy materials, or files rich with graphical content, is typically more appropriate for PCs than hand-helds.

Getting through Daily E-Mail Many high-technology companies do large amounts of their business through channel partners that are managed by groups of business development staff. One software manufacturer armed its staff with hand-helds for keeping up with the daily flood of e-mail. When staff members are traveling, they typically recover 1 to 2 hours of productivity each day. After returning to their hotel rooms each night, they have prescreened all their e-mail, and don't have to wade through 50 to 100 messages before getting on to the real work of responding to key communications.

Sales Force Automation For a large specialty chemical manufacturer, one device wasn't enough. Its traveling salespeople needed a laptop for creating PowerPoint presentations and answering requests for proposals. The laptop keyboard, large display, and ability to manipulate graphics make it the required tool for the job. The sales people also benefit from a hand-held-based SFA application that allows easy access to basic customer and order information. With users rarely in the office, long battery life proved to be a key criterion for the hand-helds, and Palm-powered devices were chosen.

Healthcare Application Finally, the small form factor of hand-held devices made them ideal for busy healthcare workers who visit patient after patient—moving throughout a large hospital in London. They can carry along a wealth of reference material and directly record patient information, eliminating the need for all those illegible paper charts. In this case, the users stop by their desks to start and end each day, so battery life was less of an issue since they could recharge overnight. However, the amount of local storage and quality of display were critical, given the large volume of reference material being accessed, and the importance of acting on correctly read information. A Pocket PC device was chosen.

Conclusion

In each of the preceding cases, the usage pattern and processes being automated made some of the device characteristics previously listed more or less critical. The business drivers should thus lead you toward the most appropriate device for the task. In some cases, you will need to support multiple devices for different user communities.

Finally, in addition to supporting different types of devices for different classes of users, industry analyst META Group predicts that by 2005, each corporate knowledge worker will have four to five different computing and information access devices that will be used to access various wireless data applications. So, do not be overly concerned on this point. Instead, look to the wireless data application development tools and mobile middleware solutions you choose to remove a majority of the challenges of integrating multiple mobile device platforms.

References

1. *The CIO Wireless Resource Book,* Synchrologic, 200 North Point Center East, Suite 600, Alpharetta, GA 30022, 2002.
2. John R. Vacca, *i-mode Crash Course,* McGraw-Hill, 2002.

CHAPTER 21

Configuring Residential Wireless Data Access Technology

The meaning of residential (home) networking configuration is changing as a result of the introduction of new wireless data access technologies that are allowing for more advanced applications. Today, the term *home networking* means much more than having a house wired with a phone line, Internet connection, and the latest computers. At the very least, families want the ability to share a printer or free themselves from the jumble of wires and cords that typically plague home offices.

The wireless data access technology available in home networking today is much more sophisticated and practical than most families realize. It can help them, simply and affordably, create and configure a home that evolves with their ever-changing communications needs, wirelessly network their entertainment and computing devices, and actually increases the level of safety and security at home. (The Glossary defines many terms, abbreviations, and acronyms used in the book.)

Transforming a Home

Almost every house in America comes equipped with a standard phone line that runs throughout the structure on twisted pair copper wire, delivering reliable voice and data service to users. But this same, inexpensive, ubiquitous wiring, in combination with wireless data access technology, can also be used to network a home's printer and entertainment devices and to add a new level to home security.

By simply installing this twisted pair wiring, in combination with wireless data access to all rooms in the home, you can virtually provide any type of wireless data access technology service required to each room. These wireless data access services include 75- to 300-ohm video feeds for cable, satellite, VCR, and DVD; 550-kHz to 108-MHz audio feeds for AM/FM and CD; 300- to 3400-kHz telephony feeds for phones, conferencing, and intercoms; and 10/100BaseT wireless data access feeds for computer networking, Internet access, printer sharing, and fax machines.

This home networking setup is of a simple design. By terminating these different media into a sophisticated multiplexing device that can split audio/video and voice/data input signals without degradation, a home owner will have the ability to branch these signals across the twisted pair wire to any location in the home. Wireless data access home networking technology will allow central location of satellite and cable receivers, AM/FM antennas, computer resources, etc., that plug into these multiplexers. Then, by simply making patch cord changes to a small rack, located in a closet, the homeowner can control what piece of information is fed to each room.

By implementing this new technology, homeowners will no longer need sloppy, expensive coaxial cabling, generally used for video. Nor will they need to install new cabling to meet the changing needs of a particular room in the home. By simply making a change in the small patch panel rack, families can deliver wireless data access to a single, centrally located device, such as a DVD player, stereo, or printer, from any room in the house. Instead of buying expensive equipment for every room where they desire access to movies, music, and printers, this technology allows families to share just one unit.

Wireless data access home networking technology essentially transforms the home into a flexible series of rooms that can easily be readapted as the family unit changes over time. For example, the same standard RJ 45 outlet that can be used for an audio/video baby monitor when a child is young can later be easily used to network the room for a computer and Internet and DVD wireless data access when that child grows up. That bedroom can later use the same connectors and be turned into a wireless data home office when the child leaves home.

Safety and Security Features

These same twisted pair wires and RJ 45 outlets can be used to increase the level of safety and security in a home. By sending a standard 75-ohm video stream signal from a wireless camera outside or inside the home to the multiplexer, any TV in the home can become a surveillance monitor. This can also be set up to provide surveillance to a pop-up window for TVs equipped with P-I-P. Once it is installed, families can see images of anyone standing in front of their homes via their television sets.

Wireless security features can also be accessed from outside of the home, via the Internet. While away, parents can use features such as the wireless baby camera to check up on their child and the baby sitter, making sure that everything is going well.

Market Outlook

The future of wireless data access home networking technology is strong (see sidebar, "Wireless Data Access Residential Networking Technology Future"). In fact, the Yankee Group estimates that at least 32 million households in the United States are interested in wireless data access home networking technology and that 23.5 million would like to implement systems within the next year. Of those interested, the most popular uses for a wireless data access home networking system include communicating with friends and family and, for entertainment purposes, viewing movies and listening to music in any room in the house.

Wireless Data Access Residential Networking Technology Future

People have been talking about the connected home for years now, so why hasn't home networking finally penetrated the market on a broad scale? There is not one answer to this question, as a number of factors have contributed to the gradual adoption of home networking. Home networking has had to overcome expensive technology, limited access to broadband connections, and consumer confusion. Now, with the lowering costs of technology, the increase in high-speed connections, and the proliferation of easy-to-use devices, consumers have a real need for home networking.

Step 1: Following Broadband Adoption

The adoption of high-speed Internet connections may have gotten off to a slow start, but most industry analysts agree that broadband is the next true killer app. With the increase of Internet usage, consumers do not have the time for anything but broadband. Broadband and home networking go hand in hand because a user can maximize a single high-speed connection and share it with all the connected devices, which enables access to bandwidth-intensive applications and files like multimedia streaming, VoIP, and multiple-player gaming. In fact, according to Parks Associates (http:// www.parkassociates.com), of the 40 million U.S. households predicted to have broadband Internet connections by 2005, 28 million of them are projected to have wireless data home networks.

The number of multiple-PC homes is growing at a similarly astounding rate. The number of households with more than one computer will grow at double-digit rates through 2003, while the growth for single-PC homes remains the same. Wireless data home networking adoption depends not only on the penetration of broadband and multiple PCs, but also the pervasiveness of devices that drive consumer demand for wireless data networking technology. As more and more appliances penetrate the market, consumers are interested in a way to simplify and coordinate the abundance of products.

Step 2: The Proliferation of Devices

The second factor driving the demand for wireless data home networking is the vast market penetration of devices. In fact, according to Strategy Analytics (http://www.strategyanalytics.com/cgi-bin/gsearch.cgi?dr=01&restrict=25), 86 percent of U.S. homes will be on line by 2006, with at least 76 percent of that group using multiple devices. Wireless data home networking enables all devices (PCs, stereos, household appliances, TVs, printers, PDAs, electronic games, etc.) to operate together, sharing data over one unanimous information source. By enabling these devices to communicate, users can perform the wireless data networking basics like sharing peripherals and Internet access. They also can utilize the wireless data network to enable the printer in the office to print documents stored on the computer in the bedroom, while the bedroom PC is supplying the MP3 files that are played on the stereo downstairs.

Tomorrow's Wireless Data Home Networking Solution

At the convergence of these trends, there is a clear demand within the technology industry for a wireless data home networking solution. Currently, there are several ways to network the home:

- Phone-line networking uses the existing home phone wiring connections to transmit information. This technology has been around the longest. Phone-line connections are compatible with other wireless data networking technologies and require no additional networking.
- Ethernet enables flexible networks, meaning a user can simply use two network interface cards and a Category 5 cable to create an Ethernet connection, or the user can set up multiple hubs, routers, and bridges to create a multifaceted network. With this flexibility, there is room to select the speeds at which the network runs, which range from 10 to 100 Mbps. Yet Ethernet requires new wiring to be installed, often requiring reconstruction of homes. It is also hard to use and often requires a lot of technical support.
- Wireless data networking enables computers and appliances to communicate through radio signals. This provides added mobility, which is particularly convenient for laptops and hand-held devices.
- Power-line networking uses the home's existing power lines to send data at Ethernet-class rates to and from computing and household appliances. Power-line networking can coexist with already-popular devices that use residential power lines to communicate, including X-10, Cebus, and LonWorks.

Step 3: Choosing the Best Home Network

It is evident that there are a number of viable home networking solutions, from Ethernet to phone lines. The most effective network depends primarily on the products being connected. Each solution presents unique benefits, but some of the available networks do not provide a comprehensive solution that meets all of a consumer's demands, which is why using power line and wireless data simultaneously makes sense in many cases.

One networking option may lie in the marriage of wireless data and power-line networking. This solution combines the ubiquity and pervasiveness of power line with the utility and mobility of wireless data. With power line as the backbone of the network, consumers can plug in their appliances to any power outlet in the home. Users can then set up a wireless data network that allows the mobile appliances[2] to communicate with the static appliances.

Ensuring interoperability and coexistence between networks are the industry alliances that have brought companies together to develop standards-based wireless data networking solutions. One alliance that is bringing power line to the forefront of home networking is the HomePlug Powerline Alliance (http://www.homeplug.org/), which comprises more than 90 companies that are developing an open specification that leverages the wide availability of residential power lines. Industry alliances, like HomePlug, help reduce consumer confusion by creating one industry standard for its specific home networking solution. Operating with one industry standard helps bring the vision of the connected home to a reality by ensuring compatibility within the proliferation of home networking products and services.

While wireless data home networking is in the early-adopter stage of its lifecycle, this is one market that will experience substantial growth. Today, several wireless data networking solutions are being developed by industry alliances, which have brought companies together to create home networking options that represent all segments of the technology market. With an increasingly tech-savvy population using multiple appliances, users will demand a way to organize the mass of information stored on their computing and household devices, and, as the technology continues to advance, more devices will be developed that help people add practicality, leisure, and efficiency to their lives. With the broad adoption of wireless data home networking, you may be living in that futuristic home where your appliances are part of the family in the near future.[1]

Conclusion

Because most families want to use a wireless data home network for communication and entertainment, several different types of companies will be able to break into this burgeoning market. The companies include cable and infrastructure manufacturers such as ITT Industries and Avaya, home automation and home theater installers such as AVS and Panja, and computer and networking manufacturers such as Dell and IBM.

Eventually, this market will expand to include the use of smart appliances and controllers, but most analysts and industry professionals agree that popular use and acceptance of these technologies is still several years away. In the meantime, families can still enjoy some practical uses of home networking today—bringing the sophistication of the future into everyday living.

References

1. Alberto Mantovani, "The Future of Home Networking," Technology Marketing Corporation, One Technology Plaza, Norwalk, CT 06854, 2002.
2. John R. Vacca, *i-mode Crash Course,* McGraw-Hill, 2002.

PART 5

Advanced Wireless High-Speed Data Network Solutions and Future Directions

CHAPTER 22

Residential High-Speed Wireless Data Personal Area Networks

Wireless data personal area networks (WDPANs) enable short-range ad hoc connectivity among portable consumer electronics and communications devices. The coverage area for a WDPAN is generally within a 10-m radius. The term *ad hoc connectivity* refers to both the ability for a device to assume either master or slave functionality and the ease in which devices may join or leave an existing network. The Bluetooth radio system has emerged as the first technology addressing WDPAN applications with its salient features of low power consumption, small package size, and low cost. Wireless data rates for Bluetooth devices are limited to 1 Mbps, although actual throughput is about half this data rate. A Bluetooth communication link also supports up to three voice channels with very limited or no additional bandwidth for bursty wireless data traffic.

The next wave of portable consumer electronics and communications devices will support multimedia traffic that requires high-speed data rates (high-rate traffic). Applications include high-quality video and audio distribution and multimegabyte file transfers for music and image files. Figure 22-1 illustrates a few example devices in a high-rate WDPAN that include digital camcorders and TVs, digital cameras, MP3 players, printers, projectors, and laptops.[1] In addition, the high-rate WDPANs may find compelling applications as a cable replacement technology for residential/home entertainment systems capable of high-definition video and high-fidelity sound, and DVD or high-quality graphics-based interactive games with multiple consoles and virtual reality goggles. The need for communications between these multimedia-capable devices leads to peer-to-peer ad hoc type connections that warrant wireless data rates well in excess of 20 Mbps and quality of service (QoS) provisions with respect to guaranteed bandwidth. To accommodate the required Physical (PHY) layer wireless data rates and medium access control (MAC) layer QoS requirements, the IEEE 802.15 WDPAN Working Group initiated a new group, the 802.15.3 High-Rate WDPAN Task Group. The IEEE 802.15.3 Task Group has been chartered with creating a high-rate WDPAN standard that provides for low-power, low-cost, short-range solutions targeted to consumer digital imaging and multimedia applications. The final version of the IEEE 802.15.3 high-rate WDPAN standard is expected to be approved in 2003. The MAC and PHY layer descriptions presented in this chapter only reflect the ongoing work based on the draft version of the standard. (The Glossary defines many technical terms, abbreviations, and acronyms used in the book.)

There are several alternative wireless data local-area network (LAN) technologies, such as IEEE 802.11a and b and HiperLAN, that are also targeting the use of unlicensed spectrum at 2.4- and 5-GHz bands. Compared to existing wireless data LAN systems, the 802.15.3 high-rate WDPAN technology possesses desirable features suited for portable communications and electronic devices and their applications. The salient characteristics of the IEEE 802.15.3 high-rate WDPAN standard are:

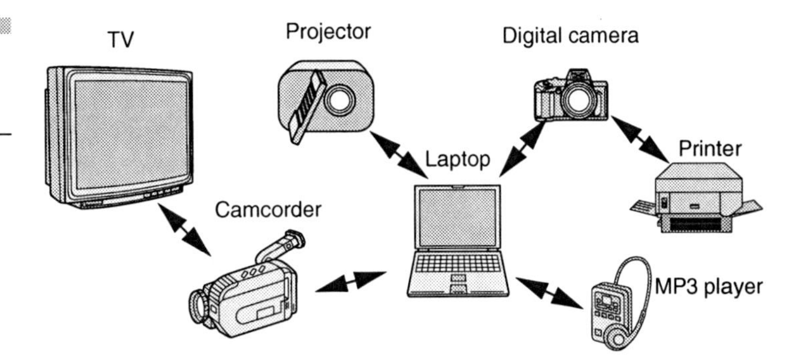

Figure 22-1 High-rate WDPAN target applications.

- Ability to form ad hoc connections with QoS support for multimedia traffic
- Ease of joining and leaving an existing network
- Advanced power management to save battery power
- Low-cost and low-complexity MAC and PHY implementations optimized for short-range (less than 10 m) communications
- Support for high-speed data rates up to 55 Mbps for video and high-quality audio transmissions[1]

Alternatives: IEEE 802.11b, e, and g

The IEEE 802.11 standard offers several wireless data LAN technologies for use in the unlicensed 2.4- and 5-GHz bands. Legacy 802.11 systems operate in the 2.4-GHz band with three different PHY layers sharing the same MAC layer. These PHY layer specifications are the seldom-used infrared (IR) technology and the more popular direct-sequence spread spectrum (DSSS) and frequency-hopping spread spectrum (FHSS) systems achieving 1- and 2-Mbps data rates. Operating under the same 802.11 MAC layer in the 2.4-GHz band, higher wireless data rates are supported by the IEEE 802.11b PHY layer specification with complementary code keying (CCK) modulation, achieving wireless data rates of 5.5 and 11 Mbps. Recently, as previously explained in earlier chapters, a new task group, IEEE 802.11g, has been formed to draft a standard that achieves data rates higher than 22 Mbps. Alternatively, in the 5-GHz band, the IEEE 802.11 standard offers the 802.11a specification that uses orthogonal frequency-division multiplexing (OFDM), achieving data rates up to 54 Mbps. A new task group, 802.11e, has been created to accommodate additional QoS provisions and security requirements at the MAC layer while supporting all of the previously mentioned legacy 802.11 PHY layers.

The IEEE 802.11 MAC protocol has some shortcomings with respect to supporting high-rate WDPAN applications. First, the 802.11 MAC layer is burdened with legacy LAN functionality and applications (not ad hoc). Second, less emphasis has been given to power management in the MAC protocol, since 802.11 wireless data LAN applications were not mainly targeted to battery-operated portable communications devices. Furthermore, the 802.11 MAC specification does not provide adequate QoS guarantees with respect to guaranteed transmission slots for isochronous traffic.

The IEEE 802.11e group is now tasked with creating a new 802.11 MAC protocol by adding on top of the legacy 802.11 MAC layer specification. The proposed 802.11e additions do not include guaranteed QoS in the ad hoc connection mode, which is paramount for high-rate WDPAN applications. Some form of QoS will be available in the coordinated network mode. Even at the completion of the new 802.11e MAC specification, the legacy burden for LAN functionality and the requirement to support numerous PHY layers will likely render the MAC implementations too complex and power-inefficient for the high-rate WDPAN applications. From a PHY layer point of view, the IEEE 802.11b specification supports wireless data rates up to 11 Mbps, which is not nearly sufficient for the WDPAN applications related to video distribution.

Support for wireless data rates higher than 22 Mbps is contemplated by the newly formed task group, IEEE 802.11g. Two leading PHY layer candidates for the 802.11g standard are single-carrier trellis-coded 8-phase shift keying (PSK) modulation and OFDM schemes. Both candidates offer considerably more costly radio and baseband implementations than the 802.15.3 PHY layer. The approval of the 802.11g specification requires an FCC rule change in the FCC Part 15.247 rules, for which a Notice for Proposed Rule Making (NPRM) process is underway.

Finally, the 802.11a technology operating in the unlicensed 5-GHz band supports data rates up to 54 Mbps. However, in addition to the MAC layer inefficiencies described earlier, the OFDM baseband processor consumes too much power and requires too expensive RF front-end implementations for WDPAN applications and devices.

IEEE 802.15.3 High-Rate WDPAN Standard

The 802.15.3 MAC layer specification is designed from the ground up to support ad hoc networking, multimedia QoS provisions, and power management. In an ad hoc network, devices can assume either master or slave functionality, depending on existing network conditions. Devices in an ad hoc network can join or leave an existing network without complicated setup procedures. The 802.15.3 MAC specification has provisions for supporting multimedia QoS. Figure 22-2 illustrates the MAC superframe structure that consists of a network beacon interval, a contention access period (CAP), and a contention-free period (CFP) reserved for guaranteed time slots (GTS).[1] The boundary between the CAP and GTS periods is dynamically adjustable.

A network beacon is transmitted at the beginning of each superframe carrying WDPAN-specific parameters, including power management, and information for new devices to join the ad hoc network. The CAP period is reserved for transmitting non-QoS data frames such as short bursty wireless data or channel access requests made by the devices in the network. The medium access mechanism during the CAP period is collision sense multiple-access/collision avoidance (CSMA/CA). The remaining duration of the superframe is reserved for GTS to carry data frames with specific QoS provisions. The type of wireless data transmitted in the GTS can range from bulky image or music files to high-quality audio or high-definition video streams. Finally, power management is one of the key features of the 802.15.3 MAC protocol, which is designed to significantly lower the current drain while being connected to a WDPAN. In the power save mode, the QoS provisions are also maintained.

The 802.15.3 PHY layer operates in the unlicensed frequency band between 2.4 GHz and 2.4835 GHz, and is designed to achieve wireless data rates of 11 to 55 Mbps that are commensurate with the distribution of high-definition video and high-fidelity audio. The 802.15.3 systems employ the same symbol rate, 11 Mbaud, as used in the 802.11b systems. Operating at this symbol rate, five distinct modulation formats are specified, namely, uncoded quadrature PSK (QPSK) modulation at 22 Mbps and trellis-coded QPSK and 16/32/64 quadrature amplitude

Figure 22-2
An IEEE 802.15.3 MAC superframe.

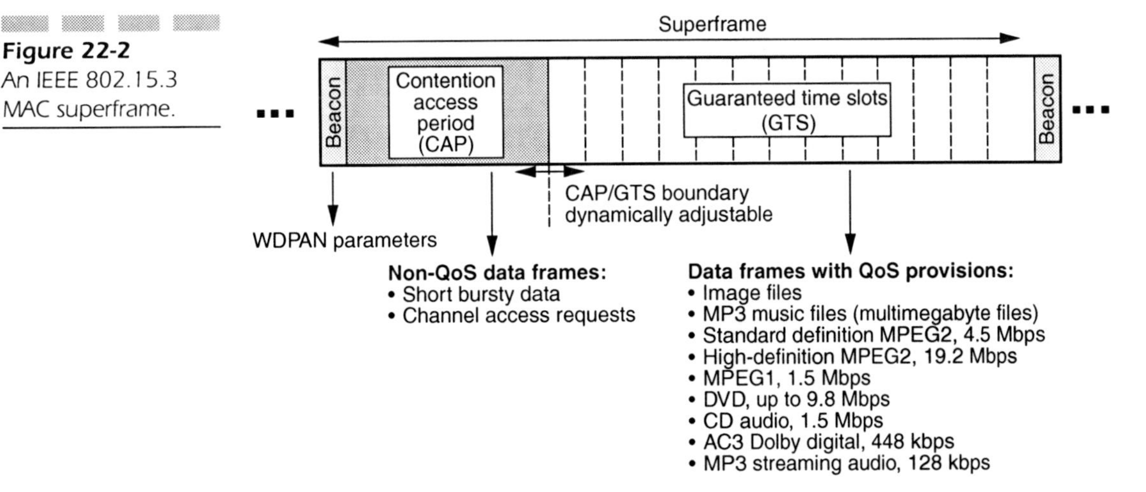

modulation (QAM) at 11, 33, 44, 55 Mbps, respectively (TCM). The base modulation format is QPSK (differentially encoded). Depending on the capabilities of devices at both ends, the higher data rates of 33 to 55 Mbps are achieved by using 16/32/64 QAM schemes with 8-state 2D trellis coding. Finally, the specification includes a more robust 11-Mbps QPSK TCM transmission as a dropback mode to alleviate the well-known hidden node problem. The 802.15.3 signals occupy a bandwidth of 15 MHz, which allows for up to four fixed channels in the unlicensed 2.4-GHz band. The transmit power level complies with the FCC 15.249 rules with a target value of 0 dBm.

The RF and baseband processors used in the 802.15.3 PHY layer implementations are optimized for short-range transmission limited to 10 m, enabling low-cost and small-form-factor MAC and PHY implementations for integration in consumer devices. The total system solution is expected to fit easily in a compact flash card. The PHY layer also requires low current drain (less than 80 mA) while actively transmitting or receiving data and minimal current drain in the power save mode.

Table 22-1 summarizes the key features of the IEEE 802.15.3 standard in comparison to existing IEEE 802.11a, b, and g and Bluetooth standards.[1] As shown, the 802.15.3 devices draw much less current while active than 802.11 implementations, due to shorter-range transmission requirements. From an ad hoc networking point of view, it is important that devices have the ability to connect to an existing network with a short connection time. The 802.15.3 MAC protocol targets connection times much less than 1 s. Reviewing the regulatory requirements, it should be noted that the operation of WDPAN devices in the 2.4-GHz band is highly advantageous since these devices cannot be used outdoors in Japan while operating in the 5-GHz band. The outdoor use of most portable WDPAN devices prohibits the use of the 5-GHz band for worldwide WDPAN applications.

IEEE 802.15.3 Physical Layer Modulation and Coding

The IEEE 802.15.3 PHY layer standard specifies the QAM signal constellations illustrated in Fig. 22-3.[1] The symbol rate is 11 Mbaud and applies to all specified modulation formats. Consequently, the raw PHY layer data rates are 22 Mbps for uncoded QPSK modulation, and 11, 33, 44, and 55 Mbps for trellis-coded QPSK and 16/32/64 QAM, respectively, due to 1-b/symbol redundancy introduced by trellis coding.

A 2D 8-state (2D-8S) trellis code is applied to the QPSK and 16/32/64 QAM signaling schemes. The implementation of the 2D-8S QPSK and 16/32/64 QAM TCM encoder is illustrated in Fig. 22-4.[1] In the 16/32/64 QAM TCM modes, the lower-order 3 bits select one of the eight subsets,

TABLE 22-1

IEEE 802.15.3 versus Others

	802.15.3	802.11b,g*	802.11a	Bluetooth 1.1
Frequency band, GHz	2.4	2.4	5	2.4
Data rate, Mbps	Up to 55	Up to 22	Up to 54	1
Current drain, mA	<80	<350	>350	<30
Number of video channels	4	2 (see QoS)	5 (see QoS)	None
Range, m	10	100	100	10–100
Regulatory:				
North America	FCC 15.249	802.11g requires FCC rule change	FCC 15.407	FCC 15.247
Europe		ETSI 300.328		
Japan		RCR-STD-T66/33A, no outdoor use at 5 GHz		
Complexity (area)	1.5X	3X	4X	1X
Connect time, s	≪1	N/A	N/A	~5
QoS	Guaranteed time slots for multimedia	802.11e patched QoS (TBD) with legacy	LAN support	No video support

*802.11g is expected to support >22 Mbps.
NA = not available.

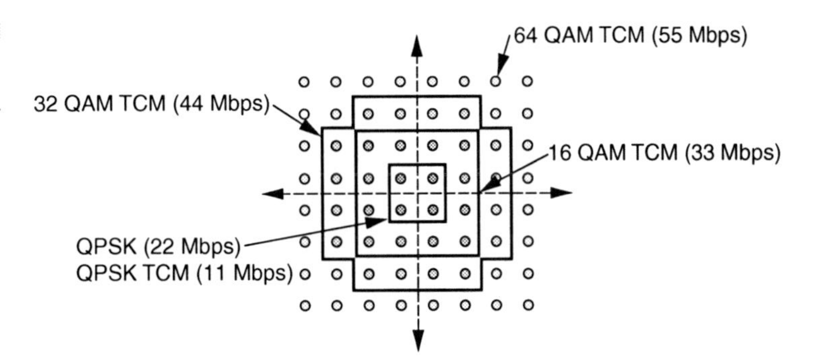

Figure 22-3 Signal constellations.

D0, D1, . . . , D7, which are generated by applying a set-partitioning rule to the signal constellations shown in Fig. 22-3. Each subset contains 2, 4, and 8 symbols for 16/32/64 QAM constellations, respectively. The remaining higher-order bits (1, 2, and 3 bits for 16/32/64 QAM, respectively) select one of the symbols from each subset. In the QPSK TCM mode, the lower-order 2 bits select one of the four subsets, D0, . . . , D3, each containing a single symbol. The set-partitioning concept is designed such that the minimum squared euclidean distance between symbols increases at each stage of the set-partitioning process. For the 16/32/64 QAM TCM schemes, the encoder determines the next symbol to be transmitted from one of the subsets, D0, D1, . . ., D7, based on a state transition rule governed by the 8-state finite state machine (FSM), shown in Fig. 22-4. Four subsets are assigned to each state. Consequently, there are four possible transitions from each state. The transmitted symbol associated with a particular transition contains 2 bits due to the number of possible transitions from a state plus 1, 2, or 3 bits due to selecting one of the 2, 4, and 8 symbols in that particular subset for the 16/32/64 QAM, respectively. Therefore, 16/32/64 QAM TCM carry 3, 4, and 5 b/symbol, respectively. For the QPSK TCM scheme, two subsets are assigned to each state, resulting in two possible transitions from each state. Since each subset contains a single symbol, the transmitted QPSK TCM symbol carries only 1 information bit.

IEEE 802.15.3 Physical Layer Frame Format

The IEEE 802.15.3 PHY layer frame format consists of a preamble, payload, cyclic redundancy check (CRC), and a trellis tail, as illustrated in Fig. 22-5.[1] The preamble contains 10 periods of a special constant-amplitude

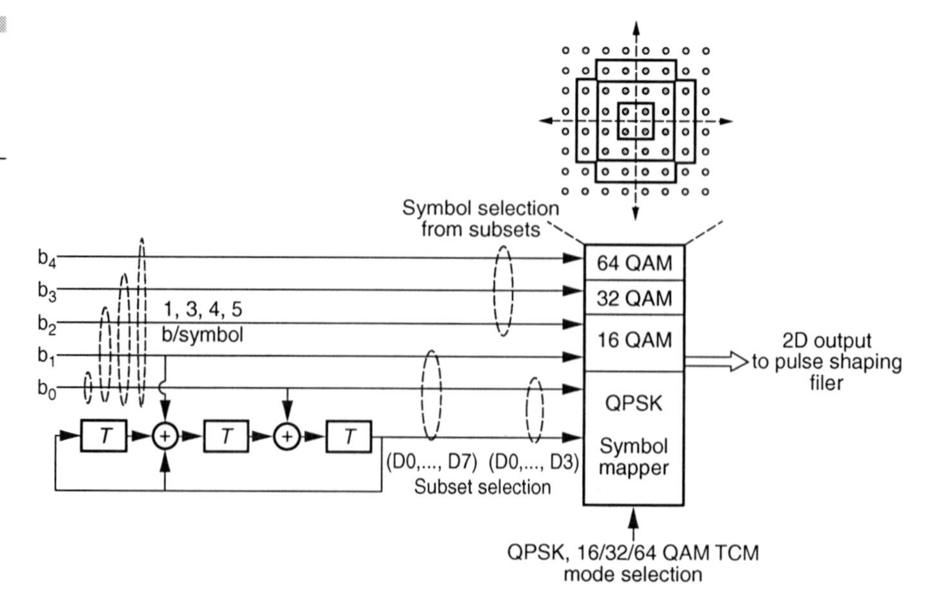

Figure 22-4 An 8-state QPSK and 16/32/64 QAM TCM encoder.

zero autocorrelation (CAZAC) sequence. Each period of the CAZAC sequence contains 16 QPSK symbols. The preamble is used for various signal acquisition functions such as gain adjustment, carrier frequency offset compensation, symbol timing adjustment, channel estimation, and calculation of equalizer coefficients. The length of the preamble and periodicity thereof depend on the length of the overall symbol response. The 16-symbol period is sufficient to handle an overall symbol response of 1.5 μs, which includes all of the signal shaping filters and the effects of the delay-spread channel. The payload contains up to 2048 bytes. It is followed by CRC bits and a trellis tail. To aid in the sequence decoding process in the receiver, a 2-symbol interval trellis tail is appended to the end of every packet to terminate the TCM code at a known state. The trellis tail extends to 3 symbols in the QPSK TCM mode.

Receiver Sensitivity

IEEE 802.15.3 receiver sensitivity is defined as the minimum received signal level that guarantees a bit error rate (BER) performance of less than 10^{-5} in the presence of additive white gaussian noise (AWGN). The 802.15.3 receiver has a noise bandwidth of 11 MHz, which determines the amount of AWGN power present in the receiver. A receiver noise figure of 12 dB is assumed for sensitivity calculations. The receiver sensitivities required by the QPSK TCM, QPSK, and 16/32/64 QAM TCM (11, 22, 33, 44, 55 Mbps) transmission modes are −82, −75, −74, −71, and −68 dBm, respectively. In terms of the signal-to-noise ratio (*S/N*) required while operating at a BER of less than 10^{-5}, *S/N* values of 5.5, 12.6, 13.5, 16.6, and 19.8 dB are necessary for the QPSK TCM, QPSK, and 16/32/64 QAM TCM schemes.

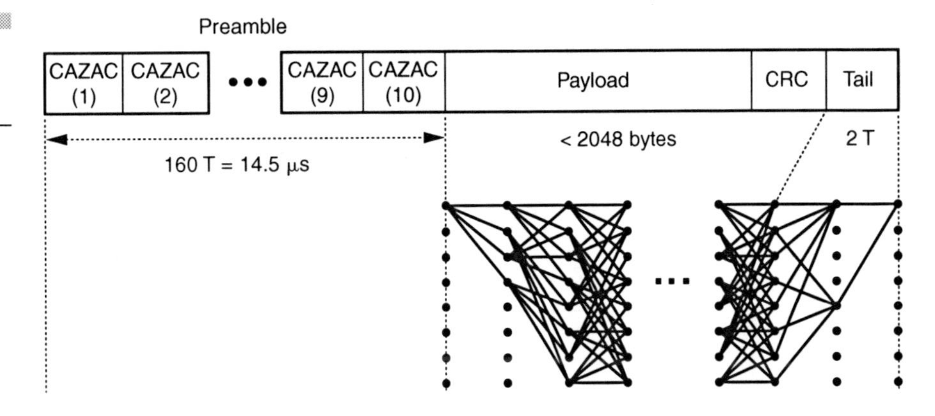

Figure 22-5 Variable-length PHY layer frame format.

Characteristics of Short-Range Indoor Propagation Channels

A short-range indoor propagation channel in the unlicensed 2.4-GHz band can be described by an exponentially decaying (delay spread) Rayleigh fading channel model. The impulse response of the channel consists of random complex samples with a uniformly distributed phase and Rayleigh distributed magnitude, where the average magnitude response decays exponentially with time. The channel characteristics can change drastically from one transmission burst to another because of both the movement of the communication devices and the objects within the transmission range. The frequency response of the channel may exhibit deep and wide spectral notches due to multipath propagation. Within the 2.4-GHz industrial, scientific, and medical (ISM) band, the 802.15.3 signal occupies a fixed bandwidth of 15 MHz. Several spectral notches may be present within the 15-MHz signal bandwidth in certain portions of the unlicensed 2.4-GHz band. As noted earlier, the placement and severity of these spectral notches change from one transmission burst to another because of the movement of objects in the medium.

Figure 22-6 illustrates a typical overall symbol response of an 802.15.3 signal.[1] The overall symbol response is the combination of a raised cosine pulse with 11-Mbaud symbol rate and 30 percent excess bandwidth, and a single instantiation of the exponentially decaying Rayleigh fading indoor propagation channel. The effects of the Rayleigh fading delay spread channel on the overall symbol response are clearly noticeable from the two large spectral notches within signal bandwidth.

IEEE 802.15.3 Receiver Performance

Indoor propagation channels pose serious challenges in establishing reliable high-data-rate communications. As described previously, these challenges manifest themselves in the form of severe fading and multipath propagation, resulting in multiple spectral notches within the signal spectrum. Compounding the challenge is the time-varying nature of the channel from one transmission burst to another. Use of equalization in the receiver proves to be an effective means to mitigate the effects of distortion introduced on the signal by the indoor propagation channel. Furthermore, equalization needs to be performed on a burst-by-burst basis because of the time-varying nature of the channel. To facilitate rapid equalization in the receiver, the 802.15.3 standard specifies the use of a periodic CAZAC preamble, as shown in Fig. 22-5.

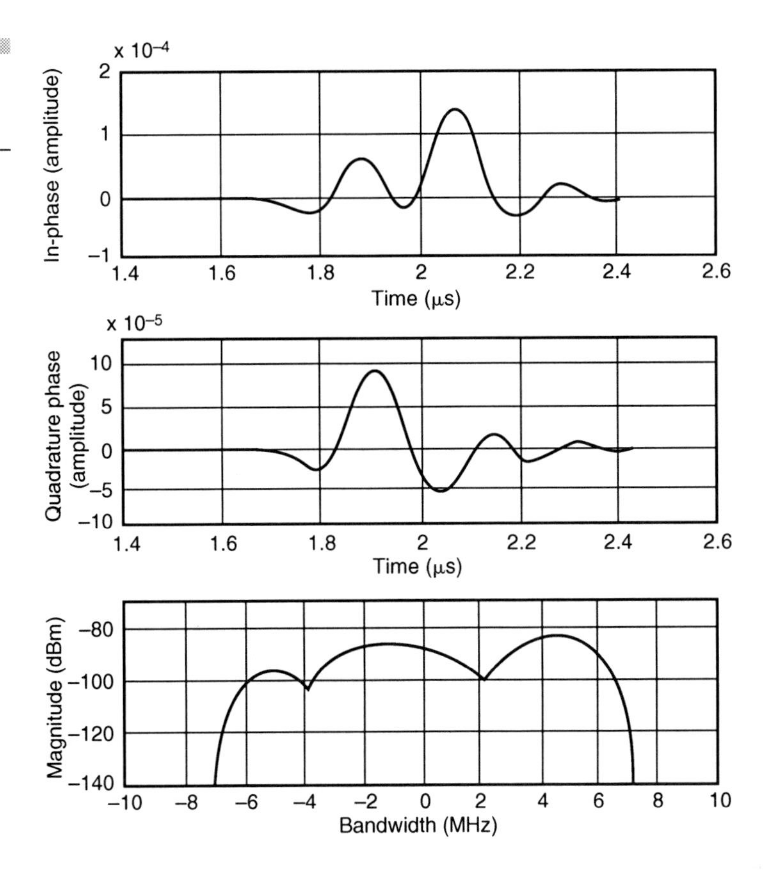

Figure 22-6 802.15.3 overall symbol response.

A receiver based on a minimum mean square error decision feedback equalizer (MMSE-DFE) is simulated to evaluate the performance of an 802.15.3 system. The MMSE-DFE architecture is modified appropriately in order to decode the 2D-8S trellis-coded transmitted signal. Simulations used in the evaluation of the system incorporate the effects of carrier frequency and symbol timing offsets. The carrier frequency and symbol timing offsets included in the simulations are −300 kHz and −25 ppm, respectively.

First, the performance of the MMSE-DFE is evaluated in terms of the *S/N* observed at the equalizer output. Since the channel under consideration is time-varying on a burst-by-burst basis, the probability distribution of *S/N* performance is obtained against 1000 instantiations of the Rayleigh fading channel with an RMS (root mean square) delay spread of 75 ns. The average received signal level is −59 dBm, which is measured over 1000 simulation runs using the Rayleigh fading delay spread channel. The combined simulation results for an MMSE-DFE with eight feedforward taps and four and six feedback taps indicated that, while eight taps are sufficient for the feedforward filter, the feedback filter requires at least six taps to ensure good performance. The 64 QAM/TCM modulation scheme requires an *S/N* of 19.8 dB to operate at a data rate of 55 Mbps with a BER of 10^{-5} in the presence of AWGN. An *S/N* of 20 dB or better was achieved for 98 percent of the Rayleigh fading channels simulated with up to 75 ns RMS delay spread. In practice, short-range indoor propagation channels typically introduce an RMS delay spread of 25 ns. Therefore, you can conclude that even at the highest data rate, 55 Mbps, supported by the 802.15.3 standard, a reliable communication link can be established for 98 percent of the channels encountered.

Another meaningful performance measure in the receiver is to determine the frame error rate (FER) performance with respect to various received signal levels relative to the receiver sensitivities. Due to the time-varying nature of the Rayleigh fading channel on a burst-by-burst basis, the FER results have been obtained by using thousands of instantiations of the randomly generated Rayleigh fading channel response. The frame size used in the simulations is 8192 bits (1024 bytes). The FER performance is measured against several received signal levels that are 7 to 15 dB higher than the receiver sensitivity required for the particular modulation scheme simulated. In the simulations, an RMS delay spread of 25 ns is used.

Figure 22-7 plots the FER performance of the 64 QAM/TCM scheme operating at 55 Mbps against average received signal levels of −61, −58.9, −57.1, −54.9, and −53.2 dBm.[1] The vertical axis represents the probability that the FER is less than the corresponding amount shown in the horizontal axis. The 802.15.3 standard targets an FER of better than 1 percent for greater than 95 percent of the channels encountered at received signal levels of 10 dB above the receiver sensitivity. Figure 22-7 shows that an FER of 1 percent is ensured for greater than 97 percent of the channels simulated at a received signal level of 10 dB higher than the −68-dBm receiver sensitivity required by the 64 QAM/TCM scheme (55 Mbps).

NOTE For all modulation formats simulated, an FER of 1 percent was ensured for almost all of the channels simulated when the received signal level is 15 dB higher than the receiver sensitivity required by the specific modulation scheme under consideration.

Conclusion

This chapter presented an overview of high-rate wireless data personal area networks, their targeted applications, and a technical overview of medium access control and physical layers, and system performance.

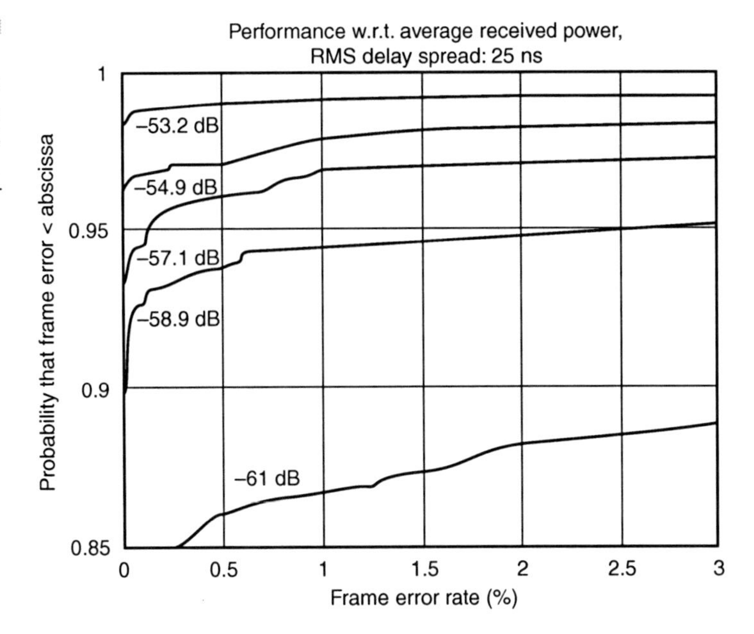

Figure 22-7 Probability distribution of FER versus average received power for 64 QAM/TCM (55 Mbps).

The high-rate WDPANs operate in the unlicensed 2.4-GHz band at data rates up to 55 Mbps that are commensurate with distribution of high-definition video and high-fidelity audio. An industry effort to create a MAC and PHY layer standard specification for high-rate WDPANs has been ongoing in the IEEE 802.15.3 High-Rate WPAN Task Group.

High-rate wireless data personal area networks enable compelling multimedia applications by establishing connectivity among consumer electronics and portable communications devices. These devices and supported applications require certain technology features, such as the ability to form ad hoc connections with QoS guarantees, support for data rates commensurate with video distribution, power management capability, and low current drain. The IEEE 802.15.3 High-Rate WDPAN Standard addresses these features from QoS to high wireless data-rate support while maintaining both power- and cost-efficient implementations.

Establishing high-data-rate WDPANs can be challenging because of the nature of time-varying indoor communications channels that cause significant delay spreads and severe fading. The use of equalization and coding mitigates the adverse effects of the indoor communications channels to a large extent. Simulation results showed that even at a data rate of 55 Mbps, reliable communications can be established for 98 percent of the Rayleigh fading channels simulated with an RMS delay spread of 25 ns.

References

1. Jeyhan Karaoguz, "High-Rate Wireless Personal Area Networks," *IEEE Communications Magazine,* 445 Hoes Lane, Piscataway, NJ 08855, 2002.
2. John R. Vacca, *The Essential Guide to Storage Area Networks,* Prentice Hall, 2002.

CHAPTER 23

Summary, Recommendations, and Conclusions

Wireless data communications has emerged as one of the largest sectors of the telecommunications industry, evolving from a niche business in the last decade to one of the most promising areas for growth in the twenty-first century. This book explored the key technological advances and approaches that are now emerging as core components for wireless data solutions of the future.

Summary

The 1990s were a period of tumultuous growth for the wireless data communications industry, and few could have predicted the rapid rise of many of today's key players that chose "winning" approaches and technologies. Likewise, there were some amazing and startling failures in the wireless data sector, despite the brilliant engineering and technological efforts that went into their formations.

One of the most successful wireless data communications technologies of the previous decade was code-division multiple access (CDMA), pioneered by Qualcomm, Inc. Qualcomm introduced its CDMA concept for mobile radio in 1990, at a time when the U.S. cellular industry was selecting its first digital mobile telephone standard.

To appreciate the growth of the wireless data sector, it is worth noting that in 1990, there were only 10 million cell phone subscribers worldwide, mostly using analog FM (first-generation) technology. Today, there are approximately 800 million subscribers, and this is expected to increase to more than 3 billion subscribers in the 2007–2008 time frame. In China alone, more than 26 million cell phone subscribers are being added each month, more than the cumulative number of wireless data subscribers that existed throughout the entire world in 1991.

Just prior to Qualcomm's introduction of its wideband digital CDMA mobile radio standard in 1990, now known as IS-95, the U.S. cellular industry was poised to select TDMA (which became IS-136) as the digital successor to the analog AMPS standard. The European community had already adopted GSM for its own pan-European digital cellular standard a couple of years earlier, and Japan's popular second-generation digital TDMA standard, PDC (Pacific Digital Cellular), was introduced shortly after IS-136's acceptance in the United States. As cellular telephone service caught on with consumers, governments across the world auctioned additional spectrum [the personal communications system (PCS) spectrum] to allow new competitors to support even more cellular telephone subscribers. The PCS spectrum auctions of the mid-1990s created a vast increase in frequencies for cellular telephone providers across the globe, thereby providing the proving ground for the second generation of cellular technology (2G, the first generation of digital modulation technologies). (The Glossary defines many technical terms, abbreviations, and acronyms used in the book.)

While the pioneering design of GSM, which included international billing, short messaging features, and network-level interoperability, now enjoys the lead in today's global wireless data market, it is also evident that wireless data CDMA was a breakthrough technology, offering increased wireless data capacity by increasing channel bandwidth and moving complexity in the handset to low-cost baseband signal processing circuits. All proposed third-generation wireless data standards (except for EDGE) use some form of CDMA (see Fig. 23-1), and the number of subscribers using the major second-generation technologies (see Fig. 23-2) clearly show CDMA and GSM as the two leading worldwide technology standards. In fact, within the past year major wireless carriers in Japan and the United States announced they were abandoning IS-136/PDC technology in favor of newer third-generation standards that have a core wideband CDMA component. While CDMA was an example of a breakthrough technology of the past decade, there were many other brilliant system concepts that ultimately failed.

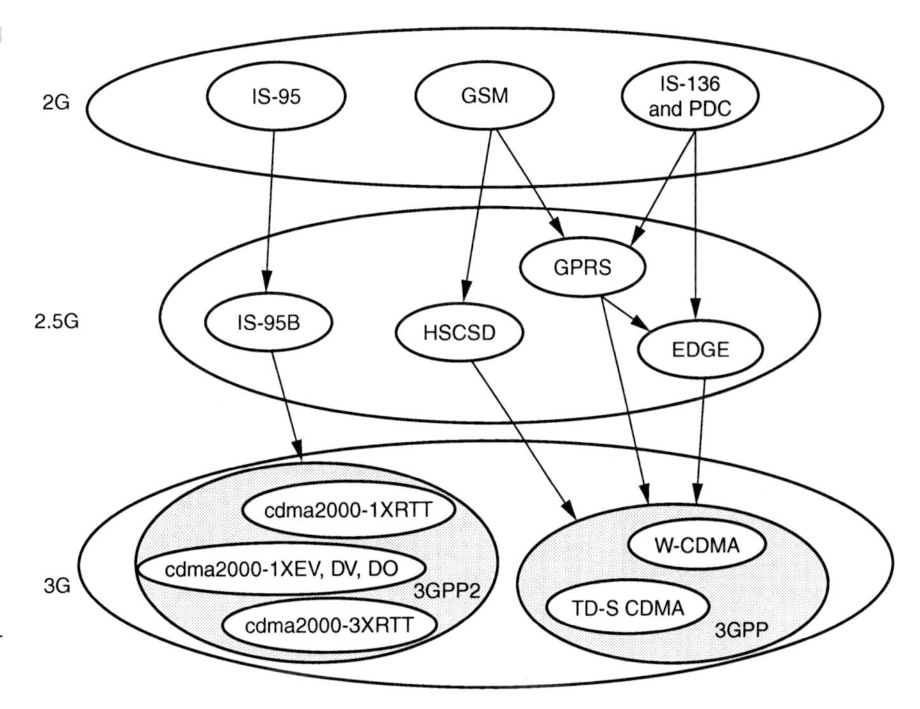

Figure 23-1 Cellular/PCS technologies and their evolution to 3G. The "alphabet soup" of wireless standards continues into the third generation of cellular phones. First-generation analog FM systems of the 1980s gave way to second-generation (2G) standards in the 1990s. Today, 2.5G standards are being rolled out, and 3G is in its infancy, waiting for better economic conditions.

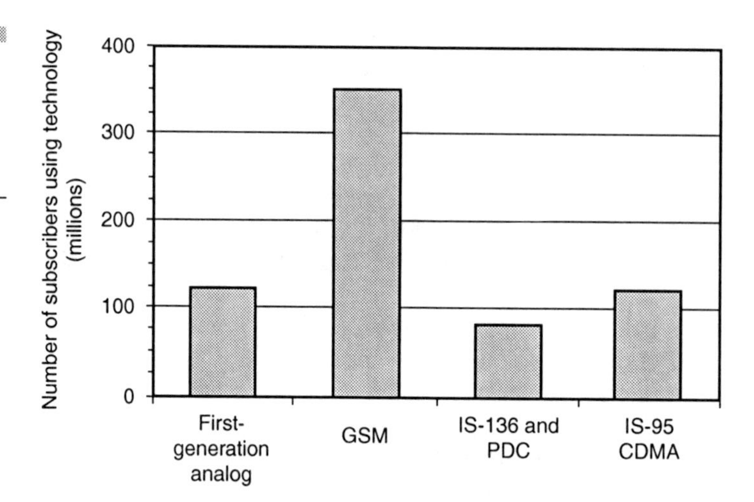

Figure 23-2 Number of subscribers of key wireless technologies in late 2001.

The vision of anytime, anywhere communications was championed by two companies that ultimately declared bankruptcy, although both companies were ahead of their time. Iridium (and companies like it) attempted to provide satellite-based wireless data communications throughout the globe, using cellular telephone concepts from space, whereas Metricom attempted to provide a nationwide service of always-on wireless data in metropolitan areas using Internet Protocol connectivity over a large network of low-power devices operating in an unlicensed spectrum.

In the case of Iridium, the cost to build and deploy a complete network of medium earth orbit (MEO) satellites and ground stations was enormous, in the many billions of dollars; the relatively slow early-adoption rate of customers made it impossible to pay back the debt service for the initial infrastructure quickly enough. Pricing of the now defunct worldwide space-based global roaming telecommunications service hovered around $3 US per minute, making it prohibitively expensive for the mass consumer market. Nevertheless, the technological breakthroughs pioneered by Iridium in space-based handoffs, spot-beam antenna technology, power-efficient engineering, handset engineering, and network management were truly extraordinary.

Metricom pioneered the vision of always-on tetherless network access, and offered the first glimpse at ubiquitous wireless data Internet access for users on the move. Metricom successfully deployed its Ricochet packet-based wireless data service in many metropolitan areas, providing its customers with 64- to 128-kbps peak wireless data throughput (and even greater in some cities) by using the license-free ISM bands and an extensive network of radio repeaters, relay stations, and network servers. The Ricochet infrastructure was installed on thousands of buildings, lampposts, and broadcast towers in select cities, and provided high-quality wireless data access and Internet for mobile and portable users of personal computers. Metricom was ahead of its time, as it built and operated one of the first examples of an ad hoc wireless data network for packet-based data access, years ahead of the 2.5G cellular/PCS technologies that are just now rolling out their medium- and high-data-rate solutions. Ultimately, Metricom was forced to file for bankruptcy in 2001, unable to justify the mounting debt incurred from aggressive build-out plans. The network infrastructure and subscriber equipment were costly, and subscribers were slow to adopt the service.

Another company with an exciting public wireless data Internet vision was Mobilestar, best known for its public WLAN access deployed in Starbucks coffee shops throughout the world. Voicestream Wireless recently purchased the assets of Mobilestar and may be exploring WLAN service as an augmentation to its conventional PCS wireless data business.

There are many other examples of both successes and failures in the past decade. The wireless local-area network (WLAN) industry, for example, is an exciting and emerging bright spot for enterprise networking within and between buildings through the use of unlicensed frequencies, whereas the collapse of several promising wireless data competitive local exchange carriers (W-CLECs) and wireless Internet service providers (W-ISPs) is a further example of businesses that were ahead of their time (and which may someday stage a comeback with the IEEE 802.16 wireless data metropolitan-area network standard) or that faced difficult or expensive access to the incumbent carrier's customers, in addition to brutal capital market conditions.

As of this writing, the telecommunications industry is undergoing an economic depression. Access to capital has been extremely difficult and valuations of several telecommunication companies have sunk by 90 percent or more since 2001. Thousands of companies have either been forced to file for bankruptcy or have jettisoned slow-growth or money-losing businesses in order to survive. Many wireless experts, some of the leading contributors to the wireless data field, are out of work or are seeking jobs elsewhere. While many new technologies abound, those that are now successful were implemented at a time when capital was readily available and governments throughout the world provided spectrum for broad adoption of new services. The availability of PCS spectrum throughout the world, for example, created the opportunity for companies such as Qualcomm to gain a foothold in the worldwide market with their CDMA concept. On the flip side, the U.S. Telecommunications Act of 1996 promised a competitive landscape that proved to be financially untenable for most new entrants, after all.

Recommendations

As you consider what may influence the wireless data technology landscape in the coming decade, you must know by now that quite often the past is prologue. The winning technologies will require a new or existing spectrum allocation to allow them to be readily adopted. There must be access to capital, meaning that it is most likely that well-heeled competitors and deep-pocketed incumbents will be involved in some way in breakthrough advances. Perhaps most important, you must consider the Internet, which was not even part of the wireless data industry's thinking through most of the 1990s. The Internet, and the widespread demand for always-on access to data, is sure to be a major driver for the wireless data industry in the coming years.

The fact that the Internet is now universally popular suggests that someday wireless data networks will be made to behave in a fashion similar to today's packet-based networks and computing devices, just as early cell phones were made to emulate the functionality of wired phones. Ad hoc networking, where users and routers move randomly throughout a network, is growing as an important research field and represents a technology that is in its early stages, but which promises to extend portable access and improve emergency communications. To date, wireless data networks have been designed with distinct approaches at the lowest and highest levels of the OSI network-layer model, with the view that base stations are fixed in position with unlimited access to bandwidth. Ad hoc networks of the future, however, will merge immediate knowledge of the physical and MAC layers with adaptive strategies at the higher-level networking layers, so that future networks can be rapidly optimized for performance at specific instances of time, using resources and connection points that may be moving or limited in bandwidth.

In today's conventional wireless data networks, where the network access points are fixed and connected to broadband backbones,[2] the quest for greater data rates, as evidenced by the WLAN industry's move to IEEE 802.11a/g 54-Mbps wireless data rates, suggests that where wireless data are concerned, more is better, especially in and around homes and buildings. A number of exciting technologies in this area are evolving, and promise to make a large impact on the wireless data landscape in the coming decade. Ultra-wideband (UWB), which was just recently approved by the FCC for a number of communications and sensing applications, is an intriguing signaling method that relies on the fabrication of ultrashort baseband pulses that have enormous bandwidths, on the order of several gigahertz. Unlike conventional wireless data systems that upconvert baseband signals to radio-frequency (RF) carriers, UWB can be used at baseband and can be thought of as a baseband transmission scheme that happens to propagate at RF frequencies. UWB has been demonstrated to provide reliable wireless data rates exceeding 100 Mbps within buildings, with extremely low power spectral densities.

Another exciting development, particularly applicable to home or campus wireless data distribution, is the commercialization of orthogonal frequency-division multiplexing (OFDM). OFDM offers multiple access and signal processing benefits that have not been available in previous modulation methods. It allows wireless data networks to pack high spectral efficiency into relatively small spectrum bandwidths. This is similar to how digital subscriber line (DSL) technology allows high wireless data rates to be passed through low-bandwidth copper cables. IEEE 802.16 point-to-multipoint MAN wireless data networks certainly could provide tetherless broadband access in the local loop, and are already doing so in developing nations.

New discoveries in the 1990s have shown us how to exploit the spatial dimension of wireless data channels through the use of multiple antennas at the transmitter and receiver, where significant gains in either energy efficiency or (more important, perhaps) spectral efficiency can be obtained. Pioneering work showed that the theoretical wireless data rates obtained with such systems in an independent Rayleigh scattering environment increase linearly with the number of antennas, and these rates approach 90 percent of the theoretical maximum Shannon capacity. New space-time methods have been shown to offer more than an order of magnitude of increase in spectral efficiency over today's modulation and coding techniques used in current WLANs and cell phone systems, and these methods hold promise for wireless data networks of the future. As an example, Lucent's V-BLAST laboratory prototype system was demonstrated to provide spectral efficiencies of 20 to 40 bps/Hz at average signal-to-noise ratio ranging from 24 to 34 dB in an indoor environment, and potential capacities on the order of 60 to 70 bps/Hz were demonstrated at 30-dB *S/N* using 16 antennas at both the transmitter and receiver.

Now, let's explore in more detail some of the exciting technologies previously listed, and postulate how they may be deployed in networks of the future. Some of these new technologies will require new spectrum allocations in order to succeed, and some may exploit already congested spectrum through the promise of greater capacity. Yet, some of these ideas may still be ahead of their time, and may need to wait another decade or so to gain widespread acceptance.

Indoor Access: The Wireless Data Frontier

It is only when sitting, studying, or concentrating that human beings are most able to use large bandwidths, and this activity happens primarily inside buildings. Just like watching a movie or television, the absorption of wireless data is primarily a passive activity, occurring at home or at work while you sit or stand in a pseudostationary position. Yet, the entire wireless data industry, as you know it today, was originally developed for mobile voice users, for people traveling in cars between home and work, before the Internet was even available to the public.

Internet usage has exploded because of consumer and business adoptions inside buildings using fixed connectivity provided by Internet service providers (ISPs) that team with a local exchange carrier, a long-distance company, or a cable company to gain access to each home. By stark contrast, wireless data carriers have spent huge amounts of capital to purchase spectrum licenses and to deploy infrastructure for outdoor mobile coverage, and have historically had difficulty penetrating their signal into buildings or homes. Furthermore, all current second-generation digital wireless data technologies were developed with a voice-centric architecture, before the widespread acceptance of the Internet, leaving all wireless data carriers vulnerable to each other and to alternative providers who can provide reliable voice and wireless data service into buildings. The battle for indoor wireless data access, where broadband data will be most needed and wanted, is shaping up to be one of the most important industry issues in the coming decade. Cellular and PCS operators desperately need third-generation Web-centric wireless data equipment that can provide Internet-like capabilities in the hands of its consumers inside buildings, as much to reduce subscriber churn as to offer new services, yet most carriers do not have existing infrastructure to provide indoor coverage or capacity reliably for today's more primitive cellular technology. This offers an opening for a new type of competitor that can exploit the availability of low-cost, license-free wireless LAN (WLAN) equipment.

By using the existing wired Ethernet infrastructure within a building or campus, WLANs are being deployed rapidly and inexpensively today, providing tetherless computer access with wireless data rates over an order of magnitude greater than those promised by much more expensive 3G cellular equipment. As voice over IP technology is improved, it is conceivable that WLANs could offer mobile/portable wireless data service that integrates phone-like features with Internet access throughout a campus without any reliance upon the cellular infrastructure.

Today, many early stage companies are looking at ways to integrate 2.5G and 3G cellular technology with WLAN technology, in order to provide coverage and capacity distribution systems for any carrier that wishes to penetrate campuses or buildings. Phones are now being built that combine WLAN and cellular capabilities within them, as a way to ensure connectivity for either type of indoor service.

Dual-mode chip sets for cellular mobile and WLAN are already becoming available from Nokia and other sources, and Intel and Microsoft (two titans steeped in software and semiconductors) recently announced a joint venture to make a new generation of cell phone. Where in-building wireless data connectivity is concerned, WLANs and their existing, widely installed IP-based wired network infrastructure may soon become a serious contender to the radio-centric cellular/PCS carriers of today who are just now seriously addressing the need for connectivity and capacity inside buildings. Moreover, WLANs are extending to campus-size areas and in outdoor venues such as tourist attractions and airports.

Multiple Access: The Universal Acceptance of CDMA

Code-division multiple access (CDMA) allows multiple users to share the same spectrum through the use of distinct codes that appear as noise to unintended receivers, and which are easily processed at baseband for the intended receiver. The introduction of CDMA seemed to polarize service providers and network system designers. On the one side, there were those who saw CDMA as a revolutionary technology that would increase cellular capacity by an order of magnitude. On the other side, there were the skeptics who saw CDMA as being incredibly complex, and not even viable. While CDMA did not immediately realize a tenfold capacity increase over first-generation analog cellular, it has slowly won over skeptics and is the clear winner in the battle of technologies, having emerged as the dominant technology in third-generation cellular standardization (see Fig. 23-1).[1] Furthermore, CDMA techniques have also been adopted for many consumer appliances that operate in unlicensed bands, such as WLANs and cordless phone systems. Early indications are that ultra-wideband technology may also rely on CDMA for multiple access, thereby completing the domination of CDMA as a wireless data technology.

Wireless Data Rates: Up, Up, and Away!

The next decade (starting in 2010) will finally see high-speed wireless data come to maturity. A key to making this a reality will be spectral efficiencies that are an order of magnitude greater than what is seen today. At the Physical layer, three technologies will play a role in achieving these efficiencies: orthogonal frequency-division multiplexing, space-time architectures, and ultra-wideband communications.

Orthogonal Frequency-Division Multiplexing and Multicarrier Communications

Orthogonal frequency-division multiplexing (OFDM) is a special form of multicarrier transmission in which a single high-speed wireless data stream is transmitted over a number of lower-rate subcarriers. While the concept of parallel wireless data transmission and OFDM can be traced back to the late 1950s, its initial use was in several high-frequency military systems in the 1960s such as KINEPLEX and KATHRYN. The discrete Fourier transform implementation of OFDM and early patents on the subject were pioneers in the early 1970s. Today, OFDM is a strong candidate for commercial high-speed broadband wireless data communications, as a result of recent advances in very large scale integration (VLSI) technology that make high-speed, large-size fast Fourier transform (FFT) chips commercially viable. In addition, OFDM technology possesses a number of unique features that make it an attractive choice for high-speed broadband wireless data communications:

- OFDM is robust against multipath fading and intersymbol interference because the symbol duration increases for the lower-rate parallel subcarriers. For a given delay spread, the implementation complexity of an OFDM receiver is considerably less than that of a single carrier with an equalizer.
- OFDM allows for an efficient use of the available radio-frequency (RF) spectrum through the use of adaptive modulation and power allocation across the subcarriers that are matched to slowly varying channel conditions using programmable digital signal processors, thereby enabling bandwidth-on-demand technology and higher spectral efficiency.
- OFDM is robust against narrowband interference, since narrowband interference only affects a small fraction of the subcarriers.
- Unlike other competing broadband access technologies, OFDM does not require contiguous bandwidth for operation.
- OFDM makes single-frequency networks possible, which is particularly attractive for broadcasting applications.[1]

In fact, over the past decade, OFDM has been exploited for wideband data communications over mobile radio FM channels, high-bit-rate digital subscriber lines (HDSL) up to 1.6 Mbps, asymmetric digital subscriber lines (ADSL) up to 6 Mbps, very high speed subscriber lines (VDSL) up to 100 Mbps, digital audio broadcasting, and digital video broadcasting. More recently, OFDM has been accepted for new wireless local-area network standards, which include IEEE 802.11a and IEEE 802.11g, providing data rates up to 54 Mbps in the 5-GHz range, as well as for high-performance local-area networks such as HiperLAN2 and others in ETSI-BRAN. OFDM has also been proposed for IEEE 802.16 MAN and integrated services digital broadcasting (ISDB-T) equipment.

Coded OFDM (COFDM) technology is also being considered for the digital television (DTV) terrestrial broadcasting standard by the Federal Communications Commission (FCC) as an alternative to the already adopted digital trellis-coded 8-T VSB (8-VSB) modulation for conveying approximately 19.3 Mbps MPEG transport packets on a 6-MHz channel. The transition period to DTV in the United States is scheduled to end on December 31, 2006, and the broadcasters are expected to return to the government a portion of the spectrum currently used for analog stations. The proponents of COFDM technology are urging the FCC to allow broadcasters to use it because of its robustness in urban environments, compatibility with DTV in other countries, and appeal in the marketplace for development of DTV.

Current trends suggest that OFDM will be the modulation of choice for fourth-generation broadband multimedia wireless data communication systems. However, there are several hurdles that need to be overcome before OFDM finds widespread use in modern wireless data communication systems. OFDM's drawbacks with respect to single-carrier modulation include OFDM and multicarrier systems.

OFDM OFDM inherently has a relatively large peak-to-average power ratio (PAPR), which tends to reduce the power efficiency of RF amplifiers. Construction of OFDM signals with low crest factors is particularly critical if the number of subcarriers is large because the peak power of a sum of N sinusoidal signals can be as large as N times the mean power. Furthermore, output peak clipping generates out-of-band radiation due to intermodulation distortion.

Multicarrier Multicarrier systems are inherently more susceptible to frequency offset and phase noise. Frequency jitter and doppler shift between the transmitter and receiver cause intercarrier interference (ICI), which degrades the system performance unless appropriate compensation techniques are implemented.

The preceding problems may limit the usefulness of OFDM for some applications. For instance, the HiperLAN1 standard completed by the European Telecommunications Standards Institute (ETSI) in 1996 considered OFDM but rejected it. Since then, much of the research effort on multicarrier communications at universities and industry laboratories has concentrated on resolving the preceding two issues. OFDM remains a preferred modulation scheme for future broadband radio area networks, because of its inherent flexibility in applying adaptive modulation and power loading across the subcarriers. Significant performance benefits are also expected from the synergistic use of software radio technology and smart antennas with OFDM systems. Several variations of multicarrier communication schemes have been proposed to exploit the benefits of both OFDM and single-carrier systems such as spread spectrum.

Ultra-Wideband (UWB) Ultra-wideband modulation uses baseband pulse shapes that have extremely fast rise and fall times in the subnanosecond range. Such pulses produce a true broadband spectrum, ranging from near dc to several gigahertz, without the need for RF upconversion typically required of conventional narrowband modulation. The ideas for UWB are steeped in original nineteenth-century work by Helmholtz and were viewed as controversial at the time (and are still viewed as such today).

UWB, also known as *impulse radio,* allows for extremely low cost, wideband transmitter devices, since the transmitter pulse shape is applied directly to the antenna, with no upconversion. Spectral shaping is carried out by adjusting the particular shape of the ultrashort-duration pulse (called a *monopulse*), and by adjusting the loading characteristics of the antenna element to the pulse. Figure 23-3 illustrates a typical bimodal gaussian pulse shape for a UWB transmitter.[1] The peak-to-peak time of the monopulse is typically on the order of tens or hundreds of picoseconds, and is critical to determining the shape of the transmitted spectrum. When applied to a particular antenna element, the radiated spectrum of the UWB transmitter behaves as shown in Fig. 23-3.

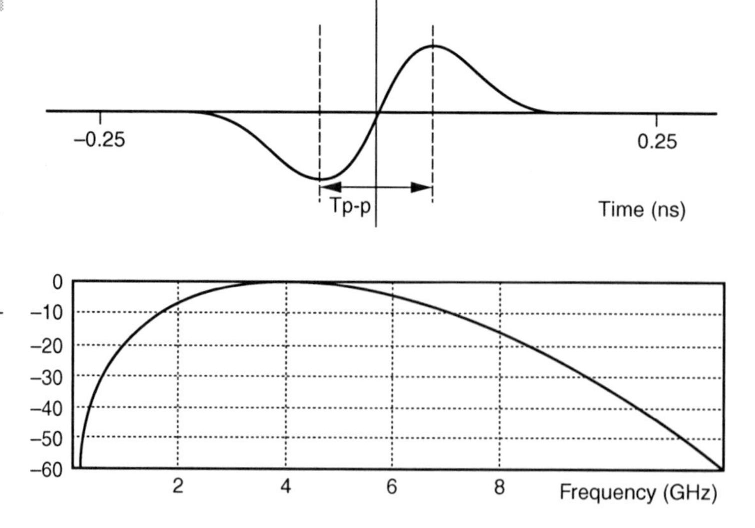

Figure 23-3 Time domain response and frequency domain response of a gaussian UWB monopulse applied to an antenna. Pulses have durations that are fractions of a nanosecond.

The UWB signals, which may be thinly populated over time as shown in Fig. 23-4, have extremely low power spectral density, allowing them to be used simultaneously with existing RF devices throughout the spectrum.[1] Because of the extremely wide bandwidths, UWB signals have a myriad of applications besides communications. On February 14, 2002, the FCC in the United States authorized the introduction of UWB for radar ranging, metal detection, and communications applications. The UWB authorization, while not completely final, is likely to limit transmitters according to FCC Part 90 or Part 15 rules. Primary UWB operation is likely to be contained to the 3.1- to 10.6-GHz band, where transmitted power levels will be required to remain below 41 dBm in that band. To provide better protection for GPS applications, as well as aviation and military frequencies, the spectral density is likely to be limited to a much lower level in the 960-MHz to 3.1-GHz band.

The ultrashort pulses allow for accurate ranging and radar-type applications within local areas, but it is the enormous bandwidth of UWB that allows for extremely high signaling rates that can be used for next-generation wireless data LANs. UWB can be used like other baseband signaling methods, in an on-off keying (OOK), antipodal pulse shift keying, pulse amplitude modulation (PAM), or pulse position modulation (PPM) format (see Fig. 23-4). Furthermore, many monopulses may be transmitted to make up a single signaling bit, thereby providing coding gain and code diversity that may be exploited by a UWB receiver.

Space-Time Processing Since the allocation of additional protected (licensed) frequency bands alone will not suffice to meet the exploding

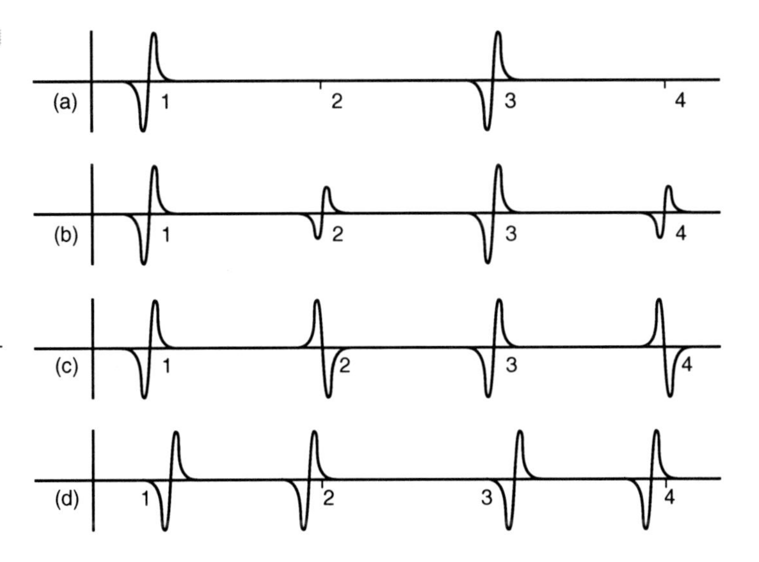

Figure 23-4 Examples of symbols sent using: (*a*) on-off keying; (*b*) pulse amplitude modulation; (*c*) binary phase shift keying; and (*d*) pulse position modulation using UWB technology.

demand for wireless data services, and frequency spectrum represents a significant capital investment (as seen from the 3G spectrum auctions in Europe), wireless data service providers must optimize the return on their investment by increasing the capacity of cellular systems. Cell-splitting can achieve capacity increases at the expense of additional base stations. However, space-time processing technology and multiple-input, multiple-output (MIMO) antenna architectures (which simultaneously exploit small-scale temporal and spatial diversity by using antennas and error-control codes in very close proximities) hold great promise to vastly improve spectrum efficiency for PCS service providers by providing capacity enhancement and range extension at a considerably lower cost than the cell-splitting approach. Moreover, space-time technology is envisioned to be used in both cellular and ad hoc network architectures. For instance, the use of smart antennas in rural areas can be effective in range improvement over a larger geographical area, resulting in lower equipment costs for a cellular system. The use of smart antennas in an ad hoc network could increase network throughput, because of suppression of the cochannel and adjacent-channel interference provided by the directional antenna gain pattern, in addition to supporting LPI/LPD features for military applications. Space-time processing could also enable 3G infrastructure to accommodate location technology in order to meet the requirements for E911.

Since multipath fading affects the reliability of wireless data links, it is one of the issues that contributes to the degradation of the overall quality of service. Diversity (signal replicas obtained through the use of temporal, frequency, spatial, and polarization spacings) is an effective technique for mitigating the detrimental effects of deep fades. In the past, most of the diversity implementations have focused on receiver-based diversity solutions, concentrating on the uplink path from the mobile terminal to the base station. Recently, however, more attention has been focused on practical spatial diversity options for both base stations and mobile terminals. One reason for this is the development of newer systems operating at higher frequency bands. For instance, the spacing requirements between antenna array elements for wireless products at 2.4-GHz and 5-GHz carriers do not significantly increase the size of the mobile terminals. Dual-transmit diversity has been adopted in 3G partnership projects (3GPP and 3GPP2) to boost the wireless data rate on downlink channels, because future wireless data multimedia services are expected to place higher demands on the downlink rather than the uplink. One particular implementation, known as open-loop transmit diversity or space-time block coding (STBC), is illustrated in Fig. 23-5.[1]

The spreading out of wireless data in time and through proper selection of codes provides temporal diversity, while using multiple antennas

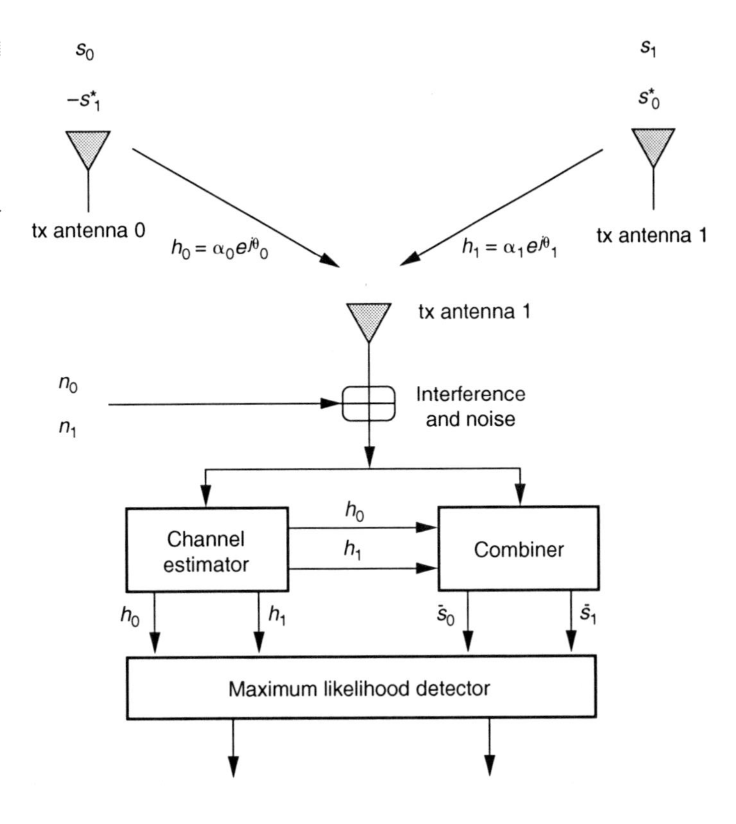

Figure 23-5 Functional block diagram of the space-time block code (STBC).

at both the transmitter and receiver provides spatial diversity. This implementation increases spectrum efficiency and affords diversity gain and coding gain with minimal complexity (all the transmit coding and receiver processing may be implemented with linear processing). Furthermore, it is shown in Fig. 23-5 that the resultant signals sent to the maximum likelihood detector are identical to those produced by a single transmit antenna with a two-antenna maximum ratio receiver combiner (MRRC) architecture. Thus, without any performance sacrifice, the burden of diversity has been shifted to the transmitter, resulting in a system and individual receiver that are more cost-effective (see Fig. 23-6).[1] It is possible to further increase the wireless data rate on the downlink by adding one or more antennas at the mobile terminal such as in Qualcomm's high-data-rate (HDR) system specification.

In a closed-loop transmit diversity implementation scheme, the receiver will provide the transmitter information on the current channel characteristics via a feedback message. It can then select the best signal or predistort the signal to compensate for current channel characteristics. Obviously, the performance of a closed-loop transmit diversity scheme will be superior to that of the simple "blind transmit" STBC

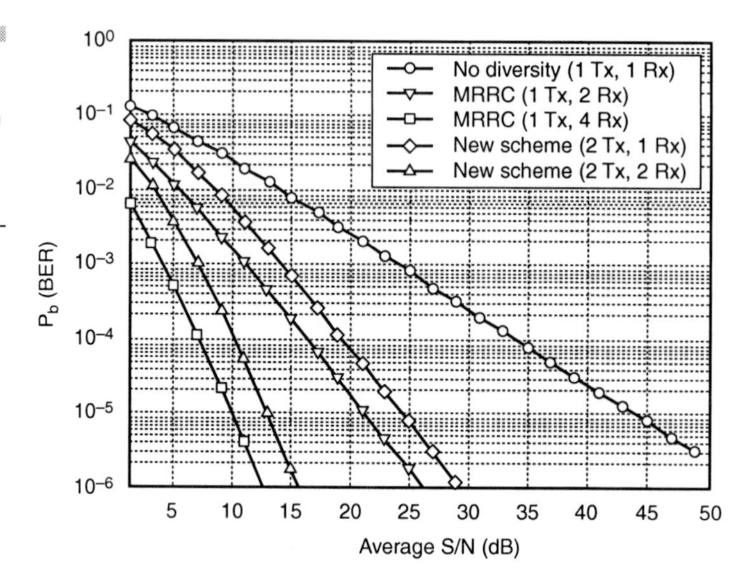

Figure 23-6 Performance comparison between STBC and MRRC for various antenna configurations.

scheme shown in Fig. 23-5. The latter approach would be preferred for small hand-held wireless data devices since the transmit power and battery life are at a premium. Besides STBC, blind transmit diversity may also be implemented by using a delay diversity architecture, where the symbols are equally distributed, but incrementally delayed among different antennas, emulating a frequency-selective channel. An equalizer at the receiver will utilize training sequences to compensate for the channel distortion, and diversity gain is realized by combining the multiple delayed versions of a symbol. A shortcoming of this approach, however, is that it suffers from intersymbol interference, if channel propagation differences are not integer multiples of the symbol periods. In this case, feedback from the receiver may be used to adjust delays.

MIMO architectures utilizing multiple antennas on both transmitter and receiver are one of the important enabling techniques for meeting the expected demand for high-speed wireless data services. Figure 23-7 illustrates the expected capacities for systems exploiting spatial diversity along with capacities of existing wireless data standards.[1] Looking at these trends, you may conclude that spatial diversity at both transmitter and receiver will be required for future-generation high-capacity wireless data communication systems.

The Bell Labs layered space-time (BLAST) approach (also known as *diagonal BLAST* or simply D-BLAST) is an interesting implementation of a MIMO system to facilitate a high-capacity wireless data communication system with greater multipath resistance. The architecture could increase the capacity of a wireless data system by a factor of m, where m

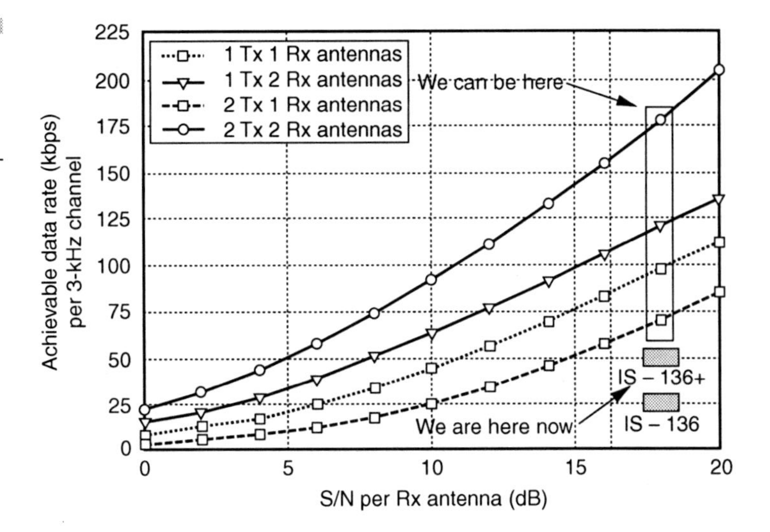

Figure 23-7 Achievable wireless data rates for several MIMO systems.

is the minimum number of transmit or receive antennas. Like the delay diversity architecture, BLAST does not use channel coding. Instead, it exploits multipath through the use of multiple transmit antennas and utilizes sophisticated processing at the multielement receiver to recombine the signals that are spread across both time and space. Figure 23-8 depicts a functional block diagram of a BLAST transmitter and receiver.[1]

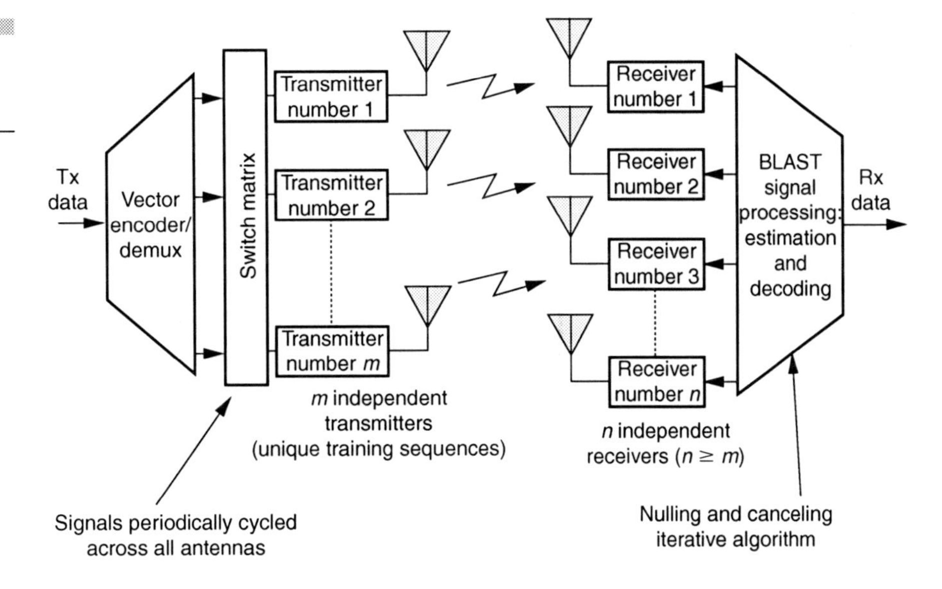

Figure 23-8 BLAST functional block diagram.

To minimize complexity, the BLAST architecture employs a recursive "divide and conquer" algorithm for each time instant, which is known as a *nulling and cancellation* process. Figure 23-9 illustrates this process over one complete cycle for one out of m processing channels (four transmit antennas are being received by one of the four receiver channels).[1] In this illustration, the receiver will receive packet A as it sequences through the transmit antennas. At the beginning of a cycle, the signal from a specific transmit antenna is isolated by canceling other signals that have already been received from other transmitters. After the first transmit antenna shift, the known, previously received signals are again subtracted from the composite signal, but now there is a "new" signal that has not been identified and must be removed. The nulling process is performed by exploiting the known channel characteristics (which are determined by the training sequences received from each transmit antenna, typically $2m$ symbols long). By projecting this new received signal vector against the transpose of the channel characteristics from the target antenna, it is effectively removed from the processing. At the same time, the known channel characteristics are used to maximize the desired signal. At the next shift of transmit antennas, this process

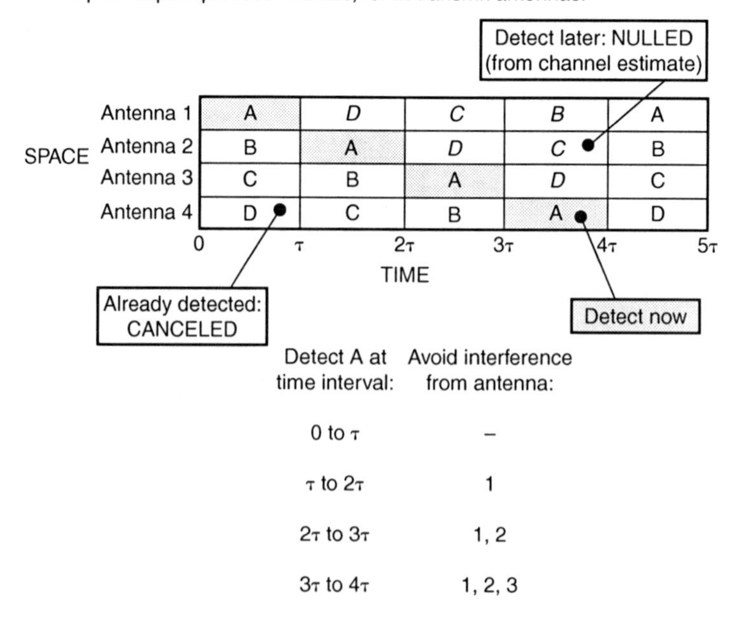

Figure 23-9 Illustration of one cycle of layered space-time receiver processing for a system with four transmit and receive antennas.

continues, with the known signals cancelled and the new signals nulled on the basis of channel characteristics.

With the promise of considerable capacity increase, there has been significant research into BLAST architectures focusing on optimized training sequences, different detection algorithms, and analysis of the benefits of combining the BLAST architecture with coding, among other topics. One of the most prevalent research areas is the development of vertical BLAST (V-BLAST), a practical BLAST architecture with considerably simpler processing. In V-BLAST, there is no cycling of codes between antennas, and therefore this simplifies the transmitter. At the receiver, the nulling and cancellation process is a recursive algorithm that orders the signals, chooses the optimum *S/N* at each stage, and linearly weights the received signals. These modifications greatly simplify the receiver processing, making V-BLAST a leading candidate for next-generation indoor and mobile wireless data applications.

Several near-future wireless data systems already plan to use space-time codes. For instance, the proposed Physical layer of the IEEE 802.16.3 broadband fixed wireless data access standard is considering using space-time codes as the inner code and a Reed-Solomon outer code. The European WIND-FLEX project is studying the "optimum" number of transmitter and receiver antennas and algorithm complexity for the design of 64- to 100-Mbps adaptive wireless data modems for indoor applications. Also, the fourth-generation (4G) cellular standards are expected to support data rates up to 20 Mbps with bandwidth efficiencies of up to 20 per cell. Space-time coding has been identified as one of the technologies needed to meet this performance requirement.

Ad Hoc Networking

Clearly, achieving higher wireless data rates at lower cost is a key for wireless data ubiquity. As previously stated, there are several Physical layer technologies that hold promise for achieving higher wireless data rates. However, another key to the future of wireless data networks is the ability to adapt and exist without substantial infrastructure. Thus, ad hoc networks are a key technology for future systems. An *ad hoc network* (also known as a *packet radio network*) is the cooperative engagement of a collection of mobile nodes that allows the devices to establish ubiquitous communications without the aid of a central infrastructure. The links of the network are dynamic in the sense that they are likely to break and change as the nodes move about the network. The roots of ad hoc networking can be traced back as far as 1968, when the work on the ALOHA network was initiated. The ALOHA protocol supports distributed channel access in a single-hop network (every node must be within reach of all other participating nodes), although it was originally employed for fixed nodes. Later in 1973, DARPA began the development of a multihop packet radio network protocol. The multihopping technique increases the network capacity by spatial domain reuse of concurrent, but physically separated, multihop sessions in a large-scale network (reduces interference); conserves transmit energy resources; and increases the overall network throughput at the expense of a more complex routing-protocol design.

In the past, ad hoc networking has been primarily considered for communications on battlefields and at the site of a disaster area, where a decentralized network architecture is an operative advantage or even a necessity. For instance, when major catastrophes happen, such as the September 11, 2001, attack, the need for a rapidly deployable, seamless communications infrastructure between public service agencies, military entities, and commercial communication systems becomes essential. Now, as novel radio technologies such as Bluetooth 1 materialize, the role of ad hoc networking in the commercial sector is expected to grow through interaction between the applications of various portable devices such as notebooks, cellular phones, PDAs, and MP3 players.

While present-day cellular systems still rely heavily on centralized control and management, next-generation mobile wireless data system standardization efforts are moving toward ad hoc operation. For instance, in the direct-mode operation of HiperLAN2, adjacent terminals may communicate directly with one another. Fully decentralized radio, access, and routing technologies are enabled by Bluetooth, IEEE 802.11 ad hoc mode, IEEE 802.16 mobile ad hoc networks (MANET), and IEEE 802.15 personal area networks (PAN). Someone on a trip who has access to a Bluetooth PAN could use a GPRS/UMTS mobile phone as a gateway to the Internet or to the corporate IP network. Also, sensor networks enabled by ad hoc multihop networking may be used for environmental monitoring (to monitor and forecast water pollution, or to provide early warning of an approaching tsunami) and for homeland defense (to perform remote security surveillance). Therefore, it is not surprising that the trends of future wireless data systems, characterized by the convergence of fixed and mobile networks, and the realization of seamless and ubiquitous communications, are both attributed to ad hoc networking.

The lack of a predetermined infrastructure for an ad hoc network and the temporal nature of the network links, however, pose several fundamental technical challenges in the design and implementation of packet radio architectures. Some of them include:

- Security and routing functions must be designed and optimized so that they can operate efficiently under distributed scenarios.

- Overhead must be minimized, while ensuring connectivity in the dynamic network topology is maintained (approaches are needed to reduce the frequency of routing table information updates).
- Fluctuating link capacity and latency in a multihop network must be kept minimal with appropriate routing protocol design.
- Acceptable tradeoffs are needed between network connectivity (coverage), delay requirements, network capacity, and the power budget.
- Interference from competing technology must be minimized through the use of an appropriate power management scheme and optimized medium access control (MAC) design.[1]

Network Optimization: Removing Boundaries

While the layered OSI design methodology (see Fig. 23-10) has served communications systems well in the past, evolving wireless data networks are seriously challenging this design philosophy.[1] Emerging networks must support various and changing traffic types with their associated quality of service (QoS) requirements as well as networks that may have changing topologies. The problem of various traffic types is typified in newly defined 3G networks. These networks must support multimedia traffic with manifold delay, error rate, and bandwidth needs. Networks

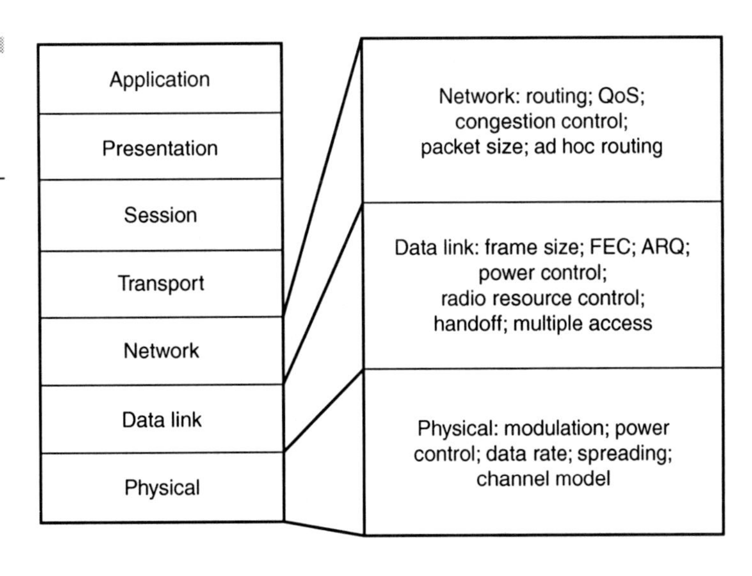

Figure 23-10 Traditional OSI communication network layers.

that experience changing topologies include ad hoc networks that lack network infrastructure and have nodes that are continuously entering and leaving the network.

In order to meet the challenges of ubiquitous wireless data access, network functions (the various OSI layers) must be considered together in designing a network. QoS requirements, which can and will vary according to application, will force the Network layer to account for the Physical layer design when the network throughput is optimized. Furthermore, different applications are better served by different optimizations. This leads to a design methodology that blurs the lines between layers and attempts to optimize across layer functionality.

As a primitive example, consider two techniques that have been proposed to improve system performance at different layers: 41 space-time block codes (STBC) at the Physical layer and a "greedy" scheduling algorithm at the MAC layer. Greedy scheduling means a simplified version of the scheduling algorithm employed in cdma2000 3G1X-EVDO, also called HDR. This scheduler is based on feedback from the mobile units, and schedules packet transmissions to the mobile that is currently experiencing the best channel conditions (highest SINR). STBC is capable of providing significant diversity advantage at the Physical layer. An even larger advantage can be provided by greedy scheduling provided that the scheduler has 20 users from which to choose. This *multiuser diversity* can provide great advantages (albeit at the sacrifice of delay, which is beyond the scope of this book). However, if you add 41 STBC on top of greedy scheduling, you obtain virtually no further advantage at a cost of quadrupling the RF cost. It can also be shown that as the number of users increases, STBC can actually degrade the SINR performance. However, in round-robin scheduling or in the case of a small number of data users, STBC helps significantly. Thus, ideally, the scheduler and the Physical layer should be optimized together to maximize performance. This simple example also shows the importance of the QoS requirements. If an application has very strict delay requirements (voice), greedy scheduling is not desirable since users experiencing bad channels must wait for service, but STBC would be an acceptable way to achieve diversity advantage. On the other hand, wireless data applications that are delay-insensitive (Web traffic) would lend themselves well to greedy scheduling rather than STBC, which requires four transmitters and RF chains.

While cross-layer network design is an important step in optimizing new multimedia networks, it is still a step below what will be necessary to truly maximize the performance of future networks. True optimization will not only require cross-layer design, but also cross-layer adaptability. Traditionally, networks have contained some ability to adapt. For example, many communications systems can adjust to changing channel conditions by using signal processing methods, or to changing traffic loads by adjusting routing tables. However, these adjustments have been isolated to a specific layer. Cross-layer adaptability will allow all network functions to pass information between functions and adapt simultaneously. Such adaptability will be required to meet the demand of changing QoS requirements, along with changing network loads and channel conditions. While the cross-layer network design requires static optimization across network layers, adaptability requires dynamic optimization across layers.

Challenges to Cross-Layer Optimization

There are several challenges and research issues associated with the vision of cross-layer optimization. First and most obviously, full network design and optimization are extremely complicated (and nearly intractable). This is particularly true when attempting real-time dynamic optimization. Some attempt must be made to determine design methodologies that encompass the incredible freedom offered to the designer when cross-layer optimization is possible.

A second serious problem involves the metrics to be used in the optimization. Network layers (and, consequently, functionalities) have traditionally had their own isolated optimization criteria. For example, Physical layer design is primarily focused on minimizing the bit error rate, while the MAC layer design is concerned with node throughput or channel availability. The network design, on the other hand, typically uses delay or routing efficiency. Thus, you should ask: What metrics represent all of these concerns? How do you optimize all concerns together or prioritize them intelligently?

A related issue arises in the context of dynamic optimization. In dynamic optimization, information is passed between the Network layers. The system designer must judiciously choose the information to be passed. It must not be overly complicated for risk of creating large delays or computationally expensive optimization routines. However, it cannot be overly simplistic and risk communicating too little information.

The design of such systems clearly requires sophisticated modeling (simulation) procedures. Traditional network simulators do not have sufficient granularity at the Physical layer to allow Physical layer design. On the other hand, adding network functionality to traditional Physical layer simulators would result in prohibitively long run times. Furthermore, network simulators embrace an event-driven methodology while Physical layer simulators use a time-driven methodology. The typical solution to this problem may be a two-tier simulation approach that uses the output of a Physical layer simulation to stimulate network simulations. However, this does not allow for interaction between the layers and precludes cross-layer optimization. Thus, hybrid approaches are necessary. Some possible options include:

- Combined simulation and semianalytic approaches that simulate high-level functionality and use semianalytic simulation approaches to approximate lower-level functionality.
- Combined simulation and hardware approaches that use hardware to perform lower-level functionality.
- Variable-granularity approaches that use a network simulator with coarse granularity (abstracting lower layers) for a majority of Physical layer links and fine granularity (possibly down to the sample level) for links of specific interest.
- Emulation and real-time processing involving all facets, from Physical layer to application, simultaneously.[1]

These hybrid approaches have yet to be firmly established and represent significant research areas.

A final research issue in the area of dynamic network optimization concerns network control. When functionality across layers is allowed to adapt, it is important that something has control of the process. Otherwise, the various adaptations can work at cross purposes. Thus, the question becomes, "Who has control?" Arguments can be made for each layer concerning the best place to locate the control, but the fact remains that this is a serious research issue that may indeed have different solutions depending on the end-user application or particular physical environment of operation.

Conclusions

This book has described many important new technologies and approaches to the wireless data communications field that are likely to evolve rapidly in the early part of the twenty-first century. In the 1990s, cellular telephone service and the Internet grew from the incubator stage to global acceptance. In the next 10 years, the Internet and wireless data communications will become intertwined in ways only imagined today.

NOTE The great new frontier for the wireless data communications industry is inside buildings, and a battle for access is emerging between cellular/PCS license holders and ad hoc networks installed by the building owners using license-free WLAN technology.

The book overwhelmingly illustrated the worldwide acceptance of CDMA as the multiple access system of choice, and presented the many challenges CDMA faces as the whole communications industry evolves to fourth-generation wireless data networks. Clearly, the need for higher data rates will lead to new modulation and coding techniques that can provide high spectral efficiencies. This book discussed three candidates for providing improved spectral efficiency at the Physical layer: orthogonal frequency-division multiplexing, ultra-wideband transmission, and space-time modulation/coding. Each of these technologies has the potential to increase the spectral efficiency of the Physical layer and will likely find its way into future systems. OFDM was highlighted as an emerging signaling method that holds promise for broadband wireless data access. The fundamentals and challenges for OFDM were given, and new applications that use OFDM were presented. Ultra-wideband, recently approved for U.S. deployment by the FCC, was highlighted in the book as an important emerging technology, and some of the fundamentals of this controversial signaling method were given. Space-time coding was also discussed in detail, with several examples given to highlight the tremendous potential of this technique.

While Physical layer advances will be a key to the future, an even more critical area for future networks exists at the higher layers. Ad hoc networks will clearly play a large role in future systems because of the flexibility that will be desired by the consumer. The book also discussed the key aspects of ad hoc networks and the research issues that must be examined to advance the use of ad hoc networks in future systems. In addition, the book discussed the idea of cross-layer optimization. The emergence of wireless data applications with diverse delay and fidelity requirements along with constantly changing topologies and requirements for future networks will require a new design methodology. Specifically, future network designs will need to consider the interaction of network layers. The book also examined a simple example as well as the key challenges associated with such a design approach.

While predicting the future is a tricky business, it is clear that wireless data will be a key technology in the future of communications. Finally, the book presented several of the technologies that will advance wireless data communications, and the challenges that must be met to make ubiquitous communications a reality.

References

1. Theodore S. Rappaport, A. Annamalai, R. M. Buehrer, and William H. Tranter, "Wireless Communications: Past Events and a Future

Perspective," *IEEE Communications Magazine,* 445 Hoes Lane, Piscataway, NJ 08855, 2002.

2. John R. Vacca, *Wireless Broadband Networks Handbook,* McGraw-Hill, 2001.

GLOSSARY

10 Base-T Basic Ethernet at 10 Mbps.

100 Base-T Ethernet running at 100 Mbps.

1000 Base-T Ethernet running at 1000 Mbps.

1G First generation. Refers to analog cellular systems.

1xEV-DO CDMA 1x evolution—data only.

1xEV-DV CDMA 1x evolution—data and voice services. Based on Qualcomm HDR.

1XRTT cdma2000 operating mode at basic chip rate (1.2288 Mbps).

2G Second generation. Refers to digital cellular and PCS wireless systems oriented to voice and low-speed data services.

2.5G Faster than today's wireless networks, but slower than 3G, 2.5G technologies face limited trial deployments in 2003, and are promoted by carriers as stepping stones to eventual 3G deployments.

2R Receive, reshape (an optical signal).

3G Third generation. Refers to the next generation of wireless systems—digital with high-speed data. Being standardized by 3GPP and 3GPP2.

3GiA 3G Internet appliance.

3GPP Third Generation Partnership Project for W-CDMA (GSM).

3GPP2 Third Generation Partnership Project for cdma2000.

3GSP 3G service provider.

3R Reshaping, retiming, reamplifying (an optical signal).

3WC Three-way call.

3XRTT cdma2000 operating mode at 3 times the basic chip rate of 1.2288 Mbps.

802.11 An IEEE wireless Ethernet replacement technology in the ISM band. Runs at approximately 10 Mbps.

A(2) An IP host address.

A3 GSM authentication algorithm.

A5 GSM data encryption algorithm.

A8 GSM voice encryption algorithm. Used to generate Kc.

AAA Authentication, authorization, and accounting entity.

AAL AAL adaptation layer.

A-bis Interface between BTS and BSC.

ABNF Augmented BNF. Defined in RFC 2234.

ABR Average bit rate.

ABS Alternate billing service.

AC Authentication center. Stores information for authenticating mobiles, and encrypting their voice and data transmissions.

ACC Analog control channel.

ACCOLC Access overload class.

ACD Automatic call distributor. Distributes incoming calls to one of a number of people equally able to handle them (for customer service).

ACELP(1) Adaptive CELP.

ACELP(2) Algebraic CELP.

ACF Authentication control function.

ACG Automatic code gapping. A method of shedding load in telecommunications systems.

ACH Access channel.

ACK Acknowledgement signal.

ACM SS7 ISUP address complete message. Response to IAM.

ACP Adjacent channel power.

ACRE Authorization and call routing equipment. Used for routing calls to cellular phones with a "cordless" mode.

AD Abbreviated dialing.

A/D Analog/digital. Usually used in the context of conversion from analog to digital (or vice versa).

ADA Advertising agent. Provides information to an MS on the services provided by a 3G network.

ADDS Application data delivery service.

ADN Abbreviated dialing numbers.

ADO ActiveX Data Objects. Microsoft's newest high-level interface for data objects.

ADPCM Adaptive differential PCM.

ADS Asynchronous data service.

ADSL Asymmetric DSL. Bit rates are higher from the network than from the client.

AEG Asian Expert Group. A WAP working group.

AES (1) Advanced Encryption Standard. The Rijndael encryption algorithm. Replacement for DES chosen by NIST. (2) Audio Engineering Society.

AH Answer hold. Service that allows an incoming call to be placed on hold without answering it first.

A-interface Interface between the MSC and BS.

AK Anonymity key.

AKA Authentication and key agreement.

A-key The primary CAVE authentication key, used to generate SSD.

AMA Automatic message accounting.

AMPS Advanced mobile phone service. A term used for the first generation of analog wireless technology. It is based on waveform transmission, unlike digital technologies, which broadcast ones and zeros.

ANI Access network identifier.

ANSI American National Standards Institute.

ANSI-41 Wireless intersystem operations standard.

AOA Angle of arrival. A technique for locating a radio by estimating the angle of signal arrival at multiple points.

AP Application part (of a protocol).

APDU Application PDU.

ARQ Automatic repeat request.

ATIS Alliance for Telecommunications Industry Solutions. Parent organization for the T1 standards committees and many telecom industry groups, such as OBF.

ATM Asynchronous transfer mode. Transmits data as 53-byte units using a connection-oriented protocol.

AUTHR Authentication response. The output of CAVE when RAND is used as a global challenge.

AWGN Additive white gaussian noise.

B8ZS Bipolar with 8-zero substitution. Replaces an all-0 octet by one containing two BPV.

Backhaul Routing trunks from a cell site to an MSC before routing to the PSTN.

Badput A cute name for wasted bandwidth. Bandwidth = goodput (throughput) + badput + unused bandwidth.

BAF Billing automatic message accounting format. The CDR/AMA format used by most U.S. wireline telecom carriers.

BAIC Barring of all incoming calls.

BAOC Barring of all outgoing calls.

Barring Refusal to allow certain types of calls.

BATS Broadcast air-interface transport service used by TIA/EIA-136.

BCCH Broadcast control channel.

BCD Binary coded decimal. Digits 0 to 9 are encoded as 4-bit numbers.

BCE Base station central equipment.

BCH (1) Broadcast channel. A channel transmitted by one (BS) and received by many (MS). (2) Bose, Chaudhuri, and Hocquengham error detection and correction methodology.

BCM Basic call manager.

BCSM Basic call state model. An IN concept.

BDN Barred dialing number.

Bearer capability A capability of a transport protocol (a maximum bit rate or message latency). A teleservice may be able to use any facility that can provide a specified bearer capability.

BEG Billing Expert Group. A WAP working group.

BER (1) Bit error rate. (2) Basic encoding rule.

BGP Border Gateway Protocol.

B/I Busy/idle bit.

BIB Backward indicator bit. Indicates when a received MTP frame is out of sequence.

BICC Bearer independent call control. ISUP adapted for use over IP-based transport. Compare with SIP.

BIC-Roam Barring of incoming calls while roaming.

BID A SID allocated for accounting purposes. BIDs are allocated by Cibernet.

BLAST An architecture for realizing a very high bit data rate. This multiple transmit/receive antenna system [referred to as multiple input multiple output (MIMO) or BLAST for the Lucent Bell Labs version] is being touted mainly in the United States as a means for achieving very high capacities in wireless systems.

BNF Backus-Naur form. A precursor to ASN.1 and other metalanguages.

BPV Bipolar violation. Transmission of two 1-bits in a PCM channel (DS0) with the same polarity (both positive or both negative). May be deliberately used to indicate all-zero octets.

BS Base station. Includes BTS (base transceiver system; radio portion of BS) and BSC (base station controller; the "brains" of a base station, controlling the radio equipment in the BTS).

BSC Base station controller. The "brains" of a base station, controlling the radio equipment in the BTS.

BSS BS subsystem.

BWDA Broadband wireless data access. The creation of efficient end-user terminals capable of delivering a full range of multimedia data services.

CA Certificate of authority.

CAC Carrier access code. Identifies a long-distance carrier. 101 + CIC.

CALEA U.S. Communications Assistance for Law Enforcement Act. Requires that telecommunications carriers provide for surveillance (i.e., wiretaps) at the switch site.

CAMA Centralized automatic message accounting.

CAMEL Customized applications for mobile networks enhanced logic, based on CAP. IN capabilities for GSM. Compare with WIN.

Candidate MSC An MSC being considered as the target MSC of a handoff.

CANID Current ANI.

CAP (1) CAMEL AP. (2) Carrierless amplitude and phase modulation. (3) Competitive access provider.

CAPCS Cellular Auxiliary Personal Communications Service.

CAR Committed access rate. An IP method to achieve higher QoS.

CARE Customer account record exchange. Sent from an LEC to an IXC to establish a long-distance account for a customer.

CAS Call-associated signaling. Contrast with NCAS.

CAT Smart Card Application Toolkit.

CATPT CDMA UIM Card Application Toolkit Protocol Teleservice.

CAVE TIA cellular authentication and voice encryption algorithm.

CB Cell broadcast.

CBC CB center.

CBMI CB message identifier.

CBR Constant bit rate.

CBS GSM/W-CDMA cell broadcast SMS.

CC (1) E.164 country code. (2) GSM Call Control Protocol. (3) Content of communications (for LAES).

CCA Common cryptographic algorithm.

CCAT CDMA Card Application Toolkit. Specifies communications between a CDMA R-UIM and the ME.

CCB U.S. FCC Common Carrier Bureau.

CCH Control channel. A cellular or PCS channel that broadcasts information about a cell to mobiles that are not currently in a call.

CDMA Code-division multiple access. Refers to any of several protocols used in so-called second-generation (2G) and third-generation (3G) wireless communications. Implemented in AMPS-compatible systems by IS-95.

CDPD Cellular Digital Packet Data. A protocol that uses 30-kHz AMPS channels to transmit packets of data. Standardized in TIA/EIA/IS-732.

CDR Call detail record.

CELP Card edge low profile. Socket into which COAST circuit boards are plugged.

CGL Calling geodetic location. The position of a mobile phone, as transmitted through various signaling protocols.

CHTML Compact HTML. Used by iMode.

C/I Carrier-to-interference ratio.

CIC Carrier identification code.

Circuit-switched data Data transmitted over a dedicated (although usually virtual) channel. The destination address is implicitly defined by the (virtual) circuit that is selected.

CLASS Custom local-area signaling services. AT&T developed a set of 1A ESS revenue-generating features called LASS (local-area signaling services). Pacific Bell requested customized software enhancements for some of the features, and will refer to them as CLASS. Consists of number-translation services, such as call forwarding and caller identification, available within a local exchange of a LATA.

CLEC Competitive LEC. A new entrant in a market previously limited to one carrier. Some wireless carriers may qualify for this designation.

CN Core network. Protocols for this include GSM MAP and ANSI-41.

CTIA Cellular Telecommunications Industry Association.

D-AMPS Digital AMPS (IS-54 and IS-136 TDMA).

DataTAC Data TAC. A Motorola wireless data system. Formerly known as Ardis.

dB Decibel; 10 times the logarithm of the value in base 10.

DBm Decibels referenced to 1 milliwatt.

DCC Digital color code. A number assigned to a control channel used to limit erroneous accesses.

DCCH Digital control channel. The control channel used by IS-136 and TIA/EIA-136 D-AMPS systems.

DCE Data communications equipment (a computer).

DCN Data communications network.

DCS (1) Data coding scheme. (2) Digital cross-connect system.

D digit The fourth digit of an NANP phone number. Currently restricted to the values 2 to 9 to allow 7-digit dialing.

DECT Digital enhanced cordless telephony.

DES Data encryption standard. A commonly used encryption method, usually used with 56-bit keys.

DF (1) UIM dedicated file. Compare with EF and MF (2). (2) Delivery function (for LAES).

DFP Distributed functional plane NRM.

DHCP Dynamic host control protocol. Allows automatic assignment of IP addresses on a network.

DHKE Diffie-Hellman key exchange. A method of securely exchanging encryption keys over an insecure interface.

DID Direct inward dialing. Directs all calls to a block of numbers to a PBX.

Diffie-Hellman A secure key exchange mechanism.

Diffserv Differentiated services. Different QoSs for different types of traffic (voice, video, e-mail).

Digital Transmission of information through a signal that can take on only certain discrete values (bits with values 0 or 1). Compare with analog.

Disconnection The end of a call. Not to be confused with termination of a call or the release of a trunk.

DL Downlink. Radio link from network "down" to terminal. Compare with UL.

DLC Digital loop carrier. A single digital facility (T1 or T3) carrying multiple lines to a business or other large customer.

DN Directory number. The number dialed to terminate a call to a phone.

DS0 Digital signal level 0. A 64-kbps digital link used to carry a single voice conversation or signaling traffic for multiple calls/trunks. In ANSI networks, 8 kbps is usually reserved for in-band signaling (on-hook/off-hook, etc.), reducing the bandwidth to 56 kbps.

DSL Digital subscriber line. DSL technologies use sophisticated modulation schemes to pack data onto copper wires.

DTC Digital traffic channel.

DTE Data terminal equipment. A device that controls data flowing to or from a computer.

DVB Digital Video Broadcasting. A set of standards that define digital broadcasting using existing satellite, cable, and terrestrial infrastructures.

DVD Digital versatile disc. An optical disc technology that is expected to rapidly replace the CD-ROM (as well as the audio compact disc) over the next few years.

E1 A digital link carrying 32 DS0 channels, with two used for signaling purposes. Used mostly outside North America. Compare with T1.

E.164 ITU-T dialing plan standard. Numbers are composed of CC + NSN (NDC + NDC).

E.212 ITU-T mobile identification number standard.

E.214 ITU-T standard that allows an E.212 IMSI number to be mapped onto an E.164 number to allow routing through SS7 networks. Unfortunately, this mapping does not work in North America.

E911 Enhanced 911 service. Provides the identity and the approximate location of the calling phone.

EACC Emergency area congestion control.

EAP IETF Extensible Authentication Protocol.

EAR Export administration regulations. Replace ITAR for control of export of encryption technologies.

EAS Emergency Alert System. A U.S. government system that transmits audio or text information about emergencies (mostly weather) to radio and TV stations. There has been some talk about extending this to wireless data phones via broadcast SMS.

Eb Energy of an information bit.

EBCDIC Extended binary coded decimal interchange code. IBM's byte code for letters, numbers, and special characters. Not as good as ASCII because, for example, letters are not all in a single group, making software more awkward.

EBNF Extended BNF. Used to define XML, for example.

EC Exchange carrier.

ECC Elliptic curve cryptography.

ECDLP Elliptic curve discrete logarithm problem.

ECMA European Computer Manufacturer's Association.

ECSA Exchange Carriers Standards Association. Renamed ATIS several years ago.

ECT Explicit call transfer.

EDAC CDMA error detection and correction coding.

EDGE Enhanced data rates for GSM (or global) evolution. Use of a new modulation scheme to increase data rates within the 200-kHz RF bandwidth to 384 kbps, although per-user rates will be significantly lower.

EF UIM elementary file.

E-GGSN Enhanced GGSN.

EIA Electronics Industry Alliance.

E-OTD Enhanced observed time difference. A positioning technology for wireless data phones. Compare with AOA, TOA, TDOA.

ESCON Enterprise Systems Connection. A marketing name for a set of IBM and vendor products that interconnect S/390 computers with each other and with attached storage, with locally attached workstations, and with other devices using optical fiber technology and dynamically modifiable switches called ESCON directors.

E-SMR Enhanced SMR.

ETSI European Telecommunications Standards Institute.

EUT Equipment under test.

FA (1) Foreign agent. (2) Flexible alerting. A badly named feature that is really an extension phone service for wireless data calls.

FAC (1) FA challenge. (2) Final assembly code. The 2-digit IIMEI ME manufacturer identity.

FACCH Fast associated control channel.

FAQ Frequently asked questions.

F-BCCH Fast broadcast control channel.

FBI U.S. Federal Bureau of Investigation.

FC Feature code. An asterisk followed by digits indicating the invocation of a feature (*73 may be used to disable call forwarding). Feature codes should be sent to the HLR for interpretation. There is no standardization of feature codes.

FCC U.S. Federal Communications Commission.

FCI Forward call indicator. Used to indicate whether a number portability query has occurred for this call (to prevent looping).

FCS Frame check sequence. A checksum for a transmitted frame.

FDCCH Forward DCCH.

FDD Frequency-division duplex.

FDDI Fiber distributed data interface.

FDM Frequency-division multiplex.

FDMA Frequency-division multiple access. Compare with TDMA and CDMA.

FDN Fixed dialing number.

FDTC Forward DTC.

FE Functional entity. A logical element of a network. Not necessarily realized as a physically distinct device.

FEC (1) Forward error correction. (2) Forwarding equivalence class. An identity for packets that all get routed via MPLS in the same way.

FE-NTS Feature-enhanced NTS.

FER Frame error rate. The number of frames in error divided by the total. These frames are usually discarded, in which case this can be called the *frame erasure rate.*

FET Field-effect transistor.

FEXT Far-end cross talk.

FFPC CDMA fast forward power control.

FFT Fast Fourier transform.

FICON Fiber connectivity. A high-speed input/output (I/O) interface for mainframe computer connections to storage devices.

FIFO First in, first out. A queueing methodology similar to lining up for a bank teller.

FLEX A Motorola one-way paging protocol that runs at 1600, 3200, or 6400 bps.

FM^2 Frequency modulation squared or to the second power.

FPLMTS Future public land mobile telecommunications systems. Now IMT-2000.

F-SCH Forward SCH (from BS).

FSK Frequency shift keying.

FSOP Free-space optics photonics. Also called free-space photonics (FSP), this refers to the transmission of modulated visible or infrared (IR) beams through the atmosphere to obtain broadband communications.

G3FAX Group III facsimile service.

G95 A subgroup of GGRF investigating roaming between GSM and CDMA systems based on TIA/EIA-95 and cdma2000.

GAD Geographic area description. GSM 101.109 and 3GPP 23.032 define this. Related to CGL in T1.628.

GAIT GSM/ANSI-41 interoperability trial.

GAP SS7 ISUP generic address parameter.

Gateway MSC An MSC designed to receive wireless calls from the PSTN.

GbE Gigabit Ethernet. A transmission technology based on the Ethernet frame format and protocol used in local-area networks (LANs); provides a data rate of 1 billion bits per second (1 gigabit).

GDP ISUP generic digits parameter.

GECO Global emergency call origination.

GEO Geostationary orbit (for a satellite).

GERAN GSM/EDGE RAN.

GETS U.S. Government Emergency Telecommunications Service.

GFP Generic framing procedure. A standard that has been devised to offer more bandwidth-efficient ways of packing Ethernet traffic into a SONET/SDH transport network.

GFSK Gaussian FSK.

GGRF GSM Global Roaming Forum within the GSM Association. Composed of GAIT, G95, and groups considering interworking with IDEN and TETRA.

GGSN Gateway GSN. Connects to other packet data networks (Internet).

GHOST GSM hosted short message teleservice. Allows TIA/EIA-136 messages to be delivered to a GSM MS.

GHz One thousand million hertz. A measure of radio frequency.

Giga Prefix to indicate one billion (as in gigabit). Abbreviated G.

GII Global information infrastructure.

GIWU GSM Interworking unit. An interface to various networks for data communications.

Global challenge Method of authentication using the same random number (RAND) for every mobile in a cell site or sector.

GMM GPRS Mobility Management Protocol.

GMSK Gaussian MSK.

GPRS General Packet Radio Service. A GSM-based packet data protocol using up to all 8 of the time slots in a GSM channel. It supports data rates up to 115 kbps, although 10 to 40 kbps is more likely in practice.

GPS Global Positioning System. A system for determining location based on comparing signals from several U.S. military satellites. Compare with AOA, TOA.

GR Telecordia generic requirements document.

GSM Global System for Mobile Communication. A digital mobile telephone system that is widely used in Europe and other parts of the world.

GTT Global title translation. A method of routing in SS7 networks based on global titles and not point codes.

GUI A program interface that takes advantage of the computer's graphics capabilities to make the program easier to use.

H.248 ITU-T protocol to control the MG.

HA MIP home agent.

HAAA Home AAA.

HAC HA challenge.

HANDMREQ TIA/EIA-41 HandoffMeasurementRequest INVOKE message. Being replaced by HANDMREQ2.

HANDMREQ2 TIA/EIA-41 HandoffMeasurementRequest2 INVOKE message. Used by the serving MSC to determine the signal strength being received by a neighboring candidate MSC. Replaces HANDMREQ.

Handoff The process of an MS changing from one frequency in one cell or sector to a different frequency in a neighboring cell or sector.

HARQ Hybrid ARQ.

HC Home country.

HCM Handset configuration management.

HDLC High-level data link control.

HDML Hand-held markup language.

HDR High data rate. High-speed data-only version of CDMA, standardized as 1XEV-DO. Now called HRPD.

HDSL High-bit-rate DSL. Symmetric DSL used for corporate networking.

HDTV High-definition television. A television display technology that provides picture quality similar to that of 35-mm movies with sound quality similar to that of today's compact disc.

HE Home environment.

HF Hyperframe.

HFC (1) HF counter. (2) Hybrid fiber-coax.

HFR Hybrid fibre radio. The capacity of optical networks combined with the flexibility and mobility of wireless access networks.

HFSS High-frequency structure simulator.

HI Handover interface.

HINFO Host information.

HIPPI High-performance parallel interface.

HLPI Higher layer protocol identifier.

HLR Home location register.

HMAC Hashed MAC.

Home agent IP node that receives packets at a mobile node's static address and forwards them to a foreign agent for delivery at its current (care-of) address.

Home MSC The MSC to which the PSTN routes based on an MDN.

HomeRF Promoters of SWAP for in-home wireless data networking.

HPC High probability of completion.

HSCSD High-speed circuit-switched data GSM-enhanced to allow up to 57.6-kbps data rates.

HTML Hypertext markup language.

IA5 International Alphabet 5. Basically the ASCII character set.

IAM SS7 ISUP initial address message. Used to initiate a call.

IANA Internet Assigned Number Authority.

IAP Intercept access point. A point within the telecommunications network that provides voice or data for CALEA interception to a DF.

ICMP Internet Control Message Protocol. Method for reporting errors and performing loopback testing on the Internet.

ICO Intermediate circular orbit.

ICS Implementation conformance statement.

IDB ITS data bus (SAE J2366).

iDEN Integrated Digital Enhanced Network. Motorola's proprietary system for E-SMR, used by Nextel and others. Based on GSM.

IDL Interface Description Language.

IDSL ISDN digital subscriber line. Just a fancy name for ISDN.

IE Information element.

IEC (1) Internet Engineering Consortium. (2) International Electrotechnical Commission.

IEEE Institute of Electrical and Electronics Engineers.

IEEE-ISTO IEEE Industry Standards and Technology Organization.

IETF Internet Engineering Task Force. Standards-setting body for the Internet.

IF Information flow.

IFAST International Forum on AMPS Standards Technology. Allocates IRM and SID codes.

IFFT Inverse FFT.

IGP Interior Gateway Protocol.

IIF Interworking and interoperability function.

IK Integrity key. Protects (via encryption) the contents of signaling messages (in AKA).

IKE Internet key exchange.

ILEC Incumbent local exchange carrier. An ILEC is a telephone company that was providing local service when the Telecommunications Act of 1996 was enacted. Compare with CLEC.

ILM Incoming label map.

IM IP multimedia.

IMC Internet Mail Consortium.

IMEI GSM International Mobile Equipment Identity. Composed of TAC + FAC + SNR + spare digit. A check digit is also included, but not always transmitted.

IMSI International mobile subscriber identity. Formerly called international mobile station identity. Based on the ITU-T E.212 numbering plan. Composed of mobile country code (MCC), mobile network code (MNC), and MSIN.

IN Intelligent network. Wireless equivalents are WIN and CAMEL.

IntServ Integrated services.

IP Internet Protocol.

IRM International roaming MIN. A mobile subscription identifier beginning with the digit 0 or 1 to avoid conflict with NANP MINs.

IS-54 First-generation TDMA radio interface standard.

IS-136 Second-generation TDMA air interface standard.

ISDN Integrated services digital network. PSTN utilizing SS7 interfaces.

ISM Industrial, scientific, and medical applications (of radio-frequency energy). Operation of equipment or appliances designed to generate and use locally radio-frequency energy for industrial, scientific, medical, domestic, or similar purposes, excluding applications in the field of telecommunications.

ISUP ISDN user part. Defines the protocol and procedures used to set up, manage, and release trunk circuits that carry voice and data calls over the public switched telephone network (PSTN). ISUP is used for both ISDN and non-ISDN calls. Calls that originate and terminate at the same switch do not use ISUP signaling.

ITAR U.S. International Traffic in Arms Regulations. These rules used to govern export of encryption technology in the United States and, to some extent, in Canada.

ITS Intelligent transportation system (“smart highways”).

ITU International Telecommunications Union.

ITU-T ITU—Telecommunications Division.

IWU Interworking unit.

IXC Interexchange carrier. A carrier licensed to provide long-distance services.

J-STD- Joint ATIS-T1/TIA standard.

J-STD-025 LAES standard jointly produced by ATIS T1 and TIA TR-45.

J-STD-034 Wireless data E911 Phase I standard (cell site and mobile identification, callback capabilities).

J-STD-036 Wireless data E911 Phase II standard (position of caller within 125 m/400 ft).

JDC Japanese digital cellular.

JIP ISUP jurisdiction information parameter.

JPEG Joint Photographic Experts Group. Not a group, but a graphics format created for them that is suitable for digital photographs and similar images. Features variable levels of compression and quality.

JTACS Japanese TACS.

JTC Joint technical committee (with members of more than one SDO).

kbps kilobits per second.

Kc GSM cipher key. Used for voice encryption.

KG Key generator.

Ki GSM subscriber authentication key.

kilo Prefix to indicate one thousand (as in kilobit). Abbreviated k.

KSG Keystream generator.

KSI Key set identifier.

L1 Layer 1.

L2 Layer 2.

L2TP Layer 2 Tunnel Protocol. Defined in RFC 2661. Successor to PPTP.

L3 Layer 3.

LAC (1) Link access control. (2) Location area code, 16-bit number identifiying a location area (group of base stations) within a GSM PLMN.

LAES Lawfully authorized electronic surveillance.

LAI GSM location area identity. Composed of MCC + MNC + LAC.

LAK Local authentication key.

LAN Local-area network.

LAP Link Access Protocol.

LAPB Link Access Protocol, Balanced.

LAPD Link Access Protocol for the ISDN D Channel.

LATA Local access and transport area. U.S. local telephone service area.

LB Letter ballot.

LBS Location-based services.

LCAS Link capacity adjustment scheme protocols.

LCP Link Control Protocol.

LCS GSM location services.

LDAP Lightweight Directory Access Protocol.

LDP Label Distribution Protocol.

LDT Location determination technology.

LEA Law enforcement agency.

Leakage Inability to bill for a service.

LEC Local exchange carrier (wireline carrier for local calls).

LEMF Law enforcement monitoring facility (for LAES).

LEO Low Earth orbit.

LER Label edge router.

LERG Local exchange routing guide. Includes a list of valid number blocks in the NANP.

LFB Look-ahead for busy.

LIDB Line information database. Used for validating collect calls, third-party calls, etc.

LIF Location Interoperability Forum.

LIFO Last in, first out. A queueing methodology similar to stacking dishes. Compare with FIFO.

LIR Location identification restriction. Prevents the location of a mobile phone from being revealed (with legal exceptions for E911, etc.).

LLA-RHOC Link layer assisted robust header compression.

LMDS Local multipoint distribution service. Uses the 27.5- to 31.3-GHz frequency band to serve multiple points within a radius of 3 to 5 km.

LMSD Legacy MS domain. Refers to existing SS7-based telecommunications systems, when accessed from an all-IP system.

LMSI Local mobile station identity.

LMU GSM location measurement unit.

LND Last number dialed.

LNP Local number portability. Allows consumers to keep a phone number when changing carriers within the same geographical region.

LNPWG Local Number Portability Working Group of NANC.

LOS Line of sight. A direct path through the air from transmitter to receiver. Always desirable for wireless communications, although not always necessary (for cellular/PCS).

LPC Linear predictive coding.

LPDE Local PDE.

LPDU Link protocol data unit. A frame.

LPI/LPD Low probability of interception/low probability of detection.

LRF Location registration function.

LRN Location routing number. A routing number that identifies the terminating switch for a ported DN.

LSB Least significant bit or byte. Compare with MSB.

LSMS Local service management system. Telephone carrier interface that downloads information from the NPAC into one or more NPDBs.

LSP Label switched path.

LSPI Proposed SS7 ISUP local service provider identification.

LSR Label switching router.

LSSU Link status signal unit. The link status message for SS7 MTP.

LT Line termination.

LUDT Long UDT. Supports messages up to 3 kilobytes in length.

M Mandatory (for a parameter).

M3UA SS7 MTP3 user adaptation layer. Resides above SCTP and below SCCP in an IP telecom signaling protocol stack.

MABEL Major account billing exchange logistical record. Defined by Cibernet to facilitate centralized billing for large account customers.

MAC (1) Medium access control. (2) Message authentication code.

MACA Mobile Assisted Channel Allocation. Mobile unit determines the best alternate cell site to originate a call in, when no channels are available in the current site.

MAH Mobile access hunting. Termination to one of a group of mobiles.

MAHO Mobile-assisted handoff. Mobile unit measures signal strength at neighoring base stations to determine which to hand off to.

MAI Multilateral investment agreement.

MAN Metropolitan-area network. Compare with LAN, WAN.

MAP Mobile AP. Protocol that interconnects wireless telephone systems.

Mbps Megabits (millions of bits) per second (bps).

MC (1) Short message center. (2) Multicarrier.

MCC E.212 mobile country code. First 3 digits of IMSI.

MCI Malicious call identification.

Mcps Million chips per second.

MD5 Message digest 5.

MD-IS Mobile data—intermediate system. A CDPD term.

MDN Mobile directory number (DN).

ME Mobile equipment. Synonym for UE. Contrast with MS.

MEA Message encryption algorithm.

Mega Prefix to indicate 1 million (as in megabit). Abbreviated M.

MEGACO Media Gateway Control—IETF Working Group.

MEI Mobility event indicator.

MEMS Microelectronic mechanical system.

metalanguage A language that can be used to define other languages.

MeXe 3GPP TSG-T mobile execution environment.

MF (1) Multifrequency tone signaling. Used for trunk signaling. (2) UIM master file. Compare with EF and DF.

MG Media gateway. Interface between packet environment of CN and circuit environment of PSTN.

MGC MG controller.

MGCF MG control function. Controls the media gateway.

MGCP MG Control Protocol. A VoIP call control standard.

MHz Million hertz. A measure of radio frequency.

MIB Management information (data) base.

MIN Mobile identification number. The 10-digit identifier of a mobile subscription.

MIN1 The last 7 digits of a MIN encoded in a weird, but compact, 24-bit binary format.

MIN2 The first 3 digits of a MIN encoded in a very wierd, but compact, 10-bit binary format.

Minimization Removal of information that has been collected, but that is outside the scope of a court order.

MIP Mobile IP.

MIPS Million instructions per second. Measurement of processor speed.

MISP Mobile Internet service provider.

MMDS Multipoint microwave distribution system. Also known as multichannel multipoint distribution system and wireless cable, it is another wireless broadband technology for Internet access.

MMIC Monolithic microwave IC; modules with the addition of eight new modules. The MMIC modules have the industry's widest frequency range: 1.9 to 2.0, 5.8 to 7.2, 10.0 to 13.3, 13.75 to 14.5, and 14.4 to 15.4 gigahertz (GHz).

MNC E.212 mobile network code (identifies an individual carrier, or a portion of a carrier network). Part of IMSI.

MNP—10 Microcom Networking Protocol—10. The alternate protocol, as described by V.42. MNP, classes 2 through 10, implements end-to-end error correction through automatic repeat-transmission request algorithms.

MPEG2 Motion Pictures Experts Group 2. Not a group but a format for audio/video compression version 2.

MPLS Multiprotocol Label Switching. An IP (RFC 3031) designed to provide preferred treatment to priority traffic to achieve a higher QoS. Every packet labeled with the same FEC gets the same treatment.

MS Mobile station (wireless phone).

MSB Most significant bit (or byte).

MSC Mobile switching center.

MSK Minimum shift keying. FSK where peak-to-peak FM deviation is half the data rate.

MSSPRING BLSR is called MSSPRING in SDH (ITU-T G.841).

MTP Mail Transfer Protocol. A protocol for sending e-mail messages between servers.

N-1 switch The switch responsible for LNP queries.

NAA Network access application.

NACK Negative ACK.

NAI Network access identifier.

NAM Number assignment module. A portion of a mobile phone that stores a single user profile, including MIN and IMSI, but not ESN.

NAMPS Narrowband AMPS.

NANC North American Numbering Council. The organization responsible for overseeing the distribution of telephone numbers in the NANP.

NANP North American numbering plan.

NANPA North American numbering plan administrator (or area).

NAS (1) Network access servers. (2) Nonaccess stratum. Protocols between the UE and the core network that are transparent to the RAN (SMS).

NAT Network address translation. Maps an internal set of IP addresses to an external set (perhaps assigned dynamically by DHCP).

NBAP Node B application part.

NCAS Non-call-associated signaling. Compare with CAS.

NCGW Network capability gateway.

NCS National Communication System. Responsible for U.S. Government communications, including WPS and GETS.

NDC E.164 national destination code.

NE Network entity.

NEBS Network equipment building system.

Net additions Increase in new revenue-generating subscribers over a time period (usually a quarter year).

NFS Number field sieve. Used in cryptography.

NHLFE Next hop label forwarding entry.

NI (1) Network identity. (2) Number incomplete.

Nibble Four bits.

NIC Network interface card. A computer circuit board or card that is installed in a computer so that the computer can be connected to a network.

NID Network identification. A number that uniquely identifies a portion of a wireless carrier's total system. Used by TIA/EIA-95 systems.

NII National Information Infrastructure. A proposed advanced, seamless web of public and private communications networks, interactive services, interoperable hardware and software, computers, databases, and consumer electronics to put vast amounts of information at users' fingertips.

NIST U.S. National Insitute of Standards and Technology.

NPAC NP administration center. Often used to refer to the NPAC SMS.

NPDB Number portability database. A list of ported numbers and associated LRNs used for LNP routing.

NRM Network reference model (TIA/EIA/TSB-100).

NSN National significant number. All of an E.164 number apart from the CC (NDC + SN).

O Optional (for a parameter).

OAA Over-the-air activation.

OA&M Operations, administration, and maintenance.

OAM&P Operations, administration, maintenance, and provisioning.

OATS Over-the-air activation teleservice.

OBF Ordering and billing forum of ATIS.

O-BCSM Originating BCSM.

OC-1 SONET service at 51.84 Mbps. Equivalent to one DS3 (672 DS0).

OC-3 SONET service at 155.52 Mbps. Equivalent to 3 DS3 or 2016 DS0.

OC-9 SONET service at 466.56 Mbps. Equivalent to 9 DS3 or 6048 DS0.

OC-12 SONET service at 622.08 Mbps. Equivalent to 12 DS3 or 8064 DS0.

OC-18 SONET service at 933.12 Mbps. Equivalent to 18 DS3 or 12,096 DS0.

OC-24 SONET service at 1.24416 Gbps. Equivalent to 24 DS3 or 16,128 DS0.

OC-36 SONET service at 1.86624 Gbps. Equivalent to 36 DS3 or 24,192 DS0.

OC-48 SONET service at 2.48832 Gbps. Equivalent to 48 DS3 or 32,256 DS0.

OC-96 SONET service at 4.97664 Gbps. Equivalent to 96 DS3 or 64,512 DS0.

OC-192 SONET service at 9.95328 Gbps. Equivalent to 192 DS3 or 129,024 DS0.

OC-768 SONET service at 38.813 Gbps. Equivalent to 768 DS3.

OCDM Optical code-division multiplexing.

OCH Optical channel.

OCN Operating company number. U.S. identifier for a telecom carrier.

OCSP Online Certificate Status Protocol. An IETF protocol.

Octet A snooty standards term for byte.

ODB Operator-determined barring.

ODBC Open Database Connectivity. A standard database access method developed by Microsoft Corporation.

O/E Odd/even.

OET U.S. FCC Office of Engineering and Technology.

OFA Optical-fiber gateway.

OFDM Orthogonal frequency-division multiplexing. An FDM modulation technique for transmitting large amounts of digital data over a radio wave.

OFDMA Orthogonal FDMA.

OG Optical gateway.

OHG Operators Harmonization Group. A group of wireless carriers that wants to make CDMA systems work on both ANSI-41 and GSM networks.

O&M Operations and maintenance.

OTA Over-the-air programming. Uploads internal mobile tables.

PAC PPTP access concentrator. Concentrates PPP traffic on a dial access platform.

PACA Priority access and channel assignment. Channels are assigned to mobiles attempting to originate a call in order of priority, rather than first-come, first-served.

Packet data Data transmitted from multiple users in individually addressed discrete packets. Compare with circuit-switched data.

PACS Personal Access Communications System.

PAD Packet assembly/disassembly.

Page The process of telling an MS to prepare for an incoming call.

PAN Personal area network. A network that connects personal devices, such as computer, keyboard, mouse, phone, and monitor. Also known as Piconet.

PANID Previous ANI.

PAP Password Authentication Protocol.

PAS Priority access service.

PBCCH GPRS packet broadcast CCH.

PBX Private branch exchange. A privately owned switch providing wired and possibly also wireless data service for an office, factory, campus, etc.

PC Personal computer.

PCB Printed circuit board. A thin plate on which chips and other electronic components are placed.

PCCCH GPRS packet control channel.

PCF Packet control function.

PCG 3GPP project coordination group.

PCH Paging channel.

PCI Protocol capability indicator.

PCIA Personal Communications Industry Association. Rival trade association to CTIA.

PCM Pulse code modulation.

PCMCIA Personal Computer Memory Card International Association.

PCN Personal communications network.

PCS Personal communications system(s).

PCSC Personal communications switching center.

PD Packet data.

PDA Personal digital assistant. A hand-held device that combines computing, telephone/fax, Internet, and networking features.

PDA Personal digital assistant.

PDC Pacific digital cellular. A Japanese cellular standard.

PDC-P PDC packet network used.

PDP Packet Data Protocol. Used to perform signaling tasks of GPRS Tunneling Protocol (GTP).

PIM Personal information manager software. A type of software application designed to help users organize random bits of information.

PLMN Public land mobile network. A cellular, PCS, or 3G wireless network.

POP Post Office Protocol. A protocol used to retrieve e-mail from a mail server.

POS Personal operating space. A space about a person or object about 10 m in radius that envelopes the person.

PPTP Point-to-Point Tunneling Protocol. Allows PPP to be tunneled through an IP network. Uses an enhanced GRE mechanism for flow control and congestion control. Defined in RFC 2637. May be replaced by L2TP.

PSTN Public switched telephone network. The international telephone system based on copper wires carrying analog voice data.

QAM Quadrature amplitude modulation.

QCELP Qualcomm code excited linear prediction.

QoR Query on release. A method for implementing LNP.

QoS Quality of service. A list of measurable attributes such as bandwidth, delay, and jitter that should be met for a customer.

QPSK Quadrature PSK, used in CDMA. Phase can be in one of four states.

Quintet A group of five security data elements used in 3GPP AKA. Composed of RAND, expected user response (XRES), cipher key (CK), integrity key (IK), and network authentication token (AUTN).

RA Registration authority.

RAB Radio access bearer.

RAC Routing area code. An 8-bit number identifying a routing area with a location area.

RACF Radio access control function. An IN term.

RACH Random access channel. Mobile units compete to access this shared control channel.

Radio interface The interface between an MS and a BS.

RADIUS Remote Authentication Dial-In User Service. An Internet user authentication system.

RAI GSM routing area identification. Composed of LAI + RAC.

RAKE A receiver technique that uses several baseband correlators to individually process several signal multipath components. The correlator outputs are combined to achieve improved communications reliability and performance.

RAN Radio access network.

RAND Random number used for authentication purposes.

RANDBS Random number used for base station challenge authentication operation.

RANDC RAND confirmation. A portion of RAND used to try to confirm the value that was used by an MS during a global challenge.

R and R Revise and resubmit. The usual treatment for a contribution to a standards committee.

RANDSSD Random number used for CAVE SSD update operation.

RANDU The random number used for a unique challenge.

RAO Revenue Accounting Office.

R-APDU Response APDU.

RASC Radio access system controller.

Rate center The location of a phone for billing purposes. Long-distance charges are based on the distance between two rate centers (often the switch location). Rate centers and competition combine to cause very inefficient number assignment, because every carrier operating in a rate center must be assigned at least 10,000 numbers.

RBOC Regional Bell operating company. One of seven local exchange companies formed from the breakup of AT&T: NYNEX, Bell Atlantic, BellSouth, Southwestern Bell, Pacific Bell, USWest, and Ameritech.

RCF Radio control function. An IN term.

RDCCH Reverse DCCH.

RDTC Reverse DTC.

RDV Roamer database verification (IS-847). Ability of HLR to query VLR database to determine whether information needed to support roamers is correct.

RECC Reverse analog control channel.

Recipient switch The switch to which a number has been ported.

RED Random early detection.

Reed-Solomon A type of forward error correction (FEC).

ReFLEX A two-way paging protocol related to FLEX. Operates at speeds up to 9600 bps inbound and 6400 bps outbound (25-kHz channel) or 25,600 bps (50-kHz channel).

Registration The process by which an MS informs a system of its presence.

Rel Abbreviation for release of a specification (as in GSM Rel 5).

REL SS7 ISUP circuit release message.

Release Returning a trunk, transceiver, or other telephony resource to the idle list.

RETURN RESULT Message sent to successfully end TCAP transaction.

REVAL Recommendations on the Procedures for Evaluation of Radio Transmission Technologies for FPLMTS.

Revertive dialing Calling your own phone number. Often used to access a voice mail system.

RF Radio frequency.

RFC Request for comments. Internet standard.

RFI (1) Request for information. (2) Radio-frequency interference.

RFP Request for proposal.

R-SCH Reverse SCH (from MS).

R-SGW Roaming SGW. Interface between SS7 SCCP and MTP to IP protocols.

R-UIM Removable UIM.

RSVP Resource Reservation Setup Protocol. A new Internet protocol being developed to enable the Internet to support specified QoSs.

SA Security Association.

SACCH Slow associated control channel.

SAE J2366 Society of Automotive Engineers Golden Reference driver implementation for the Intelligent Transportation System Data Bus (IDB Forum) and AMI-C.

SAFER Safe and Fast Encryption Routine. Based on IBC.

SAMPS ANSI-136 system assisted mobile positioning through satellite. TDMA, MS-assisted location determination, using GPS.

SAP Service access point.

SAPI Service access point identifier.

SAR (1) Segmentation and reassembly. (2) Successive approximation converter. Type of analog-to-digital converter. (3) Specific absorbtion rate. Amount of RF energy absorbed (for RF/health studies).

SAT Supervisory audio tone. One of three tones around 6 kHz that are transmitted from a base station to an MS by analog cellular (EIA/TIA-553).

S-BCCH SMS BCCH.

SC Steering committee.

SCC (1) SAT color code. A slight variation in frequency to limit false SAT detection. (2) Service control code. A 1-, 2-, or 3-digit code followed by an asterisk (*) used for controlling switch capabilities during a call (777*). Compare with FC.

SCCP Signaling connection control part. SS7 enhanced routing and identification layer.

SCE Service creation environment for the intelligent network (IN, WIN).

SCEG Smart Card Expert Group. A WAP working group.

SCEMA Secure cellular encryption module algorithm.

SCF Service control function. Most important part of an SCP.

SCH CDMA supplemental channel. Used for transmitting higher-speed data.

SCID Subchannel identity.

SCM (1) Station class mark. Identifies the power class and transmission mode of an AMPS terminal. (2) 3GPP2 session control manager. Manages multimedia sessions.

SCP Service control point. An SS7 network element used to control call processing (by providing database or intelligent network services).

SCPT CDMA service category programming teleservice.

SCTP Stream Control Transmission Protocol.

SDH Synchronous digital hierarchy.

SDO Standards development organization. TIA and ATIS are SDOs.

SDSL Symmetric DSL. Bit rates are the same in both directions.

SFA Sales force automation. An application service provider that focuses on providing and hosting applications related exclusively to business functions.

SGW Signaling gateway.

SID System identifier. A 15-bit identifier of an AMPS wireless license or system.

SIM GSM subscriber identification module (smart card).

SINR Signal-to-noise-plus-interference ratio. SINR refers to signal tracking. It determines the angle of arrival of the desired signal with phased arrays to determine which beam to use and adjusts the weights with adaptive arrays to maximize the desired signal-to-noise-plus-interference ratio in the output signal.

SIP Session Initiation Protocol. An IETF IP for VoIP (packetized voice) call processing.

SMR Specialized mobile radio. A form of private mobile radio that allows one talker and multiple listeners in a group (for example, taxi company).

SMS Short message service. The transmission of short text messages to and from a mobile phone, fax machine, and/or IP address.

S/N Signal-to-noise ratio.

SNCP In a general ring topology, any recovery mechanism may be used. This may be the underlying transport's BLSR/MSSPRING or UPSR/SNCPRING, or it may be using control plane to set up 1 + 1 or 1 : 1. In a BLSR/MSSPRING ring, the type of protection has already been decided (BLSR is the name of the protection mechanism). Similarly for UPSR/SNCPRING.

SONET Synchronous optical network.

SONET/SDH Synchronous optical network/synchronous digital hierarchy. An international standard for synchronous data transmission over fiber-optic cables.

S/R Emergency services selective router.

SRD CTIA standards requirement document.

SS7 Signaling System 7. A telecommunications protocol defined by the International Telecommunication Union (ITU) as a way to offload PSTN data traffic congestion onto a wireless or wireline digital broadband network.

SSD Shared secret data. A secondary key used in most CAVE operations.

SSL Secure Sockets Layer. A protocol developed by Netscape for transmitting private documents via the Internet.

SUA SCCP user adaptation layer. Replaces SCCP and M3UA. Resides above SCTP and below the application in an IP telecom signaling protocol stack.

SWAP Shared Wireless Access Protocol. Operates in the 2.4-GHz ISM band.

T1 (1) Group of standards committees related mostly to wireline standards, such as SS7, but also North American GSM standards. (2) A digital link carrying 24 DS0 channels. Used mostly within North America. Compare with E1.

T1M1 OA&M standards committee for North American wireline networks.

T1P1 Personal communications (U.S. GSM) standards committee for North America.

T1S1 SS7 standards committee for North America.

T3 Physical device carrying one DS3 service.

TA Terminal adaptor

TAC (1) Total access communications. A name used by Motorola in several products, such as TACS, DataTAC. (2) Type approval code. Six-digit part of IMEI that is assigned by a central authority (national telecom regulatory body).

TACS TAC system. Basically AMPS in the 900-MHz frequency band.

TADIG Technology and Documentation Interoperability Group. A GSM organization.

Tandem Any switch that is used to receive and route traffic (phone calls), but neither originates nor terminates them.

TAP (1) Transfer account procedure. A billing record format used mainly by GSM carriers. (2) Telocator (now PCIA) Alphanumeric Protocol for paging.

TAR Temporary alternative routing.

Target MSC The destination MSC for an intersystem handoff.

TBCD Telephony BCD. Digit 0 is encoded as 10, as on a rotary dial phone.

T-BCSM Terminating BCSM.

TCAP Transaction capabilities application part. Message packaging standard used by TIA/EIA-41, LNP, 800 and other SS7-based applications. Defined in ANSI T1.114.

TCC Telephony country code.

TCH Traffic channel. Often classified as full rate (FR) or half rate (HR).

TCP Transmission Control Protocol. A protocol that provides for reliable delivery of messages over the Internet.

TCP/IP TCP used over IP.

TD-CDMA TDD mode of operation for UTRA.

TDD Telephone device for the deaf. See TTY.

TDMA Time-division multiple access. A wireless technology that allows for increased bandwidth over digital cellular networks. Similar to CDMA, the call stream is broken into fragments so that multiple calls can take place over a single frequency.

TDOA Time difference of arrival.

TETRA Terrestrial trunked radio.

TIA Telecommunications Industry Association. Develops telecommunications standards for use in the United States and in any other countries that care to adopt them or at least allow their use. The TIA spun off from the EIA (Electronics Industry Alliance) in 1988 and is ANSI-accredited.

TIA/EIA-136 ANSI version of the TDMA air interface standard. Replaces IS-136.

TOA Time of arrival. A technique for locating a radio by comparing the time of signal arrival at multiple points.

TOS Type of service.

Traffic channel A portion of a radio channel used to transmit one direction of a digital voice conversation.

TSG-T 3GPP TSG for terminal and UIM standardization.

T-SGW Trunk SGW.

TTY A device used by the deaf or hearing-impaired to communicate text messages over telephone systems. It runs at 45.45 bps.

UAProf WAP User Agent Profiles Drafting Committee.

UATI Unicast access terminal identifier.

UDH Universal data header.

UDP User Datagram Protocol. An Internet protocol providing basic services only. Compare with TCP.

UDR Usage data record.

UDT SS7 MTP unit data message. Carries a payload of about 250 octets.

UDTS UDT service message. Used to send error responses to UDT messages.

UE User equipment (phone and all peripherals such as USIM).

UGID User group identification.

UHF Ultrahigh frequency; 300 to 3000 MHz (used in the United States and Canada as television channels 14 to 83).

UICC Universal integrated circuit card.

UID User interactive dialog.

UIM User identification module (smart card).

UL Uplink. Radio link in the direction "up" to network. Compare with DL.

Um Radio interface between MS and BSS/BSC.

UMTS Universal mobile telecommunications system (a 3G initiative).

UNI User-network interface.

Unique challenge A method of encryption using a unique random number (RANDU) as a challenge.

UPR CTIA user performance requirements document. Now known as SRD.

UPT Universal personal communications.

URI Uniform resource indicator.

URL Uniform resource locator. Form of Internet address usable by a browser.

URN Uniform resource name.

USAT USIM Application Toolkit.

USCF User-selected call forwarding. Allows an incoming call to be diverted before it is answered, either to a preset number, or to a number specified at the time of diversion. Closely related to AH.

USIM User services identity module. SIM for UMTS.

USSD GSM unstructured supplementary service data.

UTC Universal Coordinated Time.

UTRA Universal terrestrial radio access.

VA Validation authority.

VAAA Visited AAA.

VAD Voice activity detection. A system that detects the absence of voice and prevents transmission of unecessary digitized voice during this time.

V and V Verification and validation. A committee review of a standard before formal ballot.

VAR Value-added reseller.

VAS Value-added service.

VC Virtual circuit.

VCI VC identifier.

VDSL Very high speed DSL, providing about 58 Mbps in both directions. Available as asymmetric (see ADSL) or symmetric (see SDSL).

VDSL Alliance Promoters of VDSL.

VGC Voice group call.

VHE Virtual home environment.

VHF Very high frequency; 30 to 300 MHz (used in the United States and Canada as television channels 2 to 13).

VLR Visitor location register.

VMAC Voice mobile attenuation code.

VMS Voice message system.

Vocoder Voice coder or codec.

VOFDM Vector OFDMA.

Voice channel A radio channel used to transmit one direction of an analog voice conversation. Compare with traffic channel.

Voice coder Converts an analog voice signal into a digitally coded representation, and vice versa. Wireless voice coders often also compress the voice into a bit rate from 8 to 13 kbps. Also called codec or vocoder.

VoIP Voice over IP.

VP (1) Voice privacy (encryption). (2) Virtual path.

VPI VP identifier.

VPLMN Visited PLMN.

VPM Voice privacy mask.

VPN Virtual private network.

W3C World Wide Web Consortium.

Walsh codes A group of $2N$ vectors or words that contain $2N$ binary elements, which with themselves and their logical inverses form a mutually orthogonal set.

WAN Wide-area network. Compare with LAN and MAN.

WAP Wireless Application Protocol. A new protocol that is supposed to provide more efficient Internet access from wireless phone.

WAP W3C WAP-W3C Coordination Committee.

WARC World Administrative Radio Convention.

WATS Wide-area telephone service

W-CDMA Physical layer of the FDD mode of operation of UTRA. A European version of CDMA and the 3G evolutionary step planned for GSM. Operates in pairs of 5-MHz channels at 3.84 Mbps.

WCMP Wireless Data Control Message Protocol. Protocol for reporting errors and performing loopback testing in WAP. Based on ICMP.

WCS Wireless Communications Service. The WCS band consists of two separate 15-MHz chunks of bandwidth in the 2.3-GHz frequency range.

WDAE Wireless data application environment.

WDAG Wireless Data Applications Group.

WDASP Wireless data application service provider.

WDASU Wireless data access subscriber unit.

WDATM Wireless data ATM.

WDBMP Wireless data bit map.

WDEMT CDMA wireless data enhanced messageing teleservice.

WDM Optical wavelength-division multiplexing. Refers to a type of multiplexing developed for use on optical fiber.

WDP Wireless Datagram Protocol.

WFQ Weighted fair queueing.

WG Working group.

WGS-84 World Geodetic System 1984.

WI Work item.

WID 3GPP WI description.

WDIF Wireless Data Interconnect Forum (managed by Telcordia).

WIDMS Wireless data multimedia message service.

WIG WAP Interoperability Group.

WIM WAP identity module.

WIN Wireless Intelligent Network. Protocol with similar goals as IN, AIN, and CAMEL.

WLL Wireless local loop. Wireless devices or systems that are situated in fixed locations, such as an office or home, as opposed to mobile, such as cell phones and PDAs.

WML Wireless markup language. An HTML-like page description and scripting language that is an essential part of WAP.

WPS Wireless priority service.

X.25 An ITU/CCITT-defined general-purpose packet switching protocol.

XDSL Refers to all the DSL variants.

XHTML Extensible HTML. A version of HTML defined by XML and designed to be extensible. This may also be the evolutionary path for cHTML and WML.

XMAC Expected MAC.

XML Extensible Markup Language. A metalanguage that can be used to define languages like HTML and WML.

XOR Exclusive OR. A boolean operation, often used in cryptography. The exclusive OR of two bits is 1 if they have the same value and 0 if they have different values. This operation has the nice property that no information is lost. For example, (A xor B) xor B = A and (A xor B) xor A = B. Try that on any other boolean operation!

XRES Expected response to authentication challenge for 3GPP AKA. Compare with AUTHR.

XUDT Extended UDT. Supports message segmentation.

ZBTSI Zero byte time slot interchange (TSI). Bits 2 through 193 of each DS1 frame are scrambled to minimize the possibility of an all-zero octet. If all-zero octets are still found, a group of all-ones is substituted. Because of its complexity this method is not often used.

ZCS Zero code suppression. The seventh bit of an all-zero octet is replaced by a 1. Only applicable to voice because there is no way to remove this 1-bit error.

INDEX

Index

About the Author

John R. Vacca is an information technology consultant and internationally known author. Since 1982, he has published more than 440 articles and 36 books, including the leading technical reference in this field, McGraw-Hill's *Wireless Broadband Networks Handbook.* He is also the author of McGraw-Hill's *i-mode Crash Course* and *Net Privacy: A Guide to Developing and Implementing an Ironclad ebusiness Privacy Plan.* A configuration management specialist, Mr. Vacca was a computer security official (CSO) on NASA's Freedom space station program.

Printed in the United States
116873LV00003B/96/A